Marine Macro- and Microalgae
An Overview

Editors

F. Xavier Malcata

LEPABE – Laboratory of Process Engineering
Environment, Biotechnology and Energy
College of Engineering
University of Porto
Porto, Portugal

Isabel Sousa Pinto

Interdisciplinary Centre of Marine
and Environmental Research (CIIMAR)
University of Porto
Matosinhos, Portugal

A. Catarina Guedes

Interdisciplinary Centre of Marine
and Environmental Research (CIIMAR)
University of Porto
Matosinhos, Portugal

CRC Press
Taylor & Francis Group
Boca Raton London New York

CRC Press is an imprint of the
Taylor & Francis Group, an **informa** business

A SCIENCE PUBLISHERS BOOK

Cover illustrations reproduced by kind courtesy of Dr. Alejandro Buschmann

CRC Press
Taylor & Francis Group
6000 Broken Sound Parkway NW, Suite 300
Boca Raton, FL 33487-2742

First issued in paperback 2021

© 2019 by Taylor & Francis Group, LLC
CRC Press is an imprint of Taylor & Francis Group, an Informa business

No claim to original U.S. Government works

Version Date: 20180704

ISBN 13: 978-0-367-78066-1 (pbk)
ISBN 13: 978-1-4987-0533-2 (hbk)

Library of Congress Cataloging-in-Publication Data
Names: Malcata, F. Xavier, editor.
Title: Marine macro- and microalgae : an overview / editors, F. Xavier Malcata, LEPABE Laboratory of Process Engineering, Environment, Biotechnology and Energy, College of Engineering, University of Porto, Porto, Portugal, Isabel Sousa Pinto, Interdisciplinary Centre of Marine and Environmental Research (CIIMAR), University of Porto, Matosinhos, Portugal, A. Catarina Guedes, Interdisciplinary Centre of Marine and Environmental Research (CIIMAR), University of Porto, Matosinhos, Portugal.
Description: Boca Raton, FL : CRC Press, [2018] | "A science publishers book." | Includes bibliographical references and index.
Identifiers: LCCN 2018031105 | ISBN 9781498705332 (hardback)
Subjects: LCSH: Marine algae.
Classification: LCC QK570.2 .M36 2018 | DDC 579.8/177--dc23
LC record available at https://lccn.loc.gov/2018031105

Visit the Taylor & Francis Web site at
http://www.taylorandfrancis.com

and the CRC Press Web site at
http://www.crcpress.com

Preface

//

The marine environment accounts for most of the biodiversity of our planet, while offering a huge potential for citizens' well-being. Its extensive resources constitute the basis of several economic activities—yet many more are expected in coming years.

This book intends to cover the state-of-art on uses of marine algae to obtain bulk and fine chemicals, with an emphasis on optimization of the underlying production and purification processes. Major gaps and potential opportunities in this field are discussed in a critical manner.

This book is accordingly divided into 16 chapters, along two sections. The first chapter entails an overview of marine algae—covering both the current situation and unfolding opportunities; while Chapters 2 and 5 address the ecological importance of algae. Chapter 3 focuses on microalgal metabolic features and general utilization thereof; algal biomass production and downstream processing features are also addressed in Chapters 4 and 6, respectively.

The next section focuses on economic products obtained from algae, including: (i) marine alga bioactivities (Chapter 7); (ii) macroalgal phycocolloids (Chapter 8); (iii) microalgae for feed (Chapter 9); (iv) seaweed extracts against plant pathogens (Chapter 10); (v) cosmeceutical properties of algal compounds (Chapter 11); (vi) biotechnological and pharmacological applications of biotoxins from dinoflagellates (Chapter 12); (vii) integrated multi-trophic aquaculture; (viii) alternative green biofuels from microalgae; and (ix) genetic engineering approaches to exploit marine algae as potential sources of biofuels and added-value products (Chapter 15). Finally, present and future economic and environmental impacts of algal technology, based on critical questions arising from this book, are presented in the last chapter (Chapter 16).

The information provided in this book – co-authored by some of the most recognized experts in the field, should be of interest for graduate students, research scientists, technologists, and industrialists—in their quest for new technological and business opportunities in this exciting world of marine algae and microalgae.

Acknowledgments

First and foremost, we would like to acknowledge the several colleagues who have agreed to this book upon our invitation—for the time and effort they allocated in preparing the various chapters, and for their continuing interest and commitment.

We are grateful for financial support through projects: DINOSSAUR—PTDC/BBB-EBB/1374/2014 - POCI-01-0145-FEDER-016640, coordinated by author F.X.M., and funded by FEDER funds through COMPETE2020—Programa Operacional Competitividade e Internacionalização (POCI), and by national funds through FCT—Fundação para a Ciência e a Tecnologia, I.P.; and the project NORTE-01-0145-FEDER-000035-INNOVMAR—Innovation and Sustainability in the Management and Exploitation of Marine Resources, namely in research line INSEAFOOD – Innovation and valorization of seafood products, supported by the Northern Regional Operational Programme (NORTE2020), through the European Regional Development Fund (ERDF). A postdoctoral fellowship (ref. SFRH/BPD/72777/2010) for author A.C.G., granted by Fundação para a Ciência e Tecnologia (FCT, Portugal, Lisbon), under the auspices of ESF and Portuguese funds (MEC) is also acknowledged.

F. Xavier Malcata
Isabel Sousa Pinto
A. Catarina Guedes

Contents

Preface   iii

Acknowledgements   v

1. **Introduction to the Marine Algae: Overview**   1
   Ana Isabel Neto and *Isabel Sousa Pinto*

2. **Importance of Seaweed in the Climate Change—Seaweed Solution**   20
   Ik Kyo Chung, Jung Hyun Oak, Jin Ae Lee, Tae-Ho Seo, Jong Gyu Kim and *Kwang-Seok Park*

3. **Microalgal Metabolism and their Utilisation**   43
   Michael A. Borowitzka

4. **Microalgal Biomass Production**   63
   Rathinam Raja, Hemaiswarya Shanmugam, Ramanujam Ravikumar and *Isabel S. Carvalho*

5. **Advances and Constraints in Seaweed Farming as a Basal Constituent for the Environmentally Sustainable Development of Aquaculture in Chile**   78
   Alejandro H. Buschmann, María C. Hernández-González and *Bárbara S. Labbé*

6. **Microalgal Downstream Processing: Harvesting, Drying, Extraction, Separation, and Purification**   90
   Xavier C. Fretté

7. **Marine Algal Bioactivities**   115
   Catarina Vizetto-Duarte, Carolina Bruno de Sousa, Maria Rodrigues, Luísa Custódio, Luísa Barreira and *João Varela*

8. **Macroalgal Phycocolloids**   146
   Leonel Pereira

9. **Microalgae for Feed**   161
   Vítor Verdelho, João Navalho and *Ana P. Carvalho*

10. **Potential Use of Extracts of Seaweeds Against Plant Pathogens**   177
    Jatinder Singh Sangha, Robin E. Ross, Sowmyalakshmi Subramanian, Alan T. Critchley and *Balakrishnan Prithiviraj*

11. **The Cosmeceutical Properties of Compounds Derived from Marine Algae**   198
    Snezana Agatonovic-Kustrin and *David W. Morton*

12. **Dinoflagellates and Toxin Production**   216
    Joana Assunção and *F. Xavier Malcata*

13. **Integrated Multitrophic Aquaculture: An Overview**   235
    Isabel C. Azevedo, Tânia R. Pereira, Sara Barrento and *Isabel Sousa Pinto*

14. Alternative Green Biofuel from Microalgae: A Promising Renewable Resource   **248**
Katkam N. Gangadhar, João Varela and *F. Xavier Malcata*

15. Genetic Engineering Approaches for the Exploitation of Marine Algae as   **274**
Potential Sources of Organic and Inorganic Biofuels
Maria Gloria Esquível, Julia Marín-Navarro, Teresa S. Pinto and *Joaquín Moreno*

16. Present and Future Economic and Environmental Impacts of Microalgal Technology   **300**
Miguel Olaizola, Rob C. Brown and *Elizabeth D. Orchard*

Index   **331**

About the Editors   **333**

1

Introduction to the Marine Algae

Overview

Ana Isabel Neto[1,2,*] *and Isabel Sousa Pinto*[3,4]

Definition

The term algae has no formal taxonomic standing but it is routinely used to indicate a large, polyphyletic, artificial, and diverse assemblage of mostly photosynthetic nonvascular organisms that contain chlorophyll *a* and have simple reproductive structures (Barsanti and Gualtieri 2014). As opposed to higher plants, algae do not have true roots, stems, leafy shoots, and transporting tissues; like mosses and ferns, algae do not form flowers and seeds, but reproduce and complete their life cycle by releasing gametes or spores into the environment. The algae concept encompasses a number of life forms that are not closely related; they are included in four kingdoms: Plantae, Chromista, Protozoa, and Bacteria (Ruggiero et al. 2015) including both prokaryotic and eukaryotic organisms (Lee 2008). The former, often classified as Cyanobacteria, are in fact photosynthetic bacteria, lacking the membrane-bounded organelles (plastids, mitochondria, nuclei, Golgi bodies, and flagella) that are present in the latter group.

The profound morphological diversity of forms and sizes, and the cytology, ecology, and reproductive biology of algae make it difficult to clearly define them (Bold and Wynne 1985). In a simplistic approach, the term *algae* refer to both macroalgae and a highly diversified group of microorganisms known as microalgae.

Species and environments

A conservative approach estimates the number of algal species in 72,500, most of them being microalgae (Guiry 2012).

[1] cE3c - Centre for Ecology, Evolution and Environmental Changes/Azorean Biodiversity Group, 9501-801 Ponta Delgada, Açores, Portugal.

[2] Department of Biology, Faculty of Sciences and Technology, University of Azores, Rua da Mãe de Deus, 9500-321 Ponta Delgada, Açores, Portugal.

[3] Interdisciplinary Centre of Marine and Environmental Research (CIIMAR), University of Porto, Terminal de Cruzeiros do Porto de Leixões, Avenida General Norton de Matos, S/N, P-4450-208 Matosinhos, Portugal.

[4] Department of Biology, Faculty of Sciences, University of Porto (FCUP), Rua do Campo Alegre, s/n, 4050 Porto, Portugal.

* Corresponding author

Algae most commonly occur in water—marine, freshwater, and brackish—but they can also be found in almost every other environment on earth, for example, growing on snow in some high mountains and in the Arctic or living in lichen associations on bare rocks, in desert soils, and in hot springs (Lee 2008). In most habitats, algae are extremely important as they function as primary producers in the food chains and produce the oxygen necessary for their metabolism and that of the consumer organisms. Algae also form mutually beneficial partnerships with other organisms in which they provide oxygen and organic compounds and receive protection and nutrients. This is the case of the zooxanthellae that live inside the cells of reef-building corals, and of the green algae or cyanobacteria that associate with fungi to form lichens (Barsanti and Gualtieri 2014).

Algae cells contain the green pigment chlorophyll that captures the sun's energy for photosynthesis, that is, the process of building energy-rich compounds from water and carbon dioxide. Photosynthetic algae will therefore grow only where there is light. In marine environments below low tide level, the amount and quality of light decreases with increasing depth to a point where the light level is not enough for algal growth. Some species contain additional pigments that enable them to absorb different light wavelengths and to use faint levels of light. This is the case of a coralline red algae collected at 268 m deep, the depth record for marine macrophytes (Littler et al. 1985). Due to these additional pigments, algae can exhibit different colors, the more common being red or brown. These shades are themselves extremely variable, with some brown species appearing olive-green and some red species almost black.

Many algae grow attached to the firm substratum (benthic). Benthic algae can grow attached on stones (epilithic), on sand or mud (epipelic), on animals (epizoic), or on other algae or plants (epiphytic). In marine environments, they can grow from the littoral zone (encompassing the intertidal area) to the shallow subtidal to around 200 m in very clear waters (Barsanti and Gualtieri 2014).

Most algae, however, are small single celled and filamentous organisms floating freely in the sea and constitute the phytoplankton, which forms the base of the marine food chains (Harris 1986; Sournia 2008). Most planktonic species are independent from the coastal and benthic processes, except for the temporary microscopic stages of the life cycle of the macroalgae (South and Whittick 1987; Falkowski and Knoll 2007).

Size and shapes

The size of algae is highly variable ranging from the tiny picoplankton which is only 0.2–2.0 μm in diameter to giant kelps with fronds up to 60 m in length.

Algae occur in dissimilar forms such as microscopic single cells (Fig. 1A), filaments (Fig. 1B), macroscopic multicellular loose or filmy conglomerations (Fig. 1C), colonies matted or branched (Fig. 1D), and more complex forms encompassing thin foliose sheets (Fig. 1E), tubes (Fig. 1F), sacs or bulbs (Fig. 1G), leather sheets (Fig. 1H), and gelatinous forms (Fig. 1I), cartilaginous (Fig. 1J) or calcareous forms, these including bushy (Fig. 1K), foliose (Fig. 1L) and crustose habits (Fig. 1M).

The unicells may or may not be solitary, and may or may not be motile through flagella. They can also exist as aggregates of single cells in colonial morphs, which can be more or less organized and have a variable number of cells. A colony is termed coenobium when the number and arrangement of cells are determined and remain constant. A good example of a motile coenobium is provided by the green alga *Volvox* (Fig. 2A).

The filamentous algae are formed by cells divisions along a plane perpendicular to the main axis in a way that all daughter cells are connected by their end wall. Filaments can be simple, as in *Spirogyra* (Fig. 2B), or branched. Branching can be further classified as false (as in the Cyanobacteria *Tolypothrix*, Fig. 2C) or true (as in the green *Cladophora*, Fig. 2D). Filaments can also consist of a single layer of cells (uniseriate) as in the red *Lejolisia* (Fig. 2E) or made up of multiple layers (multiseriate) as in *Polysiphonia* (Fig. 2F).

A peculiar form occurs in the siphonous algae, which have a coenocytic construction, consisting of tubular filaments that grow as their nuclei undergo repeated nuclear divisions but in which transverse cell walls never form. The resulting filament is therefore unicellular but multinucleate (coenocytic). A classic example of a branched coenocyte thallus is found on the green alga *Bryopsis* (Fig. 2G).

Fig. 1. Examples of algae with different shapes: (A) unicellular as in *Haematococcus*, (B) filamentous, (C) cyanophyceae film, (D) colonies as in *Haematococcus pluvialis*, (E) thin foliose sheets as in *Porphya*, (F) foliose tubes as in *Ulva intestinalis*, (G) bulbs as in *Colpomenia sinuosa*, (H) leader sheets as in *Laminaria,* (I) Gelatinous, (J) cartilageneous as in *Cystoseira abiesmarina*, (K) calcareous bushy as in *Ellisolandia elongata*, (L) calcareous foliose as in *Padina pavonica*, and (M) crustose habits as in *Peyssonnelia squamaria*. Photos from Island Aquatic Research, Azorean Biodiversity Group/Center for Ecology, Evolution and Environmental Changes (IAE/GBA/cE3c).

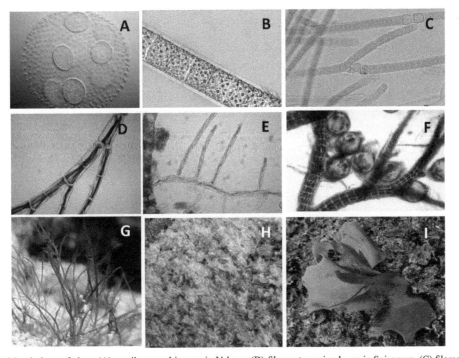

Fig. 2. Morphology of algae (A) motile coenobium as in Volvox, (B) filamentous simple, as in Spirogyra, (C) filamentous with false branches as in the Cyanobacteria *Tolypothrix*, (D) filamentous with true branches as in *Cladophora*, (E) uniseriate filaments as in the red *Lejolisia*, (F) multiple layers (multiseriate) as in *Polysiphonia*. Photos from Island Aquatic Research, Azorean Biodiversity Group/Center for Ecology, Evolution, and Environmental Changes (IAE/GBA/cE3c).

The more complex types and forms (cartilaginous) have a parenchymatous or pseudoparenchymatous construction, which originates from a meristem with cell divisions in three dimensions. In the parenchymatous thallus, cells of the primary filament divide in all directions and the essential filamentous structure is lost (e.g., *Ulva*, Fig. 2H). The pseudoparenchymatous thallus is made by the aggregation, in a loose or compact way, of numerous, intertwined, branched filaments held together by mucilages to collectively form the thallus, which therefore has little internal differentiation. A good example is found in the red alga *Schizymenia* (Fig. 2I).

Reproduction

The reproduction in algae is highly variable. Asexual reproduction, arguably the most widespread, can be vegetative (by the division of a single cell or by the fragmentation of a colony or filament) or by the production of motile spores. This mode of reproduction allows stability of an adapted genotype from one generation to the next, but restricts genetic variability. Sexual reproduction, involving the union of gametes, results in the genetic recombination needed for adaptation to changing environments (Lee 2008).

The binary fission of unicellular algae is the simplest form of asexual reproduction, in which the parent organism divides into two equal parts, each containing the same hereditary information as the parent. The autocolony formation is a specific mode in which each cell of the colony can produce a new colony similar to the one to which it belongs. Cell divisions produce a sort of embryonic colony smaller than the parent colony but equal in cell number (Fig. 2A).

In the fragmentation mode, the filaments and/or non-coenobic colonies break into two or several parts, each having the capacity of developing into new individuals.

Asexual spores can be flagellate motile (zoospores) and produced by a parental vegetative cell, aflagellate (aplanospores) that begin their development within the parent cell before being released, and autospores that are aflagellate daughter cells that will be released by the rupture of the parent cell wall. These later are almost perfect replicas of the mother cell as in the green alga *Chlorella* (Fig. 3). Spores may be produced within ordinary vegetative cells or within specialized structures called sporangia.

Under unfavorable conditions many algae can produce thick-walled resting cells, such as hypnospores, hypnozygotes, statospores, and akinetes, and enter a dormancy period. The former two are produced by protoplasts that were previously separated from the walls of the parental cells. They enable the species to survive temporary drying out of water bodies. Statospores are endogenous cysts formed within the vegetative cell. Their silica enriched walls, often ornamented with spines and other projections, make them good fossil records. Akinetes are of widespread abundance in the blue-green algae and used in Alfa taxonomy. They are essentially enlarged vegetative cells that develop a thickened wall in response to limiting environmental nutrients or limiting light.

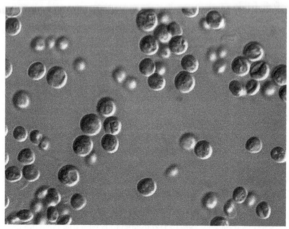

Fig. 3. *Chlorella* spores resembling mother cells. Photos from Island Aquatic Research, Azorean Biodiversity Group/Center for Ecology, Evolution and Environmental Changes (IAE/GBA/cE3c).

The various types of gametes can be classified according to relative size and motility. When both are motile and the same size (isogametes) the process is called isogamy; when gametes differ in size (heterogametes) they can undertake the anisogamy process, when both are motile but one is small (male, by convention) and the other is large (female); or the oogamy process, in which only the male gamete is motile and fuses with a very large and non-motile female gamete.

Gametes may also be morphologically identical or markedly different from vegetative cells, depending on the algal group. When similar they can only be distinguished based on the DNA ploidy.

The meiosis occurrence in a sexual life cycle and the type of cells it produces determines different types of reproductive cycles. If the meiosis occurs immediately after the zygote formation and the cycle is predominantly haploid it is called the haplontic or zygotic cycle (Fig. 4). If the cycle is predominantly diploid and the meiosis give rise to haploid gametes that are short lived and fuse to form a new zygote, the cycle is called diplontic or gametic (Fig. 4). On the diplohaplontic or sporic life cycle (Fig. 4) there is an alternation of generations between two phases, one haploid, represented by the gametophyte, and the other diploid, represented by the sporophyte. The former produces gametes by mitosis, the later produces spores through meiosis. If the two generations are morphologically identical, the alternation of generations is isomorphic. If they are different, the alternation of generations is hetermorphic.

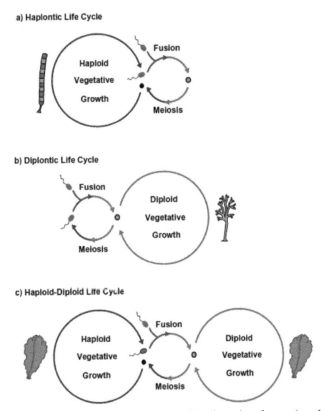

Fig. 4. Adapted from 20 Bell, G. 1994. The comparative biology of the alternation of generations: Lectures on mathematics in the life sciences: Theories for the evolution of haploid–diploid Life Cycles. 25: 1–26.

Evolution and classification

Evolution

Biotic interactions between algae and other eukaryotes are very common in aquatic and terrestrial ecosystems. These interactions can be intracellular and extracellular/surface interactions in the phycosphere, which is the ecologically and physiologically integrated neighborhood inhabited by the alga (Brodie et al. 2017). Epibiosis (surface colonization of one organism, by other organisms, epibionts) occurs on all immersed surfaces, including those of micro- and macroalgae. Epibiotic interactions (e.g., alga-alga, alga-bacterium, alga-virus) play key roles in nutrient acquisition and recycling, metabolic flux, energy flow, and developmental processes. Together with herbivory, epibiosis represents one of the most important interactions for algae and has been shown to shape entire marine communities (Brodie et al. 2017).

A very important contribution to our knowledge of the evolution of life on earth was the endosymbiotic theory by Lynn Margulis (1981) according to which single-cell creatures without complex internal structures achieved evolutionary success through increasing cellular complexity. This complexity emerged through symbiosis between organisms of different species, the most effective forms occurring when two cell types merged, each contributing to the creation of a harmonious living whole, the eukaryotic cell.

This theory assumes a monophyletic origin of eukaryotes from a common ancestor (progenote), which through various symbiosis events acquired the various organelles typical of eukaryotes. A large evolutionary radiation happened very early at the cellular level, resulting in celled organisms that differed in their cell structure. Some variants acquired chloroplast and become photoautotrophic. Others did not and remained heterotrophic.

Chloroplasts, flagella, and mitochondria have descended from free-living prokaryotes; the chloroplasts derived from cyanophytes and the latter from true bacteria. When the Earth's atmosphere already had oxygen, a primitive phagotrophic eukaryote with amoeboid movements incorporated an aerobic bacterium. It was not digested, and started functioning as a respiratory organelle (mitochondrion). As a result, the primitive eukaryote became aerobic. Later, this aerobic eukaryote incorporated another prokaryotic saprophyte, mobile by microtubules in a 9 + 2 arrangement. The incorporation was not total and the flagellum started acting as the host locomotion organelle, evolving later into the typical flagellum of eukaryotes. At about 1500 Ma ago, some of these heterotrophic flagellates incorporated a cyanobacterium (Palmer 2003; Keeling 2004) which was not digested and turned into a photosynthetic organelle (chloroplast), giving rise to the first photoautotrophic eukaryotes which subsequently gave rise to the various groups of algae and land plants.

Three major eukaryotic photosynthetic groups (possessing primary plastids) have descended from a common prokaryotic ancestor, whereas the remaining algal groups have acquired their plastids via secondary or even tertiary endosymbiosis. The three major algal lineages with primary plastids are the Glaucophyta, the Rhodophyta, and the Chlorophyta. Multicellular green algae then successfully invaded the land environment and originated mosses, ferns, and vascular plants. Other algae underwent processes of secondary endosymbiosis (Keeling 2004), for example, green algae evolved into Euglenophyta and red originated the Chromalveolata, a diverse group of algae whose chloroplasts possess chlorophyll a and c, including the Cryptophyta, the Haptophyta (coccolitophoroids), the Ochrophyta (=Heterokontophyta), and the dinoflagellates.

In this view, green algae and higher plants form a group having a common ancestor with the red algae, and the chloroplasts of brown algae resulted from a secondary endosymbiosis with a red alga. Although our understanding of the evolution of algae has improved in the last years, first by phylogenetic analyses of nuclear ribosomal sequence data (mainly 18S), and more recently by analyses of multi-gene and chloroplast genomic data, the phylogenetic relationships are still not well established (Fang et al. 2017).

Classification

The classification of living organisms has undergone a major evolution over time as a result of an increased phylogenetic knowledge. In this text we will follow the higher classification, proposed by Ruggiero et al. (2015), in two superkingdoms (Prokaryota and Eukaryota) and seven kingdoms, namely the prokaryotic Archaea (Archaebacteria) and Bacteria (Eubacteria), and the eukaryotic Protozoa, Chromista, Fungi, Plantae, and Animalia. This classification is neither phylogenetic nor evolutionary but instead represents a consensus view that accommodates taxonomic choices and practical compromises among diverse expert opinions, public usages, and conflicting evidence about the boundaries between taxa and the ranks of major taxa, including kingdoms. It reflects a hierarchical system on the basis of similarities between organisms, each group consisting of a set of organisms that are more closely related to each other than to organisms of a different group. It considers similarities in the organism's genomes as a result of a common ancestral.

There is no easily definable classification system acceptable to all algae, since taxonomy is under constant and rapid revision at all levels following new genetic and ultrastructural evidence. Keeping in mind that the polyphyletic nature of the algal group is somewhat inconsistent with traditional taxonomic groupings, and that taxonomic opinion may change as information accumulates, we will use in this text the classification adopted by Guiry and Guiry (2017) which considers 10 phyla: Cyanobacteria (=Cyanophyta) in kingdom Eubacteria; Euglenophyta in Protozoa; Glaucophyta, Rhodophyta, and Chlorophyta in Plantae; and Ochrophyta, Haptophyta, Cryptophyta, Miozoa (Dynophyceae), and Cercozoa (Chlorarachniophycea) in the Chromista. From these, only Glaucophyta is restricted to freshwater habitats. Rhodophyta and Haptophyta are predominantly marine, and the remaining phyla can be found in both marine and freshwater systems (Graham et al. 2009).

Unlike vascular plants, there are few common names for algae. Species are named according to the binomial system of nomenclature proposed by Linnaeus (1758, 1759) and in accordance to the rules of the International Code of Botanical Nomenclature (McNeill et al. 2012).

Algae phyla

Cyanobacteria

The Cyanobacteria is the most widely distributed group of algae and the dominant one in the ocean ecosystems, with about 2,000 known species (Graham et al. 2009). It includes organisms that are unicellular, colonial, and filamentous branched or unbranched (also known as trichomes). Some filamentous colonies have the ability to differentiate into several different cell types: vegetative cells; akinetes (normal, photosynthetic cells formed under favorable growing conditions); climate-resistant spores that are formed in harsh environmental conditions; and thick-walled heterocysts, which are responsible for nitrogen fixation.

Cyanobacteria are the most successful group of algae on earth. They are an important component of the picoplankton in both freshwater and marine systems and they can grow symbiotically in ferns, lichens, diatoms, sponges and other algae. Some even live in the fur of sloths, providing a form of camouflage. They also have benthic representatives that can grow as dense mats on soils, mud flats, on the seashore, and in hot springs (Barsanti and Gualtieri 2014).

The Cyanobacteria contain chlorophyll *a*, *β* carotenes, zeaxanthin, and blue and red phycobilins (phycoerythrin, phycocyanin, allophycocyanin, and phycoerythrocyanin). Their cell wall is characterized by a peptoglycan layer and the food reserve is cyanophycean starch (Dawes 1998). Some marine species also contain gas vesicles used for buoyancy regulation. A few species produce potent hapatotoxins and neurotoxins. The reproduction is strictly asexual by simple cell division or by fragmentation of colonies or filaments (Barsanti and Gualtieri 2014).

Cyanobacteria produced the Earth's first oxygen atmosphere, which fostered the rise of eukaryotes, and remain important today. They produce organic compounds used by other organisms, fulfill vital ecological functions in the world's oceans, being important contributors to global carbon and nitrogen budgets, contribute significantly to global ecology and the oxygen cycle, stabilize sediments and soils, increase water fertility and foster the growth of certain plants and fungi in symbiotic associations. Some have potential biotechnological application. Aquatic Cyanobacteria are probably best known for the extensive and highly visible blooms they can form. Some of the species involved in this process produce toxins and cause the occurrence of harmful blooms well known by their negative effects on the aquatic systems (Graham et al. 2009).

Euglenophyta

The about 900 known Euglenophyta include unicellular flagellates and colonial forms, which are widely distributed in marine, brackish, and freshwater systems, soils, and mud (Graham et al. 2009). They are especially abundant in heterotrophic environments and one of the best-known groups of flagellates.

In the flagellate forms, the flagella arise from a cavity called reservoir, located in the anterior end of the cell. Cells can also move by a series of flowing movements made possible by the presence of the pellicle, a proteinaceous wall which lies inside the cytoplasm (Lee 2008).

The Euglenophyta possess chlorophyll a, and b, β and γ carotenes, and various xanthophylls, although plastids can be colorless or absent in some species, mainly in the ones that use phagocytosis as the mode of nutrition. The reserve polysaccharide is paramylon, β-1,3-glucan stored in granules scattered inside the cytoplasm and not in the chloroplasts. Only asexual reproduction has been reported (Bold and Wynne 1985).

Glaucophyta

The Glaucophytes are simple unicellular flagellates with a dorsiventral construction and bearing two unequal flagella inserted in a depression just below the apex of the cell. The 13 known species are strictly form freshwater habitats and characterized by containing chlorophyll a, β carotenes, zeaxanthin, and phycocyanin, allophycocyanin, and phycoerythrocyanin. The food reserve is a starch (Graham et al. 2009).

Rhodophyta

Red algae, together with many representatives of the Phyla Ochrophyta (Phaeophyceae, brown algae) and Chlorophyta (green algae), are mostly considered seaweeds and, very frequently, referred to as macroalgae. Although morphologically closely related, these three phyla have fundamental differences that are evident when comparing their photosynthetic pigments, reserve foods, cell wall, mitosis, flagellar construction, morphology, and life history cycles.

The red algae are one of the oldest groups of eukaryotic algae, and also one of the largest, the number of species ranging from 5,000 to as many as 20,000 (Norton et al. 1996; Guiry and Guiry 2017).

The Rhodophyta are predominantly marine and predominantly found in warm temperate to tropical latitudes. These organisms occur both intertidally and subtidally and some species have adapted to grow at depths of 200 m and more. They encompass both benthic and free living forms and cover almost all types of growth and sizes. The Bangiophycidae (Bangiales) range from unicells to multicellular filaments of sheet-like thalli and retain morphological characters that are found in the ancestral pool of red algae. The Floridophyceae includes the more complex red algae and a higher variety of growth forms (Barsanti and Gualtieri 2014).

The Rhodophyta contain chlorophyll *a*, *α*, and *β* carotenes, zeaxanthin, and the phycobilins phycocyanin, phycoerythrin, and allophycocyanin pigments that allow these organisms to grow at depths where no other photosynthetic organisms can adapt. Their food reserve is floridean starch, a highly branched amylopectin insoluble in boiling water. The cell wall has an inner layer of randomly arranged microfibrils and an outer layer that may contain sulfated galactan polymers, some economically important such as agar, carrageenan, funoran, and furcellarin. The inner microfibrils are composed of cellulose polymers except in genus as *Porphyra* and *Bangia* (Bangiales), the more primitive red algae, where the polymers are of xylose and mannose. Calcification of cell walls, enriched by the calcite crystalline form of calcium carbonate, is characteristic in the coralline algae (Corallinales), which play a critical but often neglected role in coral reef development (Dawes 1998).

Reproduction can be sexual and asexual. In the great majority of red algae, cytokinesis is incomplete and daughter cells are linked by pit connections. A proteinaceous plug fills the junction between cells and can appear refractive under the light microscope, being a distinctive feature in alpha taxonomy (Pueschel 1989). The sexual life cycle is generally diplohaplontic, iso-, or heteromorphic (Bold and Wynne 1985).

Red algae are probably best known for their economic and ecological importance. There are over 300 species of economic importance, either for direct consumption (e.g., nori), or extraction of commercial colloids for laboratory cell-culture media or industrial additives (Graham et al. 2009).

Chlorophyta

The Chlorophyta are a large group of algae (with about 6,000 species recorded) exhibiting a wide range of somatic differentiation, from flagellates to complex multicellular thalli and including unicellular, colonial, filamentous, siphonous, and parenchymatous organisms (Thomas 2002). The different level of thallus organization is so distinct and characteristic that it has traditionally been used as the basis of the classification within this phylum. Green algae are dominated by unicellular freshwater species, the marine species being mostly macroalgae.

Most green algae are uninucleated. Exceptions occur in the orders Siphonocladales (e.g., *Ernodesmis*, *Valonia*), Cladophorales (*Anadyomene*, *Cladophora*), and Bryopsidales (e.g., *Caulerpa*, *Codium*, *Halimeda*). This later order includes coenocytic forms which are multinucleated and single-cell organisms in which there are no cell walls or cell membranes separating the nuclei of the entire plant.

The Chlorophyta contain chlorophyll *a* and *b*, responsible for the typical green colour, β and γ carotene, and various xanthophylls. The cell structure is eukaryotic and the main food reserve is starch. The structure and composition of green algal cell walls range from cellulose microfibrils (Ulvales) to highly crystalline cellulose (Siphonocladales, Cladophorales), or to polymers of xylan and mannan, which form microfibrils (Dasycladales). Several tropical green algae of the orders Bryopsidales and Dasycladales have calcified walls consisting of the aragonite form of calcium carbonate crystals. These calcified species give a significant contribution to the calcium carbonate sediments in tropical waters (Hillis-Colinvaux 1980).

The sexual life cycle can be haplontic (e.g., Bryopsidophyceae) or diplohaplontic (e.g., Cladophorophyceae) and is predominantly isomorphic (Bold and Wynne 1985).

Like other groups, some green algae are important economically and used as a direct food source, and additives for several purposes (Graham et al. 2009).

Ochrophyta

With more than 100,000 known species (van den Hoek et al. 1995), this phylum comprises mostly marine organisms and the whole variety of forms of growth and sizes, including the giant kelps (large Phaeophyceae responsible for the submerged forests that characterize the rocky sea bottom of cold and temperate regions, van den Hoek et al. 1995), the single-celled and colonial Chysophyceae, the unicellular, filamentous, and siphonous Xantophyceae, the coccoids unicells of the Eustigmatophyceae, and the well-known diatoms (Bacillariophyceae). Important members of this phylum are the brown algae belonging to the single class Paeophyceae (Silva 1980), which includes about 1,500 to 2,000 species (van den Hoek et al. 1995). Almost exclusively marine and dominant in temperate to arctic latitudes, these organisms occur both intertidally and subtidally down to depths of 100 m and more. Most require stable hard substrata for attachment but some *Sargassum* species occur only as drift populations (Dawes 1998).

The Phaeophyceae are the most complex algae and encompass filamentous, syntagmatic, or parenchymatous organisms ranging from small filamentous forms (*Ectocarpus*) and crustose forms (*Ralfsia*), to massive cartilaginous thallus (*Fucus*, *Ascophyllum*, the giant kelps). The major taxa are separated on the basis of life history and gene sequence information, but taxonomic classification is still in flux. The order Ectocarpales, which encompasses about 30 genera including *Ectocarpus*, includes the most primitive species of brown algae in which growth occurs by intercalary cell divisions of the uniseriately branched filaments. On the opposite side, the Laminariales include the largest and more complex of all algae (*Laminaria*, *Macrocystis*), which can exhibit substantial tissue differentiation including an epidermis, outer and inner cortex, and a central medulla that contains sieve like cells called trumpet hypae with a conduction function similar to sieve cells in higher plants (Schmitz and Srivastava 1980).

The Ochrophyta possess chlorophyll *a* and *c* (with the exception of the Eustigmatophyceae that only have chlorophyll *a*), *β* carotene, and various xanthophylls responsible for the typical greenish-brown coloration. The main food reserve is the complex carbohydrate polymer laminarin. Mannitol, a low molecular sugar-alcohol, can also be present and is thought to serve also as a reserve food and as an osmoticant (Bold and Wynne 1985). The cell structure is eukaryotic, with a distinct cell wall containing alginate compounds and cellulose. Typically, cells are uninucleated with the exception of the medullary cells in *Durvillea* and *Laminaria*.

The sexual life cycle can be haplontic (Chysophyceae), diplontic (Bacillariophyceae), and diplohaplontic (Paeophyceae) and can be iso- or hetermorphic (Bold and Wynne 1985).

As in the Rhodophyta, there are many brown algae species important economically, used as direct food source, commercial colloids extracts, and additives for several purposes (Graham et al. 2009).

Haptophyta

The 300 known Haptophyta (Graham et al. 2009) encompass simple organisms including unicellular, palmelloid, or coccoid forms, and also a few colonies and short filaments. The flagellate cells bear two naked flagella with different lengths, inserted apical or laterally. The algae of this phylum possess a haptonema, a long thin organelle somehow similar to a flagellum but with a distinct ultrastructure (Barsanti and Gualtieri 2014).

Haptophytes are predominantly marine, with only a few records from freshwater and terrestrial systems. The best-known are coccolithophores, some of the most abundant marine phytoplankton, especially in the open ocean. These organisms have an exoskeleton of calcareous plates called coccolith and are extremely abundant as microfossils. Other well-known planktonic haptophytes are *Chrysochromulina* and *Prymnesium*, which periodically form toxic marine algal blooms (Graham et al. 2009).

The Haptophyta possess chlorophyll *a* and *c*, *β* carotene, and various xanthophylls responsible for the typical golden yellow-brown coloration. The main food reserve is the polysaccharide chrysolaminarine. The cell wall is covered with tiny cellulosic or calcified scales bearing spot-like fibrils which are radially arranged.

Their sexual life cycle is hetermorphic diplohaplontic, in which a diploid planktonic flagellate stage alternates with a haploid benthic filamentous stage.

Cryptophyta

This cryptophytes are simple unicellular flagellate asymmetric organisms that bear two unequal hair flagella, subapically inserted parallel to one another, covered by bipartite hairs, and emerging form a deep gullet located on the ventral side of the cell. The 200 known species are typically free-swimming organisms in both marine and freshwater systems. Some members are known to be zooxanthellae in host invertebrates or within marine ciliates.

Cryptophyta possess chlorophyll *a* and *c*, *α* carotene, and various phycobilins. The main food reserve is starch granules. The cell is enclosed in a stiff, proteinaceous periplast made of polygonal plates. Cryptomonads are distinguished by the presence of characteristic extrusomes called ejectosomes, which consist of two connected spiral ribbons held under tension (Graham et al. 2009).

Although a sexual life cycle has been recently reported, the main method of reproduction is by longitudinal cell divisions (Barsanti and Gualtieri 2014).

Miozoa (Dinophyceae)

This phylum includes about 3,000 to 4,000 species, the most common being unicellular flagellates. Ameboid, coccoid, palmelloid, filamentous, and non-flagellate unicellular forms, although less common, can also occur (Graham et al. 2009). The dinoflagellates have two flagella with an independent beating pattern and a distinct location, one girdling and conferring the cell the rotator swimming motion, the other trailing. The cell is surrounded by a layer of flat, polygonal vesicles, which can be empty or filled with cellulose plates. In the typical dinoflagellates, these plates generally form a bi-partite armor, consisting of an upper anterior half and a posterior half, separated by a grove known as cingulum where the transversal flagellum is located (Barsanti and Gualtieri 2014).

The organisms of this phylum are important components of the microplankton of marine and freshwater systems. Some are parasites in invertebrates; others are endosymbiotic (zooxanthellae) of typical corals. Some are notorious for nuisance blooms and toxin production and many exhibit bioluminescence (Graham et al. 2009).

The Miozoa possess chlorophyll a, b, and c, α carotene, and various xanthophylls and phycobilins. The principal reserve polysaccharide is starch granules, but some genera also have oil droplets. These organisms have a haplontic sexual life cycle and some produce resting stages, called dinoflagellate cysts or dinocysts (Bold and Wynne 1985).

Cercozoa (Chlorarachniophyceae)

The members of this phylum are naked uninucleated cells that form a net-like plasmodium. Cercozoa is a small group of algae that occur in temperate and tropical marine waters, growing among sand grains, on mud, in tidepools, on seaweeds, or in plankton. They are typically mixotrophic, ingesting bacteria and smaller protists as well as conducting photosynthesis (Graham et al. 2009).

The Cercozoa only include few genera. They possess chlorophyll a and b. The reserve polysaccharide is paramylon, β-1,3-glucan. The reproduction is predominately asexual by cell divisions or zoospore formation. Sexual reproduction is reported for two species. The basic life cycle comprises ameboid, coccoid, and flagellate cell stages (Barsanti and Gualtieri 2014).

Distribution and major communities

Algae are very important organisms, especially in aquatic ecosystems constituents of both, planktonic and benthonic systems.

Phytoplankton

The phytoplankton is composed of autotrophic and auxothrophic organisms that float freely and let themselves be carried by water movement. It is widely distributed in the marine and freshwater environments and is probably the best studied and known group of algae that can be even studied with satellite images. It plays a key role in the primary production and marine food chains (Whittaker and Likens 1975) and is crucially dependent on minerals such as nitrate, phosphate or silicic acid, whose availability is governed by the proximity of land and rivers and the balance between the so called biological pump and the upwelling of deep, nutrient-rich waters. Many phytoplankton species also depend on organic water-soluble molecules that they can't synthesize like vitamin B. Since large areas of the world ocean are devoid of B vitamins, this limitation could be important for phytoplankton growth and for efficiency of carbon and nitrogen fixation in those regions (Sañudo-Wilhelmy et al. 2014; Bertrand et al. 2015).

The phytoplankton is dominated by unicellular forms but also includes some colonial and filamentous organisms (South and Whittick 1987), ranging in size from the smaller forms named picoplankton 0,2–2 µm, to nanoplankton 2–20 µm and up to macrophytoplankton, of 2 mm (Sournia 2008). The Sargasso Sea, which covers an area of about 4,000 km in diameter, is the only situation in which relatively large macroalgae, similar in everything to their benthic counterparts, have a planktonic life. Important constituents of phytoplankton are the coccolithophores (Haptophyta), the cyanobacteria, the euglenoids (Euglenophyta), the diatoms (Bacillariophyceae), the dinoflagellates (Dinophyceae), and reproductive stages of Chlorophyceae and Phaeophyceae.

Smayda (1980) reviewed the distribution and succession of phytoplankton species in the marine environment and concluded that, in general, the cold, nutrient rich (eutrophic) waters are dominated by diatoms and cyanophytes (Bacillariophyceae and Cyanophyceae), while in warm and oligotrophic waters (poor in nutrients), the coccolithophores predominate. Dinoflagellates are abundant in both situations. The cyanobacteria are most commonly found in tropical and/or subtropical environments (Round 1981).

Benthic algae

The benthic algae comprise micro- and macroalgae and both groups are more abundant on rocky shores. In general, the multicellular benthic macroalgae are commonly called seaweeds because of their size, construction, and attachment to firm substrata (Dawes 1988). Because they require a stable substrate for attachment and light, they are confined to marginal areas of continents and islands, where they occupy a relatively narrow range and some seamounts with tops near the surface (Ramos et al. 2016).

The macroalgae comprise the largest and most complex group of algae on rocky shores encompassing organisms which are phylogenetically distinct but morphologically similar and adapted to the same type of habitat. The morphological parallelism displayed by macroalgae of different phyla appears to be an adaptation to life in similar conditions (Norton et al. 1981).

Macroalgae contribute to primary production and oxygen release, as well as to the formation of siliceous and calcareous deposits. Some are small, forming productive turfs on coral reefs, while others, such as kelps can form large marine forests. Benthic Cyanobacteria are widespread on temperate rocky and sandy shores (Whitton and Potts 1982) and in the tropics, can form large macroscopic tufts (for example, Oscillatoriaceae) and smaller but abundant nitrogen-fixing Nostocaceae that are major components of the reef flora (Charpy et al. 2012). Most of the macroalgal biomass enters the food chains in the form of debris (Fenchel and Jörgensen 1977), and is consumed by scavengers (bacteria and fungi), which are considered structuring bodies in coastal communities (especially at the subtidal level). Other rocky shore organisms feed on macroalgal debris in suspension.

Seaweeds can also have a major role in carbon sequestration (Krause-Jensen and Duarte 2016). The fate of seaweed net primary productivity and export to the deep sea, is still not well known. But some studies (Krause-Jensen and Duarte 2016 and references therein) help to provide a first estimate of the contribution of seaweeds to carbon sequestration and export to the deep sea (> 1,000 m depth), where the carbon is precluded from exchanging with the atmosphere over long periods of time even after being remineralized.

Seaweeds offer a habitat for many different species of animals and their morphology influences the composition of the macrofauna community (Veiga et al. 2014; Matias et al. 2015; Torres et al. 2015).

Most benthic algae display seasonal differences in growth in response to environmental changes in factors such as radiation, photoperiod, temperature, nutrient availability, and herbivory (Little and Kitching 1996).Temperature is an important factor in the distribution of algae so that, for the same type of substrate, rocky shorelines located at different latitudes exhibit distinct communities. Usually, with increased latitude there is an increase in the abundance of brown algae and a reduction in the red algae (Lüning 1990). The tropical flora is devoid of large fronds and dominated by calcareous forms, including crusts, which are more resistant to the high herbivory typical of the tropics (Lubchenco and Gaines 1981). In tropical waters the predominant leafy forms are Dictyotales and Siphonales, while in northern Europe dominate the large Fucales and Laminariales.

Benthic communities in temperate and cold waters

The European temperate and cold water intertidal algal communities are dominated by brown algae, notably *Ascophyllum nodosum*, *Fucus*, and *Laminaria*. Less abundant *taxa* are *Porphyra* and *Ulva* at higher levels, and *Mastocarpus stellatus*, *Palmaria palmata*, and *Chondrus crispus* lower in the shore. Other species usually present under the Laminariales canopies include crustose coralline algae, *Polysiphonia* spp., *Dilsea carnosa*, *Ulvaria urceolata*, *Ahnfeltia plicata*, and *Rhodomela confervoides* (Dawes 1998).

The typical intertidal zonation at these latitudes is characterized by the presence of three distinct zones. The upper zone (high shore) is dominated by lichens (e.g., *Verrucaria maura*), littorinids (e.g., *Littorina*), cyanophytes (e.g., *Calothrix crustacea*), some Chlorophyta (e.g., *Prasiola*), and some red algae (e.g., *Bangia*). The intermediate zone (mid shore) is characterized by three horizontal bands: the upper one containing barnacles (*Chthamalus*, *Semibalanus*), mussels (*Mytilus*), some Chlorophyta (*Blidingia*), and some brown algae (e.g., *Pelvetia*, *Fucus distichus*, *Ralfsia*); the mid band still containing littorinids and barnacles, but dominated by *Fucus vesiculosus*, *Ascophyllum nodosum*, *Hildenbrandia*, and *Porphyra*; and finally the lower band which is totally dominated by algae, namely *Fucus serratus*, *Mastocarpus stellatus*, *Chondrus crispus*, corallines, *Ulva*, *Cladophora rupestris*, and seasonally *Dumontia* and *Scytosiphon*. The lower intertidal zone (low shore) is dominated by the red *Palmaria palmata* and several *Laminaria* species. At these northern latitudes, the Fucales distribution follows a gradation with *Pelvetia* appearing higher on the coast, followed by *Fucus spiralis*, *F. vesiculosus*, and *F. serratus*, and then *Laminaria digitata* and *L. hyperborea* (Johnson et al. 1998).

The dense mats of fucoids influence the structure and functioning of communities because these large algae provide available colonization area to numerous epiphytes; they also offer these epiphytes protection against desiccation and hydrodynamism. Consequently, fucoids dominated communities support multiple herbivores, including mollusks, barnacles, and echinoderms. In the rocky shores of northern Europe, algae and barnacles share dominance and the algal growth is the main factor influencing the coastal ecology (Hawkins and Hartnoll 1983; Jenkins et al. 2004). The transition between the intertidal and subtidal communities in temperate and cold North Atlantic is marked by the presence of kelps, species of Laminariales, predominantly *Laminaria*, *Alaria*, *Ecklonia*, *Nereocystis*, and *Macrocystis*. These algae extend their distribution to lower subtidal levels, where they can form extensive submerged forests, depending on water clarity. Kelps are globally important foundation species that occupy coastlines of all continents except Antarctica. Kelps are among the most productive primary producers on the planet, and greatly enhance diversity and secondary productivity locally by forming habitat for many other species, including those that support commercial fisheries (Bertocci et al. 2016). Sea urchins are often the structuring organisms of these forests and can cause deforestations (called sea urchin barrens) when their populations expand, eliminating all large seaweeds from considerable areas. Kelp forests are receding in many coastal areas in Europe (Araújo et al. 2016) and globally (Krumhansl et al. 2016) mostly due to climate change and other stresses.

Benthic communities of warm temperate and subtropical waters

The intertidal algal communities in warm temperate and subtropical waters are characterized by some basic differences in relation to the ones from temperate and cold waters especially with regard to the absence of the large Laminariales. This is the situation of the Macaronesian Region composed of the Atlantic islands of Azores, Madeira, the Canaries, and Cape Verde (Fig. 5).

The Macaronesia seems to represent a bridge between Northern Europe, the Mediterranean, and the tropical and subtropical coasts further south. Many cold-water species in this region have their southern limit of distribution, while others have their northern limit of occurrence here.

The typical intertidal zonation at these latitudes is characterized by the presence of three main zones (Neto et al. 2005). Higher in the shore is a zone characterized by animals (winkles, *Littorina* spp., Fig. 6A) with the marine lichen *Lichina confinis* and the Cyanobacteria *Rivularia* sporadically present. The upper mid shore zone is characterized by a patchy mosaic of animals (mainly barnacles) and algae (principally *Fucus spiralis*, *Gelidium microdon*, and *Caulacanthus ustulatus*, Fig. 6B). The lower part of this zone is characterized by algal turf, a growth of either diminutive algae or diminutive forms of larger species that create a dense, compact mat 20–30 mm thick often covering large areas of rock (Fig. 6C). Turfs may be monospecific or contain numerous species and may be characterized by articulated calcareous algae such as *Elisolandia* and *Jania*, or by soft algae such as *Ceramium*, *Chondracanthus*, and *Osmundea*. The lower level (low shore) is dominated by the larger, erect, corticated macrophytes (*Pterocladiella capillacea*, *Cystoseira abies-marina*, *Halopteris scoparia*) and articulated calcareous algae.

As in Northern latitudes, seaweed communities create a suitable habitat for many small invertebrates that live among algal thalli or within the algal turf. The more common herbivorous are the limpets that graze on the microscopic germling algal stages, and the sea-urchins.

North-East ATLANTIC

Fig. 5. Map of the Macaronesia Islands in the North-East Atlantic: Azores, Madeira, Canary Islands and Cape Verde.

Fig. 6. Intertidal zonation typical from the Macaronesean Islands. (A) High intertidal zone, dominated by animals (winkles, *Littorina* spp.) with the marine lichen *Lichina confinis* and the Cyanobacteria *Rivularia* sporadically present. (B) The upper mid shore zone is characterized by a patchy mosaic of animals and algae. (C) Low intertidal zone characterized by algal turf. Photos from Island Aquatic Research, Azorean Biodiversity Group/Center for Ecology, Evolution and Environmental Changes (IAE/GBA/cE3c).

The transition between the intertidal and subtidal is characterized by the presence of frondose and cartilaginous algae (e.g., *Elisolandia*, *Cystoseira*, *Dictyota*, *Halopteris*, *Pterocladiella*, *Stypocaulon*), although the crusts and turfs remain present.

At lower subtidal levels, rocks are often covered by filamentous forms and large foliose species (mainly Dictyotales). Crustose corallines constitute the first strata.

Benthic communities in tropical waters

Coral reefs are important components in tropical and subtropical regions characterized by clear and saline waters with high circulation and an average temperature of 22°C (Dawes 1998). They are known for their structural complexity and high productivity, for which algae are primarily responsible. This observation is counterintuitive, given that at first glance reefs seem to be depleted of algae and that species diversity of benthic macroalgae in tropical areas is about half of that in northern temperate and cold regions (Keith et al. 2014).

The main groups of algae present in coral reefs are: dinoflagellates endosymbionts (e.g., *Symbiodinium*); perforating filamentous algae of classes Cyanophyceae and Chlorophyceae that inhabit living and dead corals; crustose corallines; multispecific turf; and several erect species (*Dictyota*, *Padina*, *Sargassum*, *Udotea*, *Halimeda*). The calcareous algae act as builders of the structure that supports the whole system due to its ability to fix calcium carbonate (highly abundant in the water surrounding the coral reefs), depositing calcite ($CaCO_3$) in their cell wall (Littler 1976). Many reefs are essentially formed by these organisms. The fragmentation of these calcareous algae, by drilling organisms and scrappers, increases the amount of sediment and calcium carbonate that occurs on the reef floor. The genus *Halimeda*, for example, is responsible for about 77% of the sediments that make up some tropical reefs (Littler and Littler 1988).

The reefs are characterized by a huge quantity and diversity of animal life being among the more productive marine systems (Dawes 1998). The herbivores (mainly fish, sea urchins, and mollusks) dominate and keep algal growth under control. Without this intense grazing pressure algae would develop rapidly, covering the available substrate and inhibiting the colonization of the area by corals (Paddack and Cowen 2006).

Chemical defenses

Algae, as many other organisms, have a set of metabolic pathways that produce molecules essential for its basic operation. They also produce other, non-essential, molecules. Primary metabolites are those essential compounds without which the organism cannot survive (e.g., aminoacids, cofactors, lipids, sugars, pigments) and secondary compounds (also known as secondary metabolites or natural products) are the ones that are not involved in the development and maintenance of the organism. The latter have a limited biological distribution (often occur only in a given species) and are normally produced in response to an ecological intervention (Maschek and Baker 2009), for example, excessive solar radiation, epiphytism, or herbivory (Williams et al. 1989). Scheuer (1990) proposed that all secondary metabolites evolved from primary ones and that their action in algae is mainly protective.

Life in intertidal environments involves periods of emersion in which there is direct exposure to solar ultraviolet radiation (UV), with significant adverse effects on algal tissue (Holzinger and Lutz 2006). Seaweeds subjected to these conditions have defense mechanisms involving the action of metabolites such as phenols and pigments. Some mechanisms include: (i) changes in the amount and composition of carotenoids and xanthophylls (Goss and Torsten 2010); (ii) the mobilization of non-enzymatic antioxidants such as phenols (Bischof et al. 2006); and (iii) the synthesis of compounds which absorb UV such as aminoacids (Solovchenko and Merzlyak 2008).

The competition for space and resources is very intense in marine environments (McClintock and Baker 2001) and macroalgae are an ideal substrate for the attachment and growth of endo- and epibionts (Lane and Kubanek 2009). Some associations between algae and guests are mutual but most are competitive and the algae can be subjected to a number of infections caused by various microorganisms (Harvell et al. 1999). The effect of pathogens and herbivores can even cause the destruction of algae populations, causing imbalances in the associated communities and ecosystems (Potin 2009). Survival in an environment with these constraints led the algae to develop metabolic strategies and physiological mechanisms that enable them to eliminate or impede the deleterious action of herbivores and parasites (Armstrong et al. 2001). Among these mechanisms allelopathy, the chemical inhibition of growth of other organisms, has been the object of much attention in recent years (Hellio et al. 2000; Pereira and Gama 2009). One of the best examples of allelopathy is the production of chemicals by crustose corallines to destroy zoospores of other species, preventing their attachment and the subsequent overgrowth of upright epiphytes (Gross 2003). Kim et al. (2004) reported the destruction of spores of 14 species of macroalgae by allelopathic substances produced by the crustose *Lithophyllum yessoense*.

As the epibionts have a very diverse nature, the algae have to produce different metabolites for different guests. For a product to be a chemical inhibitor of the growth of other organisms and therefore constitute a natural and effective antiepibiont (antifouling), it must ensure the algal thallus resistance against colonization by epibiontic organisms. This is only possible through a continuous production and release of the metabolite (Jormalainen and Honkanen 2009). Each type of antiepibiont has a specific target and therefore prevents the colonization of the algal thallus by a specific organism. The revisions of Amsler et al. (2009), Jormalainen and Honkanen (2009), and Pereira and Gama (2009) on the work performed, respectively, in polar, tropical, and temperate regions, show that the production of secondary metabolites and their manifestation varies according to the geographical area. They also concluded that algae in tropical areas produce a greater variety of natural products, particularly terpenoids, as a defensive response against the higher effect of herbivory in those systems.

Natural algae products have also been related to the success of invasive species, allowing them to colonize and establish populations in areas far away from their area of origin (Liu and Pang 2010). *Caulerpa taxifolia*, for example, a species with known colonization success, produces a sesquiterpene which makes it inedible by sea urchins and toxic to fish (Uchimura et al. 1999). Also, the stronger chemical antifouling properties and competitiveness of *Sargassum* spp. compared to, for example, *F. vesiculosus*, might contribute to the invasion success of *S. muticum* (Schwartz et al. 2017).

Research on the applications of algae secondary metabolites is intense and has resulted in the isolation of numerous active compounds important for pharmaceutical uses such as antimicrobials, antivirals, antitumor and cytotoxics, hemagglutinating, and antioxidants. Hundreds of articles are published and patents are granted each year (see revision in, e.g., Blunt et al. 2011; Guedes et al. 2011). On a more personal level, several researchers recommend the integration of edible seaweed in the human diet as a way to prevent and even fight disease (Aguilera-Morales et al. 2005; Li et al. 2011; Mohamed et al. 2012).

Acknowledgements

This chapter was partially funded by Project NORTE-01-0145-FEDER-000035-INNOVMAR – Innovation and Sustainability in the Management and Exploitation of Marine Resources, namely in research line ECOSERVICES – Assessing the environmental quality, vulnerability and risks for the sustainable management of NW coast natural resources and ecosystem services in a changing world, supported by the Northern Regional Operational Programme (NORTE2020), through the European Regional Development Fund (ERDF).

References

Aguilera-Morales, M., M. Casas-Valdez, B. Carrillo-Dominguez, B. Gonzalez-Acosta and F. Pérez-Gil. 2005. Chemical composition and microbiological assays of marine algae *Enteromorpha* spp. as a potential food source. J. Food Compos. Anal. 18: 79–88.

Amsler, C.D., J. McClintock and B.J. Baker. 2009. Macroalgal chemical defenses in polar marine communities. pp. 91–104. *In*: C.D. Amsler (ed.). Algal Chemical Ecology. Springer-Verlag, Berlin.

Armstrong E., L.M. Yan, K.G. Boyd, P.C. Wright and J.G. Burgess. 2001. The symbiotic role of marine microbes on living surfaces. Hydrobiologia 461: 37–40.

Araújo, R.M., J. Assis, R. Aguillar, L. Airoldi, I. Bárbara, I. Bartsch, T. Bekkby, H. Christie, D. Davoult, S. Derrien-Courtel, C. Fernandez, S. Fredriksen, F. Gevaert, H. Gundersen, A. Le Gal, I. Lévêque, N. Mieszkowska, K.M. Norderhaug, P. Oliveira A. Puente, J.M. Rico, E. Rinde, H. Schubert, E.M. Strain, M. Valero, F. Viard and I. Sousa-Pinto. 2016. Status, trends and drivers of kelp forests in Europe: an expert assessment. Biodivers Conserv. 25: 1319–1348.

Barsanti L. and P. Gualtieri. 2014. Algae-Anatomy, Biochemistry and Biotechnology. CRC Press, Boca Raton. 361 pp.

Bertocci, I., R. Araújo, P. Oliveira and I. Sousa-Pinto. 2015. Review: Potential effects of kelp species on local fisheries. J. Appl. Ecology 52: 1216–1226.

Bertrand, E.M., J.P. McCrow, A. Moustafa, H. Zheng, J.B. McQuaid, T.O. Delmont and A.E. Allen. 2015. Phytoplankton–bacterial interactions mediate micronutrient colimitation at the coastal Antarctic sea ice edge. P. Natl. Acad. Sci. USA 112: 9938–9943.

Bischof, K., I. Gómez, M. Molis, D. Hanelt, U. Karsten, U. Lüder, M.Y. Roleda, K. Zacher and C. Wiencke. 2006. Ultraviolet radiation shapes seaweed communities. Rev. Environ. Sci. Bio. 5: 141–166.

Blunt, J.W., B.R. Copp, M.H.G. Munro, P.T. Northcote and M.R. Prinsep. 2011. Marine natural products. Nat. Prod. Rep. 28: 196–268.

Bold, H.C. and M.J. Wynne. 1985. Introduction to the Algae: Structure and Reproduction. Prentice-Hall, Englewood Cliffs, New Jersey, 720 pp.

Brodie, J., S.G. Ball, F.-Y. Bouget, C.X. Chan, O. De Clerck, J.M. Cock, C. Gachon, A.R. Grossman, T. Mock, J.A. Raven, M. Saha, A.G. Smith, A. Vardi, H.S. Yoon and D. Bhattacharya. 2017. Biotic interactions as drivers of algal origin and evolution. New Phytol. 216: 670–681.

Cavalier-Smith, T. 2004. Only six kingdoms of life. Proceedings of the Royal Society of London 271: 1251–1262.

Charpy, L., B.E. Casareto, M.J. Langlade and Y. Suzuki. 2012. Cyanobacteria in coral reef ecosystems: a review. J. Mar. Biol., Article ID 259571.

Dawes, C.J. 1998. Marine Botany. John Wiley & Sons, New York, USA, 480 pp.

Fang, L., F. Leliaert, Z.-H. Zhang, D. Penny and B.-J. Zhong. 2017. Evolution of the Chlorophyta: Insights from chloroplast phylogenomic analyses. J. Syst. Evol. 55: 322–332.

Falkowski, P.G. and A.H. Knoll. 2007. Evolution of Primary Producers in the Sea. Elsevier Academic Press Publications, 458 pp.

Fenchel, T.M. and B.B. Jörgensen. 1977. Detritus food chains of aquatic ecosystems: the role of bacteria. Adv. Microb. Ecol. 1: 1–58.

Goss, R. and J. Torsten. 2010. Regulation and function of xanthophyll cycle-dependent photoprotection in algae. Photosynthesis Research 106: 103–122.

Graham, L.E., J.M. Graham and L.W. Wilcox. 2009. Algae. Benjamin Cummings. San Francisco. xviii+616 pp.

Gross, E.M. 2003. Allelopathy of aquatic autotrophs. Crc. Cr. Rev. Plant Sci. 22: 313–339.

Guedes, A.C., H.M. Amaro and F.X. Malcata. 2011. Microalgae as sources of high added-value compounds—a brief review of recent work. Biotechnol. Progr. 27: 597–613.

Guiry, M.D. 2012. How many species of algae are there? J. Phycol. 48(5): 1057–1063.

Guiry, M.D. and G.M. Guiry. 2017. AlgaeBase. World-wide electronic publication, National University of Ireland, Galway. http://www.algaebase.org; searched on 22 December 2017.

Harris, G.P. 1986. Phytoplankton Ecology: Structure, Function and Fluctuation. University Press, Cambridge. 384 pp.

Harvell, C.D., K. Kim, J.M. Burkholder, R.R. Colwell, P.R. Epstein, D.J. Grimes, E.E. Hofmann, E.K. Lipp, A. Osterhaus, R.M. Overstreet, J.W. Porter, G.W. Smith and G.R. Vasta. 1999. Review: Marine ecology—Emerging marine diseases—Climate links and anthropogenic factors. Science 285: 1505–1510.

Hawkins, S.J. and R.G. Hartnoll. 1983. Grazing of intertidal algae by marine invertebrates. Oceanogr. Mar. Biol. 21: 195–282.

Hellio, C., G. Bremer, A.M. Pons, Y. Le Gal and N. Bourgougnon. 2000. Inhibition of the development of microorganisms (bacteria and fungi) by extracts of marine algae from Brittany, France. Appl. Microbiol. Biot. 54: 543–549.

Hellio, C., D. de La Broise, L. Dufosse, Y. Le Gal and N. Bourgougnon. 2001. Inhibition of marine bacteria by extracts of macroalgae: potential use for environmentally friendly antifouling paints. Mar. Environ. Res. 52: 231–247.

Hillis-Colinvaux, L. 1980. Ecology and taxonomy of Halimeda: primary producer of coral reefs. Adv. Mar. Biol. 17: 1–327.

Holzinger, A. and C. Lütz. 2006. Algae and UV irradiation: effects on ultrastructure and related metabolic functions. Micron. 37: 190–207.

Jenkins, S.R., T.A. Norton and S.J. Hawkins. 2004. Long term effects of Ascophyllum nodosum canopy removal on mid shore community structure. J. Mar. Biol. Assoc. UK 84: 327–329.

Johnson, M.P., S.J. Hawkins, R.G. Hartnoll and T.A. Norton. 1998. The establishment of fucoid zonation on algal-dominated rocky shores: hypotheses derived from a simulation model. Funct. Ecol. 12: 259–269.

Jormalainen, V. and T. Honkanen. 2009. Macroalgal chemical defenses and their roles in structuring temperate marine communities. pp. 57–90. *In*: Amsler, C.D. (ed.). Algal Chemical Ecology. Springer-Verlag, Berlin.

Keeling, P.J. 2004. Diversity and evolutionary history of plastids and their hosts. Am. J. Bot. 91: 1481–1493.

Keith, S.A., A.P. Kerswell and S.R. Connolly. 2014. Global diversity of marine macroalgae: environmental conditions explain less variation in the tropics. Global Ecol. Biogeogr. 23(5): 517–529.

Kim, J., J.S. Choi, S.E. Kang, J.Y. Cho, H.J. Jin, B.S. Chun and Y.K. Hong. 2004. Multiple allelopathic activity of the crustose coralline alga *Lithophyllum yessoense* against settlement and germination of seaweed spores. J. Appl. Phycol. 16: 175–179.

Krause-Jensen, D. and C.M. Duarte. 2016. Substantial role of macroalgae in marine carbon sequestration. Nat. Geosci. 9: 737–742.

Krumhansl, K.A., D.K. Okamoto, A. Rassweiler, M. Novak, J.J. Bolton, K.C. Cavanaugh, S.D. Connell, C.R. Johnson, B. Konar, S.D. Ling, F. Micheli, K.M Norderhaug, A. Pérez-Matus, I. Sousa-Pinto, D.C. Reed, A.K Salomon, N.T. Shears, T. Wernberg, R.J. Anderson, N.S. Barrett, A.H. Buschmann, M.H. Carr, J.E. Caselle, S. Derrien-Courtel, G.J. Edgar, M. Edwards, J.A. Estes, C. Goodwin, M.C. Kenner, D.J. Kushner, F.E. Moy, J. Nunn, R.S. Steneck, J. Vásquez, J Watson, J.D. Witman and J.E.K. Byrnes. 2016. Global patterns of kelp forest change over the past half-century. P. Natl. Acad. Sci. USA 113: 13785–13790.

Lane, A.L. and J. Kubanek. 2009. Secondary metabolite defenses against pathogens and biofoulers. pp. 229–243. *In*: Amsler, C.D. (ed.). Algal Chemical Ecology. Springer-Verlag, Berlin.

Lee, R.E. 2008. Phycology. Cambridge University Press, Cambridge. 560 pp.

Li, C., Y. Gao, Y. Xing, H. Zhu, J. Shen and J. Tian. 2011. Fucoidan, a sulfated polysaccharide from brown algae, against myocardial ischemia-reperfusion injury in rats via regulating the inflammation response. Food Chem. Toxicol. 49: 2090–2095.

Linnaeus, C. 1758. Systema Naturae per Regna tria Naturae. Stockholm 1: 823.

Linnaeus, C. 1759. Systema Naturae per Regna tria Naturae. Stockholm 2: 825–1384.

Little, C. and J.A. Kitching. 1996. The Biology of Rocky Shores. Oxford University Press New York.

Littler, M.M. 1976. Calcification and its role among the macroalgae. Micronesica 12: 29–41.

Littler, M.M., D.S. Littler, M.B. Stephen and N.N. James. 1985. Deepest known plant life discovered on an uncharted seamount. Science 227(4682): 57–59.

Littler, M.M. and D.S. Littler. 1988. Structure and role of algae in tropical reef communities. pp. 29–56. *In*: Lembi, C.A. and J.R. Waaland (eds.). Algae and Human Affairs. Cambridge University Press, Cambridge.

Liu, F. and S.J. Pang. 2010. Stress tolerance and antioxidant enzymatic activities in the metabolisms of the reactive oxygen species in two intertidal red algae *Grateloupia turuturu* and *Palmaria palmata*. J. Exp. Mar. Biol. Ecol. 382: 82–87.

Lubchenco, J. and S.D. Gaines. 1981. A unified approach to marine plant-herbivore interactions. I. Populations and communities. Annu. Rev. Ecol. Syst. 12: 405–437.

Lüning, K. 1990. Seaweeds: their Environment, Biogeography, and Ecophysiology. John Wiley & Sons, New York.

Margulis, L. 1981. Symbiosis in Cell evolution: Life and its Environment on the Early Earth. WH Freeman and Co, San Francisco.

Maschek, J. and B. Baker. 2009. The chemistry of algal secondary metabolism. pp. 1–24. *In*: Amsler, C. (ed.). Algal Chemical Ecology. Springer-Verlag, Berlin.

Matias, M.G., F. Arenas, M. Rubal and I. Sousa-Pinto. 2015. Macroalgal composition determines the structure of benthic assemblages colonizing fragmented habitats. PLOS ONE 10: e0142289.

McClintock, J.B. and B.J. Baker (eds.). 2001. Marine Chemical Ecology. CRC Press, New York, USA. xiv+610 pp.

McNeill, J., F.R. Barrie, W.R. Buck, V. Demoulin, W. Greuter, D.L. Hawksworth, P.S. Herendeen, S. Knapp, K. Marhold, J. Prado, W.F. Prud'Homme Van Reine, G.F. Smith, J.H. Wiersema and N.J. Turland. 2012. International Code of Nomenclature for algae, fungi and plants (Melbourne Code) adopted by the Eighteenth International Botanical Congress Melbourne, Australia, July 2011. Regnum. Veg. 154: 1–240.

Mohamed, S., S.N. Hashim and H.A. Rahman. 2012. Seaweeds: a sustainable functional food for complementary and alternative therapy. Trends Food Sci. Tech. 23: 83–96.

Neto, A.I., I. Tittley and P.M. Raposeiro. 2005. Flora Marinha do Litoral dos Açores. Secretaria Regional do Ambiente e do Mar, Horta, Açores.

Norton, T.A., A.C. Mathieson and M. Neushul. 1981. Morphology and environment. pp. 421–452. *In*: Lobban, C.S. and M.J. Wynne (eds.). Biology of Seaweeds. Blackwell Scientific Publications, Oxford.

Norton, T.A., M. Melkonian and R.A. Anderson. 1996. Algal biodiversity. Phycologia 35: 308–326.

Paddack, M.J. and R.K. Cowen. 2006. Grazing pressure of herbivorous coral reef fishes on low coral-cover reefs. Coral Reefs 25: 461–472.

Palmer, J.D. 2003. The symbiotic birth and spread of plastids: how many times and whodunit? J. Phycol. 39: 4–11.

Pereira, R.C. and B.A.P. da Gama. 2009. Macroalgal chemical defenses and their roles in structuring tropical marine communities. pp. 25–55. *In*: Amsler, C.D. (ed.). Algal Chemical Ecology. Springer-Verlag, Berlin.

Potin, P. 2009. Oxidative burst and related responses in biotic interactions of algae. pp. 245–271. *In*: Amsler, C.D. (ed.). Algal Chemical Ecology. Springer-Verlag, Berlin.

Pueschel, C.M. 1989. An expanded survey of the ultrastructure of red algae pit plugs. J. Phycol. 25: 625–636.

Ramos, M., I. Bertocci, F. Tempera, G. Calado, M. Albuquerque and P. Duarte. 2016. Patterns in megabenthic assemblages on a seamount summit (Ormonde Peak, Gorringe Bank, Northeast Atlantic). Mar. Ecol. 37: 1057–1072.

Round, F.E. 1981. The Ecology of the Algae. Cambridge University Press, Cambridge.

Ruggiero, M.A., D.P. Gordon, T.M. Orrell, N. Bailly, T. Bourgoin, R.C. Brusca, T. Cavalier-Smith, M.D. Guiry and P.M. Kirk. 2015. A higher level classification of all living organisms. PLoS ONE 10(4): e0119248. doi:10.1371/journal.pone.0119248.

Ruggiero, M.A., D.P. Gordon, T.M. Orrell, N. Bailly, T. Bourgoin, R.C. Brusca, T. Cavalier-Smith, M.D. Guiry and P.M. Kirk. 2015. Correction: A higher level classification of all living organisms. PLoS ONE 10(6): e0130114. doi:10.1371/journal.pone.0130114.

Sañudo-Wilhelmy, S.A., L. Gómez-Consarnau, C. Suffridge and E.A. Webb. 2014. The Role of B Vitamins in Marine Biogeochemistry. Annu. Rev. Mar. Sci. 6: 339–367.

Schwartz, N., S. Rohde, S. Dobretsov, S. Hiromori and P.J. Schupp. 2017. The role of chemical antifouling defence in the invasion success of Sargassum muticum: a comparison of native and invasive brown algae. PLoS ONE 12(12): e0189761.

Scheuer, P.J. 1990. Some marine ecological phenomena - chemical basis and biomedical potential. Science 248: 173–177.

Schmitz, K. and L.M. Srivastava. 1980. Long distance transport in Macrocystis integrifoli. III. Movement of THO. Plant Physiol. 66: 66–69.

Silva, P.C. 1980. Names of classes and families of living algae: with special reference to their use in the Index Nominum Genericorum (Plantarum). Regnum. Veg. 103: 1–156.

Smayda, T.J. 1980. Phytoplankton species succession. pp. 493–570. *In*: Morris, I. (ed.). The Physiological Ecology of Phytoplankton. Blackwell Scientific Publications, Oxford.

Solovchenko, A.E. and M.N. Merzlyak. 2008. Screening of visible and UV radiation as a photoprotective mechanism in plants. Russ. J. Plant Physl.+ 55: 719–737.

Sournia, A. 2008. Form and function of marine phytoplankton. Biol. Rev. 57: 347–394.

South, G.R. and A. Whittick. 1987. Introduction to Phycology. Blackwell Scientific Publications, Oxford. 341 pp.

Thomas, D. 2002. Seaweeds. The Natural History Museum, London.

Torres, C., P. Veiga, M. Rubal and I. Sousa Pinto. 2015. The role of annual macroalgal morphology in driving its epifaunal assemblages. J. Exp. Mar. Biol. Ecol. 464: 96–106.

Uchimura, M., R. Sandeauz and C. Larroque. 1999. The enzymatic detoxifying system of a native Mediterranean scorpion fish is affected by *Caulerpa taxifolia* in its environments. Envir. Sci. Tech. 33: 1671–1674.

Veiga, P., M. Rubal and I. Sousa-Pinto. 2014. Structural complexity of macroalgae influences epifaunal assemblages associated with native and invasive species. Mar. Environ. Res. 101: 115–123.

Whitton, B.A. and M. Potts. 1982. The Biology of Cyanobacteria, Blackwell Sci. Public. Oxford.

Whittaker, R.H. and G.E. Likens. 1975. The Biosphere and Man. Springer-Verlag, Berlin. 305–328 pp.

Wijesinghe, W.A.J.P. and Y.J. Jeon. 2011. Biological activities and potential cosmeceutical applications of bioactive components from brown seaweeds: a review. Phytochem. Rev. 10: 431–443.

Williams, D.H., M.J. Stone, P.R. Hauck and S.K. Rahman. 1989. Why are secondary metabolites (natural products) biosynthesized. J. Nat. Prod. 52: 1189–1208.

Zechman, F.W., H. Verbruggen, F. Leliaert, M. Ashworth, M.A. Buchheim, M.W. Fawley, H. Spalding, C.M. Pueschel, J.A. Buchheim, B. Verghese and M.D. Hanisak. 2010. An unrecognized ancient lineage of green plants persists in deep marine waters 1. J. Phycol. 46: 1288–1295. doi:10.1111/j.1529-8817.2010.00900.x.

Importance of Seaweed in the Climate Change

Seaweed Solution

Ik Kyo Chung,[1,*] *Jung Hyun Oak,*[2] *Jin Ae Lee,*[3] *Tae-Ho Seo,*[4] *Jong Gyu Kim*[5] and *Kwang-Seok Park*[6]

Introduction

Approximately 71% of the earth's surface is covered by seawater. Because the sea plays a dominant role in regulating climate, it also offers great potential for removing and fixing atmospheric carbon dioxide (De Vooys 1979; Raven and Falkowski 1999; Falkowski et al. 2000; Pelejero et al. 2010). The estimated contribution of marine vegetation, which covers less than 2% of the ocean surface, is about 210 to 244 Tg C yr^{-1}, or almost half of the total carbon sequestered in the oceans (Duarte et al. 2005). In coastal areas, mangroves, seagrasses, and salt marshes, large amounts of "blue" carbon are accumulated and retained in sediments (Laffoley and Grimsditch 2009; Nellemann et al. 2009). Although they do not have sedimentary substrates, these kelp forests and seaweed beds can be considered blue carbon (Smith 1981; Ritschard 1992; Chung 2007; Chung et al. 2011, 2013, 2017; N'Yeurta et al. 2012; Duarte et al. 2017).

Although no complete survey has been done, the global standing kelp crop, allowing for seasonal and spatial variances, is as much as 20 Tg C, with a biomass density of 500 g C m^{-2} (Reed and Brzezinski 2009). A conservative estimate of global net primary production is up to 39 Tg C yr^{-1} for deep tropical kelp forests

[1] Dept. of Oceanography, Pusan National University, Busan 46241, R. Korea.
[2] Marine Research Institute, Pusan National University, Busan 46241, R. Korea.
  Email: oakjh@pusan.ac.kr
[3] School of Environmental Science and Engineering, Inje University, Gimhae 50834, R. Korea.
  Email: envjal@inje.ac.kr
[4] Coastal Production Institute, Yeosu 59699, R. Korea.
  Email: eelis@hanmail.net
[5] Response to Climate Change Co. Ltd., Pohang 37673, R. Korea.
  Email: jkkim@rcc-posco.co.kr
[6] Material Research Division, Research Institute of Industrial Science & Technology, Pohang 37673, R. Korea.
  Email: kspark@rist.re.kr
* Corresponding author: ikchung@pusan.ac.kr

(Graham et al. 2007; Reed and Brzezinski 2009). Based on these data, seaweeds could account for about 16.0 ~ 18.7% of the total marine-vegetation sink (Duarte et al. 2005; Krause-Jensen and Duarte 2016).

Because space is scarce on islands in the open ocean or in countries with relatively short coastlines, the concept of "multiple use" must be addressed. The massive expansion of wind farms in offshore areas of the North Sea have introduced the possibility of combining those turbines with the installation of extensive facilities for shellfish and seaweed aquaculture (Buck and Buchholz 2004; Buck et al. 2004). This might include the integration of fish cages within the foundation of wind farms as well as the extractive components of integrated multi-trophic aquaculture (IMTA) systems (McVey and Buck 2008). Offshore wind farms provide areas that are free of shipping traffic and are appropriately sized for farming. Such sites present an ideal opportunity for devising and implementing a multiple-use concept (Buck et al. 2004; Michler-Cielucha et al. 2009). Furthermore, recent advances in mariculture techniques have led to greater supplies of seaweeds as a "marine crop" (Chritchley and Ohno 1998).

The concept of Ocean Macroalgal Afforestation (OMA) has been introduced to reduce atmospheric CO_2 and to produce biomethane and biocarbon dioxide. In fact, OMA could remove 53 billion tons of CO_2 per year based on a calculation if macroalgae forests cover 9% of the ocean surface (N'Yeurta et al. 2012). According to a Korean survey from a small-scale experiment, seaweed farms could hold up to 16 tons of CO_2 ha^{-1} yr^{-1} (Chung et al. 2013).

Generally, the impact of climate change on the functioning of the world's marine ecosystems includes decreased ocean productivity, altered food web dynamics, reduced abundance of habitat-forming species, shifting species distributions, and a greater incidence of disease (Hoegh-Guldberg and Bruno 2010). Consequently, the structure of this ecosystem directly and indirectly changes, and researchers do not yet have a comprehensive understanding of those effects. For example, fluctuations in temperature also influence general metabolic activities, the ratio of photosynthesis to respiration, and net primary productivity by seaweed communities.

Seaweeds are the fundamental biota in building marine food webs, serving as habitat and nursery and removing organic pollutants from the seawater. They are also affected by ocean warming and acidification (Wernberg et al. 2011; Koch et al. 2013). Although changes in the assemblage and distribution of seaweeds have been reported (Müller et al. 2009; Wernberg et al. 2011), it is sometimes difficult to conclude that such events are caused by climate change due to natural variations and anthropogenic alterations (Merzouk and Johnson 2011). Even though there are ever increasing concerns about these matters, they are beyond the scope of this chapter.

Phycologists in Korea have initiated a project for possible applications of algae in carbon sequestration and the Asian Pacific Phycological Association (APPA) launched a working group, "Asian Network for Using Algae as a CO_2 Sink," at the 4th Asian Pacific Phycological Forum at Bangkok in 2005. Since then, network members have cooperated in Side Events at the Conference of the Parties (COP) to the United Nations Framework Convention on Climate Change (UNFCCC) (http://unfccc int). After the Paris Agreement at the COP21 was adopted in December 2015, we are now in a new global climate regime. In the same context, the APPA network has adopted a new paradigm including adaptation measures of seaweeds and shifted to the Asian Network of Algae as Mitigation and Adaptation Measures (ANAMAM). The second Wando Seaweed Expo hosted a special workshop on "ANAMAM" and "Carbon Zero Seaweed Town (CØST)" on Apr. 14–15, 2017. The results of the workshop on the seaweed solution as mitigation and adaptation measures are summarized in Fig. 1.

When the Korean project 'GHG emissions reduction using seaweeds' was conducted from 2006 to 2011, all the concept and purposes of the project were focused on the Kyoto Protocol and the Kyoto Mechanism. Although we are in a new climate regime now, the concept and process of Korean project in mitigation and adaptation measures should be the same as implementation of Nationally Determined Contributions (NDCs) in the context of the Paris Agreement. As capacity-building of NDCs is a key element for implementing the Paris Agreement, the seaweed solution should be considered in improving and broadening knowledge to meet the goals of the Paris Agreement.

In addition, there has been ever increasing attention on seaweeds in the context of climate change as blue carbon (Krause-Jensen and Duarte 2016; Duarte et al. 2017) including seaweed aquaculture beds (Sondak et al. 2017). The ecosystem services of seaweed beds and kelp forests become critical in

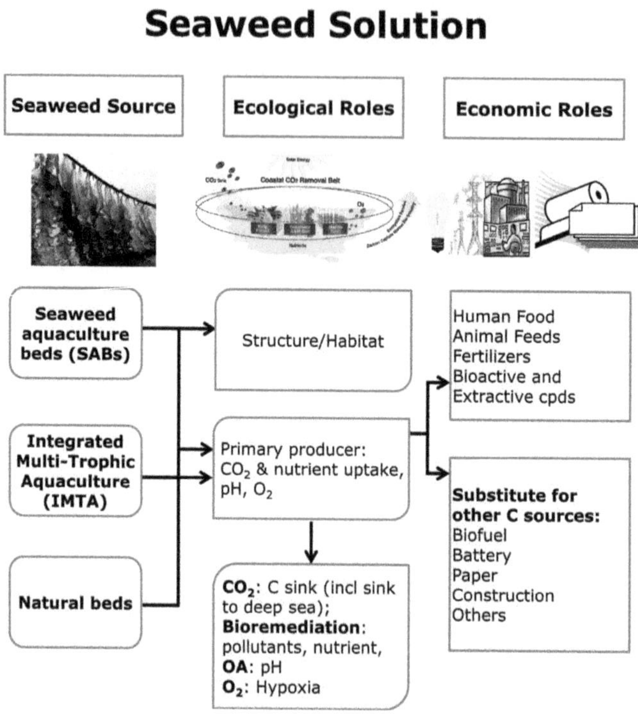

Seaweed Solution

Fig. 1. Conceptual diagram of the seaweed solution discussed at the first workshop of the Asian Network of Algae as Mitigation and Adaptation Measures (ANAMAM) and Carbon Zero Seaweed Town (CØST) summarized by Prof. Put O. Ang, Jr. in 2017.

regulating capabilities against ocean acidification, deoxygenation, and eutrophication (Chung et al. 2017; Kang and Chung 2017).

We briefly introduce the Korean project (Chung et al. 2011, 2013) in the present study. For the first time, we developed a draft of Seaweed CDM-AR-PDD. Hopefully, the work could provide valuable information for countries preparing the NDCs with seaweeds (Sondak et al. 2017).

Korean initiative

Concept of a Coastal CO_2 Removal Belt (CCRB)

At their annual meeting in 2005, the Korean Society of Phycology decided to recruit a task force to evaluate algae-related projects for combating the effects of climate change. From those proposals, the Project 'Greenhouse Gas (GHG) Emissions Reductions using Seaweeds' was funded by the Korean Ministry of Maritime Affairs and Fisheries (MOMAF, now Ministry of Oceans and Fisheries) in 2006 (http://agw-seaweed.org). The purpose of that project was to develop a seaweed Clean Development Mechanism (CDM). The objective of the Coastal CO_2 Removal Belt (CCRB; Fig. 2) was to accomplish CO_2 removal via forests or beds in coastal waters, implementing those tasks along various spatial-temporal scales (Chung 2007; Chung et al. 2013). The CCRB was to be established and managed by CDM Projects. To achieve this goal efficiently, the project was conducted in two phases. The first, from 2006 to 2008, involved selecting suitable species for the CCRB and developing methods for monitoring the baseline of seaweed communities. The second phase, from 2009 to 2011, focused on conducting a pilot CCRB farm study to devise a new method for measuring the actual, net removal of GHG by seaweeds.

As a newly developed concept, the CCRB had several facets that required open discussion. Intended for coastal regions, it was to be a natural and/or man-made community that achieved CO_2 removal in the manner of a forest, and it was defined according to various spatial-temporal scales. Its operational

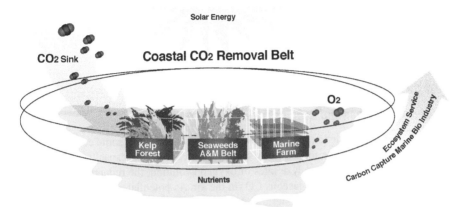

Fig. 2. Conceptual schematic diagram of the Seaweed Adaptation & Mitigation belt within CCRB. A, Adaptation; M, Mitigation (Chung et al. 2013).

definitions stipulated that (1) it should be a man-made marine plant community managed by CDM project participants; (2) it had a definite scale of area or volume, as designated in the Project Design Document (PDD) and approved by the CDM Executive Board (EB); and (3) it would operate during the proposed crediting period.

Strategic goals

The project focused on mapping short- to long-term strategies to encourage the UNFCCC's approval of seaweed CDM methodology, while also assessing the economic impacts of reducing GHG emissions (http://agwseaweed.org). It included several important activities: (1) collecting data to verify that the seaweed mechanism is effective as a CO_2 sink, (2) identifying the mechanism by which CO_2 is removed through the metabolism of seaweeds and other marine organisms, (3) developing management technologies to enhance marine-environment conservation while conducting research on sustainable CO_2 removal, (4) drawing up a plan to participate in CO_2-removal technology sales and emissions-trading in response to enactment of the Kyoto Protocol, and (5) uniting with other international parties to create a favorable environment for the approval of methodologies employed in seaweed CDM project activities.

Project tasks

Our project entailed the following tasks:

- Develop CDM methodology
- Assess its environmental and economic impacts
- Set up a national/international network to deal with relevant policies
- Design management technologies for the establishment and conservation of CCRB in coastal waters
- Estimate CO_2 removals by sinks and evaluate their efficiency
- Lay the groundwork for the practical use of marine biotechnologies that employ CO_2-fixing mechanisms
- Take a lead role in conducting research on seaweed physiology and ecology
- Establish a foundation for seaweed studies and the development of professional human resources, taking into account the fact that Korea has one of the top three seaweed-production technologies in the world
- Contribute to the recovery and increase in marine-life resources via seaweed afforestation in Korean coastal waters
- Add new value to seaweed studies that help to increase Korea's total seaweed stocks

- Favor local and national growth through the sale of seaweed CO_2-removal technologies, and participate in emissions trading
- Develop baseline/monitoring methodologies and new techniques for the management of seaweed CDM projects in coastal zones

Results of the Korean project

Selection of suitable seaweed species for CCRB

We investigated the photosynthetic capacities of wild seaweed species and some cultivar strains to select those that would be most suitable for CCRB in Korea. To measure their capacity for CO_2 removal, we applied three different methods to analyze rates of oxygen evolution, the reduction of inorganic carbon in seawater, and the accumulation of gaseous CO_2 in the air. As potential candidates, we recommended six seaweeds for the removal of atmospheric CO_2—*Ulva pertusa* ('Galparae' in Korean), *Saccharina* (=*Laminaria*) *japonica* ('Dashima'), *Undaria pinnatifida* ('Meeyeok'), *Ecklonia cava* ('Gamtae'), *E. stolonifera* ('Gompee'), and *Grateloupia lanceolata* ('Gaedobak').

Methods for monitoring baseline populations for seaweed communities

We monitored seaweed communities along the Korean coast, including Jeju Island, to estimate a baseline for natural beds. Ten core sites were evaluated for three years (2006–2009) and another 20 satellite sites were surveyed for one year only. A total of 64 sites were seasonally surveyed in July, October, January, and April from the subtidal to the splash zone. Conventional methods were applied. To estimate the seaweed biomass or standing crop, we adopted a nondestructive technique. Based on the percent coverage and weight of dominant species within a community, we were able to obtain species-specific regression equations between coverage and biomass for each dominant species (Ko et al. 2008). In the case of kelp with long and tall fronds, data for density and individual sizes were required. At each sampling site, five replicates of 50×50 cm quadrats were assessed in the intertidal, upper, middle, and lower zones, at subtidal depths of 1, 5, and 10 m. Each community was categorized overall into five size classes, based on average standing stocks: Class 1, > 2400 g m^{-2}; Class 2, 1800 ~ 2400 g m^{-2}; Class 3, 1200 ~ 1800 g m^{-2}; Class 4, 600 ~ 1200 g m^{-2}; and Class 5, < 600 g m^{-2}. From this analysis (densities shown in Fig. 3), we determined that the coast of Jeju Island had the largest average size class at 1.8, followed by 2.9 for the southern coast (Choi et al. 2008a; Kim et al. 2008), 3.2 for the eastern coast (Kang et al. 2008; Shin et al. 2008), and 5.0 for the western coast (Choi et al. 2008b; Wan et al. 2009).

Fig. 3. Regional average biomass densities (by fresh weight) for five dominant species in eastern, western, and southern coastal regions of Korea, as well as Jeju Island, based on seasonal assessments. Values are means ± standard deviations.

Establishment and maintenance of CCRB with artificial seaweed communities

To develop a seaweed A/R (afforestation/reforestation) CDM PDD, we established and maintained a pilot CCRB farm during the second phase of the project period. The concept promoted the removal of CO_2 via marine vegetation when sites were populated with the perennial brown alga *Ecklonia*. Our farm drew down approximately 10 tons of CO_2 ha^{-1} yr^{-1}. This success was manifested as an increment in biomass accumulations and a decrease in the amount of dissolved inorganic carbon in the water column (Chung et al. 2013).

Development of components for a seaweed A/R CDM PDD application

Project Algae and Global Warming (AGW) utilizes innovative research on seaweed biology and ecology to establish a new baseline and to monitor methodologies with the goal of creating international consensus that seaweed should be recognized as a GHG sink. Bench-marking of forests and CDM-A/R on land is a good starting point. From this, a CCRB and Coast Use Coast-Use Change and Aquatic Vegetation (tentative abbreviation: CUCUCAV) can then be developed. Paradigms and concepts can also be provided for these new CDM methodologies, with respect to both lifespan and scale.

In the beginning, baseline data must be gathered in order to devise a protocol. As presented in guidelines by the UNFCCC, certain information are required for the PDD, including the following:

- General description of the proposed project activity
- Application of a baseline methodology
- Description and plan for a monitoring methodology
- Declaration of the duration for project activities/crediting
- Estimations of the net anthropogenic GHG removals by sinks and GHG emissions by sources
- Environmental and socio-economic impacts of the proposed A/R CDM project activity
- Stakeholders' comments

The required content differs slightly between regular- and small-scale projects. For example, our seaweed A/R CDM project was to be conducted on a small scale because seaweeds were not yet recognized as a carbon sink. Therefore, we submitted a form for a Simplified Project Design Document for Small-Scale project activities (CDM-SSC-PDD) that would describe our new small scale methodologies (F-CDM-SSC-NM) in order to be validated and registered (http://cdm.unfccc.int/methodologies/SSCAR/index.html). As of December 2012, only seven small-scale A/R methodologies have been approved (http://cdm.unfccc.int/methodologies/SSCAR/approved).

A draft version of seaweed CDM-AR-PDD

Section A: General description of the seaweed CDM project activity

A.1 Title of the proposed seaweed CDM project

Title: Seaweed A/R CDM Project (Seaweed CO_2 Sink Farm Project)

Version of the document: Draft

Date of the document: December 2012 (prepared in 2011)

A.2 Description of activities in the seaweed CDM project

We established a 0.5-ha farm for perennial brown seaweed in the coastal waters of Korea to develop a new method for an A/R CDM PDD. Activities included removing CO_2 from the seawater and storing it, via photosynthesis, in carbon pools within the project boundary. Carbon was to be sequestered within this ocean farm, where plants of perennial seaweed species would serve as a CO_2 sink.

Background: The pilot project was supported by the Korean Ministry of Oceans and Fisheries, formerly the Ministry of Marine Affairs and Fisheries (MOMAF). It was managed by the Marine Research Institute

(MRI) of the Institute for Research and Industry Cooperation (IRIC) at Pusan National University (PNU), through an initiative for "Greenhouse Gas (GHG) Emissions Reduction Using Seaweeds". This seaweed CO_2 sink farm was built as a mid-water long-line culture system along the southern coast of Korea (Chung et al. 2013). The designated area, within a coastal environment, had muddy-sand sediment at the bottom and previously lacked aquatic vegetation. The site did not involve any land- or coastal-based uses related to fisheries products.

Purpose of the pilot project: Using carbon credits from CDM projects, we hope to secure the economic status of this business interest while also contributing to the sustainable development of our country by preserving and restoring coastal ecosystems. In doing so, we can take steps toward coping with climate change and implement a green growth policy that utilizes seaweed resources. Activities within this pilot project should improve water quality and preserve fisheries resources in the southern coastal environment of Korea. Measures can be developed for reducing GHG emissions and mitigating global warming. Projects such as these are reasonably conducted through policies that are fully compliant with standards recommended by various environmental groups. Moreover, we can identify ways to minimize any conflicts within those groups. Our two primary objectives are (1) to mitigate climate change by reducing GHG emissions through greater biomass production and (2) to increase incomes in local communities through short-term job creation, while improving the sustainable harvest of additional, subsidiary seaweed farm products in the mid-term.

To attain these objectives, we included the following components: installing a plantation of *Ecklonia cava*, *Saccharina japonica*, and *E. stolonifera* for ecosystem restoration that would not involve harvesting; increasing awareness and improving technical assistance to local residents by operating a seaweed farm for conservation and sustainability; and monitoring and managing the implementation of this project over its entire life span.

Contribution to sustainable development: This project should contribute not only to environmental protection, but also to socio-economic improvement in coastal areas by building capacity for residents and creating opportunities for employment. Several scheduled events would enhance the knowledge base for local communities on environmental conservation, seaweed restoration, and management. Short-term employment would be available during the implementation of the project and throughout its phase of monitoring. A seaweed farm was expected to improve water quality, leading to a potential ecosystem service and better fisheries resources that might increase incomes for local fishermen. Various project activities were planned to provide several benefits, as outlined below.

Establishing and restoring kelp forests through pilot A/R farms can help maintain and gradually improve marine biodiversity in several ways. First, the successful, intensive cultivation of massive seaweed communities provides sufficient food and habitat for marine animals. This leads to better fisheries stocks and eventually increases the potential for marine biodiversity in waters near the project site. Second, the effects of several threats to the marine environment can be mitigated or neutralized, so that the project areas are no longer barren. This is manifested in the natural and spontaneous enhancement of ocean resources and may attract other marine life to the area, ultimately increasing local fisheries stocks. Third, greater production of seaweed biomass means that more CO_2 can be removed, thereby reducing the acidity of surrounding waters through photosynthetic activity. Lowering the levels of ocean acidification also promotes marine biodiversity and abundance.

This farming project should encourage the development of new streams of income and increased revenues for fishermen and other participants. The gradual rise in incomes is accomplished in several ways. First, fisheries resources are enhanced through improvements to the marine environment. This contributes to the maintenance of a stable income for fishermen's households and can help develop the local economy by increasing yield landings and reducing deviations in the catch. Second, the farms provide a sink for absorbed CO_2, allowing participants to trade carbon credits achieved through reductions in GHG emissions. Therefore, business conducted via this CDM can be a source of economic development. Shared profits can also be allocated to participating economic agents, which will help raise the incomes of local communities.

To create more profits, we suggested that this pilot program be combined with the construction/restoration of a seaweed (kelp forest) A/R CDM project. This would involve the following:

- Constructing/restoring an approximately 0.5-ha pilot farm with perennial brown seaweeds, such as *Saccharina* (=*Laminaria*) *japonica*, *Ecklonia cava*, and *E. stolonifera*
- Establishing a legal organization that supports the trade of carbon credits produced through this construction and restoration. Such an organization would analyze those credit purchases in detail while simultaneously gaining practical and technical experience in related CDM businesses
- Observing and evaluating how CDM project activities impact the environment and matters related to socio-economics
- Designing and managing the build-up of project funds for pilot A/R CDM farms
- Developing, investigating, and promoting the best approach for managing the marine environment, while also strengthening training programs and giving technical assistance to communities

Individual operators might not have been interested in this pilot study if they did not see that an additional source of income could be generated from these efforts. However, if new revenues were simultaneously obtained by trading carbon credits and managing a pilot farm, then local fishermen and CDM business sectors would recognize the benefits that can ensue. This profit created by the additional carbon credit trading through the seaweed CDM operation could subsidize incomes of individual operators directly and be advantageous in both social and environmental service aspects of local communities.

Construction and restoration of the 0.5-ha seaweed farm, which began in May 2009, was scheduled to take between 24 and 60 months and be completed to a depth of 5 to 10 m. Basic data needed for the financial analysis model and the selection of target seaweed species were determined after discussions with experts and local fishermen. Some considerations included the rates of carbon absorption, enhancement of biodiversity, seawater purification, and additional value from byproducts while the project was active. This farm was managed to minimize the risk of biomass losses caused by pests, diseases, and typhoons, and was aimed at maximizing profits, both environmental and economical, through the cultivation of numerous seaweed species and a multi-stage, integrated aquaculture system.

A.3 Project participants

The host country was Korea, which ratified the Kyoto Protocol in November 2002. Project participants were from MRI/PNU, located at Busan, Korea (Busandaehak-ro 63beon-gil 2, Busan Metro City 46241, Korea). In addition, the stakeholders' group includes individual fishermen (residents in the CDM region, i.e., Pyeongsan-1-li, Namhae-gun, Gyeongsangnam-do), local municipal governments in Gyeongsangnam-do Province, the Fisheries Cooperation Federation, the Fisheries Association Co-op, other stakeholders responsible for emissions reductions, private companies interested in the CDM project (e.g., POSCO), and countries or local governments who may be responsible for delegated duties of reduction obligations.

A.4 Technical description of the seaweed A/R CDM project and community surveys

The project activities took place at Pyeongsan-1-li, Namhae-gun, Gyeongsangnam-do (in front of Sojukdo), with coordinates of N 34°45'49" E 127°50'10". In July 2009, characteristics of seaweed ecology were surveyed in the vicinity to provide baseline information (Table 1). From this, 13 seaweed species were identified: two greens, three browns, and eight reds. The greatest biomass was measured from *Gelidium amansii*, followed by *Ulva pertusa* and *Ecklonia cava*. In October 2009, 12 species were identified (two greens, six browns, and four reds). *Sargassum fusiforme* (=*Hizikia fusiformis*) produced the most biomass, followed by *Ishige okamurae* and *G. amansii*. In January 2010, 18 species were recorded (one green, seven browns, and 10 reds). Biomass was greatest from *Sargassum horineri*, *Undaria pinnatifida*, and *G. amansii*. The final survey, in April 2010, revealed 16 species (one green, eight browns, and seven reds), dominated by *S. horineri*, *U. pinnatifida*, and *I. okamurae*. During the survey period, most of the biomass was produced at a 1-m depth, between the lower intertidal zone and the subtidal zone.

Table 1. Estimates of carbon levels within dominant species based on CHN analyses of samples taken from natural seaweed communities near the project area. Each sampling date covered 800 m^2.

Dominant species*	Jul. 2009		Oct. 2009		Jan. 2010		Apr. 2010	
	Ul. p	G. a	S. f	I. o	S. h	Un. p	S. h	Un. p
Dry wt. (g m^{-2})	21.6	40.2	15.7	14.7	127.1	62.6	66.3	61.7
C content in fronds (%)	27.9	27.4	27.0	29.8	24.9	22.4	24.0	23.8
Areal C content (g m^{-2})	6.0	11.0	4.2	4.3	31.6	14.0	15.9	14.7
Total C removed (kg)	4.8	8.8	3.4	3.4	25.3	11.2	12.7	11.7

*Ul. p = *Ulva pertusa*, G. a = *Gelidium amansii*, S. f = *Sargassum fusiforme* (=*Hizikia fusiformis*), I. o = *Ishige okamurae*, S. h = *Sargassum horneri*, Un. p = *Undaria pinnatifida*

The PDD included a description of current environmental conditions in the area, for example, climate, hydrology, soils, ecosystems, and the possible presence of rare or endangered species and their habits. Data such as temperature (°C), salinity, pH, dissolved oxygen (mg L^{-1}), Chemical Oxygen Demand (mg L^{-1}), Dissolved Inorganic Nitrogen (mg L^{-1}), Dissolved Inorganic Phosphorus (mg L^{-1}), Suspended Solids (mg L^{-1}), and the concentration of Chlorophyll a (μg L^{-1}) were obtained from an environmental report by the National Institute of Fisheries Research (NIFS; http://www.nifs.re.kr).

Seaweeds used in this project were chosen by a group of experts from academia, research institutes, and industries. The NIFS, PNU, Chonnam National University, and the Research Institute of Industrial Science and Technology (RIST) participated in the selection process, during which rates of CO_2 uptake, survival, and growth, as well as other related characteristics, were discussed. Opinions from the stakeholders' community were also considered when making decisions. Table 2 presents various characteristics for these selected species.

Specifications for GHG emissions that would be part of this project were based on the assumption that CO_2 would be captured. This gas is emitted by human activities, seaweed spore production, and the transportation of materials and personnel needed for installing the project. Selected carbon pools are shown in Table 3.

Because this pilot project was the first attempt to recognize seaweeds as a carbon sink, we needed to provide new terminology for various aspects. For example, 'Seaweeds' are collectively defined as photosynthetic marine algae for which the mature fronds are ≥ 30 cm long and/or wide. Here, the focus was on perennial macro-algae. Seaweeds are categorized according to morphology, frond texture, and photosynthetic activity (Littler and Littler 1980; Littler et al. 1983). They include groups that are sheet, filamentous, coarsely branched, thick-leathery, jointed-calcareous, or crustose.

Table 2. Characteristics of seaweed species selected for the pilot project.

Species	*Saccharina japonica* (Areschoug) C.E. Lane, C. Mayes, Druehl & G.W. Saunders*	*Ecklonia cava* Kjellman	*Ecklonia stolonifera* Okamura
Korean name	Dashima	Gamtae	Gompee
Classification**	Ochrophyta Phaeophyceae Laminariales Laminariaceae	Ochrophyta Phaeophyceae Laminariales Lessoniaceae	Ochrophyta Phaeophyceae Laminariales Lessoniaceae
Distribution	Korea, Japan, Kamchatka, Sakhalin in the Pacific Coast	Korea (southern coast and Jeju Island), Japan	Korea (eastern and southern coasts), Japan
Size	Length: 1.5 ~ 3.5 m Width: 25 ~ 40 cm	Length: 1 ~ 2 m	Length: 30 cm ~ 1 m Width: 5 ~ 30 cm
Life span	Perennial (2–4 yrs)	Perennial (5–7 yrs)	Perennial (5–7 yrs)
Usage	Food	Food	Food, mouthwash, cosmetics (raw materials)

* =*Laminaria japonica* Areschoug; ** http://www.algaebase.org

Table 3. Carbon pools.

Pool	Application (yes/no)	Explanation
Above-ground biomass	Yes	Biomass situated above the sediment at the bottom of the water column
Below-ground biomass	No	Root and rhizome biomasses in the sediment
Dead algae	No	Usually disintegrated in the water column
Seabed organic carbon	No	Equivalent to the organic soil content in the sediment

Generally, seaweeds cannot grow on soft and fine sediments. Because their habitats have been degraded to barren rocky grounds, the term 'kelp (seaweed) forests (or beds)' was used here to encompass communities that occur on artificial substrates, for examples, blocks, ropes, man-made reefs, abandoned ships, Triton (made of slag from the iron-making process, etc.), or a complex of such elements. These substrates were installed so that seaweeds could be attached for construction and/or restoration. One method, developed by RIST, enabled a crew to construct reefs after a hard bottom forms. This improved substrate property on the soft floor, making it possible to cultivate or grow seaweeds in challenging locations where it is difficult to apply more conventional methods.

The minimum area that can be populated by a seaweed forest (i.e., the crown cover in the A/R) differs among substrates. For example, when ropes are used, the coverage is over 70% whereas that with artificial reefs is 30%. This estimate is based on a community in which plants of the target species grow longer than 30 cm.

To utilize the marine space for these project activities, an operator must obtain a license and permission from relevant authorities, such as local governments and stakeholders as well as other agencies. According to the Fisheries Act in Korea, these types of fisheries licenses are valid for 10 years.

The technology employed for this CDM project involved mid-water rope construction, which is appropriate for cultivating *Ecklonia* and *Saccharina* (Chung et al. 2013). Seedlings of both were produced at the NIFSNFRDDI, per their well-established nursery method for culturing young sporophytes. Transplantation occurred at the seaweed forest farm, which was established during the pilot survey. Growth data, standing stock, and plant sizes were recorded from July 2009 to April 2011. After their seeding, *E. cava* and *E. stolonifera* grew steadily from July 2009 to May 2010 before their growth rates gradually decreased after the temperature began to rise above 20°C in June (Fig. 4). Although *E. cava* grew faster than *E. stolonifera*, the highest growth rate was found with an annual cultivar of *S. japonica* during May 2010. The average frond density was calculated by counting the number of attached fronds

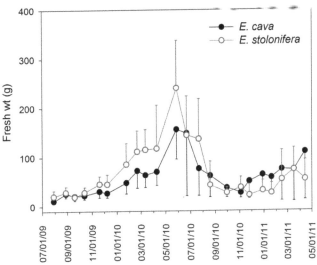

Fig. 4. Growth of *Ecklonia* in pilot seaweed farm along southern coast of Korea.

per meter of rope. For example, in July 2009, *E. cava* had 92 fronds m^{-1} versus 103 fronds m^{-1} for *E. stolonifera*. Those respective densities decreased to 54 and 58 in December 2009 before rising to 81 and 109 in May 2010. This increase in density may have been due to the growth of young fronds that matured under suitable conditions as well as the outgrowth of fronds from new holdfast propagation.

Management activities could be visualized within the framework of an integrated system that included a detailed location of the project boundaries via GPS and GIS, utilization of marine spatial resources, the range of marine space, the duration of the project period leased from the owner, permanent monitoring, afforestation, thinning, harvesting, etc. A task force team for marine affairs was responsible for overseeing the entire process of preparation and implementation. This involved training personnel, providing management assistance, and giving instruction in maintaining quality control. Such an approach clearly delineated the project boundaries and activities that occurred. For example, by using the GIS system, the floating buoys could be remotely monitored by satellite.

Over the specified credit period, the amount of GHG emissions to be reduced through implementation of this new carbon sink was expected to be > 12.1 tCO_2eq ha^{-1} yr^{-1}. As a factor in risk analysis, the sea urchin is a major grazer but it does not affect plants at greater depths. However, because this current project site was shallow and plants were prostrate during low tides, some sea urchins moved from the rocky substrate to graze. This action accounted for approximately 50% of the damage and loss to *Saccharina japonica*, 20% to *Ecklonia cava*, and just under 10% to *E. stolonifera*. Furthermore, the larvae of the blue mussel appeared from April to June and matured in September and October. Therefore, the space available for plant growth was diminished by the influx of mussel settlements. This affected the attachment of *Ecklonia cava* and *E. stolonifera* when they regenerated in October and November as the water temperature dropped, leading to a decrease in their biomass. Some blue mussels died due to low salinity during seasons of heavy rain. When they became detached from the substrate after death, seaweeds were also lost with them. During the first five years of the project, we noted that over 90% of the seaweed became detached in this manner.

Section B. Application of a baseline methodology

B.1 Title of reference for the approved baseline methodology applied to this project

No corresponding baseline methodology could be applied to the Seaweed A/R CDM Project as a proven method. Instead, we proposed a new methodology for constructing a seaweed community on an artificial reef.

B.2 Justification for the choice of methodology and its applicability to the project

The following methodology was applied in accordance with the conditions stipulated below:

- A direct A/R seaweed plantation with selected species was installed in the marine substrate.
- The site for constructing this pilot farm with perennial species was selected because no seaweed community had previously existed there. Moreover, the plantation substrate for project activities had to be open; current coverage by seaweeds did not already qualify at the level described for this pilot farm.

This project could be implemented as one of afforestation. We provided evidence that, in the past 30 years, no such activity of planting seaweed vegetation on artificial reef substrates (blocks, ropes, reefs, Triton, etc.) had occurred in this devastated coastal area, no nitrogen fertilizer was used, and the proposed cultivation practices did not interfere with natural seaweed habitats.

Reforestation would be a possible alternative for this environment using seaweed forests restored via artificial plantation into previously forested areas. In addition, seeding might be a viable option, with the goal of promoting natural restoration on sites where communities have been destroyed and areas used for other purposes. Appropriate survey data were provided to prove the conditions stated above.

- It was not possible to consider this project activity as an attractive investment factor. No other source of income existed here other than that from certified emission reduction (CER) credits because, based on an economic analysis, the carbon sink produced by the pilot seaweed plantation was expected to be a deficit. In terms of that analysis, the Internal Return Rate (IRR) would be very small and meaningless if this project did not exist.
- Even though state and local governments set the purpose for this project area as the construction and growth of seaweed forests that use algae, barriers arose in methods of financing, the design of key technologies for reducing GHG emissions with seaweed, and various institutional aspects of technology development. These interfered with our ability to achieve the objectives described in this report.
- When carbon contents sequestered within the biomass, above- or belowground, fluctuated in the marine substrate, the reservoir (pool) was to be considered within the guidelines for this project.
- If no other existing business activities were comparable to this project, certain aspects had to be considered when predicting its success. That is, under current circumstances, the marine substrate needed for the construction of seaweed forests was more likely to be devastated within the next several decades if its quality continued to be degraded, economic analyses revealed unattractive results, barriers were presented that hindered investments (e.g., the loan fund proved impossible to use, bank loan financing encountered many obstacles, operators were unable to achieve successful seaweed plantation or manage those new forests, the support system was inactive in these construction efforts), or risks were shown in the CER credit market.

B.3 Description of how the methodology was applied to this project

It was previously difficult to establish a natural seaweed community within the project area discussed here. Habitats for several benthic invertebrate animals and surfaces sufficient to support plant growth had long been devastated and reduced to barren ground. Thus, we suggested the use of a new, artificial substrate for forest construction. The history of project activities is listed in Table 4.

Table 4. History of project activities.

Period	Conditions/activity
Past ~ Oct 2007	No seaweed vegetation
Oct 2007 ~ Nov 2007	RIST/POSCO installed an artificial reef facility with Triton for seaweed forests (0.5 ha)
Dec 2007 ~ Jun 2009	The NIFS, South (West) Branch, RIST/POSCO, and local Co-op co-managed the facility
Jul 2009 ~ present	Construction of a pilot seaweed A/R CDM farm

B.4 Description of how actual removal of net GHG by carbon sinks can be increased above levels that would have occurred in the absence of this registered project

We considered two factors—substrate eligibility and additionality—when implementing this project. Ocean (marine) substrate eligibility is equivalent to Land eligibility in terms of A/R. The marine area (substrate) used in this project had not shown any sign of recent seaweed growth and did not satisfy the standards for defining natural growth of seaweeds or seaweed forests. We determined its eligibility based on marine substrate utilization and a soil coverage map. As our second factor, the fact that a generally explained method could be adopted as an additive means proved that this project was additive and did not depict a baseline scenario.

An overview and the assessment procedures for additive approval are outlined below:

Step 0: Preliminary screening based on the starting date of A/R project activities

In the case of new afforestation, evidence should be presented that the coastal area is degraded and that no seaweed vegetation has been promoted by artificial substrates over the past 30 years. Seaweed vegetation and the project area had been gradually devastated by cultivation, harvesting of crops, and ranching.

Currently, property upon which the ocean substrate was to be used for new afforestation along the project boundary was similar to degraded coastal areas that contained seaweed, grass, and shrub species. This pilot project began on 1 May 2009; no construction of seaweed vegetation on artificial substrates had occurred before that date.

Step 1: Identification of alternative scenarios consistent with a legal and regulatory framework

Because this project was economically unattractive and entailed several financial, technical, and institutional barriers, as well as market risks, one could realize its reality, feasibility, and applicability only through consistent use of a marine substrate. Otherwise, the quality of this marine environment would be degraded to the point that it was classified as barren ground. Due to grazing by species such as sea urchins, snails, and sea stars, it had been difficult for seaweed to grow and reproduce naturally. Thus, a native seaweed forest similar to one obtained via this project could not otherwise occur. Some national and regional programs have already been implemented for six or seven years. Whether the goals of many such programs can be met depends upon the availability of financial resources. Domestic funding for construction and restoration of seaweed forests has long been limited, and has been concentrated in areas considered more economically viable and efficient rather than being used for addressing the conditions found in this project area. The baseline scenario satisfies current and foreseeable-future requirements that are entirely applicable, legitimate, and open to regulation.

Step 2: Investment Analysis

Sub-step 2a. Determine the appropriate method for analysis. In the case of proven alternatives (such as maintaining the current situation), additional investments are not required and do not provide economic benefits. We selected IRR as an economic indicator for analyzing the benchmarks in this project.

Sub-step 2b. Option III: Apply benchmark analysis. The required return on stocks within the agriculture investment business is 12%, based on standard values issued by Chinese special agencies in 2012. This means that the government will approve a project only when the IRR is above this minimum return. These standard values are typically treated as a reference point for private investment and commercial afforestation.

Sub-step 2c. Calculate and compare financial indicators. The task of "nursery operations" occurred between 2009 and 2010, during the pilot survey. Costs associated with establishing planting space are presented in Table 5.

Other operating costs consisted of those for harvesting, product transportation, post-harvest re-plantation, maintenance, management, and pest control, all of which would begin four years after

Table 5. Cost of establishment with transplantation on a 0.5-ha site in 2012.

Task		Cost
Establishment	Site preparation	KRW 20,000,000 (about US $20,000)
	Seedling care	n.a.
	Planting	*Ecklonia cava* and *E. stolonifera*: KRW 15,000 m^{-1} x 400 m = KRW 6,000,000 *Saccharina japonica*: KRW 12,000 m^{-1} x 200 m = KRW 2,400,000
	Fertilization	n.a.
	Fire and disease control	n.a.
	Weeding	n.a.
Equipment		KRW 25,000,000
Other (transportation, etc.)		n.a.
Total		KRW 53,400,000

planting and continue until credits were generated. Income includes only the proceeds from CER credits. The success of a product depended upon regional conditions.

Sub-step 2d. Sensitivity analysis. The most important factors influencing the IRR for this project were the quantity of product derived, its price, and the cost of operation.

Step 3: Barrier Analysis—The proposed seaweed A/R CDM project activities were not considered "additive".

Sub-step 3a: Identify barriers that would prevent the implementation of project tasks: (a) Investment barriers other than economic/financial. The chances were almost zero that this type of project could receive a bank loan for implementation because it carried great market risk and was economically unattractive; and (b) Technology barriers—this project was the first to apply such new technology with associated risks.

Sub-step 3b: Show that these identified barriers would not prevent the implementation of at least one alternative to the project as proposed. Alternative utilization of marine substrates might not have faced the obstacles mentioned above.

Step 4: Common Practice Analysis—The proposed seaweed A/R CDM project activities are considered "additive".

Although national and local programs determined the overall purpose in developing this project, achieving our objectives would most likely depend upon how we utilized funds. In the target region, the level of financial support for the construction and restoration of seaweed forests had been very limited for several years. No domestic monies were available for investing in its implementation. In addition, this CDM could not replace reforestation activities, but had to be included within the scale and range of other programs because vast regions require such reforestation efforts.

Step 5: Impact of CDM Registration

This A/R CDM project was created to ease economic and financial burdens and satisfy perceived barriers. Thus, the following benefits could be generated by its initiation:

- Revenue from sales of CER credits would increase because of the removal of carbon from the atmosphere. Without this project, the amount of sequestered carbon reserves would either decline or remain stable at very low levels in the target region as environmental degradation continued.
- As a pilot venue for assessing the financial activities related to future carbon businesses, the stakeholders who observed this project would be motivated. Direct experiences would provide some incentive to participate in a carbon market that would involve verification, confirmation, exchange, and the willingness to develop new projects.
  - Investment cost: KRW 53,400,000 (for 0.5 ha).
  - Expected CER profit (EUR9/tCO$_2$eq): EUR576 ha^{-1}yr^{-1}.
  - Expected IRR (economic analysis): Currently, CER revenue is very small compared to the level of investment, making IRR calculations meaningless. Thus, a larger project must be considered (500 ha or more).
  - Addition: This is secured with economic additions, including CER credits.

Because one could recognize this project area as a carbon sink only if no mature seaweed had been harvested when the seaweed pilot farm was constructed, approval of the current project and any additions related to problems associated with such harvesting could not be made in principle.

B.5 Detailed baseline information

The baseline was set for a devastated coastal area without a history of seaweed vegetation. This information included the date of completion for baseline research and the participants and organizations involved in conducting that research.

Section C: Application of monitoring methodology and monitoring plan

C.1 Title and reference for approved monitoring methodology applied to project activities

Title: Construction and Restoration of Seaweed A/R CDM Project

No currently approved (authorized) observation method was available for the artificial construction and restoration of seaweed forests. Therefore, we presented a new observation method in this report.

C.2 Justification for the choice of methodology and its applicability to the project (Refer to B.1)

The new method presented here included all of the factors concerning measuring, observing, and calculating mean values. These factors were related to the exact estimation of net removal of anthropogenic GHG by sink agents within marine substrates, where the environment was previously degraded and neglected.

To address the overall capability of researchers to perform these tasks, this report has covered the scope of the project, description of the establishment of those forest facilities, and corresponding operational activities. The task of checking feasibility within the set of scenarios adopted in the baseline methodology was related to the duration of carbon-credit production in the next stage of development. During the construction and restoration period of this project, some GHG were leaked in the course of transporting workers, seaweed seedlings, and other goods not produced within the marine substrate. Information was included that assured that the quality of the project contents was maintained and that the control plan could be fully executed.

This report comprises our field surveys, verification of data collection, and data entry file storage. It is part of the observational plan for 'construction and restoration of seaweed forest CDM project using seaweeds', which integrated the collected data and guaranteed that this collection would be effective.

For the baseline, we recorded the net amount of GHG removed by the sink at hourly intervals. Because environmental conditions were degraded at the project site, we concluded that the probability was very low for natural recruitment of seaweed to occur in this marine substrate based on conditions set by the testing methodology. If no seaweed vegetation was already present, then the amount of GHG removed from the substrate was estimated to be zero, and would fluctuate within the upper and lower parts of the ocean surface where seaweeds do exist. To maintain and execute these processes of observation in the most cost-effective manner, we focused on measuring the amount of carbon that accumulates only in the upper and lower layers of the water column. We assumed that other sink reservoirs were not likely to remove more than the amount designated as the baseline value.

The observation method integrated local climate data, existing vegetation, geographical classification of locations, seaweeds used for plantation, satellite images, soil maps, and project areas based on GPS and field surveys. This approach used a permanent sampling configuration to observe the carbon sink in the upper and lower layers of the water column. Configuration via GPS assured that measurements and observations were consistent over time.

- The marine substrate was constructed with a forest that resulted from direct plantation of seaweeds on the artificial attachment facility, as proposed by the "Seaweed A/R CDM Project".
- Because the environment for this marine substrate had been seriously degraded over time by human activity, and conditions were worsened by the addition of nutrients, it was necessary that we plant seaweeds rather than relying on natural seeding or recruitment.
- In some parts of this project area, fish populations were widely distributed several years ago. Currently, however, only a small number of fishermen are working there. They are not concerned with the seaweed forest project being conducted there.
- If the use of natural seaweed spores had failed or the project was not successful because of other environmental influences, the construction of natural seaweed forests could have been interrupted.
- Information was provided about the characteristics of each seaweed species, the length of time over which they were to be grown, and the no-harvest policy that would be followed during the carbon-credit period.

- In the baseline scenario of this 'Seaweed A/R CDM Project', it was assumed that the amount of carbon sequestered would remain stable or decrease due to the degradation of the marine environment, pollution, and the accumulation of substances such as soil organic matter from fragments of seaweed litter or dead seaweed in the substrate. Therefore, those reserves would not be considered when issuing *l*CER or *t*CER credits. Instead, we would account for only the upper and lower levels of biomass in the water column.
- We were allowed to conduct limited investigations into no-credit reservoirs to determine relative losses. If extraction was required, that amount might then be used during the second period of credit generation.
- The establishment of a seaweed forest was uniformly implemented to consider genetics, species, plantation period, and the management system.

C.3 Determining GHG removals by sinks

Monitoring baseline net GHG removals: The purpose of this project was to restore a degraded marine substrate. Its baseline scenario was written according to a new methodology. Variability in the carbon sink was set to zero when no seaweed vegetation was present; the expected carbon sink of seaweeds in the upper and lower layers of the water column would fluctuate according to the amount of vegetation that grew. The project participants were to use five-year fixed-term credits and reforestation would not be required. Therefore, we could not record net GHG removal by sinks at the baseline.

Monitoring the physical boundaries for project activities: A field survey was conducted at each location along the project boundaries where reforestation/plantation activities were to be performed. Geographical locations were established for each study area via GPS (each corner of the polygon with latitude and longitude) after any alterations were verified by the DOE. The actual project borders were confirmed as matching those described in Section *A*. When the actual border was located outside of the boundaries, additional information was provided (see Section *A*).

The legal justification was recognized for why this particular marine substrate was selected for the project, and the baseline scenario demonstrated how it was implemented in the marine space. Changes to the boundaries were reported to the DOE during the project period for verification. If those boundaries were to change during the crediting period, any damage to the seaweed forests had to be reported and the boundaries for that location were to be modified. Subsequent verification was to be done by the DOE. Those damaged areas would be removed from project consideration and their corresponding CER credits withdrawn. If seaweed forest construction at the marine substrate failed in any project area, it was to be replaced, and that region would be immediately superseded.

Monitoring the establishment and management of the coastal forest: To ensure the quality and execution of plantation activities, the following observations were to be made within three years of constructing the seaweed forests. This was to include harvest (location, size, and species), fertilization (species, location, and amount and type of fertilizer), and inspection and verification. Reforestation or re-sowing would occur after the marine substrate was harvested, and factors that promoted the best natural recruitment of seaweeds were to be considered.

Data for actual net GHG removals by sink, obtained through the pilot survey: To achieve pre-stratification, the devastated coastal areas were divided into regions of poor vegetation versus fertile areas, based on estimates of biomass production within the project boundaries. Buoys were installed within the farm structure in order to calculate the actual area according to on-board GIS coordinates. However, some modifications were made to the shape of the structure due to the influence of tides. The goal in sampling was to count and measure all of the fronds along a 1-m rope. When more than 30 individuals were found, only the 30 largest were measured. Physiological changes, in terms of photosynthetic activity, were monitored monthly using the Diving-PAM.

Over the period, divers were to collect and determine the wet weights of seaweed specimens, using an electronic balance (up to 0.1 g accuracy). Total leaf lengths and widths were measured with a tape (up to 1 mm accuracy). Using elemental analyses for each seaweed species, the carbon sink (CO_2 equivalent) was estimated based on wet and dry weights. The composition of elements is not shown.

During the construction of the pilot seaweed farm, CO_2 in the ocean was taken up through photosynthesis. Because the atmosphere and ocean are in a state of equilibrium, any carbon sequestered by these plants would contribute to the reduction in atmospheric CO_2. This was based on the following principle: if the movement of CO_2 between the atmosphere and ocean surface is in balance, then its removal from the ocean (water column) is the same as the amount removed from the atmosphere.

Calculations of carbon levels and emissions (Table 6)

1) Baseline Emissions (BE), calculated on a seaweed-biomass basis
   a) Obtain wet weights of samples. Vegetation along a 1-m-long rope was measured three times. The wet biomass of 30 specimens was weighed; if the rope contained more than 30 plants, then the largest 30 were taken for measurement.
   b) Either measure dry weights directly or else estimate the value based on wet weight (about 13% of wet weight).
   c) Total length of rope measures (extrapolated to the total area of 1.0 ha and obtained by the pilot survey).
      • *Ecklonia cava*: 49 lines × 100 m per rope (2-m interval) = 4900 m
      • *E. stolonifera*: 49 lines × 100 m per rope (2-m interval) = 4900 m
      • *Saccharina japonica*: 24 lines × 100 m per rope (4-m interval) = 2400 m
   d) Estimate the carbon content per unit length of rope structure (kg C m^{-1}). The carbon sink per 100 m of rope was calculated as 43.5 kg C for *Ecklonia cava*, 88.9 kg C for *Ecklonia stolonifera*, and 105.0 kg C for *Saccharina japonica*, based on the highest frond weight recorded in May 2009 during the initial pilot study period.
   e) Estimate BE for each seaweed species.
      • Dry weight (DW g) = wet weight (g) × 0.13. For *Ecklonia cava*, DW = 156.0 × 0.13 = 20.28 g per plant; *E. stolonifera*, 240.6 × 0.10 = 31.28 g per plant; and *Saccharina japonica*, 323.9 × 0.15 = 42.10 g per plant.
      • Dry weight per unit length (DWM, g DW m^{-1}) = frond dry weight (g) × No. of fronds per meter. For *Ecklonia cava*, DWM = 20.28 × 81 = 1642.28 g each; *E. stolonifera*, 31.28 × 109 = 3409.52 g each; and *Saccharina japonica*, 42.10 × 99 = 4167.90 g each.
      • Carbon content per unit length (CWM, g C m^{-1}) = dry weight per unit length (DWM, g DW m^{-1}) × leaf tissue carbon content (%). For *Ecklonia cava*, CWM = 1642.28 × 0.265 = 435.2 g C m^{-1}; *E. stolonifera*, 3409.52 × 0.261 = 889.9 g C m^{-1}; and *Saccharina japonica*, 4167.9 × 0.252 = 1050.3 g C m^{-1}.

Table 6. Growth characteristics and estimates of carbon contents from *Ecklonia* specimens sampled in May 2010 from the pilot seaweed farm.

	E. cava	*E. stolonifera*
Frond wet weight (g)	156.00	240.60
Frond dry weight (g) (13% of wet wt)	20.28	31.28
No. of fronds per meter of rope	81	109
Dry weight per unit rope length (g DW m^{-1})	1642.68	3409.52
Carbon content in frond tissue (%)	26.50	26.10
Carbon content per unit rope length (g C m^{-1})	435.20	889.90
Baseline emissions per unit area* (tCO$_2$eq ha^{-1})	7.82	15.99

* The framework of the mid-water rope-culture system could accommodate 49 lines of 100-m-long rope, placed at 2-m intervals to create a 1-ha substrate space overall.

- Baseline emissions of seaweed (BEi, tCO_2eq ha^{-1}) = carbon content per unit length (g C m^{-1}) × total rope length (m) × 44/12 ÷ 10^6. For *Ecklonia cava*, BEi = 435.2 × 4900 × 44/12 ÷ 10^6 = 7.82 tCO_2eq ha^{-1}; *E. stolonifera*, 889.9 × 4900 × 44/12 ÷ 10^6 = 15.99 tCO_2eq ha^{-1}; and *Saccharina japonica*, 1,050.3 × 2,400 × 44/12 ÷ 10^6 = 9.24 tCO_2eq ha^{-1}.
  f) Baseline emissions per unit area (BE, tCO_2eq ha^{-1}) = 7.82 ~ 15.99 tCO_2eq ha^{-1}
  g) Maximum baseline emissions (BE, tCO_2eq ha^{-1}) = ~ 16.00 tCO_2eq ha^{-1}
2) Project Emissions (PE): PE was defined as the GHG emissions produced by the installation of the farm structure and transportation of vehicles (trucks) and vessels (work boat) needed for the plantation project. PE, tCO_2e = Σ {fuel consumption x fuel calorific value × conversion factor of calorific value by fuel consumption (4.1868 TJ × 10^{-9} kcal^{-1}) × CO_2 emission factor (tCO_2 TJ^{-1})}. Here, we assume the amount of emissions to be zero because the amount of project emission is relatively small in comparison with baseline emission. Project emission is based on transport for algae cultivation. It is shown in Table 7.
3) Leakage (L) was assumed to be zero here. Actually there is no leakage in this project.
4) Estimation of greenhouse gas reductions (tCO_2eq) = BE − PE − L.

Table 7. Project emissions by CDM activities estimated to have occurred during the pilot survey.

Fuel consumption	Diesel (l)
Calorific value	liquid (kcal L^{-1}) = 8,540 kcal L^{-1}
Conversion factor	1 kcal = 4.1868 kJ, 1 ton = 1,000 kg, 1 TJ = 10^9 kJ
Emission factor	CO_2 emission factor = 74.1 tCO_2 TJ^{-1}

C.4 Treatments of leakage in the monitoring plan

If applicable, we included a description of the data and information that were to be collected in order to monitor leakage associated with the proposed project activities. We provided details about the formulae and/or models to estimate L (for each GHG, sources, carbon pool, in units of equivalent), the procedures for periodic review of the implementation of activities and measures to minimize leakage, and descriptions of the formulae and/or models used to estimate net anthropogenic GHG removals by sinks (for each GHG, carbon pool, in units of equivalent). Wet (fresh) weights were determined with an electronic balance and frond (leaf) widths and lengths were measured with a tape. Because of the possibility of instrument error when sampling convenience was a factor, we undertook procedures for quality control and assurance.

We also presented a description of the operational and management structure(s) that the project operator would implement in order to monitor actual GHG removals by sinks, as well as any leakage generated by project activities. The names of persons/entities that determined the monitoring methodology were recorded.

Persons and agencies associated with PNU, CNU, NIFS, RIST, and RCC were involved in making baseline decisions.

Section D. Estimation of net anthropogenic GHG removals by sinks

D.1 Estimate baseline net GHG removals by sinks

Under the baseline scenario, we assumed that the accumulation of carbon would be due to an increase in seaweed biomass. If seaweeds were not present above or below the sea surface, the accumulation of carbon would be zero. Any other reservoir was not considered if it was related to the project scenario and might produce values lower than our baseline readings.

D.2 Estimate of actual net GHG removals by sinks

GHG emissions were estimated under the assumption that changes in carbon storage, project-associated emissions, and leakage were zero. Estimations of actual net GHG removals by sinks were presented with the following terms (as shown in Table 8): Baseline net GHG removals by sinks ($tCO_2eq\,yr^{-1}\,ha^{-1}$), Project emissions ($tCO_2eq\,ha^{-1}yr^{-1}$), Leakage ($tCO_2eq\,ha^{-1}yr^{-1}$), Annual carbon stock change ($tCO_2eq\,ha^{-1}yr^{-1}$), and Cumulative net GHG removals by sinks ($tCO_2eq\,ha^{-1}yr^{-1}$).

Table 8. Estimation of actual net GHG removals by sinks.

Year No.	Year	Baseline net GHG removals by sinks ($tCO_2eq\,ha^{-1}\,yr^{-1}$)	Project emissions ($tCO_2eq\,ha^{-1}\,yr^{-1}$)	Leakage ($tCO_2eq\,ha^{-1}\,yr^{-1}$)	Annual carbon stock change ($tCO_2eq\,ha^{-1}\,yr^{-1}$)	Cumulative net GHG removals by sinks ($tCO_2eq\,ha^{-1}\,yr^{-1}$)
1	Year 1	16.0	0.0	0.0	0.0	16.0
2	Year 2	16.0	0.0	0.0	0.0	32.0
3	Year n	16.0	0.0	0.0	0.0	$16.0 \times n$

D.3 Estimated leakage

It was assumed that no leakage would occur due to project activities.

D.4 The sum of D.I minus D.2 minus D.3 represented the net anthropogenic GHG removals by sinks of the proposed project activities

The final CO_2 reduction was estimated to be ~16 tons of $CO_2eq\,ha^{-1}\,yr^{-1}$.

Section E. Environmental impacts of the seaweed A/R CDM project activity

This project does not harm the environment. Therefore, documentation is not applicable concerning the analysis of environmental impacts, including biodiversity and natural ecosystems, and impacts outside the project boundary; a statement confirming that project participants have undertaken an environmental impact assessment in accordance with the procedures required by the host Party, including conclusions and all references to supporting documents if any negative impact is considered significant by the project participants or the host party; and descriptions of planned monitoring and remedial measures to address significant impacts referred to in the previous section.

Section F. Socio-economic impacts of the seaweed A/R CDM project activity

F.I Documentation on the analysis of socio-economic impacts, including impacts outside the project boundary

This project is expected to have two major benefits.

1) Improving the marine environment will help maintain the diversity of life within it. The effects of establishing and restoring kelp forests with these pilot farms will appear gradually. First, successful cultivation of intensive, massive seaweed communities will provide abundant food and habitats for marine animals, increase fisheries stocks, and eventually elevate the potential for marine biodiversity in waters around the project area. Second, the establishment and restoration of kelp forests with pilot farms will mitigate or neutralize the effects of several threats to the marine environment, such as by protecting terrain that would otherwise be barren. Our efforts will help the natural and spontaneous enhancement of fisheries resources and attract other marine life into the area, ultimately improving local fisheries stocks. Third, the increase in seaweed biomass that is associated with the removal of CO_2 will eventually reduce the acidity of surrounding waters through photosynthetic activity, and

will mitigate ocean acidification of the coastal area. It will also promote the enhancement of marine biodiversity and abundance.

2) Implementing this project will encourage the development of new income and increase revenues for fishermen and other project participants. This effect will be manifested gradually. First, fisheries resources will be enhanced through improvements to the marine environment. This will contribute to maintaining stable incomes for fishermen's households, and will develop the local economy by increased landings and reduced deviations in their catches. Second, by providing a sink for absorbed CO_2, this seaweed farm project will allow participants to trade in carbon credits achieved by reducing GHG. Therefore, the CDM business will be a source of development. Shared profits that are thus obtained can also be allocated to participating economic agencies, which will bolster the incomes of local communities.

F.2 If any negative impact is considered significant by the project participants or the host Party, a statement that project participants have undertaken a socioeconomic impact assessment, in accordance with the procedures required by the host Party, including conclusions and all references to support documentation

not applicable

F.3 Description of planned monitoring and remedial measures to address significant impacts referred to in Section F.2 above

not applicable

Section G. Stakeholders' comments

G.1 Brief description of how comments by local stakeholders have been invited and compiled

Public hearings, consultations, and a review process have been instituted, including discussions on how this project concerns business types, facilities, locations, availability of funds, participating organizations, budgets, and other related topics (Table 9). A brief history was presented on the initiation of the pilot project and steps in consultation, the submission of proposals and the subsequent approval process, selection of the study location, and public hearings. During the period of operation, we reported on production and quality control, activities at on-site work facilities and those affiliated with post-operative research and evaluation, on-site facilities management, expansion, and the sharing of results.

- Purpose of the project: Implementation of the technology for "Construction of kelp forest using slag" developed by POSCO/RIST will contribute to improvements in the coastal water environment and increase the incomes of fishermen.
- Project overview

History of progress on project: Date (event schedule): Purpose of event/Organization and Agents

- 3 April 2007: Proposal of pilot projects and consultations/POSCO/RIST, Namhae-gun, SMFO.
- 7 June 2007: Submission of pilot project and business promotion; Namhae-gun's approval of proposal and beginning of candidate selection/POSCO/RIST, Namhae-gun.
- 11 September 2007: Selection of project activity location and public hearings → Business Promotion Agreement (Pyeongsan-1-li Co-op)/POSCO/RIST, Namhae-gun, SMFO.
- 11 September–20 October 2007: Survey of the project location in detail, production and quality control, production of on-site work facilities/POSCO/RIST, Namhae-gun, SMFO/Pyungsan-1-li Co-op.
- Since 20 October 2007: Post-operative research and evaluation, on-site facilities management, expansion and sharing of results/POSCO/RIST, Namhae-gun, SMFO/Pyungsan-1-li Co-op.

Table 9. List of topics considered for public hearings, consultations, and a review process.

Topic	Main contents
Model applied	Kelp forest construction model for abalone habitat type (Patent No. 10-0650268)
Area/Volume of the facility	Project location/size: 1 (Pyeongsan-1-li/approx. 0.5 ha (5,000 m²) Slag structures: kelp forest-type artificial reef (3 tons), 250 ea; tetrapod (2.6 tons), 300 ea; small block for filling, 1000 ea Slag for reinforcement of substrate: 12,000 tons (13- to 30-mm nodules)
Location of the facility	Co-op fisheries ground, Pyeongsan-1-li, Nam-myeon, Namhae-gun, Gyeongsangnam-do
Means of propulsion	RIST conducts the project with the support of POSCO funds (approval of Namhae-gun)
Participating organizations	Project (business) hosts: POSCO/RIST, Business subjective: POSCO/RIST, Namhae-gun/South Sea Marine Fisheries Office (SMFO) Collaborative research personnel from NIFS, Chonnam National University, and Gangneung-Wonju National University Construction of facilities: Changshin Innovative Industrial Development Co. Production of slag nodule: Hyo–Seok Co. Transportation of products and facilities: Haju Shipping Co.
Project budget	Direct facility cost: 33 billion KRW Post-project monitoring and research fund: 3.5 billion KRW

6.2 Summary of the comments received

The reactions by stakeholders included an interest in vessel usage, job creation, and expected increases in incomes gained through these project activities. They did not express any significant feelings against the project.

Prospects: New paradigm of mitigation and adaptation—seaweed solution

Verification of this seaweed mechanism as an effective CO_2 sink, thereby leading to seaweed CDM methodologies, has been the ultimate objective of this project (Chung et al. 2013). As the global climate changes, the structure and function of the marine ecosystem will be significantly altered. In response, we should develop appropriate measures that exploit the role of marine vegetation as a tool for mitigation and adaptation (Turan and Neori 2010).

Our seaweed CDM project, here in its first trial, is a major challenge. Sustainable production, in the form of a seaweed-integrated multi-trophic aquaculture system (Chopin et al. 2001), is the essential prerequisite to achievement. For this the relevant regions must be appropriately controlled under management schemes for Ecosystem-Based Aquaculture and Integrated Coastal Zones (Chung et al. 2002).

When new seaweed communities are cultivated and restored in coastal waters and also expanded offshore with artificial substrates, a tremendous amount of greenhouse gases can be sequestered. Harvesting this biomass for subsequent conversion to biofuel (Chynoweth 2002) or paper (Seo et al. 2010) is a promising measure to counteract global warming. When these circumstances are considered, actual approval of a seaweed CDM project by the UNFCCC will lead to enhanced, sustainable management of marine environment and marine resources, thereby greatly benefitting coastal communities.

Acknowledgements

This work has been supported by a "GHG emissions reduction using seaweeds" Project funded by the Korean Ministry of Oceans and Fisheries, a grant from the National Research Foundation of Korea (NRF-2016R1A2B1013637) to IKC, and the APPA Working Group, "Asian Network for using algae as a CO_2 sink."

References

Buck, B.H. and C.M. Buchholz. 2004. The offshore-ring: a new system design for the open ocean aquaculture of macroalgae. J. Appl. Phycol. 16: 355–368.

Buck, B.H., G. Krause and H. Rosenthal. 2004. Extensive open ocean aquaculture development within wind farm in Germany: the prospect of offshore co-management and legal constraints. Ocean Coast. Manag. 47: 95–122.

Choi, C.G., J.H. Kim and I.K. Chung. 2008a. Temporal variation of seaweed biomass in Korean coasts: Yokjido, Gyeongnam Province. Algae 23: 311–316.

Choi, H.G., K.H. Lee, X.Q. Wan, H.I. Yoo, H.H. Park, J.H. Kim and I.K. Chung. 2008b. Temporal variations in seaweed biomass in Korean coasts: Woejodo and Jusamdo, Jeonbuk. Algae 23: 335–342.

Chopin, T., A.H. Buschmann, C. Halling, M. Troell, N. Kautsky, A. Neori, G.P. Kraemer, J.A. Zertuche-Gonzalez, C. Yarish and C. Neefus. 2001. Integrating seaweeds into marine aquaculture systems: a key towards sustainability. J. Phycol. 37: 975–986.

Chritchley, A.T. and M. Ohno. 1998. Seaweed Resources of the World. Japan International Cooperation Agency, p. 431.

Chung, I.K., Y.H. Kang, C. Yarish, G.P. Kraemer and J.A. Lee. 2002. Application of seaweed cultivation to the bioremediation of nutrient-rich effluent. Algae 17: 1–10.

Chung, I.K. 2007. Seaweed coastal CO_2 removal belt in Korea. The United Nations Framework Convention on Climate Change and the 13th Conference of the Parties, Side Event of Seaweed Coastal CO_2 Removal Belt in Korea and Algal Paper and Biofuel, 6–10 Dec. 2007, Bali, Indonesia.

Chung, I.K., J. Beardall, S. Mehta, D. Sahoo and S. Stojkovie. 2011. Using marine macroalgae for carbon sequestration: a critical appraisal. J. Appl. Phycol. 23: 877–886.

Chung, I.K., J.H. Oak, J.A. Lee, J.A. Shin, J.G. Kim and K.-S. Park. 2013. Installing kelp forests/seaweed beds for mitigation and adaptation against global warming: Korean Project Overview. ICES J. Mar. Sci. 70: 1038–1044.

Chung, I.K., F.A. Calvyn and J. Beardall. 2017. The future of seaweed aquaculture in a rapidly changing world. Eur. J. Phycol. 52: 495–505.

Chynoweth, D.P. 2002. Review of Biomethane from Marine Biomass. Department of Agricultural and Biological Engineering, University of Florida. http://abe.ufl.edu/chyn/Webpagecurrent/publications dc.htm.

De Vooys, C.G.N. 1979. Primary production in aquatic environment. In: B. Bolin, E.T. Degens, S. Kempe and P. Ketner (eds.). SCOPE 13 – The Global Carbon Cycle. Scientific Committee on Problems of the Environment (SCOPE). (http://www.icsu-scope.org/downloadpubs/scope13/chapter10.html).

Duarte, C.M., J.J. Middelburg and N. Caraco. 2005. Major role of marine vegetation on the oceanic carbon cycle. Biogeosciences 2: 1–8.

Duarte, C.M., J. Wu, X. Xiao, A. Bhrun and D. Krause-Jensen, D. 2017. Can seaweed farming play a role in climate change mitigation and adaptation? Frontiers in Marine Science 4: 1–8.

Falkowski, P., R.J. Scholes, E. Boyle, J. Canadell, D. Canfield, J. Elser, N. Gruber, K. Hibbard, P. Högberg, S. Linder, F.T. Mackenzie, B. Moore III, T. Pedersen, Y. Rosenthal, S. Seitzinger, V. Smetacek and W. Steffen. 2000. The global carbon cycle: a test of our knowledge of Earth as a system. Science 290: 291–296.

Graham, M.H., B.P. Kinlan, L.D. Druehl, L.E. Garske and S. Banks. 2007. Deep-water kelp refugia as potential hotspots of tropical marine diversity and productivity. Proc. Natl. Acad. Sci. 104: 16576–16580.

Hoegh-Guldberg, O. and J.F. Bruno. 2010. The impact of climate change on the world's marine ecosystems. Science 328: 1523–1528.

Kang, J.W. and I.K. Chung. 2017. The effects of eutrophication and acidification on the ecophysiology of *Ulva pertusa* Kjellman. J. Appl. Phycol. 29: 2675–2683.

Kang, P.J., Y.S. Kim and K.W. Nam. 2008. Flora and community structure of benthic marine algae in Ilkwang Bay, Korea. Algae 23: 317–326.

Kim, M.S., M. Kim, M.H. Chung, J.H. Kim and I.K. Chung. 2008. Species composition and biomass of intertidal seaweeds in Chuja Island. Algae 23: 301–310.

Ko, Y.W., G.H. Sung and J.H. Kim. 2008. Estimation for seaweed biomass using regression: a methodological approach. Algae 23: 289–294.

Koch, M., G. Bowes, C. Ross and X.-H. Zhang. 2013. Climate change and ocean acidification effects on seagrasses and marine macroalgae. Global Change Biol. 19: 103–132.

Krause-Jensen, D. and C.M. Duarte. 2016. Substantial role of macroalgae in marine carbon sequestration. Nature Geoscience 9: 737–742.

Laffoley, D. d'A. and G. Grimsditch. 2009. The Management of Natural Coastal Carbon Sinks. IUCN, 53 pp.

Littler, M.M. and D.S. Littler. 1980. The evolution of thallus form and survival strategies in benthic marine macroalgae: field and laboratory tests of a functional form model. Am. Natur. 116: 25–43.

Littler, M.M., D.S. Littler and P.R. Taylor. 1983. Evolutionary strategies in a tropical barrier reef system: functional form groups of marine macroalgae. J. Phycol. 19: 229–237.

McVey, J.P. and B.H. Buck. 2008. IMTA-Design within an offshore wind farm. In: Aquaculture for Human Wellbeing—The Asian Perspective. The Annual Meeting of the World Aquaculture Society, 23 May 2008, Busan (Korea).

Merzouk, A. and L.E. Johnson. 2011. Kelp distribution in the northwest Atlantic Ocean under a changing climate. J. Exp. Mar. Biol. Ecol. 400: 90–98.

Michler-Cielucha, T., G. Krause and B.H. Buck. 2009. Reflections on integrating operation and maintenance activities of offshore wind farms and mariculture. Ocean Coast. Manag. 52: 57–68.

Müller, R., T. Laepple, I. Bartsch and C. Wiencke. 2009. Impact of oceanic warming on the distribution of seaweeds in polar and cold-temperate waters. Bot. Mar. 52: 617–638.

N'Yeurta, A.d.R., D.P. Chynoweth, M.E. Capron, J.R. Stewart and M.A. Hasan. 2012. Negative carbon via ocean afforestation. Process Safety Environ. Prot. 90: 467–474.

Nellemann, C., E. Corcoran, C.M. Duarte, L. Valdés, C. De Young, L. Fonseca and G. Grimsditch. 2009. Blue Carbon. A Rapid Response Assessment. UNEP, 80 pp.

Pelejero, C., E. Calvo and O. Hoegh-Guldberg. 2010. Paleo-perspectives on ocean acidification. Trends Ecol. Evol. 25: 332–344.

Raven, J.A. and P.G. Falkowski. 1999. Oceanic sinks for atmospheric CO_2. Plant Cell Environ. 22: 741–755.

Reed, D.C. and M.A. Brzezinski. 2009. Kelp forest. pp. 31–37. *In*: Laffoley, D. d'A. and G. Grimsditch (eds.). The Management of Natural Coastal Carbon Sinks. IUCN.

Ritschard, R.L. 1992. Marine algae as a CO_2 sink. Water Air Soil Pollut. 64: 289–303.

Seo, Y.-B., Y.-W. Lee, C.-H. Lee and H.C. You. 2010. Red algae and their use in papermaking. Bioresource Technol. 10: 2549–2553.

Shin, J.D., J.K. Ahn, Y.H. Kim, S.B. Lee, J.H. Kim and I.K. Chung. 2008. Temporal variations of seaweed biomass in Korean coasts: Daejin, Gangwondo. Algae 23: 327–334.

Smith, S.V. 1981. Marine macrophytes as a global carbon sink. Science 211: 838–840.

Sondak, C.F.A., P.O. Ang, J. Beardall, A. Bellgrove, S.M. Boo, G.S. Gerung, C.D. Hepburn, D.D. Hong, Z. Hu, H. Kawai, D. Largo, J.A. Lee, P.-E. Lim, J. Mayakun, W.A. Nelson, J.H. Oak, S.-M. Phang, D. Sahoo, Y. Peerapornpis, Y. Yang and I.K. Chung. 2017. Carbon dioxide mitigation potential of seaweed aquaculture beds (SABs). J. Appl. Phycol. 20: 2363–2373.

Turan, G. and A. Neori. 2010. Intensive seaweed aquaculture: a potent solution against global warming. pp. 359–372. *In*: Israel, A., R. Einav and J. Seckbach (eds.). Seaweeds and Their Role in Globally Changing Environments. Springer.

Wan, X.Q., H.H. Park, H.I. Yoo and H.G. Choi. 2009. Temporal variations in seaweed biomass and coverage in Korean coasts: Ongdo, Chungnam. Fish. Aquat. Sci. 12: 130–137.

Wernberg, T., B.D. Russell, M.S. Thomsen, C.F.D. Gurgel, C.J.A. Bradshaw, E.S. Poloczanska and S.D. Connell. 2011. Seaweed communities in retreat from ocean warming. Current Biol. 21: 1828–1832.

Wylie, L., A.E. Sutton-Grier and A. Moore. 2016. Keys to successful blue carbon projects: lessons learned from global case studies. Marine Policy 65: 76–84.

Yarish C. and R. Pereira. 2008. Mass production of marine macroalgae. pp. 2236–2247. *In*: Jorgensen, S.E. and B.D. Fath (eds.). Ecological Engineering Vol. 3. Encyclopedia of Ecology. Oxford. Elsevier.

3

Microalgal Metabolism and their Utilisation

Michael A. Borowitzka

Introduction

The diverse metabolic pathways and the variety of expression of these underlie the interest in microalgae as sources of potentially valuable chemicals and products, and the use of microalgae for a range of processes such as waterwater treatment, soil bioremediation and as animal feed. Microalgae, defined as those unicellular, colonial, and filamentous species of eukaryotes and the prokaryotic cyanobacteria (blue-green algae) can be found in most algal phyla. Microalgae are already important commercial sources of high-value chemicals including β-carotene, astaxanthin (Borowitzka 2010), phycocyanin (Eriksen 2008), and long-chain polyunsaturated fatty acids such as eicosapentaenoic acid, docosahexaenoic acid, and arachidonic acid (Ratledge 2010; Wynn et al. 2010), as well as being used as human nutritional supplements (Belay 1997; Iwamoto 2004) and for animal nutrition in aquaculture (Borowitzka 1997; Neori 2011). Microalgae are continuing to be developed as potential sources of a range of other chemicals and products (Borowitzka 2013a), and are receiving intense attention as potential sources of renewable fuels (Wijffels and Barba 2010; Borowitzka and Moheimani 2013, Unkefer et al. 2017).

Irrespective of the application of the microalgae and of the culture system which may be used, successful culture and product formation requires a good understanding of algal biology, physiology, and biochemistry. An understanding of those factors which affect effective utilization of light, photosynthesis, respiration, nutrient uptake and utilization, and the regulation of metabolism in microalgae assists in the selection of the most suitable strains (Borowitzka 2013b), the design of culture systems, and the optimisation of culture conditions to produce the desired product (Borowitzka 2016).

Photosynthesis

The particular attraction of microalgae for practical and commercial applications is that they can use light energy to fix inorganic carbon in photosynthesis ultimately to produce a wide variety of organic molecules. There are also species which can take up organic carbon molecules in the light as sources

Algae R&D Centre, Murdoch University, Murdoch, Western Australia 6150, Australia.
Email: m.borowitzka@murdoch.edu.au

of both energy and carbon building blocks (i.e., mixotrophy) and some can utilise organic carbon as an energy source in the dark (i.e., heterotrophy).

Photosynthesis is the process leading to primary production and can be described simply as a composite chain of cascading events starting with photon capture by the photosynthetic pigments and extending through O_2-evolution to C-fixation (Kroon et al. 1993). Antennae pigments catch photon energy and funnel this towards a transmembrane structure in the thylakoid, called Photosystem II (PSII), via resonance transfer. It is at PSII that primary charge separation occurs and electron transport continues from PSII to Photosystem I (PSI) via the cytochrome b_6/f complex. Photons are absorbed at PSI also, to provide the reducing power needed to produce NADPH. During these 'light reactions' of photosynthesis, hydroxyl ions are released into the thylakoid lumen resulting in the development of an electrochemical gradient. ATPsynthase complexes bound within the thylakoid membrane utilise this electrochemical gradient to synthesize ATP in the stroma. Thus, simple reductants are the product of the light reactions of photosynthesis, with NADPH potentially yielding one molecule of CH_2O (Jumars 1993). For a more detailed account of the photosynthetic reactions, the reader is referred to Falkowski and Raven (2007) and the book edited by Papageorgiou and Govindjee (2004).

The stoichiometry of carbon fixation is not as clearly fixed as the simple equation often shown (as below) might lead one to believe:

$$CO_2 + H_2O \xrightarrow{hv + Chl\,a} CH_2O + O_2$$

The energy comes from light (hv) and four photons are required to excite one electron. If the electron reduces the primary electron acceptors within the open reaction centres, it is eventually converted into chemical energy, where 4 electrons (= 12 photons) are needed to fix one mole of CO_2. Carbon fixation leading to the construction of carbon skeletons is essential for growth and for energy storage to be used for nocturnal metabolism. In the light, however, reductants formed by photosynthesis are distributed between carbon fixation, nitrogen assimilation, photorespiration, inorganic carbon accumulation, chlororespiration, pseudocyclic electron transport (e.g., Mehler reaction), and respiratory phosphorylation (Björkman and Demmig-Adams 1995; Behrenfeld et al. 2004). Temporal separation of metabolic events governs the prominence of any particular pathway and the fraction of photosynthate allocated to carbon fixation changes with growth conditions and on time scales from seconds to generations (Behrenfeld et al. 2004).

The theoretical limit of photosynthetic use of solar radiation is about 11.9% (Walker 2009). However, the potential efficiency which it is possible to achieve is less that this and has been calculated to be at about 4–5% (Benemann and Oswald 1996; Zhu et al. 2008; Grobbelaar 2009). As Vonshak and Torzillo (2004) have pointed out, outdoor algae cultures are exposed to a variety of changes in environmental conditions which occur at several time scales. There is the circadian cycle in light and temperature over 24 h, and the seasonal cycle which varies over the year according to the geographical and climatic location where the algae are being grown. Mixing of the cultures imposes a third cycle, a light-dark cycle which fluctuates in the order of seconds to fractions of seconds depending on the culture and mixing system. The efficient use of light by microalgae outdoors is further affected by both the high irradiance and its variability over the day and due to clouds. Microalgae have evolved a range of photoacclimation and photoprotective mechanisms to cope with this variable environment. Over long periods of days to weeks, algae acclimate to changes in irradiance, and this photoacclimation has been demonstrated in natural phytoplankton in many studies (Brown and Richardson 1968; Prezelin 1976; Meeson and Sweeney 1982; Olaizola and Yamamoto 1994), as well as in outdoor algal cultures (Moheimani and Borowitzka 2007). As the levels of Rubisco seem to be relatively constant (Sukenik et al. 1987), the major regulation must be in the light reactions, mainly in PS II. Regulation of PS II can be by one of two ways—modulation of the PS II light harvesting capacity, or by changes in the number of PS II reaction centres. In order to cope with the variable time scales of the changes in irradiance, algae have evolved both fast-responding and slow-responding acclimation mechanisms. Over a scale of days algae will increase their photosynthetic pigments in light-limiting conditions and reduce them under supra-optimal irradiance (MacIntyre et al. 2002). Increasing growth irradiances leads to changes in the chlorophyll a/b ratio in the green alga *Dunaliella tertiolecta*, the chlorophyll a/c ratio in the diatoms *Thalassiosira pseudonona* and *Skeletonema costatum* (Kolber et al. 1988), and the haptophyte *Emilianea huxleyi* (Harris et al. 2005). This increase in the pigment ratios and decline in pigment content is consistent with decreases in the size and number of PS I and PS II reaction centers protecting the photosynthetic apparatus from excess energy during growth under sustained saturating irradiances.

Relatively rapid fluctuations in light as would be expected, for example, by the passage of a cloud during the day, expose the cells to rapid changes between high and low irradiances. Microalgae have evolved a number of photoprotective mechanisms such as the PS II and PS I electron cycles, state-transitions, fast repair of the D1 protein of the PS II reaction centre, and the scavenging of reactive oxygen species (Falkowski and Raven 2007; Lavaud 2007; Minagawa 2011) to cope with rapid changes in irradiance. Amongst these mechanisms the xanthophyll cycle and the dependent thermal dissipation of the excess light energy (non-photochemical quenching—NPQ) plays a central role (Brunet and Lavaud 2010). NPQ (allows the thermal dissipation of the excess of energy and is one of the faster photoprotective processes activated by algal cells). NPQ decreases the flow of excitation energy to PS II reaction centers and helps to minimize the production of harmful oxygen radicals in the PS II antenna. NPQ has three components: qE which is a fast quenching component regulated by the build-up of a transthylakoid ΔpH and the operation of the xanthophyll cycle, qI which is an intermediate quenching component due to state transition, and qI which is a slowly relaxing quenching component due to photoinhibition (Rascher and Nedbal 2006). qE involves mostly the xanthophyll cycle activity (Brunet and Lavaud 2010), which corresponds to the reversible de-epoxidation of violaxanthin into zeaxanthin through antheraxanthin for green algae. In most chlorophyll c–containing microalgae, the photoprotective cycle involves the conversion between diadinoxanthin and diatoxanthin. Diatoms are one of the groups with the greatest NPQ and xanthophyll cycle activity (Dimier et al. 2007; Goss and Jakob 2010) and this may be the mechanism by which they can better cope with variable irradiances (Wagner et al. 2006).

Photosynthesis and photorespiration are also influenced by temperature, with higher temperatures increasing oxygenase and PSII activities (Belay and Fogg 1978; Verity 1981; Morris and Kronkamp 2003). Different algae species have different temperature optima and therefore their response to high irradiance will vary with temperature. For example, Fig. 1 shows the effect of temperature at different irradiances on the effect of increasing $[O_2]$ for *Isochrysis galbana*. The gross photosynthetic rate decreased with increasing $\{O_2\}$ above 50% sat_{air} for all conditions except for the 23°C treatment at 1200 $\mu mol_{photons}.m^{-2}.s^{-1}$ which was inhibited at concentrations above 80% sat_{air}. The rate of decrease in gross photosynthetic rate (O_2 inhibition rate) was affected by both irradiance and temperature. The gross photosynthetic rate at 1200 $\mu mol_{photons}.m^{-2}.s^{-1}$ was highest at 23°C whereas at 2500 $\mu mol_{photons}.m^{-2}.s^{-1}$ the highest temperature specific maximum gross photosynthetic rate was observed at 26°C.

Respiration

Algae cultures can lose a significant part of their biomass at night due to respiration (Grobbelaar and Soeder 1985) and it is therefore important to minimise these respiratory losses in order to maximise productivity. Dark respiration is cell size dependent (Banse 1976) and is strongly influenced by the growth irradiance and temperature (Raven 1981; Torzillo et al. 1991; Ogbonna and Tanaka 1996) with suboptimal temperatures and high irradiances resulting in higher night-time respiration. Dark respiration is also influenced by the nitrogen source (Vanlerberghe et al. 1992; Post 1993). More on dark respiration can be found in Raven and Beardall (2016).

Inorganic carbon utilisation

Algae, being aquatic organisms, face a different inorganic carbon environment than land plants. In water, inorganic carbon exists in three forms (CO_2, HCO_3^- and CO_3^{2-}), and the inorganic carbon system is the main buffering system in water with the relative proportion of these three inorganic carbon species dependent on pH, salinity, and temperature. For example, less than 1% of dissolved inorganic carbon (DIC) is present as CO_2 at normal seawater pH (8.1–8.2) and more than 90% is in the form of HCO_3^-. Furthermore, the rate of diffusion of CO_2 in water is 10^4 times slower than that in the atmosphere, and the rate of conversion of HCO_3^- to CO_2 is very slow at alkaline pH. These factors cause a limitation in the

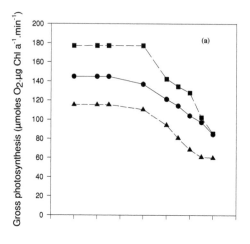

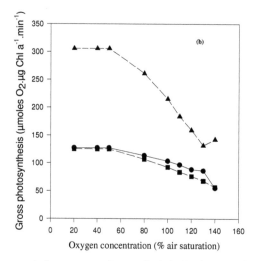

Fig. 1. The effect of oxygen concentration on gross photosynthesis in *Isochrysis galbana* at different irradiances and temperatures. (a) 1200 $\mu mol_{photon}.m^{-2}.s^{-1}$ (b) 2500 μmol_{photon} m^{-2} s^{-1}, ● = 20°C, ■ = 23°C, ▲ = 26°C.

supply of CO_2 and may therefore severely restrict photosynthesis in microalgae. Most microalgae studied can take up both CO_2 and HCO_3^-, but a few species can take up only CO_2 (e.g., *Nannochloris atomus, Nannochloris maculata, Amphidinium carterae, Heterocapsa oceanica, Monodus subterraneus* and some synurophyte algae), whereas *Nannochloropsis gaditana* and *Nannochloropsis oculata* can take up only HCO_3^- and not CO_2 (Colman et al. 2002; Bhatti and Colman 2008).

Most microalgae fix inorganic carbon via the Calvin-Benson cycle directly via ribulose bisphosphate carboxylase (Rubisco). However, Rubisco has a relatively low affinity for CO_2 and it also has a dual role as an oxygenase so that CO_2 and O_2 compete at the active binding site:

Ribulose-1,5-bisphosphate + CO_2 + H_2O → 2 x glycerate-3-P

Ribulose-1,5-bisphosphate + O_2 → glycerate-3-P + glycolate-2-P

The extent to which the competitive reactions of Rubisco occur depends on the O_2 and CO_2 concentrations at the Rubisco active site and the type of Rubisco in the particular algae taxon (Giordano et al. 2005; Beardall and Raven 2016). Actively photosynthesising algae cultures raise the pH due to CO_2 uptake, thus effectively reducing the amount of free CO_2 available as well as producing O_2 which inhibits C-fixation because of the oxygenase activity of the Rubisco. In large-scale outdoor cultures O_2

concentrations of over 150% of air saturation are usual during the day and pH can rise to above pH 9 and this can inhibit photosynthesis by 40% or more.

Most algae have evolved a diverse range of mechanisms collectively called carbon concentrating mechanisms (CCMs) that overcome the deficiencies in Rubisco (Beardall and Raven 2016). These mechanisms may involve the primary use of HCO_3^- or active CO_2 uptake. Cyanobacteria have a CCM based on CO_2 or HCO_3^- transport either at the plasmalemma or the thylakoid membrane delivering HCO_3^- to the cytosol (Ogawa and Kaplan 2003). The HCO_3^- then diffuses into the carboxysomes where carbonic anhydrase generates the CO_2 for Rubisco (Price et al. 2002). In eukaryotic algae with CCMs, the active transport mechanisms of dissolved inorganic carbon (CO_2, HCO_3^-) are located on the plasmalemma or the inner plastid envelope or both (Giordano et al. 2005). The equilibration of the CO_2 and HCO_3^- in the various cell compartments (periplasmic space, thylakoid lumen, chloroplast stroma) also involves a number of carbonic anhydrases. Diatoms such as *Thalassiosira weissflogii* appear to show C_4-type of photosynthesis where the inorganic carbon is added to a C_3 carrier to form a C_4 intermediate that is then decarboxylated at the site of Rubisco generating CO_2 at the active site; however there is still some conflict in the experimental data available (Raven 2010; Valenzuela et al. 2012).

Irrespective of the CCM mechanisms, almost all intensive algae cultures are carbon-limited and the addition of CO_2 and/or bicarbonate will stimulate growth (Olaizola et al. 1991; White et al. 2013).

Organic carbon utilisation

The ability to utilize organic carbon sources mixotrophically and/or heterotrophically is widespread amongst the algae, especially the green algae (Neilson and Lewin 1974; Perez-Garcia et al. 2011). Some algae, such as the thraustochytrids used commercially for the production of long chain polyunsaturated fatty acids, are obligate heterotrophs (Barclay et al. 1994, 2010). Mixotrophic growth overcomes some of the C-limitation usually found in algae grown photoautotrophically.

The uptake and metabolism of organic C sources such as glucose or acetate by several microalgae (e.g., *Chlorella, Chlamydomonas, Haematococcus, Phaeodactylum, Nannochloropsis*, etc.), has been extensively studied with different algae utilising different organic C sources (Chu et al. 1995; Tanner 2000; Perez-Garcia et al. 2011).

In heterotrophic culture, cell densities of more than 100 g.L^{-1} cell dry weight, were achieved with *Chlorella, Crypthecodinium*, and *Galdieria* species (de Swaaf et al. 2003; Graverholt and Eriksen 2007; Doucha and Lívanský 2012). Not surprisingly, heterotrophic growth has an effect on the composition of the algal cells. For example, heterotrophically grown *Chlorella saccharophila* (=*Chloroidium saccharophilum*) had about 3x the lipid content of autotrophically grown cells (Isleten-Hosoglu et al. 2012). The lipid composition can also be manipulated further by changing the C/N ratio. For example, in *C. sorokiniana*, low C/N ratios favoured a high proportion of trienoic fatty acids at the expense of monoenoic acids (Chen and Johns 1991) and in the diatom, *Nitzschia laevis* heterotrophically grown cells had a higher eicosapentaenoic acid content (Wen and Chen 2000).

Several studies have shown that the optimum concentration of organic C for mixotrophic growth varies between strains and that too high concentrations of the organic substrate can be inhibitory (Chu et al. 1995; Liang et al. 2009), probably because of substrate inhibition and possibly also because of osmotic effects. In many, but not all algae, mixotrophic growth is the sum of photoautotrophic growth and heterotrophic growth (Martínez and Orús 1991; Cheirsilp and Torpee 2012). Species and strain variation in the types of sugars which can be used and their effects on mixotrophic and heterotrophic growth has been reported in the genera *Chlorella* and *Auxenochlorella* (Shihira and Krauss 1965; Kessler 1976).

Glucose has been found to inhibit photosynthetic CO_2 fixation in *Chlorella* in experiments where pH was not controlled (Lalucat et al. 1984), and this is probably because glucose inhibits carbonic anhydrase induction (Shiraiwa and Umino 1991; Villarejo et al. 1997). However, specific growth rates are still doubled by the addition of glucose, and further enhanced by the addition of 2% CO_2 (Martínez and Orús 1991). Excess CO_2 may however inhibit mixotrophic growth (Sforza et al. 2012). In the light glucose also affects N uptake, stimulating nitrate uptake in *Ankistrodesmus*, especially at low CO_2 concentrations (Eisele and Ullrich 1977), and also in *Chlorella* (Schlee et al. 1985).

Nutrients

Algae require about 30 different elements for growth (Kaplan et al. 1986). In this section, however, I will only consider those which are known to influence growth and photosynthesis most. Ever since Redfield (Redfield 1934, 1958) proposed the C:N:P ratio of 106:16:1 (known as the Redfield Ratio) as representing the average atomic ratio of phytoplankton there has been discussion on how representative it is. It is clear that there are significant variations from this ratio depending on growth conditions, between taxa and with cell size (Falkowski 2000; Geider and Roche 2002; Ho et al. 2005), but 'healthy' phytoplankton cells appear to have a composition close to this ratio. Quigg et al. (2003) extended the Redfield ratio to include trace elements with the average composition of phytoplankton in this study found to be $C_{124}N_{16}P_1S_{1.3}K_{1.7}$ $Mg_{0.56}Ca_{0.5}Fe_{0.0075}Zn_{0.0008}Cu_{0.00038}Cd_{0.00021}Co_{0.00019}$. This ratio serves as a good working approximation of the nutrient requirements of microalgae and serves as a start for medium design and optimisation.

Nitrogen

As it is a key component of amino acids and proteins, nitrogen is critical for the production of biomass. The nitrogen content of algal biomass ranges from about 1% to more than 10%, and although it varies between taxa, the available nitrogen in the medium is a major determinant. Algae can utilize nitrate (NO_3^-), ammonia (NH_4^+), urea, and a range of organic N sources. Although ammonium is generally the N-source most rapidly utilized by microalgae, probably because of a lower energy demand for its metabolism, high concentrations of ammonium are detrimental, if not lethal, to most algal species because of the speciation of ammonia (NH_3)/ammonium (NH_4^+) in water which is temperature and pH dependent:

$$NH_3(aq) + H_2O \leftrightarrow NH_4^+(aq) + OH^-(aq) \; pK_a \; (25°C) = 9.25$$

Ammonia is toxic to algae (whereas ammonium is a valuable N-source) but one which may be toxic at high concentrations (Abeliovich and Azov 1976). The mechanisms of this toxicity is not well understood (Britto and Kronzucker 2002), but there is some variation in the sensitivity of different algae to ammonia. Alkaline pH and high temperatures lead to the formation of ammonia rather than ammonium (see equation above). High rates of algal photosynthesis lead to alkalinisation of the medium and can result in high concentrations of ammonia; however uptake of ammonium by algae can lead to acidification of the medium and either of these can lead to algal death (Borowitzka and Borowitzka 1988b). For example, Levine (2011) reported toxic levels of NH_4^+ at 100 mg L^{-1} for a culture of *Neochloris oleoabundans*. Other species can acclimate to high levels of nutrients and can more efficiently utilize ammonium as a nitrogen source (typically *Chlorella* and *Scenedesmus* sp. in wastewater treatment), but the exact limitations are very specific to individual algal strains and culture conditions. No significant effect on the growth of *Chlorella sorokiniana* was observed at 100 mg$_{NH4}$.L^{-1} while *Spirulina platensis* was nearly completely inhibited at 200 mg$_{NH4}$L^{-1} (Ogbonna et al. 2000). González et al. (2008) reported successful culture of a *Chlorella* sp. at concentrations as high as 373 mg.L^{-1} but Przytocka-Jusiak et al. (1984) reported that *C. vulgaris* cell growth was reduced by 50% at 330 mg.L^{-1} ammonium, while 700 mg.L^{-1} completely inhibited growth at pH 8–9. *Scenedesmus* will grow at 150 mg$_{NH4}$.L^{-1} but not at 300 mg$_{NH4}$.L^{-1} (De Godos et al. 2010). However, the effects of high ammonia/ammonium concentrations in large-scale cultures is little understood as ammonium-N salts are not normally used as the N-source in large-scale algal cultures and laboratory studies have paid little attention to the speciation of ammonia/ammonium and therefore the data in the studies cited above must be interpreted with caution.

Control of pH is important when using ammonia. Gonzalez et al. (2008) was also able to demonstrate that a much higher NH_4^+ load (~ 1100 mg L^{-1}) could be tolerated by an enclosed *Chlorella* culture when the pH was maintained at pH 7 where most of this N will be in the form of NH_4^+. Tam and Wong (1996) found very similar results with *Chlorella* able to maintain a specific growth rate of 0.2 d^{-1} at ammonia concentrations of 1000 mg.L^{-1} when pH was maintained at neutral.

Uptake of nitrate is by an H$^+$/nitrate co-transport and is stimulated by light and enhanced by CO_2 (Grant and Turner 1969; Eisele and Ullrich 1977; Tischner 2000). Although the uptake of nitrate leads to alkalinisation of the medium, high nitrate concentrations are not toxic to algae.

Algae can also use urea as an N source. Algae either have a urease which catalyses the conversion of urea to ammonium generating CO_2 in the process:

$$CO(NH_2)_2 + H_2O \rightarrow CO_2 + 2NH_3$$

or urea amidolyase which is a two enzyme, ATP requiring, system:

a urea carboxylase: $Urea + ATP + HCO_3^- \rightarrow Allophanate + ADP + P_i$

and allophanate hydrolase: $Allophanate + 3H_2O + H^+ \rightarrow 2NH_4^+ + 2HCO_3^-$

Urea amidolyase is only found in green algae, and all other algae have urease (Leftley and Syrett 1973; Syrett 1988).

Further details on Nitrogen and microalgae can be found in (Raven and Giodano 2016).

Phosphorous

Phosphorous is essential for many cellular processes such as energy transfer, nucleic acid synthesis, etc. (Dyhrman 2016). Orthophosphate (PO_4^{2-}) is the preferred form for microalgae although many microalgae have inducible extracellular phosphatases allowing the utilization of organic P sources (Burczyk and Loos 1995). Many algae also accumulate excess P intracellularly mainly in the form of polyphosphates bodies when an excess of P is available (Eixler et al. 2006). High concentrations of phosphate may inhibit algal growth and there is great variation between species in their tolerance to high phosphate.

Silicon

Silicon is generally sufficiently abundant in natural waters so as not to be limiting to microalgae; however, diatoms present a special case with a high Si requirement because of their silica valves and for most, Si must be added to the growth medium. Insufficient Si can lead to cessation of DNA synthesis and cell division (Darley and Volcani 1969; Sullivan and Volcani 1981) and, in some species such as the diatom *Skeletonema costatum*, to cell death.

Metabolism and the production of some compounds of commercial interest

Microalgae are currently used for the commercial production of carotenoids, specifically β-carotene and astaxanthin.

Dunaliella salina and β-carotene

The halophilic green alga, *Dunaliella salina*, accumulates up to 14% of dry weight as β-carotene in oil droplets located in the chloroplast (Borowitzka and Borowitzka 1988a). The β-carotene in the globules is mainly in the form of two stereoisomers: all-*trans* and 9-*cis*, with the rest a few other mono-*cis* and di-*cis* stereoisomers of the β-carotene and no xanthophylls. Both the amount of the accumulated β-carotene and the 9-*cis* to all-*trans* ratio depend on irradiance and on the algal division time, which is determined by the growth conditions (Ben-Amotz et al. 1988). The synthesis of this β-carotene is mainly influenced by salinity and irradiance. Salinity mainly regulates the maximum content of β-carotene possible, whereas irradiance regulates the rate of β-carotene formation (Semenenko and Abdullayev 1980; Ben-Amotz and Avron 1983; Borowitzka et al. 1990). Upon an upward shift in salinity *D. salina* rapidly accumulates β-carotene, but if the salinity is reduced the β-carotene is only very slowly metabolised (Borowitzka et al. 1985; Borowitzka et al. 1990). Nitrate limitation will also increase carotenoid accumulation (Mil'ko 1963; Ben-Amotz and Avron 1983) whereas P-limitation has little effect (Ben-Amotz et al. 1982). The source of N used also affects carotenogenesis with ammonia inhibiting β-carotene synthesis (Borowitzka and Borowitzka 1988b). In general, carotenogenesis is greatest under sub-optimal growth conditions when the specific growth rate is low (Borowitzka et al. 1984); that is, β-carotene is a typical secondary metabolite. The accumulation of β-carotene requires the concomitant synthesis of fatty acids (especially

C18 oleic acid) which make up the oily droplets within which the β-carotene is sequestered (Rabbani et al. 1998; Mendoza et al. 1999; Lamers et al. 2010). The carotenoid biosynthetic pathway of *Dunaliella* is the same as that in other plants (Britton 1988).

The exact function of these high amounts of carotenoids in the cell is not known, but two complementary hypotheses have been proposed. In nature, *D. salina* is usually found in shallow salt lakes where the algae cells are exposed to very high irradiances. *D. salina* cells with a high carotenoid content can tolerate much higher irradiances that those with a lower carotenoid content (Gômez Pinchetti et al. 1992). The β-carotene absorbs light in the blue region and *cis*-β-carotene is an effective quencher of singlet oxygen and is transformed to *trans*-β-carotene (Jimenez and Pick 1993) and thus, the β-carotene may also protect the chloroplast from damage by free oxygen radicals (Shaish et al. 1993). The second hypothesis proposes that the lipids and associated β-carotene are a carbon sink (Borowitzka and Borowitzka 1988a). The formation of high amounts of carotenoids occurs when one or more metabolic intermediate pathways are inhibited by lack of substrate; however, photosynthesis still continues, albeit at a reduced rate. It is essential for the survival of the cell that photosynthesis continues in order to supply sufficient energy for essential metabolic processes for survival at the high salinity at which this alga grows, such as Na^+-efflux and glycerol synthesis. One by-product of this photosynthesis is 3-phosphoglyceric acid which is further metabolised and the products of this metabolism must be either stored or excreted. If this photosynthetically generated product is stored, then it is essential that it does not inhibit cell function. Triacylglycerols and β-carotene are suitable 'neutral' compounds that could serve this function. In fact, many algae accumulate triacylglycerols (lipids) when their growth is inhibited (Roessler 1988; Ördög et al. 2013) or, alternatively, they excrete large quantities of organic C once growth limitation sets in (Arad et al. 1992; Myklestad 1995).

The high salinity brines that *D. salina* grows also mean that CO_2 solubility is reduced compared to seawater and the inorganic C equilibrium is shifted towards HCO_3^- (Lazar et al. 1983). The high temperatures where *D. salina* is grown commercially also reduce the solubility of CO_2 further. Although *D. salina* has an active CCM mechanism (Booth and Beardall 1991) and an unusual α-type external carbonic anhydrase that is active over a salinity range of 0–4 M NaCl (Fisher et al. 1996; Premkumar et al. 2003), large-scale cultures are still severely carbon limited. Some of this limitation can be overcome by the addition of bicarbonate.

The fact that β-carotene is a secondary metabolite means that conditions to achieve maximum growth rate (i.e., lower salinities, high nutrients) and the conditions to achieve a high cell β-carotene content (i.e., high salinities, low nutrients) are incompatible. In practical terms, this means two possible culture strategies for the production of β-carotene by the culture of *D. salina*. One strategy is to initially grow the algae at an optimal salinity for biomass generation (i.e., a salinity of less than 20% w/v NaCl + sufficient nutrients) and then transferring the cells to a higher salinity under nutrient limiting conditions. Such a 2 stage process however has the disadvantage that it requires a larger pond area and takes longer, thus increasing production costs. The alternative strategy is to grow the algae at high irradiances and at a salinity where β-carotene productivity is highest. Figure 2 shows the effects of salinity on growth and β-carotene content and the optimum salinity for β-carotene productivity. However, in commercial production, the algae are usually grown at a higher salinity to ensure reliable production and minimise the risk of protozoan invasions which can result from a reduction in salinity due to dilution by rainfall (Borowitzka and Borowitzka 1989). The irradiance received by the algae cells is managed by managing the areal density of the algae (Grobbelaar 1995).

Haematococcus pluvialis and astaxanthin

The freshwater green alga *Haematococcus pluvialis* is the best natural source of the oxygenated carotenoid, astaxanthin, which is accumulated in lipid droplets in the cytoplasm in the aplanospore stage of the alga (Lang 1968; Wayama et al. 2013). Aplanospore formation and astaxanthin accumulation are induced by nutrient limitation and other growth-inhibiting induction factors (often called 'stress' factors in the literature) such as high temperature, high light, or increased salinity (Borowitzka et al. 1991; Boussiba and Vonshak 1991; Vidhyavathi et al. 2008). *Haematococcus* produces only the 3*S*,3'*S* stereoisomer of

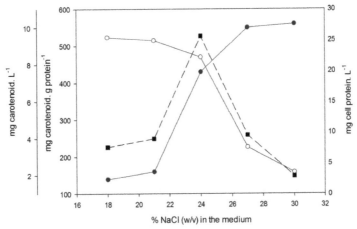

Fig. 2. Changes in biomass production (●) and carotenoid content in the cells (○) in *Dunaliella salina* W5 at different salinities. The production of carotenoid per L culture is also shown (■). Cultures were grown in modified Johnson's medium (J/1) (Borowitzka 1988a) at an irradiance of 400 $\mu mol_{photons}.m^{-2}.s^{-1}$ with a 12:12 h light:dark cycle at 33°C during the day and 28°C at night. Early stationary phase cultures were sampled. Redrawn from Borowitzka et al. (1984).

astaxanthin, both in free and esterified forms linked to 16:0, 18:1, and 18:2 fatty acids (Grung et al. 1992) and the astaxanthin pool of encysted *Haematococcus* is approximately 70% monoesters, 25% diesters, and 5% free astaxanthin. Astaxanthin formation requires the synthesis of fatty acids (Schoefs et al. 2001; Zhekisheva et al. 2002).

Astaxanthin formation and aplanospore formation generally occur together (Droop 1955); however, some astaxanthin formation may occur in the flagellate cell stage under specific nutrient limitation (Grünewald et al. 1997; Del Rio et al. 2005). *H. pluvialis* also can use organic C such as acetate either mixotrophically or heterotrophically (Kobayashi et al. 1992) and although astaxanthin formation can be observed in heterotrophic cultures, it is very much higher in mixotrophic cultures in the light (Orosa et al. 2001; Kang et al. 2005). There is a high degree of variation between strains in their response to acetate concentration and their ability to use other organic C sources such as glucose, glycerol, and glycine (Borowitzka 1992, 1995).

Astaxanthin formation in *H. pluvialis* and the complex mechanisms of the various stress factors has been extensively studied and the reader is referred to a recent review paper (Lemoine and Schoefs 2010). Like *D. salina* β-carotene, *Haematococcus* astaxanthin appears to have multiple roles in the cell. Several studies have attributed a photoprotective role to astaxanthin, especially to UV-B irradiation (Yong and Lee 1991; Qiu and Li 2006; Li et al. 2010), but Fan et al. (1998) found that astaxanthin-rich red cysts are not better protected than astaxanthin-free green cells when exposed to high light. The astaxanthin biosynthetic pathway has been shown to lower reactive oxygen species (ROS) production by lowering cellular oxygen via an electron transport from the carotenogenic desaturation steps to the plastoquinones and then to the plastid terminal oxidase (PTOX) (Li et al. 2008). The synthesis of triacylglycerol (TAG) and of the fatty acids molecules needed for esterification of astaxanthin molecules also serves as an electron sink under photo-oxidative stress (Hu et al. 2008) as the formation of a C18 fatty acid requires approximately 24 NADPH derived from the photosynthetic electron transport chain and thus relaxes the over-reduced electron transfer chain under high-light conditions. As with *Dunaliella* β-carotene formation, the rate of astaxanthin formation is light-dependent (Li et al. 2010).

The optimum growth temperature of *H. pluvialis* is between 10–20°C and, in batch culture, the flagellated green motile stage predominates during the active phase of growth and the carotenogenic aplanospores are formed during the stationary phase. This change in cell morphology—from the rapidly dividing, shear sensitive, flagellated cells to the extremely robust, thick-walled, non-motile aplanospores—means that generally a two-stage production process is used (Olaizola 2000; Cysewski and Lorenz 2004). Here, the first stage is optimised to produce growth on the flagellated 'green' cells to maximise biomass and is usually carried out in closed photobioreactors to minimise contamination. This is then followed by a second stage, often in open ponds, in which the cells are 'stressed' (high light, higher temperatures) to induce aplanospore and concomitant astaxanthin formation.

Lipids, fatty acids, and biofuels

Microalgae can accumulate high levels of lipids and are existing and potential future sources of a number of fatty acids with applications in human and animal nutrition. The fatty acids of most interest are the very long-chain polyunsaturated fatty acids arachidonic (AA; C20:4n-6), eicosapentaenoic acid (EPA; C20:5n-3), docosahexaenoic (DHA; C22:6n-3). The high lipid content and productivity of some microalgae has also made them attractive targets for the development of renewable biofuels (Griffiths and Harrison 2009; Fon Sing et al. 2013). Lipid and fatty acid composition and metabolism in algae and their manipulation has been extensively reviewed in recent years (Guschina and Harwood 2006; Harwood and Guschina 2009; Cohen and Ratledge 2010; Guschina and Harwood 2013) in much greater detail than can be covered here.

The total lipid content of microalgae ranges from about 10–50% of organic dry weight. Since the original finding that nitrogen limitation increased the lipid content in *Chlorella* (Aach 1952), many studies have shown that nutrient limitation, especially N limitation can lead to an increase in lipid content, especially triacylglycerol (TAG) content in many species of microalgae. In diatoms, Si limitation also leads to increased TAG formation. In most cases the increase in TAG is accompanied by reduced growth. Some species of algae such as members of the genus *Tetraselmis*, usually have their highest total lipid content in the exponential phase of growth and the lipid content declines in the stationary phase (Mercz 1994). *Tetraselmis* spp. therefore tend to have higher lipid productivity than many other microalgae (Barclay et al. 2005; Huerlimann et al. 2010). The effect of P limitation on lipid accumulation varies. For example, in *Phaeodactylum tricornutum* and *Chaetoceros* sp. (Bacillariophyceae), *Isochrysis galbana* (clone T-Iso) and *Pavlova lutheri* (Prymnesiophyceae), P-limitation increased the lipid content, whereas in *Nannochloris atomus* (Chlorophyceae) and *Tetraselmis* sp., it decreased (Reitan et al. 1994). Lipid content can be increased by providing additional CO_2 to algae cultures which are carbon limited (Muradyan et al. 2004; Tang et al. 2011; Moheimani 2012). High lipid productivities also have been obtained in a number of microalgae grown heterotrophically (Xiong et al. 2008) or mixotrophically (Li et al. 2011; Kong et al. 2012; Zhao et al. 2012).

The fatty acid composition of microalgae varies between taxa and, to a lesser degree, also between species (Borowitzka 1988b). Thus, the lipids of diatoms, haptophtes, and eustigmatophytes and the red unicell *Porphyridium cruentum* are particularly rich in EPA, whereas the highest content of DHA is found in thraustrochytrid algae such as *Crypthecodinium cohnii* and *Schizochytrium* spp., and the freshwater trebouxiophyte *Parietochloris incisa* contains exceptionally high amounts of AA (Bowles et al. 1999; Bigogno et al. 2002; Guschina and Harwood 2006). The content of these long-chain polyunsaturated fatty acids (PUFAs) are affected by environmental factors and can be manipulated by growth conditions and the composition of the medium (e.g., Carvalho et al. 2006), however these effects are very variable between species and the results of studies on the same species are sometimes inconsistent making it difficult to generalise. For example, Chrismadha and Borowitzka (1994) found that CO_2 addition reduced the level of EPA in the cells of *Phaeodactylum tricornutum*, as did Carvalho and Malcata (2005) in *Pavlova lutheri*, whereas Yongmanitchai and Ward (1991) found the opposite. These differences in the findings might be due to culture conditions such as available light, or due to strain differences. However, despite this reservation, some broad generalisations can be made. Nutrient limitations (especially N as well as Si in diatoms) generally lead to an increase in non-polar storage lipids with a concomitant increase in C16 and C18 fatty acids (Arisz et al. 2000; Lynn et al. 2000). High light generally leads to the production of TAG in many microalgae (Roessler 1990; Sukenik 1999) and higher temperatures increase the degree of unsaturation (Materassi et al. 1980). However, in *Parietochloris incisa* FA synthesis, mainly as TAG, was lower under high light in N replete and more so in N-starved cultures. For optimum lipid synthesis, intermediate irradiances are required (Solovchenko et al. 2008). The red unicell *Porphyridium cruentum* shows a similar response (Cohen et al. 1988).

Growth-limiting conditions, especially N-limiting conditions in the light where continuing photosynthetic carbon fixation result in an increased cellular C/N ratio mean that the photosynthetically-fixed carbon cannot be metabolised to N-containing compounds such as proteins. Therefore, this photosynthetically fixed carbon either must be stored or excreted from the cell. Clearly, there are advantages in storing the energy rich organic carbon for later re-metabolism when growth conditions improve (Pohl and Zurheide 1979; Harwood and Jones 1989), but the carbon storage product must be in a form which does not negatively affect other cellular processes such as enzyme activities or osmotic potential. Two main options are possible: storage as starch or storage in the form on neutral lipids (TAGs), or a combination of both. For biofuels production, algae with a high content of TAG are desirable. Saturated acyl moieties are preferred for this purpose because they require less energy to synthesise than PUFA and provide more energy on oxidation (Cohen and Khozin-Goldberg 2010). However, some algae such a *P. incisa* and *P. cruentum* also store appreciable amounts of PUFAs in the TAG. Cohen and Khozin-Goldberg (2010) have hypothesised that this is so because these algae species live in ecological niches subject to rapid short-term fluctuations such as in temperature or salinity where significant and rapid alterations in the fatty acid and molecular species composition of chloroplast membrane lipids to maintain membrane fluidity are required. In these situations, the TAG may also have a role as a 'buffering' agent, where the PUFA moieties are mobilised for the construction and rapid alteration of chloroplast membranes.

The synthesis of TAGs under conditions of N-deprivation requires photosynthesis and an adequate supply of inorganic carbon. Providing either additional CO_2 or bicarbonate leads to stimulation of TAG accumulation (Gardner et al. 2012; White et al. 2013). For example, in *D. salina*, a 1-day long increase in CO_2 concentration from 2% to 10% increased the total fatty acids on a dry weight basis by 30% (Muradyan et al. 2004). This was due mainly to *de novo* fatty acid synthesis with the elongation and desaturation of the fatty acids being inhibited, leading to an increase in the relative content of saturated fatty acids at high CO_2. In *Dunaliella viridis* however, the addition of CO_2 increased the total lipid content only under N-limiting conditions (Gordillo et al. 1998).

The challenge to commercially viable production of lipid-rich algae for biofuels or PUFA-rich algal lipids for nutritional use however is still a significant one, and will require the selection of strains best suited for commercial-scale culture and the manipulation of the algal physiology and metabolism by the use of appropriate culture management to maximise lipid productivity. Genetic modification of metabolic pathways to enhance lipid productivity is also an option.

Optimising productivity of outdoor cultures

Commercial production of microalgae is generally carried out in outdoor cultures, mainly in open raceway ponds, although there are a few producers using closed photobioreactors for the production of very high value products. The most important limiting factors to growth and productivity are light, carbon supply, oxygen, and temperature (Borowitzka 1998). Furthermore, the factors affecting growth (biomass production) and those affecting the production of the desired products may not be the same. In outdoor cultures, optimisation is complicated by the fact that all these parameters vary both over the day and also over the year and, unlike most laboratory cultures, algae cultures outdoors are never in a steady state. The level of control of the growth environment in large-scale cultures is limited. The provision of optimum nutrient levels is relatively easy to achieve and inorganic C can be provided in the form of CO_2 or bicarbonate, although this is often uneconomical. Temperature generally cannot be regulated, but the irradiance received by the algae cells can be managed to some degree.

The productivity of outdoor cultures is strongly correlated with the available irradiance and the efficient use of the available irradiance is a key target for optimising culture productivity (Richmond 1996). In dilute cultures (i.e., cultures with low cell densities) at optimal growth conditions of optimum temperature and non-limiting nutrients, the photosynthetic rate of the algae with relation to irradiance follows the classical text-book light response curve (P/E curve). However, dilute cultures have a low productivity as the productivity is a function of the specific growth rate (μ, time^{-1}) times the cell density; that is, Productivity = μ x cell density (g.L^{-1}). Thus, in order to achieve high productivities, high cell density cultures are required. However, not all cells in the culture receive the same amount of light in high cell density cultures because of *mutual shading* (Tamiya 1957). Light passing through the culture is absorbed by the cells so that it rapidly attenuates and cells deeper in the culture receive less or no light. The higher the cell density, the shorter the depth that the light penetrates in the culture. Oswald (1988) empirically found that the maximum depth that light penetrated in a pond of green algae was $6000/C_c$, where C_c is the cell density in mg.L^{-1}. The average amount of light, sometimes also called the average irradiance, received by the cells in a well-mixed culture is a function of the cell density (Myers and Graham 1958) and there is an optimum cell density (OCD) at which photosynthetic efficiency and productivity are at a maximum (Richmond and Vonshak 1978; Hu and Richmond 1994; Hu et al. 1998). The optimum average irradiance is equivalent to the saturating irradiance (E_s) as measured in a photosynthesis/irradiance curve using dilute algae suspensions. The OCD and light utilization efficiency, and therefore also productivity, can potentially be improved by altering the photosynthetic antenna size in the algae (Benemann 1989; Polle et al. 2002). This has been shown in the laboratory for several species of algae (Nakajima and Ueda 1997; Melis et al. 1999; Nakajima and Ueda 1999, 2000) but whether such algae can be practically grown in large-scale cultures is not yet known.

In order to accommodate the seasonal differences in irradiance, the average irradiance received by the algae cells can be altered either by changing the cell density by varying the harvesting frequency and/ or the proportion of the culture harvested, or by changing the culture depth (in the case of pond cultures) (Vonshak 1997).

However, the irradiance outdoors also changes over the day as does the culture temperature and the oxygen concentration in the water. In the morning, outdoor cultures are often too cool to use light efficiently (Vonshak et al. 2001), and by noon on sunny days, algae cultures are generally photoinhibited, even in very dense cultures (Vonshak and Guy 1992; Hu et al. 1996; Torzillo et al. 1996; Sukenik et al. 2009). The high oxygen concentration leads to photorespiration and the production of active oxygen species, such as the superoxide radical (O_2^-), the hydroxyl radical ($^{\cdot}OH$), hydrogen peroxide (H_2O_2), and singlet oxygen ($^1\Delta g\ O_2$), which can cause photoinhibition of photosynthesis (Belay and Fogg 1978; Demmig-Adams and Adams 1992; Singh et al. 1995; Cadenus 2005). Not only does photoinhibition reduce photosynthesis, but the repair of the photodamage requires energy and thus has a metabolic cost to the alga (Raven 2011), thus reducing productivity. Supraoptimal temperatures also increase photorespiration and the degree of photodamage (Kromkamp et al. 2009) and this can lead to increased dark respiration and biomass loss at night (Raven 1981; Torzillo et al. 1991; Raven 2011). Temperature and irradiance also interact and affect the chemical composition of the algae (Payer et al. 1980; Carvalho et al. 2009).

Since different algae species and strains have different temperature optima, their response to high irradiance will vary with temperature (Fig. 1). It is therefore important to select a strain which is best able to utilize high light at the temperature range which will be encountered at the site of culture (Borowitzka 2013b) and maintain the culture cell density near the OCD. This means that the cell density needs to be varied between summer and winter if operating at a constant pond depth. Alternatively, the average irradiance received by the algae can also be adjusted by varying the pond depth while maintaining the cell density constant. It is potentially possible to use different strains with different temperature optima in summer and winter (Belay 1997).

High O_2 concentrations in open ponds and photobioreactors arising from algal photosynthesis are a significant factor in limiting productivity (Ogawa et al. 1980; Kliphuis et al. 2011) and selection of algae strains whose Rubisco is less sensitive to oxygen and/or strains which have a more efficient CCM and thus an increased level of CO_2 at the site of Rubisco are highly desirable (Borowitzka 2013b).

Optimisation of growth and the cell content of specific products requires a good understanding of algal physiology (Borowitzka 2016). Furthermore, scaling-up to the very large volumes required for commercial production is not easy and is a major challenge for the transition from the laboratory to an industrial process (Borowitzka and Vonshak 2017).

Conclusion

In order to develop a reliable and highly productive large-scale culture process for microalgae and high value products from microalgae, a good understanding of algal physiology and metabolism is essential. This is important in the process of strain selection and also in the optimisation of culture conditions and for culture management. Recent new developments in genome scale metabolic reconstruction and metabolic flux analysis (Boyle and Morgan 2009; Chang et al. 2011; de Oliveira Dal'Molin et al. 2011) and analysis of gene expression (e.g., Lei et al. 2012; Valenzuela et al. 2012) are providing new tools for analysing the complex metabolic interactions in algae and should aid in the optimisation of cultures for the production of specific products of interest as well as in developing genetic engineering strategies to improve algae strains.

References

Aach, H.G. 1952. Über Wachstum und Zusammensetzung von *Chlorella pyrenoidosa* bei unterschiedlichen Lichtstärken und Nitratmengen. Arch. Mikrobiol. 17: 213–246.

Abeliovich, A. and Y. Azov. 1976. Toxicity of ammonia to algae in sewage oxidation ponds. Appl. Environ. Microbiol. 31: 801–806.

Arad, S.M., Y.B. Lerental and O. Dubinsky. 1992. Effect of nitrate and sulfate starvation on polysaccharide formation in *Rhodella reticulata*. Bioresour. Technol. 42: 141–148.

Arisz, S.A., J.A.J. van Himbergen, A. Musgrave, H. van den Ende and T. Munnik. 2000. Polar glycerolipids of *Chlamydomonas moewusii*. Phytochemistry 53: 265–270.

Banse, K. 1976. Rates of growth, respiration and photosynthesis of unicellular algae as related to cell size—a review. J. Phycol. 12: 135–140.

Barclay, W., C. Weaver and J. Metz. 2005. Development of docosahexaenoic acid production technology using *Schizochytrium*: A historical perspective. pp. 36–52. *In*: Z. Cohen and C. Ratledge (eds.). Single Cell Oils. AOCS Press, Urbana.

Barclay, W., C. Weaver, J. Metz and J. Hansen. 2010. Development of docosahexaenoic acid production technology using *Schizochytrium*: Historical perspective and update. pp. 75–96. *In*: Z. Cohen and C. Ratledge (eds.). Single Cell Oils. Microbial and Algal Oils. AOCS Press, Urbana.

Barclay, W.R., K.M. Meager and J.R. Abril. 1994. Heterotrophic production of long chain omega-3 fatty acids utilizing algae and algae-like microorganisms. J. Appl. Phycol. 6: 123–129.

Beardall, J. and J.A. Raven. 2016. Carbon acquisition by microalgae. pp. 89–99. *In*: M.A. Borowitzka, J. Beardall and J.A. Raven (eds.). The Physiology of Microalgae. Springer, Dordrecht.

Behrenfeld, M.J., O. Prasil, M. Babin and F. Bruyant. 2004. In search of a physiological basis for covariations in light-limited and light-saturated photosynthesis. J. Phycol. 40: 4–25.

Belay, A. and G.E. Fogg. 1978. Photoinhibition of photosynthesis in *Asterionella formosa* (Bacillariophyceae). J. Phycol. 14: 341–347.

Belay, A. 1997. Mass culture of *Spirulina* outdoors—The Earthrise Farms experience. pp. 131–158 *In*: A. Vonshak (ed.). *Spirulina platensis (Arthrospira)*: Physiology, Cell-Biology and Biochemistry. Taylor & Francis, London.

Ben-Amotz, A., A. Katz and M. Avron. 1982. Accumulation of ß-carotene in halotolerant algae: purification and characterisation of ß-carotene-rich globules from *Dunaliella barwdawil*. J. Phycol. 18: 529–537.

Ben-Amotz, A. and M. Avron. 1983. On those factors which determine the massive ß-carotene accumulation in the halotolerant alga *Dunaliella bardawil*. Plant Physiology 72: 593–597.

Ben-Amotz, A., A. Lers and M. Avron. 1988. Stereoisomers of ß-carotene and phytoene in the alga *Dunaliella bardawil*. Plant Physiol. 86: 1286–1291.

Benemann, J.R. 1989. The future of microalgal biotechnology. pp. 317–337. *In*: R.C. Cresswell, T.A.V. Rees and M. Shah (eds.). Algal and Cyanobacterial Biotechnology. Longman Scientific & Technical, Harlow.

Benemann, J.R. and W.J. Oswald. 1996. Systems and economic analysis of microalgae ponds for conversion of CO_2 to biomass. Final report US DOE. Pittsburgh, USA.

Bhatti, S. and B. Colman. 2008. Inorganic carbon acquisition in some synurophyte algae. Physiol. Plant 133: 33–40.

Bigogno, C., I. Khozin-Goldberg, S. Boussiba, A. Vonshak and Z. Cohen. 2002. Lipid and fatty acid composition of the green oleaginous alga *Parietochloris incisa*, the richest plant source of arachidonic acid. Phytochemistry 60: 497–503.

Björkman, O. and B. Demmig-Adams. 1995. Regulation of photosynthetic light energy capture, conversion, and dissipation in leaves of higher plants. pp. 17–47. *In*: E.-D. Schulze and M.M. Caldwell (eds.). Ecophysiology of Photosynthesis. Springer, Berlin.

Booth, W.A. and J. Beardall. 1991. Effects of salinity on inorganic carbon utilization and carbonic anhydrase activity in the halotolerant alga *Dunaliella salina* (Chlorophyta). Phycologia 30: 220–225.

Borowitzka, L.J. and M.A. Borowitzka. 1989. ß-Carotene (Provitamin A) production with algae. pp. 15–26. *In*: E.J. Vandamme (ed.). Biotechnology of Vitamins, Pigments and Growth Factors. Elsevier Applied Science, London.

Borowitzka, L.J., M.A. Borowitzka and T. Moulton. 1984. The mass culture of *Dunaliella*: from laboratory to pilot plant. Hydrobiologia 116/117: 115–121.

Borowitzka, L.J., T.P. Moulton and M.A. Borowitzka. 1985. Salinity and the commercial production of beta-carotene from *Dunaliella salina*. Nova Hedwigia, Beih. 81:217–222.

Borowitzka, M.A. 1988a. Algal growth media and sources of cultures. pp. 456–465. *In*: M.A. Borowitzka and L.J. Borowitzka (eds.). Micro-algal Biotechnology. Cambridge University Press, Cambridge.

Borowitzka, M.A. 1988b. Fats, oils and hydrocarbons. pp. 257–287. *In*: M.A. Borowitzka and L.J. Borowitzka (eds.). Micro-algal Biotechnology. Cambridge University Press, Cambridge.

Borowitzka, M.A. 1992. Comparing carotenogenesis in *Dunaliella* and *Haematococcus*: Implications for commercial production strategies. pp. 301–310. *In*: T.G. Villa and J. Abalde (eds.). Profiles on Biotechnology. Universidade de Santiago de Compostela, Santiago de Compostela.

Borowitzka, M.A. 1995. The biotechnology of microalgal carotenoid production. pp. 149–151. *In*: J.P. Aubert and P.M.V. Martin (eds.). Microbes, Environment, Biotechnology. Abstracts, Colloques Internationaux de L'annee Pasteur. Institut Louis Malarde, Papeete, Tahiti.

Borowitzka, M.A. 1997. Algae for aquaculture: Opportunities and constraints. J. Appl. Phycol. 9: 393–401.

Borowitzka, M.A. 1998. Limits to growth. pp. 203–226. *In*: Y.S. Wong and N.F.Y. Tam (eds.). Wastewater Treatment with Algae. Springer-Verlag Berlin

Borowitzka, M.A. 2010. Carotenoid production using microorganisms. pp. 225–240. *In*: Z. Cohen and C. Ratledge (eds.). Single Cell Oils. Microbial and Algal Oils. AOCS Press, Urbana.

Borowitzka, M.A. 2013a. High-value products from microalgae—their development and commercialisation. J. Appl. Phycol.

Borowitzka, M.A. 2013b. Strain selection. pp. 77–89. *In*: M.A. Borowitzka and N.R. Moheimani (eds.). Algae for Biofuels and Energy. Springer, Dordrecht.

Borowitzka M.A. 2016. Algal physiology and large-scale outdoor cultures of microalgae. pp. 601–652. *In*: M.A. Borowitzka, J. Beardall and J.A. Raven (eds.). The Physiology of Microalgae. Springer, Dordrecht.

Borowitzka, M.A. and L.J. Borowitzka. 1988a. *Dunaliella*. pp. 27–58. *In*: M.A. Borowitzka and L.J. Borowitzka (eds.). Micro-algal Biotechnology. Cambridge University Press, Cambridge.

Borowitzka, M.A. and L.J. Borowitzka. 1988b. Limits to growth and carotenogenesis in laboratory and large-scale outdoor cultures of *Dunaliella salina*. pp. 371–381. *In*: T. Stadler, J. Mollion, M.C. Verdus, Y. Karamanos, H. Morvan and D. Christiaen (eds.). Algal Biotechnology. Elsevier Applied Science, Barking.

Borowitzka, M.A., L.J. Borowitzka and D. Kessly. 1990. Effects of salinity increase on carotenoid accumulation in the green alga *Dunaliella salina*. J. Appl. Phycol. 2: 111–119.

Borowitzka, M.A., J.M. Huisman and A. Osborn. 1991. Culture of the astaxanthin-producing green alga *Haematococcus pluvialis* 1. Effects of nutrients on growth and cell type. J. Appl. Phycol. 3: 295–304.

Borowitzka, M.A. and N.R. Moheimani (eds.). 2013. Algae for Biofuels and Energy. Springer, Dordrecht.

Borowitzka, M.A. and A. Vonshak. 2017. Scaling up microalgal cultures to commercial scale. Eur. J. Phycol. 52: 407–418.

Boussiba, S. and A. Vonshak. 1991. Astaxanthin accumulation in the green alga *Haematococcus pluvialis*. Plant Cell Physiol. 32: 1077–1082.

Bowles, R.D., A.E. Hunt, G.B. Bremer, M.G. Duchars and R.A. Eaton. 1999. Long-chain n-3 polyunsaturated fatty acid production by members of the marine protistan group the thraustochytrids: screening of isolates and optimisation of docosahexaenoic acid production. J. Biotechnol. 70: 193–202.

Boyle, N. and J. Morgan. 2009. Flux balance analysis of primary metabolism in *Chlamydomonas reinhardtii*. BMC Systems Biology 3: 4.

Britto, D.T. and H.J. Kronzucker. 2002. NH4+ toxicity in higher plants: a critical review. J. Plant Physiol. 159. 567–584.

Britton, G. 1988. Biosynthesis of carotenoids. pp. 133–182. *In*: T.W. Goodwin (ed.). Plant Pigments. Academic Press, NY.

Brunet, C. and J. Lavaud. 2010. Can the xanthophyll cycle help extract the essence of the microalgal functional response to a variable light environment? J. Plankton Res. 32: 1609–1617.

Burczyk, J. and E. Loos. 1995. Cell wall-bound enzymatic activities in *Chlorella* and *Scenedesmus*. J. Plant Physiol. 146: 748–750.

Cadenus, E. 2005. Biochemistry of oxygen toxicity. Annu. Rev. Biochem. 58: 79–110.

Carvalho, A. and F.X. Malcata. 2005. Optimization of ω-3 fatty acid production by microalgae: crossover effects of CO_2 and light intensity under batch and continuous cultivation modes. Mar. Biotechnol. 7: 381–388.

Carvalho, A.P., I. Pontes, H. Gaspar and F.X. Malcata. 2006. Metabolic relationships between macro- and micronutrients, and the eicosapentaenoic acid and docosahexaenoic acid contents of *Pavlova lutheri*. Enzyme. Microb. Technol. 38: 358–366.

Carvalho, A.P., C.M. Monteiro and F.X. Malcata. 2009. Simultaneous effect of irradiance and temperature on biochemical composition of the microalga *Pavlova lutheri*. J. Appl. Phycol. 21: 543–552.

Chang, R.L., L. Ghamsari, A. Manichaikul, E.F.Y. Hom, S. Balaji, W. Fu, Y. Shen, T. Hao, B.O. Palsson, K. Salehi-Ashtiani and J.A. Papin. 2011. Metabolic network reconstruction of *Chlamydomonas* offers insight into light-driven algal metabolism. Mol. Syst. Biol. 7: 518.

Cheirsilp, B. and S. Torpee. 2012. Enhanced growth and lipid production of microalgae under mixotrophic culture condition: Effect of light intensity, glucose concentration and fed-batch cultivation. Bioresour. Technol. 110: 510–516.

Chen, F. and M.R. Johns. 1991. Effect of C/N ratio and aeration on the fatty acid composition of heterotrophic *Chlorella sorokiniana*. J. Appl. Phycol. 3: 203–209.

Chrismadha, T. and M.A. Borowitzka. 1994. Effect of cell density and irradiance on growth, proximate composition and eicosapentaenoic acid production of *Phaeodactylum tricornutum* grown in a tubular photobioreactor. J. Appl. Phycol. 6: 67–74.

Chu, W.-L., S.-M. Phang and S.-H. Goh. 1995. Influence of carbon source on growth, biochemical composition and pigmentation of *Ankistrodesmus convolutus*. J. Appl. Phycol. 7: 59–64.

Cohen, Z., A. Vonshak and A. Richmond. 1988. Effect of environmental conditions on fatty acid composition of the red alga *Porphyridium cruentum*: correlation to growth rate. J. Phycol. 24: 328–332.

Cohen, Z. and I. Khozin-Goldberg. 2010. Searching for polyunsaturated fatty acid-rich microalgae. pp. 201–224. *In*: Z. Cohen and C. Ratledge (eds.). Single Cell Oils. Microbial and Algal Oils. AOCS Publihing, Urbana.

Cohen, Z. and C. Ratledge (eds.). 2010. Single Cell Oils. microbial and Algal Oils. AOCS Press, Urbana.

Colman, B., I.E. Huertas, S. Bhatti and J.S. Dason. 2002. The diversity of inorganic carbon acquisition mechanisms in eukaryotic microalgae. Funct. Plant Biol. 29: 261–270.

Cysewski, G.R. and R.T. Lorenz. 2004. Industrial production of microalgal cell-mass and secondary products—species of high potential: *Haematococcus*. pp. 281–288. *In*: A. Richmond (ed.). Microalgal Culture: Biotechnology and Applied Phycology. Blackwell Science, Oxford.

Darley, W.M. and B.E. Volcani. 1969. Role of silicon in diatom metabolism. A silicon requirement for DNA synthesis in the diatom *Cylindrotheca fusiformis*. Exp. Cell Res. 58: 334–342.

De Godos, I., V.A. Vargas, S. Blanco, M.C. Garcia Gonzalez, R. Soto, P.A. Garcia-Encina, E. Becares and R. Munoz. 2010. A comparative evaluation of microalgae for the degradation of piggery wastewater under photosynthetic oxygenation. Bioresour. Technol. 101: 5150–5158.

de Oliveira Dal'Molin, C.G., L.-E. Quek, R. Palfreyman and L. Nielsen. 2011. AlgaGEM—a genome-scale metabolic reconstruction of algae based on the *Chlamydomonas reinhardtii* genome. BMC Genomics 12: S5.

de Swaaf, M.E., L. Sijtsma and J.T. Pronk. 2003. High-cell-density fed-batch cultivation of the docosahexaenoic acid producing marine alga *Crypthecodinium cohnii*. Biotechnol. Bioeng. 81: 666–672.

Del Rio, E., F.G. Acién, M.C. García-Malena, J. Rivas, E. Molina Grima and M.G. Guerrero. 2005. Efficient one-step production of astaxanthin by the microalga *Haematococcus pluvialis* in continuous culture. Biotechnol. Bioeng. 91: 808–815.

Demmig-Adams, B. and W.W. Adams. 1992. Photoprotection and other responses of plants to high light stress. Annu. Rev. Plant Physiol. Plant Mol. Biol. 43: 599–626.

Dimier, C., F. Corato, F. Tramontano and C. Brunet. 2007. Photoprotection and xanthophyll-cycle activity in three marine diatoms. J. Phycol. 43: 937–947.

Doucha, J. and K. Lívanský. 2012. Production of high-density *Chlorella* culture grown in fermenters. J. Appl. Phycol. 24: 35–43.

Droop, M.R. 1955. Carotogenesis in *Haematococcus pluvialis*. Nature 175: 42.

Dyhrman, S.T. 2016. Nutrients and their acquisition: Phosphorus physiology in microalgae. pp. 155–183. *In*: M.A. Borowitzka, J. Beardall and J.A. Raven (eds.). The Physiology of Microalgae. Springer, Dordrecht.

Eisele, R. and W.R. Ullrich. 1977. Effect of glucose and CO_2 on nitrate uptake and coupled OH^- flux in *Ankistrodesmus braunii*. Plant Physiol. 59: 18–21.

Eixler, S., U. Karsten and U. Selig. 2006. Phosphorus storage in *Chlorella vulgaris* (*Trebouxiophyceae, Chlorophyta*) cells and its dependence on phosphate supply. Phycologia 45: 53–60.

Eriksen, N.T. 2008. Production of phycocyanin—a pigment with applications in biology, biotechnology, foods and medicine. Appl. Microbiol. Biotechnol. 80: 1–14.

Falkowski, P.G. 2000. Rationalising elemental ratios in unicellular algae. J. Phycol. 36: 3–6.

Falkowski, P.G. and J.A. Raven. 2007. Aquatic photosynthesis. 2nd edition. Princeton University Press, Princeton, NJ

Fan, L., A. Vonshak, A. Zarka, and S. Boussiba. 1998. Does astaxanthin protect *Haematococcus* against light damage? Z. Naturforsch. 53c: 93–100.

Fisher, M., I. Gokhman, U. Pick and A. Zamir. 1996. A salt-resistant plasma membrane carbonic anhydrase is induced by salt in *Dunaliella salina*. J. Biol. Chem. 271: 17718–17723.

Fon Sing, S., A. Isdepsky, M.A. Borowitzka and N.R. Moheimani. 2013. Production of biofuels from microalgae. Mitig. Adapt. Strat. Global Change 18: 47–72.

Gardner, R.D., K.E. Cooksey, F. Mus, R. Macur, K. Moll, E. Eustance, R.P. Carlson, R. Gerlach, M.W. Fields and B.M. Peyton. 2012. Use of sodium bicarbonate to stimulate triacylglycerol accumulation in the chlorophyte *Scenedesmus* sp. and the diatom *Phaeodactylum tricornutum*. J. Appl. Phycol. 24: 1311–1320.

Geider, R.J. and J.L. Roche. 2002. Redfield revisited: variability of C:N:P in marine microalgae and its biochemical basis. Eur. J. Phycol. 37: 1–17.

Giordano, M., J. Beardall and J.A. Raven. 2005. CO_2 concentrating mechanisms in algae: mechanisms, environmental modulation, and evolution. Annu. Rev. Plant Physiol. 56: 99–131.

Gômez Pinchetti, J.L., Z. Ramazanov, A. Fontes and G. Garcia Reina. 1992. Photosynthetic characteristics of *Dunaliella salina* (Chlorophyceae, Dunaliellales) in relation to ß-carotene content. J. Appl. Phycol. 4: 11–15.

González, C., J. Marciniak, S. Villaverde, P. García-Encina and R. Muñoz. 2008. Microalgae-based processes for the biodegradation of pretreated piggery wastewaters. Appl. Microbiol. Biotechnol. 80: 891–898.

Gordillo, F.J.L., M. Goutx, F.L. Figueroa and F.X. Niell. 1998. Effect of light intensity, CO_2 and nitrogen supply on lipid class composition of *Dunaliella viridis*. J. Appl. Phycol. 10: 135–144.

Goss, R. and T. Jakob. 2010. Regulation and function of xanthophyll cycle-dependent photoprotection in algae. Photosynth. Res. 106: 103–122.

Grant, B.R. and I.M. Turner. 1969. Light-stimulated nitrate and nitrite assimilation in several species of algae. Comp. Biochem. Physiol. 29: 995–1004.

Graverholt, O.S. and N.T. Eriksen. 2007. Heterotrophic high-cell-density fed-batch and continuous-flow cultures of *Galdieria sulphuraria* and production of phycocyanin. Appl. Microbiol. Biotechnol. 77: 69–75.

Griffiths, M.J. and S.T.L. Harrison. 2009. Lipid productivity as a key characteristic for choosing algal species for biodiesel production. J. Appl. Phycol. 21: 493–507.

Grobbelaar, J.U. and C.J. Soeder. 1985. Respiration losses in planktonic green algae cultivated in raceway ponds. J. Plankton Res. 7: 497–506.

Grobbelaar, J.U. 1995. Influence of areal density on β-carotene production by *Dunaliella salina*. J. Appl. Phycol. 7: 69–73.

Grobbelaar, J.U. 2009. Upper limits of productivity and problems of scaling. J. Appl. Phycol. 21: 519–522.

Grünewald, K., C. Hagen and K. Braune. 1997. Secondary carotenoid accumulation in flagellates of the green alga *Haematococcus lacustris*. Eur. J. Phycol. 32: 387–392.

Grung, M., F.M.L. D'Souza, M.A. Borowitzka and S. Liaaen-Jensen. 1992. Algal carotenoids 51. Secondary carotenoids 2. *Haematococcus pluvialis* aplanospores as a source of (3S,3'S)-astaxanthin esters. J. Appl. Phycol. 4: 165–171.

Guschina, I.A. and J.L. Harwood. 2006. Lipids and lipid metabolism in eukaryotic algae. Prog. Lipid Res. 45: 160–186.

Guschina, I.A. and J.L. Harwood. 2013. Algal lipids and their metabolism. pp. 17–36. *In*: M.A. Borowitzka and N.R. Moheimani (eds.). Algae for Biofuels and Energy. Springer, Dordrecht.

Harris, G.N., D.J. Scanlan and R.J. Geider. 2005. Acclimation of *Emiliania huxleyi* (Prymnesiophyceae) to photon flux density. J. Phycol. 41: 851–862.

Harwood, J.L. and L. Jones. 1989. Lipid metabolism in algae. Adv. Bot. Res. 16: 2–52.

Harwood, J.L. and I.A. Guschina. 2009. The versatility of algae and their lipid metabolism. Biochimie 91: 1–6.

Ho, T., A. Quigg, C.V. Finkel, A.J. Milligan, K. Wyman, P.G. Falkowsi and F.M.M. Morel. 2005. The elemental composition of some phytoplankton. J. Phycol. 39: 1145–1159.

Hu, Q. and A. Richmond. 1994. Optimising the population density in *Isochrysis galbana* grown outdoors in a glass column photobioreactor. J. Appl. Phycol. 6: 391–396.

Hu, Q., H. Guterman and A. Richmond. 1996. A flat inclined modular photobioreactor for outdoor mass cultivation of photoautotrophs. Biotechnol. Bioeng. 51: 51–60.

Hu, Q., Y. Zarmi and A. Richmond. 1998. Combined effects of light intensity, light-path, and culture density on output rate of *Spirulina platensis* (Cyanobacteria). Eur. J. Phycol. 32: 165–171.

Hu, Q., M. Sommerfeld, E. Jarvis, M. Ghirardi, M. Posewitz, M. Seibert and A. Darzins. 2008. Microalgal triacylglycerols as feedstocks for biofuel production: perspectives and advances. Plant J. 54: 621–639.

Huerlimann, R., R. De Nys and K. Heimann. 2010. Growth, lipid content, productivity, and fatty acid composition of tropical microalgae for scale-up production. Biotechnol. Bioeng. 107: 245–257.

Isleten-Hosoglu, M., I. Gultepe and M. Elibol. 2012. Optimization of carbon and nitrogen sources for biomass and lipid production by *Chlorella saccharophila* under heterotrophic conditions and development of Nile red fluorescence based method for quantification of its neutral lipid content. Biochem. Eng. J. 61: 11–19.

Iwamoto, H. 2004. Industrial production of microalgal cell-mass and secondary products—major industrial species: *Chlorella*. pp. 255–263. *In*: A. Richmond (ed.). Microalgal Culture: Biotechnology and Applied Phycology. Blackwell Science, Oxford.

Jimenez, C. and U. Pick. 1993. Differential reactivity of β-carotene isomers from *Dunaliella bardawil* toward oxygen radicals. Plant Physiology 101: 385–390.

Juinars, P. 1993. Concepts in Biological Oceanography. Oxford University Press, New York.

Kang, C.D., J.S. Lee, T.H. Park and S.J. Sim. 2005. Comparison of heterotrophic and photoautotrophic induction on astaxanthin production by *Haematococcus pluvialis*. Appl. Microbiol. Biotechnol. 68: 237–241.

Kaplan, D., A.E. Richmond, Z. Dubisnky, and A. Aaronson. 1986. Algal nutrition. pp. 147–198. *In*: A. Richmond (ed.). Handbook of Microalgal Mass Culture. CRC Press, Boca Raton.

Kessler, E. 1976. Comparative physiology, biochemistry, and the taxonomy of *Chlorella* (Chlorophyceae). Plant Syst. Evol. 125: 129–138.

Kliphuis, A.M.J., D.E. Martens, M. Janssen and R.H. Wijffels. 2011. Effect of O_2:CO_2 ratio on the primary metabolism of *Chlamydomonas reinhardtii*. Biotechnol. Bioeng. 108: 2390–2402.

Kobayashi, M., T. Kakizono, K. Yamaguchi, N. Nishio and S. Nagai. 1992. Growth and astaxanthin formation of *Haematococcus pluvialis* in heterotrophic and mixotrophic conditions. J. Ferment. Bioeng. 74: 17–20.

Kolber, Z., J. Zehr and P. Falkowski. 1988. Effects of growth irradiance and nitrogen limitation on photosynthetic energy conversion in photosystem II. Plant Physiol. 88: 923–929.

Kong, W.B., H. Song, S.F. Hua, H. Yang, Q. Yang and C.G. Xia. 2012. Enhancement of biomass and hydrocarbon productivities of *Botryococcus braunii* by mixotrophic cultivation and its application in brewery wastewater treatment. Afr. J. Microbiol. Res. 6: 1489–1496.

Kromkamp, J.C., J. Beardall, A. Sukenik, J. Kopecky, J. Masojidek, S. Van Bergeijk, S. Gabai, E. Shaham and A. Yamshon. 2009. Short-term variations in photosynthetic parameters of *Nannochloropsis* cultures grown in two types of outdoor mass cultivation systems. Aquat. Microb. Ecol. 56: 309–322.

Kroon, B., B. Prézelin and O. Schofield. 1993. Chromatic regulation of quantum yields for photosystem II Charge separation, oxygen evolution, and carbon fixation in *Heterocapsa pygmaea* (Pyrrophyta). J. Phycol. 29: 453–462.

Lalucat, J., J. Imperial and R. Parés. 1984. Utilization of light for the assimilation of organic matter in *Chlorella* sp. VJ79. Biotechnol. Bioeng. 26: 677–681.

Lamers, P.P., C.C.W. van de Laak, P.S. Kaasenbrood, J. Lorier, M. Janssen, R.C.H. De Vos, R.J. Bino and R.H. Wijffels. 2010. Carotenoid and fatty acid metabolism in light-stressed *Dunaliella salina*. Biotechnol. Bioeng. 106: 638–648.

Lang, N.J. 1968. Electron microscopic studies of extraplastidic astaxanthin in *Haematococcus*. J. Phycol. 4: 12–19.

Lavaud, J. 2007. Fast regulation of photosynthesis in diatoms: mechanisms, evolution and ecophysiology. Funct. Plant Scie. Biotechnol. 1: 267–287.

Lazar, B., A. Starinsky, A. Katz, E. Sass and S. Ben-Yaakov. 1983. The carbonate system in hypersaline solutions: alkalinity and $CaCO_3$ solubility of evaporated seawater. Limnol. Oceanogr. 28: 978–986.

Leftley, J.W. and P.J. Syrett. 1973. Urease and ATP: Urea amidolyase activity in unicellular algae. J. Gen. Microbiol. 77: 109–115.

Lei, A., H. Chen, G. Shen, Z. Hu, L. Chen and J. Wang. 2012. Expression of fatty acid synthesis genes and fatty acid accumulation in *Haematococcus pluvialis* under different stressors. Biotech. Biofuels 5: 18.

Lemoine, Y. and B. Schoefs. 2010. Secondary ketocarotenoid astaxanthin biosynthesis in algae: a multifunctional response to stress. Photosynth. Res. 106: 155–177.

Levine, R.B., M.S. Costanza-Robinson and G.A. Spatafora. 2011. *Neochloris oleoabundans* grown on anaerobically digested dairy manure for concomitant nutrient removal and biodiesl feedstock production. Biomass Bioenergy 35: 40–49.

Li, Y., M. Sommerfeld, F. Chen and Q. Hu. 2008. Consumption of oxygen by astaxanthin biosynthesis: A protective mechanism against oxidative stress in *Haematococcus pluvialis* (Chlorophyceae). J. Plant Physiol. 165: 1783–1797.

Li, Y.T., M. Sommerfeld, F. Chen and Q. Hu. 2010. Effect of photon flux densities on regulation of carotenogenesis and cell viability of *Haematococcus pluvialis* (Chlorophyceae). J. Appl. Phycol. 22: 253–263.

Li, Z., H. Yuan, J. Yang and B. Li. 2011. Optimization of the biomass production of oil algae *Chlorella minutissima* UTEX2341. Bioresour. Technol. 102: 9128–9134.

Liang, Y., N. Sarkany and Y. Cui. 2009. Biomass and lipid productivities of *Chlorella vulgaris* under autotrophic, heterotrophic and mixotrophic growth conditions. Biotechnol. Lett. 31: 1043–1049.

Lynn, S.G., S.S. Kilham, D.A. Kreeger and S. Interlandi. 2000. Effect of nutrient availability on the biochemical and elemental stoichiometry in freshwater diatom *Stephanodiscus minutulus* (Bacillariophyceae). J. Phycol. 36: 510–522.

MacIntyre, H.L., T.M. Kana, T. Anning and R.J. Geider. 2002. Photoacclimation of photosynthesis irradiance response curves and photosynthetic pigments in microalgae and cyanobacteria. J. Phycol. 38: 17–38.

Martínez, F. and M.I. Orús. 1991. Interactions between glucose and inorganic carbon metabolism in *Chlorella vulgaris* strain UAM 101. Plant Physiol. 95: 1150–1155.

Materassi, R., C. Paoletti, W. Balloni and G. Florenzano. 1980. Some considerations on the production of lipid substances by microalgae and cyanobacteria. pp. 619–626. *In*: G. Shelef and C.J. Soeder (eds.). Algae Biomass. Elsevier, Amsterdam.

Melis, A., J. Neidhardt and J. Benemann. 1999. *Dunaliella salina* (Chlorophyta) with small chlorophyll antenna sizes exhibit higher photosynthetic productivities and photon use efficiencies than normally pigmented cells. J. Appl. Phycol. 10: 515–525.

Mendoza, H., A. Martel, M. Jiménez del Río and G. García Reina. 1999. Oleic acid is the main fatty acid related with carotenogenesis in *Dunaliella salina*. J. Appl. Phycol. 11: 15–19.

Mercz, T.I. 1994. A study of high lipid yielding microalgae with potential for large-scale production of lipids and polyunsaturated fatty acids. PhD thesis, Murdoch University, Perth, Western Australia.

Mil'kó, E.S. 1963. Effect of various environmental factors on pigment production in the alga *Dunaliella salina*. Mikrobiologya 32: 299–307.

Minagawa, J. 2011. State transitions—The molecular remodeling of photosynthetic supercomplexes that controls energy flow in the chloroplast. Biochim. Biophys. Acta - Bioenergetics 1807: 897–905.

Moheimani, N.R. and M.A. Borowitzka. 2007. Limits to growth of *Pleurochrysis carterae* (Haptophyta) grown in outdoor raceway ponds. Biotechnol. Bioeng. 96: 27–36.

Moheimani, N.R. 2012. Inorganic carbon and pH effect on growth and lipid productivity of *Tetraselmis suecica* and *Chlorella* sp. (Chlorophyta) grown outdoors in bag photobioreactors. J. Appl. Phycol. 1–12.

Morris, E. and J. Kronkamp. 2003. Influence of temperature on the relationship between oxygen- and fluorescence-based estimates of photosynthetic parameters in a marine benthic diatom (*Cylindrotheca closterium*). Eur. J. Phycol. 38: 133–142.

Muradyan, E.A., G.L. Klyachko-Gurvich, L.N. Tsoglin, T.V. Sergeyenko and N.A. Pronina. 2004. Changes in lipid metabolism during adaptation of the *Dunaliella salina* photosynthetic apparatus to high CO_2 concentration. Russ. J. Plant Physiol. 51: 53–62.

Myers, J. and J. Graham. 1958. On the mass culture of algae II. Yield as a function of cell concentration under continuous sunlight irradiance. Plant Physiol. 34: 345–352.

Myklestad, S.M. 1995. Release of extracellular products by phytoplankton with special emphasis on polysaccharides. Sci. Total Environ. 165: 155–164.

Nakajima, Y. and R. Ueda. 1997. Improvement of photosynthesis in dense microalgal suspensions by reducing the content of light harvesting pigments. J. Appl. Phycol. 9: 503–510.

Nakajima, Y. and R. Ueda. 1999. Improvement of microalgal photosynthetic productivity by reducing the content of light harvesting pigments. J. Appl. Phycol. 11: 151–201.

Nakajima, Y. and R. Ueda. 2000. The effect of reducing light-harvesting pigment on marine microalgae productivity. J. Appl. Phycol. 12: 285–290.

Neilson, A.H. and R.A. Lewin. 1974. The uptake and utilization of organic carbon by algae: an essay in comparative biochemistry. Phycologia 13: 227–264.

Neori, A. 2011. "Green water" microalgae: the leading sector in world aquaculture. J. Appl. Phycol. 23: 143–149.

Ogawa, T., T. Fujii and S. Aiba. 1980. Effect of oxygen on the growth (yield) of *Chlorella vulgaris*. Arch. Microbiol. 127: 25–31.

Ogawa, T. and A. Kaplan. 2003. Inorganic carbon acquisition systems in cyanobacteria. Photosynth. Res. 77: 105–115.

Ogbonna, J.C. and H. Tanaka. 1996. Night biomass loss and changes in biochemical composition of cells during light/dark cyclic culture of Chlorella pyrenoidosa. J. Ferment. Bioeng. 82: 558–564.

Ogbonna, J.C., H. Yoshizowa, and H. Tanaka. 2000. Treatment of a high strength organic wastewater by a mixed culture of photosynthetic organisms. J. Appl. Phycol. 12: 277–284.

Olaizola, M., E.O. Duerr and D.W. Freeman. 1991. Effect of CO_2 enhancement in an outdoor algal production system using *Tetraselmis*. J. Appl. Phycol. 3: 363–366.

Olaizola, M. 2000. Commercial production of astaxanthin from *Haematococcus pluvialis* using 25,000-liter outdoor photobioreactors. J. Appl. Phycol. 12: 499–506.

Ördög, V., W.A. Stirk, P. Bálint, C. Lovász, O. Pulz and J. Staden. 2013. Lipid productivity and fatty acid composition in *Chlorella* and *Scenepdesmus* strains grown in nitrogen-stressed conditions. J. Appl. Phycol. 25: 233–243.

Orosa, M., D. Fraqiera, A. Cid and J. Abalde. 2001. Carotenoid accumulation in *Haematococcus pluvialis* in mixotrophic growth. Biotechnol. Lett. 23: 373–378.

Oswald, W.J. 1988. Large-scale algal culture systems (engineering aspects). pp. 357–394. *In*: M.A. Borowitzka and L.J. Borowitzka (eds.). Micro-Algal Biotechnology. Cambridge University Press, Cambridge.

Papageorgiou, G. and Govindjee. 2004. Chlorophyll *a* fluorescence: A signature of photosynthesis. Springer, Dordrecht.

Payer, H.D., Y. Chiemvichak, K. Hosakul, C. Kongpanichkul, L. Kraidej, M. Nguitragul, S. Reungmanipytoon and P. Buri. 1980. Temperature as an important climatic factor during mass production of microscopic algae. pp. 389–399. *In*: G. Shelef and C.J. Soeder (eds.). Algae Biomass. Production and Use. Elsevier/North Holland Biomedical Press, Amsterdam.

Perez-Garcia, O., F.M.E. Escalante, L.E. de-Bashan and Y. Bashan. 2011. Heterotrophic cultures of microalgae: Metabolism and potential products. Water Res. 45: 11–36.

Pohl, P. and F. Zurheide. 1979. Fatty acids and lipids of marine algae and the control of their biosynthesis by environmental factors. pp. 473–523. *In*: H.A. Hoppe, T. Levring and Y. Tanaka (eds.). Marine Algae in Pharmaceutical Science. Walter de Gruyter, Berlin - New York.

Polle, J.E.W., S. Kanakagiri, E.S. Jin, T. Masuda and A. Melis. 2002. Truncated chlorophyll antenna size of the photosystems—a practical method to improve microalgal productivity and hydrogen production in mass culture. Int. J. Hydrogen. Energ. 27: 1257–1264.

Post, A.F. 1993. Ammonia enhanced dark respiration in *Chlorella vulgaris* is related to collapse of a transmembrane pH gradient. FEMS Microbiol. Lett. 113: 9–13.

Premkumar, L., H.M. Greenblatt, U.K. Bagshewar, T. Savchenko, I. Gokhman, A. Zamir and J.L. Sussmann. 2003. Identification, cDNA cloning, expression, crystallization and preliminary X-ray analysis of an exceptionally halotolerant carbonic anhydrase from *Dunaliella salina*. Acta Crystallogr. Sect. D Biol. Crystallogr. D59: 1984–1086.

Price, G.D., S.-I. Maeda, T. Omata and M.R. Badger. 2002. Modes of active inorganic carbon uptake in the cyanobacterium, *Synechococcus* sp. PCC7942. Funct. Plant Biol. 29: 131–149.

Przytocka-Jusiak, M., M. Duszota, K. Motuoiolt and R. Myoiclskl. 1984. intensive culture of *Chlorella vulgaris*/AA as the second stage of biological purification of nitrogen industry wastewaters. Water Res. 18: 1–7.

Qiu, B. and Y. Li. 2006. Photosynthetic acclimation and photoprotective mechanism of *Haematococcus pluvialis* (Chlorophyceae) during the accumulation of secondary carotenoids at elevated irradiation. Phycologia 45: 117–126.

Quigg, A., Z.V. Finkel, A.J. Irwin, Y. Rosenthal, T.-Y. Ho, J.R. Reinfelder, O. Schofield, F.M.M. Morel and P.G. Falkowski. 2003. The evolutionary inheritance of elemental stoichiometry in marine phytoplankton. Nature 425: 291–294.

Rabbani, S., P. Beyer, J.V. Lintig, P. Hugeney and H. Kleinig. 1998. Induced b-carotene synthesis driven by triacylglycerol deposition in the unicellular alga *Dunaliella bardawil*. Plant Physiology 116: 1239–1248.

Rascher, U. and L. Nedbal. 2006. Dynamics of photosynthesis in fluctuating light. Curr. Opin. Plant Biol. 9: 671–678.

Ratledge, C. 2010. Single cell oils for the 21st Century. pp. 3–26. *In*: Z. Cohen and C. Ratledge (eds.). Single Cell Oils. Microbial and Algal Oils. AOCS Press, Urbana.

Raven, J.A. 1981. Respiration and photorespiration. pp. 55–82. *In*: T. Platt (ed.). Physiological Bases of Phytoplankton Ecology. Department of Fisheries and Oceans, Canada.

Raven, J.A. 2010. Inorganic carbon acquisition by eukaryotic algae: four current questions. Photosynth. Res. 106: 123–134.

Raven, J.A. 2011. The cost of photoinhibition. Physiol. Plant 142: 87–104.

Raven, J.A. and J. Beardall. 2016. Dark respiration and organic carbon loss. pp. 129–140. *In*: M.A. Borowitzka, J. Beardall and J.A. Raven (eds.). The Physiology of Microalgae. Springer, Dordrecht.

Raven, J.A. and M. Giodano. 2016. Combined nitrogen. pp. 143–154. *In*: M.A. Borowitzka, J. Beardall and J.A. Raven (eds.). Springer, Dordrecht.

Redfield, A.C. 1934. On the proportions of organic derivatives in sea water and their relation to the composition of plankton. pp. 176–192. *In*: R.J. Daniel (ed.). James Johnstone Memorial Volume. Liverpool University Press, Liverpool.

Redfield, A.C. 1958. The biological control of chemical factors in the environment. Am. Sci. 46: 205–221.

Reitan, K.I., J.R. Rainuzzo and Y. Olsen. 1994. Effect of nutrient limitation on fatty-acid and lipid-content of marine microalgae. J. Phycol. 30: 972–979.

Richmond, A. and A. Vonshak. 1978. *Spirulina* culture in Israel. Arch. Hydrobiol. 11: 274–280.

Richmond, A. 1996. Efficient utilization of high irradiance for production of photoautotrophic cell mass: a survey. J. Appl. Phycol. 8: 381–387.

Roessler, P.G. 1988. Effects of silicon deficiency on lipid composition and metabolism in the diatom *Cyclotella cryptica*. J. Phycol. 24: 394–400.

Roessler, P.G. 1990. Environmental control of glycerolipid metabolism in microalgae—Commercial implications and future research directions. J. Phycol. 26: 393–399.

Schlee, J., B.-H. Cho and E. Komor. 1985. Regulation of nitrate uptake by glucose in *Chlorella*. Plant Sci. 39: 25–30.

Schoefs, B., N. Rmiki, J. Rachadi and Y. Lemoine. 2001. Astaxanthin accumulation in *Haematococcus* requires a cytochrome P450 hydroxylase and an active synthesis of fatty acids. FEBS Lett. 500: 125–128.

Semenenko, V.E. and A.A. Abdullayev. 1980. Parametric control of ß-carotene biosynthesis in *Dunaliella salina* cells under conditions of intensive cultivation. Fiziologya Rastenii 27: 31–41.

Sforza, E., R. Cipriani, T. Morosinotto, A. Bertucco and G.M. Giacometti. 2012. Excess CO_2 supply inhibits mixotrophic growth of *Chlorella protothecoides* and *Nannochloropsis salina*. Bioresour. Technol. 104: 523–529.

Shaish, A., M. Avron, U. Pick and A. Ben-Amotz. 1993. Are active oxygen species involved in induction of beta-carotene in *Dunaliella bardawil*. Planta 190: 363–368.

Shihira, I. and R.W. Krauss. 1965. *Chlorella*. Physiology and taxonomy of forty-one isolates. University of Maryland, College Park, Maryland.

Shiraiwa, Y. and Y. Umino. 1991. Effect of glucose on the induction of the carbonic anhydrase and the change in $K_{1/2}(CO_2)$ of photosynthesis in *Chlorella vulgaris* 11H. Plant Cell Physiol. 32: 311–314.

Singh, D.P., N. Singh and K. Verma. 1995. Photooxidative damage to the cyanobacterium *Spirulina platensis* mediated by singlet oxygen. Current Microbiol 31: 44–48.

Solovchenko, A., I. Khozin-Goldberg, S. Didi-Cohen, Z. Cohen and M. Merzlyak. 2008. Effects of light intensity and nitrogen starvation on growth, total fatty acids and arachidonic acid in the green microalga *Parietochloris incisa*. J. Appl. Phycol. 20: 245–251.

Sukenik, A. 1999. Production of eicosapentaenoic acid by the marine eustigmatophyte *Nannochloropsis*. pp. 41–56. *In*: Z. Cohen (ed.). Chemicals from Microalgae. Taylor & Francis, London.

Sukenik, A., J. Beardall, J.C. Kromkamp, J. Kopecky, J. Masojídek, S. Van Bergeijk, S. Gabai, E. Shaham and A. Yamshon. 2009. Photosynthetic performance of outdoor *Nannochloropsis* mass cultures under a wide range of environmental conditions. Aquat. Microb. Ecol. 56: 297–308.

Sukenik, A., J. Bennett and P. Falkowski. 1987. Light-saturated photosynthesis—limitation by electron transport or carbon fixation? Biochim. Biophys. Acta 891: 205–215.

Sullivan, C.W. and B.E. Volcani. 1981. Silicon in the cellular metabolism of diatoms. pp. 15–42. *In*: T.L. Simpson and B.E. Volcani (eds.). Silicon and Siliceous Structures in Biological Systems. Springer-Verlag, New York.

Syrett, P.J. 1988. Uptake and utilization of nitrogen compounds. pp. 23–39. *In*: L.J. Rogers and J.R. Gallon (eds.). Biochemistry of the Algae and Cyanobacteria. Oxford University Press, NY.

Tam, N.F.Y, and Y.S. Wong. 1996. Effect of ammonia concentrations on growth of Chlorella vulgaris and nitrogen removal from media. Bioresour. Technol. 57: 45–50.

Tamiya, H. 1957. Mass culture of algae. Ann. Rev. Plant Physiol. 8. 309–344.

Tang, D., W. Han, P. Li, X. Miao and J. Zhong. 2011. CO_2 biofixation and fatty acid composition of *Scenedesmus obliquus* and *Chlorella pyrenoidosa* in response to different CO_2 levels. Bioresour. Technol. 102: 3071–3076.

Tanner, W. 2000. The *Chlorella* hexose/H⁺-symporters. Int. Rev. Cytol. 200: 101–141.

Tischner, R. 2000. Nitrate uptake and reduction in higher and lower plants. Plant Cell Env. 23: 1005–1024.

Torzillo, G., A. Sacchi, R. Materassi and A. Richmond. 1991. Effect of temperature on yield and night biomass loss in *Spirulina platensis* grown outdoors in tubular photobioreactors. J. Appl. Phycol. 3: 103–109.

Torzillo, G., P. Accolla, E. Pinzani and J. Masojidek. 1996. *In situ* monitoring of chlorophyll fluorescence to assess the synergistic effect of low tempereture and high irradiance stress in *Spirulina* cultures grown outdoors in photobioreactors. J. Appl. Phycol. 8: 283–291.

Unkefer, C.J., R.T. Sayre, J.K. Magnuson, D.B. Anderson, I. Baxter, I.K. Blaby, J.K. Brown, M. Carleton, R.A. Cattolico, T. Dale, T.P. Devarenne, C.M. Downes, S.K. Dutcher, D.T. Fox, U. Goodenough, J. Jaworski, J.E. Holladay, D.M. Kramer, A.T. Koppisch, M.S. Lipton, B.L. Marrone, M. McCormick, I. Molnár, J.B. Mott, K.L. Ogden, E.A. Panisko, M. Pellegrini, J. Polle, J.W. Richardson, M. Sabarsky, S.R. Starkenburg, G.D. Stormo, M. Teshima, S.N. Twary, P.J. Unkefer, J.S. Yuan and J.A. Olivares 2017. Review of the algal biology program within the national alliance for advanced biofuels and bioproducts. Algal Res. 22: 187–215.

Valenzuela, J., A. Mazurie, R.P. Carlson, R. Gerlach, K.E. Cooksey, B.M. Peyton and M.W. Fields. 2012. Potential role of multiple carbon fixation pathways during lipid accumulation in *Phaeodactylum tricornutum*. Biotech. Biofuels 5: 40.

Vanlerberghe, G.C., H.C. Huppe, K.D.M. Vlossak and D.H. Turpin. 1992. Activation of respiration to support dark NO_3^- and NH_4^+ assimilation in the green alga *Selenastrum minutum*. Plant Physiol. 99: 495–500.

Verity, P.G. 1981. Effects of temperature, irradiance, and day length on the marine diatom *Leptocylindrus danicus* Cleve. I. Photosynthesis and cellular composition. J. Exp. Mar. Biol. Ecol. 55: 79–91.

Vidhyavathi, R., L. Venkatachalam, R. Sarada and G.A. Ravishankar. 2008. Regulation of carotenoid biosynthetic genes expression and carotenoid accumulation in the green alga *Haematococcus pluvialis* under nutrient stress conditions. J. Exp. Bot. 59: 1409–1418.

Villarejo, A., M.I. Orús and F. Martinez. 1997. Regulation of the CO_2-concentrating mechanism in *Chlorella vulgaris* UAM 101 by glucose. Physiol. Plant 99: 293–301.

Vonshak, A. and R. Guy. 1992. Photoadaptation, photoinhibition and productivity in the blue-green alga, *Spirulina platensis* grown outdoors. Plant Cell Environ. 15: 613–616.

Vonshak, A. 1997. Outdoor mass production of *Spirulina*: the basic concept. pp. 79–99. *In*: A. Vonshak (ed.). *Spirulina platensis* (*Arthrospira*): Physiology, Cell-Biology and Biochemistry. Taylor & Francis, London.

Vonshak, A., G. Torzillo, J. Masojidek and S. Boussiba. 2001. Sub-optimal morning temperature induces photoinhibition in dense outdoor cultures of the alga *Monodus subterraneus* (Eustigmatophyta). Plant Cell Environ. 24: 1113–1118.

Vonshak, A. and G. Torzillo. 2004. Environmental stress physiology. pp. 57–62. *In*: A. Richmond (ed.). Handbook of Microalgal Culture: Biotechnology and applied Phycology. Blackwell Science, Oxford.

Wagner, H., T. Jakob and C. Wilhelm. 2006. Balancing the energy flow from captured light to biomass under fluctuating light conditions. New Phytol. 169: 95–108.

Walker, D.A. 2009. Biofuels, facts, fantasy and feasibility. J. Appl. Phycol. 21: 508–517.

Wayama, M., S. Ota, H. Matsuura, N. Nango, A. Hirata and S. Kawano. 2013. Three-dimensional ultrastructural study of oil and astaxanthin accumulation during encystment in the green alga *Haematococcus pluvialis*. PLoS One 8: e53618.

Wen, Z.Y. and F. Chen. 2000. Production potentail of eicosapentaenoic acid by the diatom *Nitzschia laevis*. Biotechnol. Lett. 22: 727–733.

White, D.A., A. Pagarette, P. Rooks and S.T. Ali. 2013. The effect of sodium bicarbonate supplementation on growth and biochemical composition of marine microalgae cultures. J. Appl. Phycol. 25: 153–165.

Wijffels, R.H. and E. Barba. 2010. An outlook on microalgal biofuels. Science 329: 796–799.

Wynn, J., P. Behrens, A. Sundararajan, J. Hansen, and K. Apt. 2010. Production of single cell oils from dinoflagellates. pp. 115–129. *In*: Z. Cohen and C. Ratledge (eds.). SIngle Cell Oils. Microbial and Algal Oils. AOCS Press, Urbana.

Xiong, W., X.F. Li, J.Y. Xiang and Q.Y. Wu. 2008. High-density fermentation of microalga *Chlorella protothecoides* in bioreactor for microbio-diesel production. Appl. Microbiol. Biotechnol. 78: 29–36.

Yong, Y.Y.R. and Y.K. Lee. 1991. Do carotenoids play a photoprotective role in the cytoplasm of *Haematococcus lacustris* (Chlorophyta). Phycologia 30: 257–261.

Yongmanitchai, W. and O.P. Ward. 1991. Growth of and omega-3 fatty acid production by *Phaeodactylum tricornutum* under different culture conditions. Appl. Environ. Microbiol. 57: 419–425.

Zhao, G.L., J.Y. Yu, F.F. Jiang, X. Zhang and T.W. Tan. 2012. The effect of different trophic modes on lipid accumulation of *Scenedesmus quadricauda*. Bioresour. Technol. 114: 466–471.

Zhekisheva, M., S. Boussiba, I. Khozina-Goldberg, A. Zarka and Z. Cohen. 2002. Accumulation of oleic acid in *Haematococcus pluvialis* (Chlorophyceae) under nitrogen startvation or high light is correlated with that of astaxanthin esters. J. Phycol. 38: 325–331.

Zhu, X., S.P. Long and D.R. Ort. 2008. Converting solar energy into crop production. Curr. Opin. Biotechnol. 19: 153–159.

Microalgal Biomass Production

Rathinam Raja,[1] *Hemaiswarya Shanmugam,*[2,*] *Ramanujam Ravikumar*[3]
and *Isabel S. Carvalho*[1]

Introduction

Microalgae are unicellular and the most abundant primary producers found in all the aquatic systems such as, freshwater, seawater, hypersaline lakes, and even in deserts and arctic ecosystems. Sub-divided into eukaryotic and prokaryotic algae, eukaryotes possess defined cell organelles such as nuclei, chloroplasts, mitochondria whereas prokaryotes (cyanobacteria or blue-green) are primitive, possessing the simpler cellular structure of bacteria. These algae convert light energy and carbon dioxide (CO_2) into biomass such as carbohydrates (Park et al. 2011), proteins (Becker 2007), and lipids (Harwood and Guschina 2009) through a process called, photosynthesis. Their photosynthetic mechanism is similar to land-based plants, but due to a simple cellular structure, and being submerged in an aqueous environment where they have efficient to access water, CO_2 and other nutrients, they are more efficient in converting solar energy into biomass. Many microalgae species are able to switch from phototrophic to heterotrophic growth conditions. As heterotrophs, the algae rely on glucose or other carbon sources for carbon metabolism and energy and some algae can grow mixotrophically. Microalgae can also grow in extreme environments; It could also be produced on agricultural and non-agricultural lands. Fresh, brackish, saline, wastewater, municipal sewage, and industrial effluents can be used to cultivate microalgae (Raja et al. 2004). Some of the important updates on the biomass of algae will be elaborated in the upcoming sessions.

Microalgae

Natural food production is directly related to the growth of microscopic plants called algae (Aaron et al. 2011). They have many inherent advantages, some of them are: higher productivity (biomass) in few days, easily adaptable to new environments and high-lipid content, being a fundamental edge. Algae are cultivated not only as a food source but also for fuel and is unlikely to interfere with food production. It is widely believed (though research has to confirm) that the use of algal-based fuel would result in a tiny fraction of the net greenhouse gases. Also, scaling up algae could lead to yields of other commercially

[1] Food Science Laboratory, Centre for Mediterranean Bioresources and Food, FCT, University of Algarve, Campus de Gambelas, Faro 8005-139, Portugal.
[2] AUKBC Research Centre, Anna University, MIT Campus, Chrompet, Chennai-600 044, Tamil Nadu, India.
[3] Aquatic Energy LLC, One Lakeshore Drive, Lake Charles, Louisiana 70629, USA.
* Corresponding author: iswaryahema@gmail.com

viable products besides fuel (Philip et al. 2011). Among the eukaryotes, green algae are the most referred oil-rich microalgae. They are ubiquitous in a variety of habitats and grow faster than other taxa, and as much as 60% of their cell dry weight is enriched with oils. However, the composition of the oils is highly dependent on the species and the conditions in which the algae grow.

Oils that are rich in neutral lipids are desirable in a biofuel context because of their potential high fuel yield. Because TAGs are made up of three molecules of fatty acids that are esterified-or altered-to one molecule of glycerol, close to 100% of their weight can be converted into fuels. With polar lipids, on the other hand, only one or two fatty-acid molecules are esterified to glycerol and the remaining components (e.g., sugars or phosphate groups) cannot be converted to fuel feedstock. As a result, these types of lipids generate lower fuel yields (Philip et al. 2011). Two thirds of the earth's surface is covered with ocean, thus algae would be an option of great potential for food including aquaculture industries, pharma products, and global energy needs. Many countries have started to grow algae on ocean beds in floating transparent light weight tubes (e.g., Malaysia, Denmark, and USA). These projects are explained in OASIS and NASA OMEGA projects. Currently these set-ups are in trial to be a huge project soon.

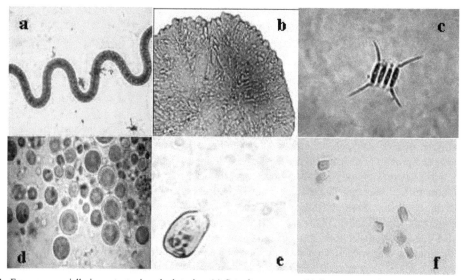

Fig. 1. Few commercially important microalgal strains. (a) *Spirulina maxima*, (b) *Botryococcus brauanii*, (c) *Scenedesmus quadricauda*, (d) *Chlorella vulgaris*, (e) *Dunaliella salina*, and (f) *Chaetoceros muelleri*.

Selection of new strains

The marine microalgae strains suitable for maximum biomass production requires a lot of important characters ideally (quantitatively) measurable. The use of locally selected strains may be of significance both for ease of management and for reasons of sustainability. An outdoor microalgal cultivation would help to achieve maximum as well optimum biomass production in large scale systems and commercialization of products such as fine chemicals, nutraceutical, and lipids. Total biomass composition includes total caloric value of the biomass, % lipids and lipid composition (for biodiesel), % starch and carbohydrate composition (for bioethanol and to identify higher value byproducts), % protein and protein composition (soluble/insoluble for food/feed purposes). Presence of heavy metals or toxins and should include relevant aspects for biorefinery, such as the cell volume, thickness/toughness of the cell wall, the presence of tough fibers (macroalgae), and the moisture content. A measure for this could be the energy input per gram of dry weight necessary for full biorefinery. It is very important to check if the organism produces any byproduct that have an intrinsic added value, such as carotenoids. This is important to reduce the costs of the final product. Here a specification of the compounds and their expected added value per gram of dry biomass should be indicated.

Maintenance of axenic culture

Marine microalgae isolated from coastal environments are unlikely to be optimally adapted to the new pond environment. Therefore, genetic selection may significantly improve productivities of the algal mass cultures. For example, when a single limiting factor exists such as the nutrient source, genetic selection will offer a greater advantage for growth. If two organisms differ through a hereditable genetic difference to utilize the same limiting substance, then the one that is better in capturing the limiting nutrient will prevail in that environment. Eventually competition for one limiting factor will result in a single type of dominant organism. There are many factors that determine the species dominance which include: (i) resource-growth rate relationships for different algae, (ii) variable environmental conditions, (iii) inhibition of one organism by another through excreted allelopathic substances, and (iv) loss of growth due to predation or sinking. Free fatty acids, cell wall degradation products, exo-metabolites produced by algae and cyanobacteria, such as cyanobacterin (Gleason and Paulson 1984; Gleason and Baxa 1986) and fischerellin (Gross et al. 1991; Hagmann and Jüttner 1996) could also have allelopathic potential. Simple mathematical competition models can be designed for two organisms that compete in the pond. The above factors can be included as terms in the model.

The effect of single or specific variables or combination variables on the growth of organisms in a laboratory scale can be made. The key issue is difficulty in simulation of the outdoor environment in the laboratory conditions or extrapolation of data from lab to mass culture condition. Maintaining a specific, genetically selected mono-algal culture of a specific inoculated strain in outdoor ponds is necessary to exhibit the high lipid productivities. The problem of species dominance and competition in outdoor mass cultures can be controlled by variations of biota, light, temperature, pH, and oxygen and nutrient supplies. For example, high ammonia or pH will inhibit most zooplankton infestations, or selecting a growth environment for specific microalgae species such as a very high alkalinity for *Spirulina* and high salinity for *Dunaliella*. These techniques are expensive and result in severe reduction of algal productivities. Severe contamination in *Chlorella* production could be solved by high density inoculations and semi-batch operations, but this resulted in reduced overall productivity. Both *Dunaliella* and *Chlorella* are dominant in their optimal environments. Contamination can be better managed in closed photobioreactors, but upon continuous cultivation, both open and closed systems become more susceptible to contaminations. Careful strain isolation and characterization, cultivation parameters of individual microalgal candidates, and the expected contaminants in the region can reduce the contamination issues.

Strain development and domestication

Increase in the production of valuable compounds of microalgae requires optimum strain development and domestication. Due to the absence of cell differentiation in microalgae represent a much simpler system for genetic manipulations compared with higher plants. Techniques to introduce DNA into algal cells with suitable promoters, new selectable marker genes, and expression vectors have to be standardized for each ideal species. Currently, all these requirements have been fulfilled for the diatom *Phaeodactylum*, the green alga, *Chlamydomonas* and the blue green algae, *Synechococcus* and *Synechocystis*. Successful genetic engineering has been achieved in the expression of mosquito larvicidal properties in blue green algae (Boussiba et al. 2000). The development of a functional transformation system can be expected in the near future for other diatoms, blue green algae, and the red alga, *Porphyridium*. The success of genetic engineering lies in the improvement of nutritional value and product yield with optimal production parameters (Raja et al. 2008). However, the following factors are to be considered to achieve the above features:

1. The accumulation of valuable substances in algae via genetic transformation can only increase up to a point where cellular metabolism starts negatively affecting the production.
2. Transgenic algae potentially pose a considerable threat to the ecosystem and will most likely to be banned from the outdoor cultivations or otherwise be under strict regulation.
3. Usually the transgenic cells exhibit less fitness than the wild type and therefore cells that lose the newly introduced gene quickly outgrow the transformants.

To prevent this, a constant selection pressure is necessary by the addition of antibiotics (a potential public health hazard).

Nutrients

Microalgae require inorganic nutrients (P, N, and C), sufficient light, and favorable temperatures to grow. Hydrogen (H) and oxygen (O) are also essential for algal growth, but water (H_2O) provides an abundance of these elements and no further discussion of H and O is necessary to algal nutrient requirements. Many other elements are needed for algal growth in trace amounts; they are collectively referred as micronutrients (calcium (Ca), magnesium (Mg), sodium (Na), potassium (K), iron (Fe), manganese (Mn), sulfur (S), zinc (Zn), copper (Cu), and cobalt (Co)). It is also important to understand that different algal species do not have identical nutritional needs (Grobbelaar 2013).

Biomass production

Algal biomass can be a source of fine chemical, proteins, pharma products, poultry feeds, feed stock and a variety of biofuels (e.g., biodiesel, hydrogen, methane, and bioethanol). Furthermore, it is being seriously considered for the removal of carbon dioxide from the flue gases (e.g., petroleum power stations). Thus it reduces global warming considerably.

Photobioreactor

Basically, photobioreactors (PBRs) have different types of tanks in which algae are cultivated (Richmond 2004). PBRs are closed cultivation systems to grow microalgae under photo-autotrophic conditions. Several types of PBRs have been experimented on since 1950s, when algal cultures were first considered the ideal solar technology to produce as a cost-effective biomass and protein on a large scale. Algal cultures consist of a single or several specific strains optimized for producing biomass as a product. Ideal growth conditions for microalgal cultures are strain specific, and the biomass productivity depends upon many factors. These include abiotic factors like temperature, pH, water quality, minerals, carbon dioxide, light cycle, and intensity. Water, nutrients, and CO_2 are provided in a controlled way, while oxygen has to be removed. Biotic factors like cell fragility and cell density, mechanical factors include mixing, gas bubble size and distribution, and mass transfer are of particular concern in photobioreactors (Schenk et al. 2008). Growth of heterotrophic algae in conventional fermentors is preferred instead of photobioreactors for production of high-value products (Jiang and Chen 1999; Wen and Chen 2003), since instead of light and photosynthesis, heterotrophic algae rely on carbon sources in the medium (Ward and Singh 2005).

Photobioreactors classified on the basis of both design and operation. The most used designs are flat-plates, tubular reactors, air-bubbled plastic bags. Some of them are artificially illuminated with fluorescent. But the recommended two types of PBRs are flat and tubular reactors (Tredici et al. 2010). Generally, PBRs are more expensive to install and operate, intensive study is still going on to reduce their cost and thus facilitate their use especially for low value products like algal oils. All parameters (nutrients, light regime, gas exchange) are maintained to realize optimal culture conditions. The contamination level is much lower compared to open systems. Engineering PBR is still a very active field of research, since closed culture systems are necessary to grow typical photosynthetic microbes and exploit them as a source of aquaculture feeds, food additives, fine chemicals, pharmaceuticals, cosmetics are preferred by industry as research tools for biofuel production. The operation costs are normally significantly higher than those of ponds (Chisti 2007; Tredici et al. 2010). Besides, there is the need for cooling, which is generally provided by water (even seawater) spraying or by insertion of a cooling serpentine in the culture.

Open raceway pond

Open pond systems are shallow ponds (a maximum of 50 cm) in which algae are cultivated. Nutrients can be provided by mixing with few liters of water nearby paddle wheels. The water is typically kept in motion by paddle wheels and some mixing can be accomplished by appropriately designed guides. Raceway ponds are being in use in Israel, USA, China, India and other countries. Fertilizer is used for mass cultivation and the culture is agitated gently by paddle wheel (Fig. 2d). The advantages of open ponds are low costs and that they are easy to operate. However, they are sensitive to contamination leading to introduction of unwanted fast growing organisms in the ponds. Especially heterotrophic organisms will graze on the autotrophic biomass and lead to loss of productivity. Therefore, the present commercial production of microalgae in open culture systems is restricted to only those organisms that can grow under extreme conditions, i.e., high pH or salinity. Thus a limited range of microalgae can be maintained as monoculture in open ponds in long-term operation. Currently there is lot of algae mass cultured and marketed commercially they are *Skeletonema* sp., *Chaetoceros* sp., *Dunaliella* sp., *Spirulina* in high alkalinity and *Chlorella* in high levels of nutrients have been successfully cultivated. Table 1 explains all the major differences between open pond and photobioreactor.

Fig. 2. (a) Seed culture scaling-up for open raceway ponds (b) A typical lab-scale photobioreactor (c) French press (side view) (d) Open raceway pond (e) Flocculated culture shows algal clumping (f) CO_2 cylinder (g) Culture storage tank (h) French press (front view) (i) Wet algal biomass collected from French press (j) Hot air dryer (Courtesy: Aquatic Energy LLC, Louisiana).

Table 1. Merits and demerits of photobioreactors and open raceway ponds (Adapted from Pulz 2001).

Parameters	Photobioreactors (PBR)	Open ponds and raceways
Space	Low	High
Water loss	Low	Very high
CO_2 loss	Low	High and completely depending on pond depth
Oxygen concentration	Build-up in closed system requires gas exchange devices (O_2 must be removed to prevent inhibition of photosynthesis and photo-oxidative damage)	Usually low enough because of continuous spontaneous out gassing
High temperature	Cooling often required (by spraying water on PBR)	Highly variable, some control possible by pond depth
Shear	High (fast and turbulent flows required for good mixing, pumping through gas exchange devices	Low (gentle mixing)
Cleaning	Required (wall-growth and dirt reduce light intensity), but causes abrasion, limiting PBR lifetime	No issue
Contamination risk	Low	High
Biomass quality	Reproducible	Variable
Biomass Concentration	High, between 2 and 8 g l–1	Low, between 0.1 and 0.5 g l–1
Production flexibility	switching possible	Few species possible, difficult to switch high
Low temperature	No issue	Risk at temperate countries

Biomass from industrial effluents and seawages

Microalgae are increasingly used in effluents and seawages for their potential to remove (and use) the excess amount of chemicals that has generated by different sources. These are removed from wastewater through direct uptake into the algae cells (Hoffman 1998). Wastewater treatment using algae has many advantages. It offers the feasibility to recycle these chemicals (or nutrients) into algae biomass as a fertilizer and thus can offset treatment cost. Oxygen rich effluent is released into water bodies after wastewater treatment using algae (Becker 2004). The addition of carbon is not required to remove nitrogen and phosphorus from wastewater. *Chlorella* (Gonzales et al. 1997), *Scenedesmus* (Martinez et al. 2000), and *Spirulina* (Olguın et al. 2003) are the most widely used algae for nutrient removal.

Municipal seawage water typically have organic and ammonia nitrogen concentrations (25 to 45 mg/L) and phosphorus concentrations (4 to 16 mg/L) (Tchobanoglous et al. 2003). At the final stage of treatment, nitrogen and phosphorus are removed by algae. Few of the algae can also be used to remove organic wastages. Some constituents of wastewater are in high concentration depending on the type of seawage that can possibly inhibit algae growth. These constituents are urea, ammonium, organic acids, phenolic compounds, and pesticides that can limit the use of seawage water to grow algae (Hodaifa et al. 2008). Kim et al. reported, almost 95 and 96% removal of nitrogen and phosphorus, respectively, by *Chlorella vulgaris* in 25% secondarily treated swine wastewater after four days of incubation (Kim et al. 1998). Travieso treated distillery wastewater from an anaerobic fixed-bed reactor in a microalgae pond and obtained 90.2%, 84.1%, and 85.5% organic nitrogen, ammonia, and total phosphorus removal, respectively (Hodaifa et al. 2008).

Nitrogen constitutes about 8% of microalgae cell dry weight and the most important nutrient for algae growth that constitutes the proteins (Richmond 2004). Ammonia, nitrite, nitrate, and urea are used as nitrogen sources in microalgae cultivation because of its lower cost (Becker 2004). Phosphorus is a macronutrient that plays an important role in growth and metabolism of algae. It is required for most cellular processes, that involving energy transfer and nucleic acid synthesis. Phosphorous is required for most of the cellular processes, that involving energy transfer and nucleic acid synthesis (Kull 1962). The two most important phosphorus forms used by algae are HPO_4^- and HPO_4^{2-}. Algae utilize the soluble form for metabolism and stores the insolubles when phosphate amounts present in the culture are limited

(Powell et al. 2008). The phosphorus consumption rate in algae depends on phosphorus concentration in both the environment and the cells, and on pH, temperature (Sancho et al. 1997).

Algae offer a great potential for performing wastewater treatment using algae. After primary and secondary treatment is fed to a race track reactor. Algae and bacteria are cultured in these reactors. Algae are continuously mixed to keep the cells in suspension and expose them periodically to light. Algae provide the dissolved oxygen required for bacterial decomposition of organic matter and bacteria provide carbon, nitrogen, and phosphorus essential for algal growth by degrading wastewater components (Garcia et al. 2000). Algae remove the nutrients directly through uptake and harvesting of the biomass. Nitrogen and phosphorus are removed indirectly by ammonia-nitrogen volatilization and orthophosphate precipitation, respectively. Directly and indirectly, the growth rate of algae controls the efficiency of nitrogen and phosphorus removal. The efficiency of nutrient removal is determined by cellular retention time, solar radiation, and temperature (Garcia et al. 2000).

Munoz and Guieysse (2006) described microalgae enhance the removal of nutrients, organic contaminants, heavy metals, and even pathogens from domestic wastewater and furnish an interesting raw material for the production of high-value chemicals or biogas. The use of algae in wastewater treatment is being well established with treatment plants in operation in California (Oswald 1988). The improvement in the quality of wastewater and the fermentation of the resulting biomass to methane were implemented (Pulz and Scheibenbogen 1998). The cultivation of algae as a source of animal feed in that respect, animal wastes represent a good source of substrates for the culture of *Spirulina*. By employing an integrated approach one can easily conceive the treatment of wastes with concomitant production of algal biomass for animal feeds. *Spirulina* deals with the recycling of animal wastes and most often with pig wastes, because intensive pig production is causing very serious problems of water pollution worldwide.

Accumulation of heavy metals especially Cd by *Scenedesmus bijugus* (Playfair) V. may show saturation in 2 h. In a similar study with Cd and Pb, *S. bijugus* could accumulate 80% of the heavy metals within the first 12 h. Mallick and Rai (1994) explained the algae-based systems for the removal of toxic minerals such as Pb, Cd, Hg, Se, Sn, Ni, As, and Br also showed great promise. The hydrocarbon rich microalga, *Botroyococcus braunii* consumed nitrate and phosphate in secondary treated water. Removals of toxic metals like As, Cr, Cd, and inorganic compounds have also reported (Sawayama et al. 1995). The use of *Dunaliella salina* as a potential test organism for the simple determination of changes in the chlorophyll content under standard conditions can be recognized as qualitative and quantitative measures of toxic residues (Yarden et al. 1993). *Dunaliella* species have also shown to be exceptionally tolerant of heavy metals such as Cu, Pb, and chlorinated hydrocarbons. *Dunaliella tertiolecta* Butcher can tolerate Cu concentration near saturation level for seawater (approx. 0.6 ug Cu mL^{-1}) (Visviki and Rachlin 1991). The effective concentration of the toxic chemicals like copper and cadmium was reduced by the marine alga, *D. minuta* Lerche, in both acute and chronic exposures. The incredible ability of the halophilic alga, *D. salina* in the treatment of salt refinery effluent was studied. Their appearance, odor, TDS, turbidity, pH, hardness, BOD, and COD concentrations of anions and cations were reduced considerably. A significant percentage of heavy metals Ba (56.5%), Al (46%), Ag (32.4), and Sr (4.8%) were also removed (Perales-Vela et al. 2006; Raja et al. 2008).

Biomass harvest

Biomass harvesting is a kind of technique and it accounts 15–20% of the production costs. The very small size of algae and their low concentration in the culture medium makes the cell recovery, harder one. The harvesting cannot be done by a single process because of the several species of algae with varying characteristics like shape, size and motility that influence to a big extent for their settling. Centrifugation is one of the most commonly used techniques to harvest microalgae in a lab scale as well in R&D laboratories. Algae are commonly used as a feed for aquaculture and they harvested by centrifuging to produce concentrates with longer shelf-life. Filter presses are used to recover fairly large microalgae like *Spirulina* sp. but are not suitable for smaller microalgae like *Scenedesmus*, *Dunaliella*, and *Chlorella*. The major costs involved in centrifugation are depreciation and maintenance of equipment (> 20000 L) (Richmond 2004). Using centrifuge in algal industries to harvest biomass is not a recommended and feasible one. The harvested algae biomass is dried so that the product can be stored without spoilage when algae are used as aquaculture feed. The drying process accounts for 20% of total algal production costs. Spray drying, freeze-drying, drum drying, and sun drying are used to dry microalgae. Spray drying is used for high-value products, but it may lead to degradation of pigments and vitamins in algae (although these may be protected by the addition of antioxidants before drying).

Flotation is a process in which a gas or air is bubbled through the liquid to be clarified. The particles are adsorbed by the rising bubbles and are removed when they reach the surface of the liquid. The bubbles can attract smaller particles easily. This process can be used for the separation of algae with particle diameter of less than 500 μm (Uduman et al. 2010). Air flotation techniques may be either dissolved or induced. This process is among several new harvesting methods proven to be efficient (Chen et al. 1998). The efficiency of dewatering microalgae by sedimentation depends on the time of removal of algae. It is influenced by the intercellular interactions of algae in the different growth phases during cultivation. The optimum harvesting time is determined by the zeta potential of the microalgal culture. It was found that algae cultures harvested during the stationary phase had a higher rate of settling than those harvested during the exponential phase. Algae cells in the exponential phase are highly stable and electrostatically repel each other. It was found that algal cultures stored in darkness settled faster compared to daylight conditions (Danquah et al. 2009).

Flocculation

Algae have high negative surface charges during the logarithmic phase of their growth. They remain widely dispersed in water because of these repulsive charges. Algae settle and forms clusters after longer residence times when they approach decay phase. This process is called, auto-flocculation. The other factors responsible for auto-flocculation include nutrient (nitrogen and phosphorus) limitation, restricted CO_2 supply, and co-precipitation of magnesium, calcium and carbonate salts (Becker 2004). Auto-flocculation is also found to take place by changing the conditions of algae cultivation. Algal flocs are formed when pond agitation stops and the CO_2 supply is cut off, causing an increase in pH. On flocculation with various pH values, *Scenedesmus* sp. showed no flocculation for pH values between 5.0 and 7.5, while at pH values above 8.5, almost 95% of the algal biomass was found to be removed (Becker 2004).

The algae cells form precipitates with the addition of chemicals. Multivalent metal salts like ferric chloride [$FeCl_3$], aluminum sulfate [$Al_2(SO_4)_3$, alum] and ferric sulfate [$Fe_2(SO_4)_3$] are used to flocculate algae (Grima et al. 2003). Alum is an efficient flocculant for *Scenedesmus* and *Chlorella* to produce fuel. The metal salts are not intended for flocculation when considering the algae for use in some aquaculture applications (Golueke and Oswald 1965). Algae can also be flocculated by using cationic polymers or polyelectrolytes (Tenney et al. 1969). The negative charge on algae is attracted to the positive charge of cationic flocculants. Algae particles are bound together by polymer flocculants through a process called bridging. This process fails to occur at high ionic strengths. Dosage between 1 and 10 mg/L of the polymer is required to flocculate fresh water algae (Bilanovic et al. 1988). Flocculation by polymeric flocculants was mainly found to depend on the molecular mass, ionic strength and dose of the polymer and also the pH and concentration of the algae culture (Chisti 1999). Chitosan has several advantages over other conventional flocculants: It does not produce any toxic effects; it is required in very low concentrations (Becker 2004). Algae can also be immobilized in a chitosan matrix and used in tertiary treatment of wastewater (Kaya and Picard 1996).

Economic importance of algal biomass

Health supplements

Microalgae can increase the larval production though the exact mechanism of action is unclear till date. Microalgae offer food for zooplankton which helps to advance the quality of the culture. For numerous freshwater and seawater animals, supplying phytoplankton to the ponds leads to much better results on survival, growth and transformation index (Muller-Feuga 2000). Due to some excreted biochemical compounds along with the induction of behavioral processes such as prey catching, regulation of bacterial community, probiotic effects and the stimulation of immunity were controlled (Hong et al. 2005; Raja and Hemaiswarya 2010). Several factors contributing nutritional value of a microalga (it includes their size and shape, digestibility, biochemical composition, enzymes, toxins and the requirements of animal feeding on the alga). Studies have tried to correlate nutritional value of microalgae with their biochemical profile (Richmond 2004; Durmaz 2007). However, results from feeding experiments that have examined microalgae which are contrary in a specific nutrient are challenging to interpret because of the confounding effects of other microalgal nutrients. Nevertheless, from examining the literature, algal foods have been enhanced with compounded diets or emulsions, some general conclusions can be reached (Knauer and Southgate 1999).

Algae grown to late logarithmic growth phase typically contain 30–40% protein, 10–20% lipid and 5–15% carbohydrate (Fujii et al. 2010). In stationary phase, composition of microalgae can change significantly for example, when nitrate is limiting, carbohydrate levels can double at the expense of protein (Liang et al. 2009). There does not appear to be a strong correlation between the proximate composition and nutritional value, though algal diets with high levels of carbohydrate are stated to produce the best growth for juvenile oysters, *Ostrea edulis* (Ponis et al. 2006). Larval scallops, *Patinopecten yessoensis* provided polyunsaturated fatty acids in adequate proportions. In contrast, high dietary protein provided best growth for juvenile mussels, *Mytilus trossulus* and Pacific oysters, *Crassostrea gigas* (Knuckey et al. 2002).

Algal pigments transferred to zooplankton may contribute to nutritional value (Lorenz and Cysewski 2000; Gagneux-Moreaux et al. 2007; Raja et al. 2008). Dominant pigments in the copepod, *Temora* sp. are lutein and astaxanthin whereas in Artemia it was canthaxanthin (Kang and Sim 2008; Gentsch et al. 2009), these prey items were fed to halibut larvae adequate amounts of vitamin A were found in halibut fed on copepods but not with halibut fed on Artemia. It was suggested that Artemia should routinely be enriched with astaxanthin and lutein to improve their nutritional value. Astaxanthin and canthaxanthin are the only pigments that can fix in the flesh of salmonids whose pinkening represents a 100 million US$, rapidly expanding market (Raja et al. 2007c). This feed additive is produced by chemical synthesis and available at a price of 3000 US$/kg approximately. The biological sources for astaxanthin are the yeast, *Phaffia rhodozyma* (Sanderson and Jolly 1994) despite its low content (0.4%), and compared to *Haematococcus pluvialis* containing 5% (Guerin et al. 2003; Kang and Sim 2008).

Some companies like Algatec-Sweden, Norbio-Norway, Biotechna-UK, Aquasearch, Cyanotech, Maricultura, Danisco Biotechnology and Oceancolor-USA have entered the astaxanthin market. In fact, microalgal astaxanthin has been approved in Japan and Canada as a pigment in salmonid feeds (Spolaore et al. 2006). Feeds including 5–20% *Arthrospira* sp. (rich in carotene pigments), to enhance the red and yellow patterns in carp. This clarity and color description increases their value. Another example is the traditional French technique so-called the greening of oysters. It consists of creating a blue-green color on the gills and labial palps of oysters using the diatom, *Haslea ostrearia*, it increases the product's market by 40% (Gagneux-Moreaux et al. 2007). For vaccine purpose *Chlamydomonas* sp. has been used using the p57 antigen, the causative agent of bacterial kidney disease. This disease is caused by the intracellular bacterium, *Reinbacterium salmoninarum* which affects wild and farmed salmonids. A huge concern from an economic point of view and the symptom of disease which develops before it can be treated with antibiotics. Fish-fed algae (4% algal dry weight of feed) fed to juvenile trout produced immunoglobulin IgM expressed in different tissues (epithelial or blood cells).

Animal feed and aquaculture

The increased shrimp farming production mainly takes place in subtropical regions of America and South-East Asia (72%, 3,718 hatcheries) (Alam et al. 2009). Microalgae are being in use to improve the nutritional quality and gives coloration to the shrimps. Among these *Chlorella* grows in nutrient-rich media, while *Spirulina* sp. requires a high pH of 9.5–11 with appropriate concentration of bicarbonate. Similarly, *Dunaliella salina* grows at the high salinity of 0.5–6 M (Raja 2007b). Several algae such as *Chaetoceros* sp., *Isochrysis* sp., *Skeletonema* sp., *Thalassiosira* sp., *Tetraselmis* sp. and *Crypthecodinium cohnii* are being useful in the aquaculture industry which do not have these selective advantages and it must be grown in closed systems. Commercial large scale systems such as the cascade system were developed in Trebon, Czech Republic in the 1970s and heterotrophic fermenters have been used for the culture of *Chlorella* sp. in Japan and Taiwan. Table 2 summarizes the commercial algal culture and their uses. Factors to be considered for production of microalgae include: the biology of the alga, cost of land, labor, energy, water, nutrients (climate if the culture is outdoors) and the type of final product. Microalgae are necessary from the second stage of larval development (zoea) and in combination with zooplankton from the third stage (myses). Naturally occurring microalgal blooms are encouraged in large ponds with low water exchange where the larvae are introduced. Sometimes fertilizers and bacteria are added to induce more favorable conditions. This production system with poor control of microalgae provides a better part of shrimp production (López Elías et al. 2003).

Non-living diets generally enhance lower growth and higher mortalities compared to those fed with live microalgae (Ponis et al. 2003). Products other than live microalgae must be exempt from contamination and nontoxic. Bacteria can provide only a part of the metabolic requirements by supplying organic molecules and vitamins. Under conditions close to those found in rearing facilities, the bacterial input represents less than 15% of the microalgal contribution for mollusks larvae and juveniles of many species (Wikfors and Ohno 2001; Knuckey et al. 2006). The uses of bacteria as food source in hatcheries seems to be invalidated, since physical and chemical treatments are often used to limit the development of bacterial contaminations which are responsible for drastic larval mortalities. However, in live microalgal culture, the natural bacterial flora was proved to enhance the health of mollusks. Antibiotic suppression of microbial flora associated with juvenile oysters fed artificially reduced growth (Durmaz 2007). Oyster larvae fed with live microalgal diets showed improved growth with the addition of some bacterial isolates. Yeast was also investigated as an alternative food source but poor results were observed (Ponis et al. 2003). Therefore, these two alternatives are not suitable to replace live microalgae.

Several factors can contribute to the nutritional value of a microalga (including its digestibility, biochemical composition, enzymes, toxins and the requirements of animal feeding on the alga). Studies have attempted to correlate the nutritional value of microalgae with their biochemical profile from feeding experiments that have tested microalgae differing in a specific nutrient are often difficult to

Table 2. Commercial algal culture and its applications (Hemaiswarya et al. 2011).

Genus	Morphology	Purpose
Nannochloropsis sp.	Small green algae	Growing rotifers and in fin fish hatcheries, used in reef tanks for feeding corals and other filter feeders, very high EPA level
Pavlova sp.	Small golden-brown flagellate, very difficult to grow so it is not produced by many hatcheries	Used to increase the DHA/EPA levels in broodstock, oysters, clams, mussels and scallops, sterol composition so it is popular with cold water fish hatcheries (cod) for enriching rotifers
Isochrysis sp.	Small golden-brown flagellate	Enrichment of zooplankton such as Artemia, used in shellfish hatcheries and used in some shrimp hatcheries, good size for feeding brine shrimp and copepods, oysters, clams, mussels, and scallops
Tetraselmis sp.	Large green flagellate	Excellent feed for larval shrimps and contains natural amino acids that stimulate feeding in marine animals, used in conjunction with *Nannochloropsis* for producing rotifers, good size for feeding brine shrimp, standard feed for oysters, clams, mussels, and scallops, excellent feed for increasing growth rates and fighting zoea syndrome
Thalassiosira weissflogii	Large diatom	Used in the shrimp and shellfish larviculture, considered by several hatcheries to be the single best alga for larval shrimps, also good for feeding copepods and brine shrimps, post-set (200 L and larger) oysters, clams, mussels, and scallops for brood stock conditioning
Dunaliella sp.	Small green flagellate	Used to increase vitamin levels in some shrimp hatcheries and also for the coloration
Chaetoceros sp.	Diatom	Used to increase vitamin levels in some shrimp hatcheries

interpret because of the confounding effects of other microalgal nutrients (Richmond 2004; Durmaz 2007). Microalgae grown to late logarithmic growth phase typically contain 30–40% protein, 10–20% lipid and 5–15% carbohydrate (Fujii et al. 2010). In the stationary phase, the proximate composition of microalgae can change significantly, e.g., when nitrate is limiting, carbohydrate levels can double at the expense of protein (Liang et al. 2009). There does not appear to be a strong correlation between the proximate composition of microalgae and nutritional value, though algal diets with high levels of carbohydrate are reported to produce the best growth for juvenile oysters, *Ostrea edulis* (Ponis et al. 2006). Larval scallops, *Patinopecten yessoensis* provided polyunsaturated fatty acids in adequate proportions. In contrast, high dietary protein provided best growth for juvenile mussels, *Mytilus trossulus* and Pacific oysters, *Crassostrea gigas* (Knuckey et al. 2002). Large sized hatcheries require highly paid technicians, multimillion dollar investments and highly controlled medium conditions. The observed trend is toward specialized production, particularly with the supply of post larvae in the hands of big centralized hatcheries. They open a pathway to new techniques especially the genetic selection of strains with stronger immunity.

Biofuel

Microalgae have high growth rates and photosynthetic efficiencies due to their simple structures. The efficiency is much higher (6–8%) than that of terrestrial plants (it is, 1.8–2.2%). The idea of using microalgae is not new, but it is now being taken seriously in several countries because of the emerging concern about global warming that is closely associated with burning of fossil fuels (Omer 2012). Microalgal biomass contains approximately 50% carbon by dry weight; therefore, it is also used to produce methane by anaerobic digestion. The process is technically feasible, but it cannot compete with many other low-cost organic substrates that are available for anaerobic digestion (Hussain et al. 2010). Depending on species, microalgae produce different kinds of lipids, hydrocarbons, and other complex oils (Guschina and Harwood 2006). Microalgal biodiesel will need to comply with existing standards. In USA, the relevant standard is the American Society for Testing and Materials (ASTM) biodiesel standard D6751 Table 3 (Knothe et al. 2005). Most of the algae are unlikely to comply with the biodiesel standards, but this may not be a significant limitation (Belarbi et al. 2000; Chisti 2007). The extent of unsaturation of microalgal oil and its content of fatty acids with more than four double bonds can be reduced easily by partial catalytic hydrogenation of the oil. The challenge lies in harvesting algal biomass and the extraction of biodiesel (Jang et al. 2005; Dijkstra 2006). The heterogeneity of algal species and growth parameters makes this bio-inspired option a technical challenge for scale-up consideration.

A sustainable and profitable biodiesel production from microalgae is possible. The biofuel can overcome the energy and environmental needs by integrating new technologies. Large quantities of algal biomass needed for the production of biodiesel could be grown in photobioreactors combined with photonics and biotechnologies. However, more precise economic assessments of production are necessary to establish with petroleum-derived fuels. The direct hydrothermal liquefaction is an energy-efficient technique for producing biodiesel from algae without the need to reduce the water content of the algal biomass (Patil et al. 2008). Although the relatedness of cyanobacteria to nonphotosynthetic bacteria allows for exploitation of genetic-engineering technologies and makes them an attractive starting

Table 3. Comparison of biodiesel and ASTM (International Trade Administration 2009).

Properties	Biodiesel from microalgae	ASTM biodiesel standard
Density (kg/L)	0.864	0.86–0.9
Viscosity (mm$_2$/s, cSt at 40°C)	5.2	3.5–5.0
Flash point (°C)	115	Minimum, 100
Solidifying point (°C)	–12	-
Cold filter plugging point (°C)	–11	Summer maximum, 0; winter maximum < –15
Acid value (mg KOH/g)	0.374	Maximum, 0.5
Heating value (MJ/kg)	41	-
H/C ratio	1.81	-

point for biofuels research, they lack one very important thing that eukaryotic microalgae can possess in abundance-neutral lipids, which are rich in triacylglycerols (TAGs).

Greenhouse gas reductions and global warming

The combustion of fossil fuel generates carbon dioxide, a major greenhouse gas that is considered as a huge threat because of its potential to cause severe global warming. Microalgae are particularly considered for bio-fixation because of their ability to grow fast and fix greater amounts of carbon dioxide. The bio-mitigation of carbon dioxide and other flue gases by microalgae have significantly gained interest in reducing the emissions from coal-fired power plants. The algae biomass thus produced by capturing carbon can be used in generating valuable products such as fuel, animal feed and fertilizer. Power plants are the major sources of CO_2 and release 5.7 giga tones of carbon dioxide per year (Kadam 2001). CO_2 is produced by both stationary and mobile sources. Microalgae offer a natural way to recycle carbon dioxide from the flue gas and thus help in reducing the effects of global warming and climate change.

Microalgal biomass is composed of 45 to 50% carbon based on dry weight measurements (Schlesinger 1991). The high carbon content of microalgae makes it suitable for storing carbon. CO_2 present is flue gas can significantly raise the growth rates of microalgae. Microalgae can be engineered in open ponds or photobioreactors to maximize CO_2 conversion to biomass thereby sequestering carbon and also producing a biofuel. The selection of microalgae is the most important factor in the bio-mitigation of carbon dioxide from flue gases generated by power plants. The algae should have high growth and CO_2 utilization rates. The other suitable characteristics for carbon dioxide bio-fixation are their ability to tolerate SOx and NOx, and thrive in mass cultures without contamination. Lastly, the algae should be chosen also considering the harvesting process. Algae with autoflocculation characteristics simplify the harvesting step and minimize the energy and cost in the downstream processing of algae production (Brennan and Owende 2010).

Algae (*Chlorella vulgaris* and *Scenedesmus obliquus*) are isolated from the water bodies near to the power plants so that they are already adapted to the flue gases generated and environmental conditions of that area were ideal for biofixation (De Morais and Costa 2007). They were able to grow in culture media with 18% (v/v) CO_2. Biological carbon sequestration using microalgae offers several advantages over geological and ocean sequestration systems. Algae can sequester carbon dioxide directly and the costs of separation of CO_2 gas are avoided. Microalgae production systems can be located near the power plant: this does not require huge costs on transportation of CO_2. The advantages of using microalgae biofixation processes are: their ability to grow on flue gas, higher carbon fixation rates than other plants and their potential to grow in wastewaters thereby minimizing the fresh water needs (Schenk et al. 2008). The Solvay process can be modified to convert CO_2 from fossil fuel power plants to bicarbonates. The carbon dioxide gas is passed through brine solution with ammonia as a catalyst under alkaline conditions to produce sodium bicarbonate according to this chemical reaction (Huang et al. 2001).

$$CO_2 + NaCl + NH_3 + H_2O \rightarrow NaHCO_3 \downarrow + NH_4Cl \text{ (Gouveia and Oliveira 2008)}$$

Sodium bicarbonate was a better carbon source for *Scenedesmus* than sodium carbonate. Carbon dioxide can be utilized in the form of carbonate salts using Solvay process. These salts can be used as a carbon source for algae growth when the power plant is not located near the algae pond. Algae can thus be used for bio-fixation of CO_2 in industrial flue gases.

Conclusion

As far as the algal biomass production has concerned, currently there are lot of projects started in many countries more particularly in India, China, United States, etc., and they are using the harvested biomass for biofuel production because it might contain heavy metals and unwanted chemicals available since the microalgae grown in municipal seawage or from effluent. This is best biological method recommended by naturalists and scientists to protect environment clean. In this way, not only the waste water has treated simultaneously the algae capturing atmospheric CO_2 thus it subsequently helps to reduce global warming.

At the same time, water quality and pH has stabilized by oxygen production during photosynthesis. It enhances waste water quality and reduces all the hazardous chemicals. Several awareness programs have been started on global warming and several countries amended new environment laws to regulate and ban the day to day issues which lead global warming.

Acknowledgement

The first author, Rathinam Raja is grateful to The Foundation for Science and Technology (FCT-SFRH/BPD/63402/2009), Portugal for funding the researcher.

References

Aaron, A., B. Amy, J. Siddharth, L. Antony and S. Sean. 2011. The potential for microalgae and other microcrops to produce sustainable biofuels. A review of the emerging industry, environmental sustainability and policy recommendations. University of Michigan, USA.

Alam, M.S., W.O. Watanabe and H.V. Daniels. 2009. Effect of different dietary protein and lipid levels on growth performance and body composition of juvenile southern flounder (*Paralichthys lethostigma*) reared in recirculating aquaculture system. J. World Aquac. Soc. 40: 513–521.

Becker, E.W. 2004. Large scale cultivation. pp. 63–171. *In*: Microalgae—Biotechnology and Microbiology. Cambridge University press, Cambridge, UK.

Belarbi, E.H., E. Molina and Y. Chisti. 2000. A process for high yield and scaleable recovery of high purity eicosapentaenoic acid esters from microalgae and fish oil. Enzyme Microb. Technol. 26: 516.

Bilanovic, D., G. Shelef and A. Sukenik. 1988. Flocculation of microalgae with cationic polymers: Effects of medium salinity. Biomass 17: 65–76.

Boussiba, S., X.Q. Wu, E. Ben-Dov, A. Zarka and A. Zaritsky. 2000. Nitrogen-fixing cyanobacteria as gene delivery system for expressing mosquiticidal toxins of *Bacillus thuringensis* ssp. *israelensis*. Journal of Applied Phycoogy 12: 461–467.

Brennan, L. and P. Owende. 2010. Biofuels from microalgae—a review of technologies for production, processing, and extractions of biofuels and co-products. Renewable and Sustainable Energy Reviews 14: 557–577.

Chen, Y.M., J.C. Liu and Y. Hsu Ju. 1998. Flotation removal of algae from water. Colloids and Surfaces Biointerfaces 12: 49–55.

Chisti, Y. 1999. Shear sensitivity. pp. 2379–406. *In*: M.C. Flickinger and S.W. Drew (eds.). Encyclopedia of Bioprocess Technology: Fermentation, Biocatalysis, and Bioseparation, vol. 5. New York: Wiley Publications, USA.

Chisti, Y. 2007. Biodiesel from microalgae. Biotechnol. Adv. 25: 294–306.

Danquah, M.K., B. Gladman, N. Moheimani and G.M. Forde. 2009. Microalgal growth characteristics and subsequent influence on dewatering efficiency. Chem. Eng. J. 151: 73–78.

De Morais, M.G. and J.A.V. Costa. 2007. Isolation and selection of microalgae from coal fired thermoelectric power plant for biofixation of carbon dioxide. Energy Conversion and Management 48: 2169–2173.

Dijkstra, A.J. 2006. Revisiting the formation of trans-isomers during partial hydrogenation of triacylglycerol oils. Eur. J. Lipid Sci. Tech. 108: 249.

Durmaz, Y. 2007. Vitamin E (a-tocopherol) production by the marine microalgae *Nannochloropsis oculata* (Fustigmatophyceae) in nitrogen limitation. Aquacul. 272: 717–722.

Fujii, K., H. Nakashima, Y. Hashidzume, T. Uchiyama, K. Mishiro and Y. Kadota 2010. Potential use of the astaxanthin-producing microalga, *Monoraphidium* sp. GK12, as a functional aquafeed for prawns. J. Appl. Phycol. 22: 363–369.

Gagneux-Moreaux, S., C. Moreau, J.L. Gonzalez and R.P. Cosson. 2007. Diatom artificial medium (DAM): a new artificial medium for the diatom, *Haslea ostrearia* and other marine microalgae. J. Appl. Phycol. 19: 549–556.

Garcia, J., R. Mujeriego and M. Hernandez. 2000. High rate algal pond operating strategies for urban wastewater nitrogen removal. Journal of Applied Phycology 12: 331–339.

Gentsch, E., T. Kreibich, W. Hagen and N. Barbara. 2009. Dietary shifts in the copepod *Temora longicornis* during spring: evidence from stable isotope signatures, fatty acid biomarkers and feeding experiments. J. Plankton. Res. 31: 45–60.

Gleason, F.K. and J.L. Paulson. 1984. Site of action of the natural algicide, cyanobacterin, in the blue-green alga, *Synechococcus* sp. Archives in Microbiology 138: 273–277.

Gleason, F.K. and C.A. Baxa. 1986. Activity of the natural algicide, cyanobacterin, on eukaryotic microorganisms. FEMS Microbiology Letters 33: 85–88.

Golueke, C.G. and W.J. Oswald. 1965. Harvesting and processing sewage grown planktonic algae. J. Water Pollution Control. Federation 37: 471–98.

Gonzales, L.E., R.O. Canizares and S. Baena. 1997. Efficiency of ammonia and phosphorus removal from a Colombian agroindustrial wastewater by the microlagae *Chlorealla vulgaris* and *Scenedesmus dimorphus*. Bioresource Technol. 60: 259–262.

Gouveia, L. and A.C. Oliveira. 2008. Microalgae as a raw material for biofuels production. J. Ind. Microbiol. Biotechnol. 36: 269–274.

Grima, E.M., E.H. Belarbi, F.G. Fernandez, R. Medina and Y. Chisti. 2003. Recovery of microalgal biomass and metabolites: process options and economics. Biotechnology Advances 20: 491–515.

Grobbelaar, J.U. 2013. Inorganic algal nutrition. Handbook of microalgal culture: Applied Phycology and Biotechnology, Chapter 8, second edition. Editors, Amos Richmond and Qiang Hu, Wiley Publications.

Gross, E., P.C. Wolk. and F. Jüttner. 1991. Fischerellin, a new allelochemical from the freshwater cyanobacterium *Fischerella muscicola*. Journal of Phycology 27: 686–692.

Guerin, M., M.E. Huntley and M. Olaizola. 2003. *Haematococcus* sp. astaxanthin; application for human health and nutrition. Trends Biotechnol. 21: 210–216.

Guschina, I.A. and J.L. Harwood. 2006. Lipids and lipid metabolism in eukaryotic algae. Prog. in Lipid Res. 45: 160–186.

Hagmann, L. and F. Jüttner. 1996. Fischerellin A, a novel Photosystem-II-inhibiting allelochemical of the cyanobacterium *Fischerella muscicola* with antifungal and herbicidal activity. Tetrahedron Letters 37: 6539–6542.

Harwood, J.L. and I.A. Guschina. 2009. The versatility of algae and their lipid metabolism. Biochimie 91(6): 679–684.

Hemaiswarya, S., R. Raja, R. Ravikumar, V. Ganesan and C. Anbazhagan. 2011. Microalgae: A sustainable source for feed in aquaculture. World Journal of Microbiology and Biotechnology 27: 1737–1746.

Hodaifa, G., E. Martinez and S. Sanchez. 2008. Use of industrial wastewater from olive-oil extraction for biomass production of *Scenedesmus obliquus*. Bioresource Technology 99: 1111–1117.

Hoffman, J.P. 1998. Wastewater treatment with suspended and non suspended algae. J. Phycol. 34: 757–763.

Hong, H.A., H.L. Duc and S.M. Cutting. 2005. The use of bacterial spore formers as probiotics. FEMS Microbiol. Rev. 29: 813–835.

Huang, H.P., Y. Shi, W. Li. and S.G. Chang. 2001. Dual alkali approaches for the capture and separation of CO_2. Energy & Fuels 15: 263–268.

Hussain, K., K. Nawaz, A. Majeed and F. Lin. 2010. Economically effective potential of algae for biofuel production. World Appl. Sci. J. 9: 1313–1323.

International Trade Administration. 2009. Energy project financing, energy projects in developing countries. Available at http://www.ita.doc.gov/td/finance/publications/Energy.

Jang, E.S., M.Y. Jung and D.B. Min. 2005. Hydrogenation for low trans and high conjugated fatty acids. Compr. Rev. Food Sci. Food Safety 4: 22.

Kadam, K.L. 2001. Microalgae Production from Power Plant Flue Gas: Environmental Implications on a Life Cycle Basis. National Renewable Energy Laboratory, technical report.

Kang, C.D. and S.J. Sim. 2008. Direct extraction of astaxanthin from *Haematococcus* culture using vegetable oils. Biotechnol. Lett. 30: 441–444.

Kaya, V.M. and G. Picard. 1996. Stability of chitosan gel as entrapment matrix of viable *Scenedesmus bicellularis* cells immobilized on screens for tertiary treatment of wastewater. Bioresource Technology 56: 147–55.

Kim, S.B., S.J. Lee, C.K. Kim, G.S. Kwon, B.D. Yoon and H.M. Oh. 1998. Selection of microalgae for advanced treatment of swine wastewater and optimization of treatment condition. Korean Journal of Applied Microbiology and Biotechnology 26: 76–82.

Knauer, J. and P.C. Southgate. 1999. A review of the nutritional requirements of bivalves and the development of alternative and artificial diets for bivalve aquaculture. Rev. Fish. Sci. 7: 241–280.

Knothe, G., J.V. Gerpen and J. Krahl. 2005. The Biodiesel Handbook. American Oil and Chemist's Society, Urbana.

Knuckey, R.M., M.R. Brown, S.M. Barrett and G.M. Hallegraeff. 2002. Isolation of new nanoplanktonic diatom strains and their evaluation as diets for the juvenile Pacific oyster (*Crassostrea gigas*). Aquaculture 211(23): 253–274.

Kull, A. 1962. Physiology and biochemistry of algae. pp. 211–229 *In*: R.A. Lewin (ed.). Academic Press, New York, USA.

Liang, H., W.-J. Gong, Z.-L. Chen, J.Y. Tian, L. Qi and G.-B. Li. 2009. Effect of chemical preoxidation coupled with in-line coagulation as a pretreatment to ultrafiltration for algae fouling control. Desalination Water Treat 9: 241–245.

López Elías, J.A., D. Voltolina, C.O. Chavira Ortega, B.B. Rodríguez Rodríguez, LM. Sáenz Gaxiola, B.C. Esquivel and M. Nieves. 2003. Mass production of microalgae in six commercial shrimp hatcheries of the Mexican northwest. Aquacultural. Eng. 29:155–164.

Lorenz, R.T. and G.R. Cysewski. 2000. Commercial potential for *Haematococcus* microalgae as a natural source of astaxanthin. Trends in Biotechnol. 18: 160–167.

Mallick, N. and L.C. Rai. 1994. Removal of inorganic ions from waste water by immobilized microalgae. World J. Microbiol. Biotechnol. 10: 439–443.

Martinez, M.E., S. Sanchez, J.M. Jimenez, F. El Yousfi and L. Munoz. 2000. Nitrogen and phosphorus removal from urban wastewater by the microalga *Scenedesmus obliquus*. Bioresource Technol. 73: 263–272.

Muller-Feuga, A. 2000. The role of microalgae in aquaculture: situation and trends. J. Appl. Phycol. 12: 527–534.

Munoz, R. and B. Guieysse. 2006. Algal-bacterial processes for the treatment of hazardous contaminants: a review. Water Res. 40: 2799–815.

Olguin, E.J., S. Galicia, G. Mercado and T. Perez. 2003. Annual productivity of *Spirulina* (*Arthrospira*) and nutrient removal in a pig wastewater recycle process under tropical conditions. J. Appl. Phycol. 15: 249–257.

Omer, A.M. 2012. Renewable energy, emerging energy technologies and sustainable development: Innovation in power, control and optimisation. ARPN J. Sci. Technol. 2: 13.

Oswald, W.J. 1988. Microalgae and wastewater treatment. pp. 305–328. *In*: M.A. Borowitzka and L.J. Borowitzka (eds). Microalgal Biotechnology. Cambridge: Cambridge University Press.

Park, J.-H., J.-J. Yoon, H-D. Park, Y.J. Kim, D.J. Lim and S.-H. Kim. 2011. Feasibility of biohydrogen production from *Gelidium amansii*. Int. J. Hydrogen. Energ. 36(21): 13997–14003.

Patil, V., K.-Q. Tran and H.R. Giselrød. 2008. Towards sustainable production of biofuels from microalgae. Int. J. Mol. Sci. 9: 1188–1195.

Spolaore, P., J.-C. Claire, Elie Duran and I. Arsène. 2006. Commercial applications of microalgae. Journal of Bioscience and Bioengineering Biotechnology 101: 87–96.

Perales-Vela, H.V., J.M. Pena-Castro and R.O. Canizares-Villanueva. 2006. Heavy metal detoxification in eukaryotic microalgae. Chemosphere 64: 1–10.

Philip, T., P.L. Laurens and A. Aden. 2011. Making Biofuel from Microalgae. So much potential coexists with so many scientific, environmental and economic challenges. American Scientist, 6.

Ponis, E., I. Probert, B. Véron, M. Mathieu and R. Robert. 2006. New microalgae for the Pacific oyster, *Crassostrea gigas* larvae. Aquacul. 253: 618–627.

Ponis, E., R. Robert and G. Parisi. 2003. Nutritional value of fresh and concentrated algal diets for larval and juvenile Pacific oysters (*Crassostrea gigas*). Aquacul. 221: 491–505.

Powell, N., A.N. Shilton, S. Pratt and Y. Chisti. 2008. Factors influencing luxury uptake of phosphorus by microalgae in waste stabilization ponds. Environ. Sci. Technol. 42: 5958–5962.

Pulz, O. and K. Scheibenbogen. 1998. Photobioreactors: Design and performance with respect to light energy input. Adv. Biochem. Eng. Biotechnol. 59: 123–151.

Pulz, O. 2001. Photobioreactors: production systems for phototrophic microorganisms. Appl. Microbiol. Biotechnol. 57: 287–293.

Raja, R., C. Anbazhagan, V. Ganesan and R. Rengasamy. 2004. Efficacy of *Dunaliella salina* (Volvocales, Chlorophyta) in salt refinery effluent treatment. Asian Journal of Chemistry 16: 1081–1088.

Raja, R., S. Hemaiswarya and R. Rengasamy. 2007c. Exploitation of *Dunaliella* for β-carotene production. Applied Microbiology and Biotechnology 74: 517–523.

Raja, R., S. Hemaiswarya, N. Ashok kumar, S. Sridhar and R. Rengasamy. 2008. A perspective on the biotechnological potential of microalgae. Critical Reviews in Microbiology 34: 77–88.

Raja, R. and S. Hemaiswarya. 2010. Microalgae and immune potential a chapter in dietary components and immune function–prevention and treatment of disease and cancer. pp. 517–529. *In*: R.R. Watson, S. Zibadi and V.R. Preedy (eds.). Humana Press/Springer, ISBN: 978-1-60761-060-1, USA.

Rathinam Raja, Shanmugam Hemaiswarya, Dakshanamoorthy Balasubramanyam and Ramasamy Rengasamy. 2007b. Protective effect of *Dunaliella salina* (Volvocales, Chlorophyta) on experimentally induced fibrosarcoma on wistar rats. Microbiological Research 162: 177–184.

Richmond, A. 2004. Basic Culturing Techniques & Downstream processing of cell-mass and products. *In*: Handbook of Microalgal Culture-Biotechnology & Applied Phycology. Blackwell publishing, Oxford, UK 40–55, 83–93 & 215–252.

Sancho, M.E., J.M. Jimenez Castillo and F. Yousfi. 1997. Influence of phosphorus concentration on the growth kinetics and stoichiometry of the microalga *Scenedesmus obliquus*. Process Biochemistry 32: 657–664.

Sanderson, G.W. and S.O. Jolly. 1994. The value of Phaffia yeast as a feed ingredient for Salmonid fish. Aquaculture 124: 193–200.

Sawayama, S., S. Inoue, Y. Dote and S.Y. Yokoyama. 1995. CO_2 fixation and oil production through microalga. Energy Convers Mgmi. 36: 729–731.

Schenk, P.M., S.R. Thomas-Hall, E. Stephens, U.C. Marx, J.H. Mussgnug, C. Posten, O. Kruse and B. Hankamer. 2008. Second generation biofuels: High-efficiency microalgae for biodiesel production. Bioenergy. Res. 1. 20–43.

Schlesinger, W.H. 1991. Biogeochemistry: An analysis of global change. 2nd ed. Academic press, California, USA.

Tchobanoglous, G., F.L. Burton and H.D. Stensel. 2003. Constituents in waste water. pp. 27–64. *In*: Wastewater Engineering-Treatment and reuse. 4th ed. Metcalf & Eddy, Inc. Tata McGraw-Hill Publishing Co. Ltd., USA.

Tenney, M.W., W.F. Echelberger, R.G. Schuessler and J.L. Pavoni. 1969. Algal flocculation with synthetic organic polyelectrolytes. Appl. Bacteriol. 18: 965–71.

Uduman, N., Y. Qi, M.K. Danquah, G.M. Forde and A. Hoadley. 2010. Dewatering of microalgal cultures: A major bottleneck to algae-based fuels. Journal of Renewable and Sustainable Energy 2: 701–715.

Visviki, I. and J.W. Rachlin. 1991. The toxic action and interactions of copper and cadmium to the marine alga, *Dunaliella minuta*, in both acute and chronic exposure. Arch. Environ. Contam. Toxicol. 20: 271–275.

Wikfors, G.H. and M. Ohno. 2001. Impact of algal research in aquaculture. J. Phycol. 37: 968–974.

Yarden, O., M. Freund and B. Rubin. 1993. *Dunaliella salina*: a convenient test organism for detection of pesticide residues in water and soil. Fresenius Environ. Bull. 2: 31–36.

Advances and Constraints in Seaweed Farming as a Basal Constituent for the Environmentally Sustainable Development of Aquaculture in Chile

Alejandro H. Buschmann,* María C. Hernández-González and
Bárbara S. Labbé

Introduction

Aquaculture development worldwide is growing continuously as an alternative to exacerbating the depletion of increasingly scarce fisheries resources (Duarte et al. 2007, 2009). In order to meet fisheries requirements around the world, considerable effort is being made in the technological and productive fields to further aquaculture development with the aim of expanding activities to include geographical zones with more extreme climatic conditions such as in the open sea (Troell et al. 2009). On the other hand, at present, aquaculture has come under strict public scrutiny in terms of ensuring compatibility between economic/social development and conservation of the environmental patrimony (Costa-Pierce 2010; Diana et al. 2013). Thus, the growing interest in world aquaculture potential and its expansion, is generating debate with regard to: (a) environmental issues; (b) the increase in the demand for resources necessary for production (principally fish meal and oil) (see Naylor et al. 2001; Olsen 2011); as well as, (c) concern over various environmental consequences generated by inorganic waste (for example, the accumulation of copper in bottom sediments; see Haya et al. 2001; Buschmann and Fortt 2005 for Chile), organics (increase of organic material and decrease of oxygen in bottom sediments below floating cages; Hargrave 2010, and for Chile, see Soto and Norambuena 2004), or biological factors, such as the increase in parasites (Carvajal et al. 1998; Sepúlveda et al. 2004) and fish escapes causing negative

Centro i-mar & CeBiB, Universidad de Los Lagos, Camino Chinquihue km 6, Puerto Montt, Chile.
* Corresponding author: abuschma@ulagos.cl

effects on the native fauna (Soto et al. 2001, 2006; Becker et al. 2007). In Chile, these aspects have also been the center of debate on the relationship between aquaculture and the environment (Buschmann et al. 2006a), resulting in a search for alternatives that contribute to reducing potentially adverse environmental externalities (Buschmann et al. 2009).

From a production efficiency perspective, in terms of use of the environment for aquaculture activities, it has been calculated that the area required to obtain all the resources necessary to maintain the productive cycle and, furthermore, to assimilate waste material produced—the "ecological footprint" —ranges from values of around 4–5 Hectares per cultivated hectare, to values as high as 50,000 hectares for each cultured hectare (Folke et al. 1998). This variability depends, principally, on the culture methods used and the species cultured. Thus, in general, carnivorous organisms have a greater ecological footprint, by one or two orders of magnitude, than extractive organisms (filterers and algae), which do not require an exogenous energetic supplement in the environment (Folke and Kautsky 1989; Folke et al. 1998). Based on this information, installation of waste recycling systems in aquaculture sites, that is, using different types of organisms that remove these harmful elements from the ecosystem has been proposed (Troell et al. 1999; Buschmann et al. 2008a; Chopin et al. 2008). These integrated aquaculture systems, using multiple species with different trophic levels, are referred to as Integrated Multi-Trophic Aquaculture (IMTA) (Chopin et al. 2001). Nevertheless, as will be mentioned later in this study, a critical evaluation is necessary as to why these systems have not been developed in Chile, in spite of the significant levels of aquaculture development reached over the past few decades (Buschmann et al. 2009).

Given this environmental context, this study outlines the function of algal cultures as organisms that extract inorganic elements, such as carbon, phosphor, and nitrogen, thus favoring coastal ecosystem health. At present, certain coastal eutrophication processes are related to the appearance of algal blooms, both of micro and macroalgae, constituting one of the biggest environmental problems affecting coastal zones globally (Clarke et al. 2006; Conley et al. 2009). Eutrophication is directly associated with emissions of inorganic elements, such as nitrogen compounds (nitrate and ammonium), and these compounds originate from urban waste, agricultural activities, and deforestation, and reach coastal zones transported by fresh water bodies, in addition to those produced by the aquaculture activity itself (Anderson et al. 2002; Buschmann et al. 2006a; Liu et al. 2012). In this study, we present an analysis of aquaculture development, emphasizing the recent developments in Chile, especially associated to advantages and restrictions of the incorporation of algae generating new business opportunities, while ensuring a more sustainable aquaculture by minimizing the effects produced by an excess of inorganic nutrients in coastal zones. The results of previously published studies have provided an exhaustive account of progress made and future challenges of algal culture in Chile, and, as a result, this line of discussion is not included in this study (see revision Buschmann et al. 2008b).

Recent aquaculture development in Chile

Marine aquaculture increased in Chile from a total production of 361,000 t in 1998 to over 870,000 t throughout 2010 (Buschmann et al. 2013). Globally, this situated Chile among the 10 largest producers and as the number one producer of marine aquaculture in the western hemisphere. During the same period, salmon culture became the principal aquaculture activity in Chile, reaching peak production levels of 630.647 t in 2006, subsequently experiencing a decline (Buschmann et al. 2013) as a result of the ISA virus (Godoy et al. 2008). The combined total of mollusk production (mainly abalones, mussels, scallops, and oysters), reached 212,210 t in 2010, the main product being muss mussels (Buschmann et al. 2013).

In the case of the exploited algal species (Fig. 1), have been increasing in time specially for different brown algae reaching above 272,000 and 313,000 t in 2010 and 2011 respectively (Table 1). Red algae reached an exploitation value of 104,000 t in 2011 (Table 1). During the last few years, commercial production of microalgae is developing with a total biomass production of over 400,000 t (Table 1). Farmed algal statistics for 2001 showed that only *Gracilaria chilensis* C. J. Bird, J. McLachlan, & E.C. Oliveira was cultivated in Chile, but in 2010 statistics indicated that in addition microalgae and the brown alga *Macrocystis pyrifera* (L.) C. Agardh are adding some biomass to the total algal farming in the country (Table 2; Fig. 2). Overall, this algal biomass contributes to the production of agar, carrageenan,

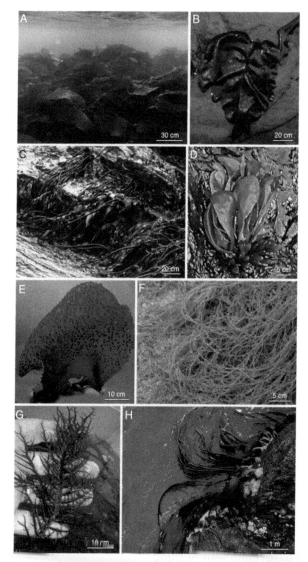

Fig. 1. Photographs of the most important seaweed species exploited and farmed in Chile. (A) *Macrocystis pyrifera*, (B) *Sarcothalia crispata*, (C) *Lessonia nigrescens*, (D) *Mazzaella laminarioides*, (E) *Gigartina skottsbergii*, (F) *Gracilaria chilensis*, (G) *Chondrachanthus chamissoii*, and (H) *Durvillaea antarctica*.

and alginates, and few other less valued products for exportation (Table 3). Algal based exportation has reached economic revenue values above 100 US$ per year for the Chilean economy.

The above-cited landings and algae cultivation values highlight the algal production potential along the Chilean coast. Specifically, the culture of the alga *Gracilaria chilensis* continues to be predominant in aquaculture landings (> 90% biomass), in spite of the fact that, at present, the technological knowhow exists to initiate aquaculture activities for various species of brown algae (e.g., *Macrocystis pyrifera*) and red algae (*Sarcothalia crispata* (Bory) Leister, *Gigartina skottsbergii* Setchell & Gardner, *Chondracanthus chamissoi* (C. Agardh) Kützing) (see revisions of Santelices 1996; Buschmann et al. 2001, 2008b, 2013).

This increased productivity has been accompanied by a progressively more intensive use of coastal zones, with aquaculture activities expanding from the Los Lagos Region to the Aysén Region (see map in Buschmann et al. 2006a). Although in many situations, mussel culture is classified as either extensive, or small-scale, according to legislation in force, it is evident that, as a result of the increased activity,

Table 1. Total algal landings in Chile over the last decade in selected years (2001, 2006, 2010, and 2011). The algal groups to which each species belong are provided in parenthesis: C = Chlorophyta; CY = Cyanophyta; R = Rhodophyta, and B = Phaeophyta. *Source of the landing data.

Species/years	Total Landing (Tones)			
Microalgae	**2001**	**2006**	**2010**	**2011**
Haematococcus pluvialis (C)	-----	1444	12	5
Spirulina sp. (CY)	-----	3189	5	22
Macroalgae	5	17	-----	-----
Callophyllis variegata (R)	402	310	209	222
Gelidium rex (R)	87508	161834	190746	241633
Lessonia nigrescens (B)	3325	1590	924	998
Chondracanthus chamissoi (R)	2098	2292	6048	6468
Durvillaea antárctica (B)	9672	9319	11735	19400
Macrocystis pyrifera (B)	18457	27552	62734	46239
Lessonia trabeculata (B)	-----	27	-----	-----
Ulva lactuca (C)	-----	215	-----	-----
Gymnogongrus furcellatus (R)	-----	4	16	41
Porphyra columbina (R)	-----	3731	1172	2096
Mazzaella laminarioides (R)	-----	17135	30194	29559
Sarcothalia crispata (R)	22717	8	19725	14616
Gigartina skottsbergii (R)	37606	8	-----	-----
Mazzaella membranacea (R)	117969	77336	52239	56732
Gracilaria chilensis (R)	32	-----	-----	-----
Non-classified algae				
Total	299,791	306,011	375,759	418,031

*http://www.sernapesca.cl/index.php?option=com_remository&Itemid=54&func=select&id=2.

Table 2. Relative (%) aquaculture contribution to total algal landings of Chile over the last decade in selected years: 2001, 2006, 2010, and 2011. *Source of aquaculture sector.

Species/year	Percentage (%)			
	2001	2006	2010	2011
Haematococcus pluvialis	-----	100	100	100
Macrocystis pyrifera	-----	-----	0.10	-----
Gracilaria chilensis	55.55	43.43	23.25	25.6
Spirulina sp.	-----	100	100	100

*http://www.sernapesca.cl/index.php?option=com_remository&Itemid=54&func=select&id=2.

Table 3. Total production of processed products over the last decade in selected years (2001, 2006, 2010, and 2011) in Chile. *Source of the data.

Products/years	Production (metric tones)			
	2001	2006	2010	2011
Agar	18	2.531	512	82
Dry algae	758	45.153	68.218	75.293
Dehydrated algal biomass	0	0	0	0
Alginate	568	-----	0	0
Carrageenan	205	751	878	307
Colagar	0	-----	5.981	10.745
Total	1549	798.68	1464.2	475.04

*http://www.sernapesca.cl/index.php?option=com_remository&Itemid=54&func=select&id=2.

Fig. 2. Photographs showing seaweed exploitation (*Lessonia* and *Sarcothalia*) of natural stocks and farming (*Gracilaria* and *Macrocystis*) in Chile.

both in terms of number of centers and biomass loads, in certain zones of southern Chile, it can no longer be considered a small-scale activity from an environmental point of view. This situation has given rise to calculations that the environmental cost, only as a result of salmon culture, reaches at least 30% of the Gross Domestic Product of fisheries activity in Chile (Buschmann and Pizarro 2001). This cost is associated mainly with cultures of high trophic level species (carnivores) that require a supply of exogenous energy (feed) that produces elevated concentrations of nutrients entering the system, even in situations where food conversion efficiency is high. It has been determined that the recuperation of inorganic dissolved wastes by harvesting the organism cultured does not exceed 75% for nitrogen or 60% of the carbon (see, as an example, mass balance in Buschmann et al. 1996a). Although nitrogen-use efficiency is comparatively greater in salmon cultures than in cattle farms, salmon activity in Chile introduces more nitrogen into the fresh and marine aquatic systems than cattle activities (Soto et al. 2007). Thus, we observe an environmental context related to this activity, where a variety of organic and inorganic waste materials are introduced into aquatic systems, with multiple and complex environmental effects (see more details in Buschmann et al. 2009). Although Chile is among the principal marine aquaculture producers worldwide, Chile's productions levels differ considerably from those of oriental countries; this is because Chile produces, predominantly, carnivorous organisms (Buschmann et al. 2009).

Evidently, the significant aquaculture activity that sustains countries such as Japan, Korea, or China, must be based on models that balance organic and inorganic waste generated by certain organisms, with organisms that enable this waste to be recycled and which, when harvested, facilitate extraction of these waste materials from the environment (Shi et al. 2011). Thus, the concept of IMTA acquires particular relevance as a means of reducing the environmental risks. In this context, algae play a vital role in reducing volumes of dissolved inorganic nutrients (Troell et al. 1999; Chopin et al. 2001, 2008, 2011; Buschmann et al. 2008a).

The effects produced by aquaculture vary significantly, generating different environmental situations that may be detrimental to the aquaculture activity itself, and that must be identified and resolved. The introduction of nitrogen in coastal zones is one of these negative effects and can induce certain eutrophication processes that favor the proliferation of green algae (green tides), as occurred in China, prior to the Beijing Olympics (Liu et al. 2012). In Chile, this association has not been verified; however, proliferations of green algae in bays with intensive aquaculture activities can present these blooms and overgrowth on other coastal organisms (see Buschmann et al. 2013). Decomposition of these algal blooms increases the oxygen demand resulting in loss of biodiversity, which can, eventually, decrease the productive potential of aquaculture activities (Peckol and Rivers 1996). Many salmon culture centers experience oxygen problems during certain seasons of the year. This phenomenon may not produce mortality directly, but can induce stressful conditions to such an extent that they provoke reduced immunological responses in fish, resulting in increased pathologies, which in turn, produce mortalities of up to 50% (Bravo and Midtlying 2007; Bustos et al. 2011).

It is not possible to resolve the environmental problems described in the preceding paragraph by merely increasing the use of drugs to control pathogens and parasites; an understanding of how the cultured organisms interact with the environment is also required. Although the elevated level of antibiotics used in Chilean aquaculture (Cabello et al. 2013) is related to the type and frequency of diseases, the high dosages can also affect natural populations of coastal organisms (Fortt et al. 2007) and increase resistance to antibiotics (Chelossi et al. 2003; Nonaka et al. 2007; Heuer et al. 2009; Buschmann et al. 2012). Triggering epidemiological responses, such as the case of the ISA virus or other pathologies, that generate an enormous economic and social impact, also implies a very significant environmental component that cannot be ignored (Mardones et al. 2009). The problem cannot be alleviated solely by veterinary solutions and the application of different therapeutics; prevalent environmental variables must also be clearly defined. As a consequence, the solution lies in obtaining independent scientific information that permits, on the one hand, clear and transparent identification of factors responsible for inducing and transferring pathogens (natural and artificial) and the environmental problems involved. Therefore, the development of production strategies is oriented towards reducing environmental consequences associated with the culture of exclusively animal species, as well as reducing the production of inorganic nitrogen and the resulting coastal eutrophication effects.

Considerable effort has been made in Chile to implement legislation that will resolve the most pressing problems. Nevertheless, the question arises as to how the regulation is implemented and whether it actively pursues and innovates in a context of establishing production models compromised with achieving greater environmental sustainability. This regulation (RAMA; Aquaculture Environmental Regulation) is based principally on the monitoring of sediments below the culture systems, using detection of anaerobiosis (absence of oxygen according to the original normative), or low oxygen concentration (according to the present normative), as a signal that generates contingency procedures (Buschmann et al. 2013). Various efforts have been carried out to maintain a regulation that is up to date and several important modifications have been undertaken from 2005 to date (see Buschmann et al. 2013 for a more comprehensive discussion). We believe that these topics require the express attention of both the governmental sector and the productive sector. Evidently, detection of anoxic conditions is not a sufficient indicator of environmental sustainability; this brings into question the quality of the current regulation and does not significantly promote the search for solutions and technological alternatives that secure a higher degree of sustainability for this activity (Buschmann et al. 2006b). Today, technological solutions exist to deal with these issues (see Buschmann et al. 2008c), but still, improvements of management practices to create more efficient are required (Diana et al. 2013). Nevertheless, the clear and transparent identification and definition of environmental problems is necessary in order to apply concrete and feasible solutions to each one of them. If these issues are addressed, Chile will be able to integrate coherently into a global market that is progressively more demanding environmentally. In the following section, we will deal with these issues in more depth, determine existing restrictions and challenges affecting the development of more sustainable aquaculture, and evaluate the importance of using algae to achieve this aim.

Future of seaweed farming in Chile

Although phycocolloid markets continue to expand on a global scale, growth over the last decade was much lower than 20 years ago (Bixler and Porse 2011). During the past years, there has been a strong competition for new markets affecting the value of seaweed and their associated products. Also, the capacity of China and other oriental countries to produce *Gracilaria* biomass lower a lower costs, has limit algal culture expansion in Chile. Only by discovering new applications that increase the aggregate value of phycocolloids, such as alginates in the production of more efficient batteries (Kovalenko et al. 2011), or other innovative uses, will necessity for algal culture increase, raising demand and improving prices (see other examples in Kim 2012). In addition to the traditional use of seaweed as a source of phycocolloids, the demand for seaweed, in particular *Macrocystis* as an ingredient in diet formulations for animals, such as the case of abalone cultures (Flores-Aguilar et al. 2007), has generated greater extraction restrictions for certain species in Chile (Vásquez 2008). On the other hand, the use of algal biomass for the production of biofuel is becoming a new energy alternative on a global level (e.g., Chisti 2008; Wargacki et al. 2012). Nevertheless, although technology must be developed that permits industrialized culture at a cost that is competitive with market prices and other biofuel sources (Maceiras et al. 2011), acceptable levels of energetic efficiency must also be ensured (Walker 2009; Clarens et al. 2010; Marquardt 2011) in order to develop an algal culture that is not only economically profitable, but also achieves positive environmental externalities. Studies presented in this paragraph show how different uses for algae are emerging, creating new demand for algal biomass, which should also improve the profitability of aquaculture practices in the near future.

From another perspective, development of algal culture is associated with its use as a bioremediator in eutrophication processes of coastal zones where aquaculture activities are undertaken, now referred to as Integrated Multi-Trophic Aquaculture (Chopin et al. 2008; Buschmann et al. 2008a). As an example, we cite the fact that an algal culture is capable of reducing nitrogen emissions released into the environment by 85% during an annual salmon production cycle; recognizing that this nutrient is the element that produces the greatest impact in coastal eutrophication processes (Buschmann et al. 1996a; Troell et al. 1999). In Chile, various studies have been carried out on integrated multi-trophic aquaculture, testing the capacity of red and brown algae to remove nitrogen and inorganic phosphorus (Troell et al. 1997; Buschmann et al. 2008c; Abreu et al. 2009). Efficiency in the removal of dissolved inorganic material requires, in relative terms, a large surface area of algae with respect to the surface area used by fish (1 hectare of fish culture activities will require at least 100 hectares of *Macrocystis* to remove 80% of the dissolved nitrogen inputs produced by the fish culture, Buschmann et al. 2008c); this relationship is based, principally on the fact that fish culture occupies a given volume that is determined by cages that reach depths of 10 or more meters, while, in the case of the algae, productivity depends on solar radiation and, thus, use of the water column is very limited (Abreu et al. 2009). Many studies have been carried out in the recent past, but still many key aspects towards the production and nutrient removal optimization requires attention (Troell et al. 2003).

A regulatory policy framework is required that permits the promotion of the use of macroalgae for bioremediation. This aspect is a key to allow that algal farming is perceived as an economically sustainable environmental alternative (Chopin et al. 2001; Neori et al. 2007). Environmental valorization of algal culture can be undertaken by internalization of the environmental costs (Buschmann et al. 1996b), or the creation of culture incentives (Buschmann et al. 2008a) still remains to be developed. As the ecological efficiency, sustainability and economics of culturing carnivorous fish are improved by growing them in an ecological balance with species from low trophic levels in IMTA conditions (Neori and Nobre 2012) a new regulatory framework seems necessary.

Although, some traditional activities have less economic vitality than in the past; new productive alternatives have emerged (see previous paragraphs) that should enhance the development of algal culture in Chile and other western countries. Production technologies are available for a variety of algae in Chile, development of algal culture is still minimal, with low levels of diversification; this would appear to be associated with a demand characterized by lower growth rates (Bixler and Porse 2011) and unattractive prices (Buschmann et al. 2008b). Thus, it seems that, without some kind of incentive that encourages changes in the structure of aquaculture production, progress will be difficult. On the other hand, from the environmental point of view, the absence of economic incentives to reduce dissolved nutrient loads in the water through use of macroalgae would also imply that such measures will not be implemented. In the meantime, the only significant demand for biomass that can be identified is the potential of certain species, especially *Macrocystis,* to be used for the production of bioethanol (Wargacki et al. 2012).

The Environmental Regulation for Aquaculture (Reglamento Ambiental de Acuicultura; RAMA), regulates part of Chile's aquaculture activities (Buschmann et al. 2013). This was approved in 2001 in the context of the General Law on Fisheries and Aquaculture (Ley General de Pesca y Acuicultura; LGPA). Since its approval in 2001, this regulation has been modified, from 2005 onwards, with the intention of adjusting the norm to its practical application. The RAMA aims to establish criteria that define the concept of "environmental damage", as well as methodologies for preliminary characterization of culture site (PCS) and environmental information (EINF), in order to evaluate (initially) and monitor (over time), environmental indicators aimed at determining whether or not the loading capacity of the water body subject to aquaculture activity has been exceeded. Thus, it will be possible to establish when the sedimentation area of the productive installations presents anaerobic conditions, equivalent to a situation where sediment or water column variables have been exceeded, i.e., $O_2 \leq 2.4$ mg L^{-1}, organic material content $\geq 9.1\%$, pH ≤ 6.7, and the absence of coats of visible microorganisms and gas bubbles. By categorizing culture centers, this regulation recognizes differences in the degree of impact generated by different aquaculture activities, according to the biomass produced, or whether they require external supply of food or fertilizers. Nevertheless, the present normative does not make explicit reference to the role of the dissolved inorganic nutrient input in environmental degradation and its effect on reducing the loading capacity. As large areas of mollusks and/or macroalgae cultivated in suspended systems, or anchored to the bottom, can reduce current velocity and, thus, increase the sedimentation rate (e.g., Buschmann et al. 1997; McKindsey et al. 2011), regulation should, contemplate appropriate measurements and indicators suitable for the massive farming of this organisms, considering species, type of culture system, and the production scale that are different than for finfish cage farming. Thus, these mentioned regulation issues could be perceived as an aspect (in addition to the commercial factors previously mentioned) that impedes seaweed aquaculture development in Chile. On the other hand, algal aquaculture must also be undertaken under conditions that minimize its environmental impact. Clearly, this normative is biased, associated with the fact that, at present, only *Gracilaria chilensis* is cultured on a commercial scale (Buschmann et al. 2001, 2013). As has been commented on previous occasions, development of culture of other species, using different production systems, is envisaged; this is case of *Macrocystis pyrifera*, where aquaculture protocols aim towards suspended production systems (Gutiérrez et al. 2006; Westermeier et al. 2006; Macchiavello et al. 2010). However, as mentioned, market factors still exist that continue to impede its development. As a consequence, it would appear necessary to incorporate the specificities of algal culture into the RAMA and aquaculture regulations in general, covering all aspects of its diversity to ensure that regulatory distortions inhibiting the future development of algal aquaculture in Chile are not produced.

Conclusions

Considering that fisheries resources are becoming increasingly more scarce, both on a national and worldwide scale (Jackson et al. 2001; Pauly et al. 2002; Worm et al. 2009), it is envisaged that the importance of aquaculture will increase significantly in the near future (Diana 2009; Hallam 2012). This will open new commercial opportunities for the global development of aquaculture (Duarte et al. 2007, 2009). However, on the one hand, it will be necessary to incorporate the explicit compromise of aquaculture activity with the sustainable management of the environment (Costa-Pierce 2010), and to focus on the development of innovative technologies to sustain this increase in production (Diana et al. 2013).

In the case of Chile, this opportunity for economic and social development cannot be sustained on the basis of a disregard for the country's environmental patrimony. Formulas and strategies must be identified, whereby productive goals can be reached in association with an explicitly described environmental component (Buschmann et al. 2009, 2013). Aquaculture diversification must be based on the incorporation of species with different ecological functions (primary producers, detritivores, herbivores, and carnivores) in order to balance the flow of material and energy in coastal systems used by aquaculture practices. In this context, generating scientific information is essential to the sustainability of aquaculture in Chile (Buschmann et al. 2009). Similarly, the technological proposals and innovations already in existence must be implemented, to resolve previously mentioned deficiencies. In particular, the ecosystemic services provided by algae must be taken into consideration, if a balance in material and energy flows is to be achieved. Incentive systems are required that encourage this activity and, thus, further progress towards the rational use of our coastal systems. The ecological efficiency, environmental sustainability, and economic profitability associated with the culture of carnivorous organisms, can be improved when the aquaculture activity is undertaken within an ecologically balanced system. This involves integrating species with a low trophic level, in accordance with the perspective of integrated multi-trophic aquaculture (Ridler et al. 2007; Neori and Nobre 2012), together with a more appropriate regulatory context. Thus, aggregate value must be generated for the algae, which will increase demand. On the other hand, efforts to internalize the environmental costs must be made to stimulate the use of technologies that reduce negative externalities and their associated costs (Buschmann et al. 1996b, 2008a; Chopin et al. 2001). If Chile tackles these issues in depth, it could export not only food products, but also sustainable production technology for world aquaculture development.

Acknowledgements

This paper is based on previous studies carried out over the past 20 years, supported, principally, by FONDECYT-Chile and Programa Basal-Conicyt (FB-001) during the past 4 years. The cooperation of Susan Angus, in the translation of the manuscript, and the text review by Matthew Lee is greatly appreciated.

References

Abreu, M.H., D.A. Varela, L.A. Henríquez, A. Villarroel, C. Yarish, I. Sousa-Pinto and A.H. Buschmann. 2009. Traditional vs. integrated multi-trophic aquaculture of *Gracilaria chilensis* C. J. Bird, J. McLachlan & E. C. Oliveira: productivity and physiological performance. Aquaculture 293: 211–220.

Anderson, D.M., P.M. Glibert and J.M. Burkholder. 2002. Harmful algal blooms and eutrophication: nutrient sources, composition, and consequences. Estuaries 25: 704–726.

Becker, L.A., M.A. Pascual and N.G. Basso. 2007. Colonization of the southern Patagonia ocean by exotic chinook almon. Conser. Biol. 21: 1347–1352.

Bixler, H.J. and H. Porse. 2011. A decade of change in seaweed hydrocolloids industry. J. Appl. Phycol. 23: 321–335.

Bravo, S. and P.J. Midtlyng. 2007. The use of fish vaccines in the Chilean salmon industry 1999–2003. Aquaculture 270: 36–42.

Buschmann, A.H., M. Troell, N. Kautsky and L. Kautsky. 1996a. Integrated tank cultivation of salmonids and *Gracilaria chilensis* (Rhodophyta). Hydrobiologia 326/327: 75–82.

Buschmann, A.H., D.A. López and A. Medina. 1996b. A review of the environmental effects and alternative production strategies of marine aquaculture in Chile. Aquacult. Eng. 15: 397–421.

Buschmann, A.H., F. Briganti and C.A. Retamales. 1997. Intertidal cultivation of *Gracilaria chilensis* (Rhodophyta) in southern Chile: long term invertebrate abundance patterns. Aquaculture 156: 269–278.

Buschmann, A.H. and R. Pizarro. 2001. El costo ambiental de la salmonicultura en Chile. Análisis de Políticas Públicas (Fundación Terram) 5: 1–7.

Buschmann, A.H., J.A. Correa, R. Westermeier, M.C. Hernández-González and R. Norambuena. 2001. Red algal farming in Chile: a review. Aquaculture 194: 203–220.

Buschmann, A.H. and A. Fortt. 2005. Efectos ambientales de la acuicultura intensiva y alternativas para un desarrollo sustentable. Ambiente y Desarrollo 21: 58–64.

Buschmann, A.H., V.A. Riquelme, M.C. Hernández-González, D.A. Varela, J.E. Jiménez, L.A. Henríquez, P.A. Vergara, R. Guíñez and L. Filún. 2006a. A review of the impacts of salmon farming on marine coastal ecosystems in the southeast Pacific. ICES J. Mar. Sci. 63: 1338–1345.

Buschmann, A.H., V.A. Riquelme, M.C. Hernández-González and L.A. Henríquez. 2006b. Additional perspectives for ecosystem approaches for aquaculture. pp. 168–176. *In*: J.P. McVey, C.-S. Lee and P.J. O'Bryan (eds.). Aquaculture and Ecosystems: An Integrated Coastal and Ocean Management Approach. The World Aquaculture Society, Baton Rouge, Louisiana.

Buschmann, A.H., M.C. Hernández-González, C. Aranda, T. Chopin, A. Neori, C. Halling and M. Troell. 2008a. Mariculture waste management. pp. 2211–2217. *In*: S.E. Jorgensen and B.D. Fath (eds.). Ecological Engineering: Vol. 3 of Encyclopedia of Ecology, Elsevier, Oxford, UK.

Buschmann, A.H., M.C. Hernández-González and D.A. Varela. 2008b. Seaweed future cultivation in Chile: perspectives and challenges. Int. J. Env. Pollut. 33: 432–456.

Buschmann, A.H., D.A. Varela, M.C. Hernández-González and P. Huovinen. 2008c. Opportunities and challenges for the development of an integrated seaweed-based aquaculture activity in Chile: Determining the physiological capabilities of *Macrocystis* and *Gracilaria* as biofilters. J. Appl. Phycol. 20: 571–577.

Buschmann, A.H., F. Cabello, K. Young, J. Carvajal, D.A. Varela and L.A. Henríquez. 2009. Salmon aquaculture and coastal ecosystem health in Chile: Analysis of regulations, environmental impacts and bioremediation systems. Ocean Coast. Manag. 52: 243–249.

Buschmann, A.H., A. Tomova, A. López, M.A. Maldonado, L.A. Henríquez, L. Ivanova, F. Moy, H.P. Godfrey and F.C. Cabello. 2012. Salmon aquaculture and antimicrobial resistance in the marine environment. PlosOne 7: e42724.

Buschmann, A.H., R. Stead, M.C. Hernández-González, S.V. Pereda, J.E. Paredes and M.A. Maldonado. 2013. Un análisis crítico sobre el uso de macroalgas como base para una acuicultura sustentable. Revista Chilena de Historia Natural (in press).

Bustos, P.A., N.D. Young, M.A. Rozas, H.M. Bohle, R.S. Ildefonso, R.N. Morrison and B.F. Nowak. 2011. Amoebic gill disease (AGD) in Atlantic salmon (*Salmo salar*) farmed in Chile. Aquaculture 110: 281–288.

Cabello, F.C., H.P. Godfrey, A. Tomova, L. Ivanova, H. Dölz, A. Millanao and A.H. Buschmann. 2013. Antimicrobial use in aquaculture re-examined: its relevance to antimicrobial resistance and to animal and human health. Env. Microbiol. (in press).

Carvajal, J., L. González and M. George-Nascimento. 1998. Native sea lice infestation of salmonids reared in netpen system in southern Chile. Aquaculture 166: 241–246.

Chelossi, E., L. Vezzulli, A. Milano, M. Branzoni, M. Fabiano, G. riccardi and I.M. Banat. 2003. Antibiotic resistance of benthic bacteria in fish-farm and control sediments of the Western Mediterranean. Aquaculture 219: 83–97.

Chisti, Y. 2008. Biodiesel from microalgae. Biotech. Adv. 25: 294–306.

Chopin, T., A.H. Buschmann, C. Halling, M. Troell, N. Kautsky, A. Neori, G.P. Kraemer, J.A. Zertuche-Gonzalez, C. Yarish and C. Neefus. 2001. Integrating seaweeds into marine aquaculture systems: a key toward sustainability. J. Phycol. 37: 975–986.

Chopin, T., S.M.C. Robinson, M. Troell, A. Neori, A.H. Buschmann and J. Fang. 2008. Multitrophic integration for sustainable marine aquaculture. pp. 2463–2475. *In*: S. Erik Jørgensen and B.D. Fath (eds.). Ecological Engineering. Vol. [3] of Encyclopedia of Ecology, Elsevier, Oxford.

Chopin, T., A. Neori, A.H. Buschmann, S. pang and M. Sawhney. 2011. Diversification of the aquaculture sector: Seaweed cultivation, integrated multi-trophic aquaculture, integrated sequential biorefineries. Global Aquaculture Advocate July/ August. 58–60.

Clarens, A.F., E.P. Resurreccion, M.A. White and L.M. Colosi. 2010. Environmental life cycle comparison of algae to other bioenergy feedstocks. Env. Sci. Technol. 44: 1813–1819.

Clarke, A.L., K. Weckström, D.J. Conley, N.J. Anderson, F. Adser, E. Andrén, V.N. De Jonge, M. Ellegaard, S. Juggins, P. Kauppila, A. Karhola, N. Reuss, R.J. Telford and S. Vaalamaa. 2006. Long-term trends in eutrophication and nutrients in the coastal zone. Limnol. Oceanogr. 51: 385–397.

Conley, D.J., H.W. Paerl, R.W. Howarth, D.F. Boesch, S.P. Seitzinger, K.E. Havens, C. Lancelot and G.E. Likens. 2009. Controlling eutrophication: nitrogen and phosphorus. Science 323: 1014–1015.

Costa-Pierce, B.A. 2010. Sustainable ecological aquaculture systems: The need for a new social contract for aquaculture development. Mar. Technol. Soc. J. 44: 88–112.

Diana, J.S. 2009. Aquaculture production and biodiversity conservation. BioScience 59: 27–38.

Diana, J.S., H.S. Egna, T. Chopin, M.S. Peterson, L. Cao, R. Pomeroy, M. Verdegem, W.T. Slack, M.G. Bondad-Reantaso and F. Cabello. 2013. Responsible aquaculture in 2050: valuing local conditions and human innovations will be key to success. BioScience 63: 255–262.

Duarte, C.M., N. Marbá and M. Holmer. 2007. Rapid domestication of marine species. Science 316: 382–383.

Duarte, C.M., M. Holmer, Y. Olsen, D. Soto, N. Marbà, J. Guiu, K. Black and I. Karakassis. 2009. Will the oceans help feed humanity? BioScience 59: 967–976.

Flores-Aguilar, R.A., A. Gutiérrez, A. Ellwanger and R. Searcy-Bernal. 2007. Development and current status of abalone aquaculture in Chile. J. Shell. Res. 26: 705–711.

Folke, C. and N. Kautsky. 1989. The role of ecosystems for a sustainable development of aquaculture. Ambio. 18: 234–243.

Folke, C., N. Kautsky, H. Berg, A. Janssson and M. Troell. 1998. The ecological footprint concept for sustainable seafood production: a review. Ecol. Appl. 8: S63–S71.

Fortt, A., F. Cabello and A.H. Buschmann. 2007. Residuos de tetraciclina y quinolonas en peces silvestres en una zona costera donde se desarrolla la acuicultura del salmón en Chile. Rev. Chil. Infect. 24: 8–12.

Godoy, M.C., A. Aedo, M.J.T. Kibenge, D.B. Groma, C.V. Yason, H. Grothausen, A. Lisperguer, M. Calbucura, F. Avendaño, M. Imilán, M. Jarpa and F.S.B. Kibenge. 2008. First detection, isolation and molecular characterization of infectious salmon anaemia virus associated with clinical disease in farmed Atlantic salmon (*Salmo salar*) in Chile. BMC Veter. Res. 4: 28.

Gutiérrez, A., T. Correa, V. Muñoz, A. Santibañez, R. Marcos, C. Cáceres and A.H. Buschmann. 2006. Farming of the giant kelp *Macrocystis pyrifera* in southern Chile for development of novel food products. J. Appl. Phycol. 18: 259–267.

Hallam, D. (ed.). 2012. Food Outlook: Global Market Analysis, May 2012. Food and Agriculture Organization of the United Nations. (17 January 2013; www.fao.org/giews/english/fo/index.htm).

Hargrave, B.T. 2010. Empirical relationships describing benthic impacts of salmon aquaculture. Aqcult. Env. Interact. 1: 33–46.

Haya, K., L.E. Burridge and B.D. Chang. 2001. Environmental impact of chemical wastes produced by the salmon aquaculture industry. ICES J. Mar. Sci. 58: 492–496.

Heuer, O.E., H. Kruse, K. Grave, P. Collignon, I. Karunasagar and F.J. Angulo. 2009. Human health consequences of use of antimicrobial agents in aquaculture. Food Safety 49: 1248–1253.

Jackson, J.B.C., M.X. Kirby, W.H. Berger, K.A. Bjorndal, L.W. Botsford, B.J. Bourque, R.H. Bradbury, R. Cooke, J. Erlandson, J.A. Estes, T.P. Hughes, S. Kidwell, C.B. Lange, H.S. Lenihan, J.M. Pandol, C.H. Peterson, R.S. Steneck, M.J. Tegner and R.R. Warner. 2001. Historical overfishing and the recent collapse of coastal ecosystems. Science 293: 629–637.

Kim, S.-K. 2012. Handbook of Marine Macroalgae. Biotechnology and Applied Phycology. Wiley-Blackwell, Chichester, UK.

Kovalenko, I., B. Zdyrko, A. Magasinski, B. Hertzberg and Z. Milicev. 2011. A major constituent of brown algae for use in high-capacity Li-ion. Science 334: 75–79.

Liu, F., S. Pang, T. Chopin, S. Gao, T. Shan, X. Zhao and J. Li. 2012. Understanding the recurrent large-scale green tide in the Yellow Sea: Temporal and spatial correlations between multiple geographical, aquacultural and biological factors. Mar. Env. Res. 83: 38–47.

Macchiavello, J., E. Araya and C. Bulboa. 2010. Production of *Macrocystis pyrifera* (Laminareales; Phaeophyceae) in Northern Chile on spore-based culture. J. Appl. Phycol. 22: 691–697.

Maceiras, R., M. Rodríguez, A. Cancela, S. Urrejola and A. Sánchez. 2011. Macroalgae: raw material for biodiesel production. Appl. Ener. 88: 3318–3323.

Mardones, F.O., A.M. Perez and T.E. Carpenter. 2009. Epidemiologic investigation of the re-emergence of infectious salmon anemia virus in Chile. Dis. Aquat. Org. 84: 105–114.

Marquardt, R. 2011. Light Acclimatation of Aquatic Photoautotrophs–Applied Aspects. Doctoral Dissertation, Universität Rostock, Rostock, Germany.

McKindsey, C.W., P. Archambault, M.D. Callier and F. Olivier. 2011. Influence of suspended and off-bottom mussel culture on the sea bottom and benthic habitats: a review. Can. J. Zool. 89: 622–646.

Naylor, R.L., R.J. Goldburg, J.H. Primavera, N. Kautsky, M.C.M. Beveridge, J. Clay, C. Folke, J. Lubchenco, H. Mooney and M. Troell. 2001. Effect of aquaculture on world fish supplies. Nature 405: 1017–1024.

Neori, A., M. Troell, T. Chopin, C. Yarish, A. Critchley and A.H. Buschmann. 2007. The need for a balanced ecosystem approach to blue revolution aquaculture. Environment 49: 37–44.

Neori, A. and A.M. Norbe. 2012. Relationship between trophic level and economics in aquaculture. Aquacult. Econ. Manag. 16: 40–67.

Nonaka, L., K. Ikeno and S. Suzuki. 2007. Distribution of tetracycline resistance gene, tet (M), in Gram-positive and Gram-negative bacteria isolated from sediment and seawater at a coastal aquaculture site in Japan. Microb. Env. 22: 355–364.

Olsen, Y. 2011. Resources for fish feed in future mariculture. Aquacult. Env. Interact. 1: 187–200.

Pauly, D., V. Christensen, S. Guénette, T.J. Pitcher, U.R. Sumaila, C.J. Walters, R. Watson and D. Zeller. 2002. Towards sustainability in world fisheries. Nature 418: 689–695.

Peckol, P. and J.S. Rivers. 1996. Contribution by macroalgal mats to primary production of a shallow embayment under high and low nitrogen-loading rates. Estuar. Coast. Shelf Sci. 43: 311–325.

Ridler, N., M. Wowchuk, B. Robinson, K. Barrington, T. Chopin, S. Robinson, F. Page, G. Reid, M. Szemerda, J. Sewuster and S. Boyne-Travis. 2007. Integrated multi-trophic aquaculture (imta): a potential strategic choice for farmers. Aquacult. Econ. Manag. 11: 99–110.

Santelices, B. 1996. Seaweed research and utilization in Chile: moving into a new phase. Hydrobiologia 326/327: 1–14.

Sepúlveda, F., S. Marin and J. Carvajal. 2004. Metazoan parasites in wild fish and farmed salmon from aquaculture sites in southern Chile. Aquaculture 235: 89–100.

Shi, J., H. Wei, L. Zhao, Y. Yuan, J. Fang and J. Zhang. 2011. A physical-biological coupled aquaculture model for a suspended aquaculture area of China. Aquaculture 318: 412–424.

Soto, D. and F. Norambuena. 2004. Evaluation of salmon farming effects on marine systems in the inner seas of southern Chile: a large-scale mensurative experiment. Journal of Applied Ichthyology 20: 493–501.

Soto, D., F. Jara and C.A. Moreno. 2001. Escaped salmon in the inner seas, southern Chile: Facing ecological and social conflicts. Ecol. Appl. 11: 1750–1762.

Soto, D., I. Arismendi, J. Gonzalez, J. Sanzana, F. Jara, C. Jara, E. Guzmán and A. Lara. 2006. Southern Chile, trout and salmon country: invasion patterns and threats for native species. Rev. Chil. Hist. Nat. 79: 97–117.

Soto, D., F.J. Salazar and M.A. Alfaro. 2007. Considerations for comparative evaluation of environmental costs of livestock and salmon farming in southern Chile. pp. 121–136. *In*: D.M. Bartley, C. Brugère, D. Soto, P. Gerber and B. Harvey (eds.). Comparative Assessment of the Environmental Costs of Aquaculture and other Food Production Sectors: Methods for Meaningful Comparisons. FAO, Rome, Vancouver.

Troell, M., C. Halling, A. Nilsson, A.H. Buschmann, N. Kautsky and L. Kautsky. 1997. Integrated open sea cultivation of *Gracilaria chilensis* (Gracilariales, Rhodophyta) and salmons for reduced environmental impact and increased economic output. Aquaculture 156: 45–62.

Troell, M., P. Rönnbäck, C. Halling, N. Kautsky and A.H. Buschmann. 1999. Ecological engineering in aquaculture: use of seaweeds for removing nutrients from intensive mariculture. J. Appl. Phycol. 11: 89–97.

Troell, M., C. Halling, A. Neori, A.H. Buschmann, T. Chopin, C. Yarish and N. Kautsky. 2003. Integrated mariculture: asking the right questions. Aquaculture 226: 69–80.

Troell, M., A. Joyce, T. Chopin, A. Neori, A.H. Buschmann and J-G. Fang. 2009. Ecological engineering in aquaculture—Potential for integrated multi-trophic aquaculture (IMTA) in marine offshore systems. Aquaculture 297: 1–9.

Vásquez, J.A. 2008. Production, use and fate of Chilean brown seaweeds: re-sources for a sustainable fishery. J. Appl. Phycol. 20: 457–467.

Walker, D.A. 2009. Biofuels, facts, fantasy and feasibility. J. Appl. Phycol. 21: 509–517.

Wargacki, A.J., E. Leonard, M.N. Win, D.D. Regitsky, C.N.S. Santos, P.B. Kim, S.R. Cooper, R.M. Raisner, A. Herman, A.B. Sivitz, A. Lakshmanaswamy, Y. Kashiyama, D. Baker and Y. Yoshikuni. 2012. An engineered microbial platform for direct biofuel production from brown macroalgae. Science 335: 308–313.

Westermeier, R., D. Patiño, M. Piel, I. Maier and D. Müller. 2006. A new approach to kelp mariculture in Chile: production of free-floating sporophyte seedlings from gametophyte cultures of *Lessonia trabeculata* and *Macrocystis pyrifera*. Aquacul. Res. 37: 164–171.

Worm, B., R. Hilborn, J.K. Baum, T.A. Branch, J.S. Collie, C. Costello, M.J. Fogarty, E.A. Fulton, J.A. Hutchings, S. Jennings, O.P. Jensen, H.K. Lotze, P.M. Mace, T.R. McClanahan, C. Minto, S.R. Palumbi, A.M. Parma, D. Ricard, A.A. Rosemberg, R. Watson and D. Zeller. 2009. Rebuilding global fisheries. Science 325: 578–585.

6

Microalgal Downstream Processing

Harvesting, Drying, Extraction, Separation, and Purification

Xavier C. Fretté

Introduction

Microalgae are now well established as crops for mankind with farms and production units spread around the world. These microalgae are produced with the aim to be sold as entire organisms (dry powder), or for specific compounds of interest to be extracted for further commercialization. Most microalgal compounds of interest fall into the following categories: lipids, carotenoids, polymers, and proteins (among others phycobilliproteins). However, there is an increasing demand for other types of compounds, for example, phenolic substances for their antioxidative properties (Goiris et al. 2012) as well as other compounds for their anti-inflammatory, antimicrobial, antiviral, and antitumoral activities (Guedes et al. 2011). Several family-owned SMEs are producing and commercializing dried microalgae worldwide, for example, *Arthrospira platensis* sold as granules and powder by more than 20 companies located around the town of Hyeres in South Eastern France. However, currently the industrial production of microalgae is mainly related to high-value compounds as the production costs are still very high. Such alga-biotech companies are producing and commercializing astaxanthin (Cyanotech, Hawai; Mera Pharmaceuticals, Hawai; Fuji Health Science, Japan), β-carotene (Betatene, Western Biotechnology and AquaCarotene, all in Australia), and the ω-3 fatty acids EPA and DHA (Cellana Inc., Hawai). Within the last decade, there has been a renewed interest in microalgal biodiesel and biofuel production. However, the production cost of these products make them noncompetitive compared to fossil fuels and therefore it has been estimated that the production costs of such microalgal low-value compounds should be reduced by at least an order of magnitude, for them to be economically viable (Greenwell et al. 2010).

The major obstacle in processing microalgae is their size and the fact that they are grown in water, which implies removal of the culture broth and, in the case of marine microalgae, desalting the harvested culture. Most common methods used today are centrifugation, flocculation, flotation, filtration, and/or a combination of these methods. Centrifugation is the most efficient method allowing recovery of over

Department of Chemical Engineering Biotechnology and Environmental Technology, Faculty of Engineering, University of Southern Denmark, Campusvej 55, 5230 Odense M - Denmark.
Email: xafr@kbm.sdu.dk

95% of algal biomass and large scale industrial equipment is readily available on the market, such as the SSD and SSE series clarifiers from GEA Westfalia; one of the proposed processes is presented in their sales documentation for the recovery of algae powder from open-ponds production (Fig. 1).

However, this method is energy consuming as it is estimated that it accounts for at least 20–30% of the total production costs (Gudin and Thepenier 1986) depending upon cell density, culture conditions, and algal species. Downstream processing also depends on the type of products to be extracted as well as on the algal species to be treated. Therefore costs and energy consumption for algal cultures harvest and processing are to be addressed in order to guarantee economic viability of these new crops. In this chapter, the downstream processes used to harvest, dry, disrupt, and extract the analytes of interest from the algae cells will be reviewed.

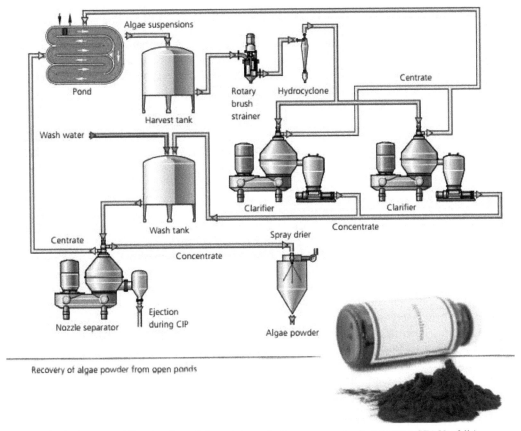

Fig. 1. Recovery of *Spirulina* (now *Arthrospira*) powder from open pond culture (source: GEA Vestfalia).

Harvesting methods

Microalgae vary considerably in shape and size. The individual algal cell size ranges from ca. 2 to 200 μm and displays shapes such as spheres, rods, or filaments. The density of microalgae consequently is very disparate. Some are buoyant while others will naturally sediment due to high density, as much as 1,150 kg/m³ for diatoms. In order to prevent sedimentation, microalgae usually have a negatively charged surface at physiological conditions, with a zeta potential ranging from –40 to –5 mV. The inter-particular repulsive force thus generated by the negatively charged cells contributes to maintaining the algal cells in suspension, but inevitably the cells will sediment at a velocity which can best be described for spherical non-interacting particles in ideal dilute conditions by the following equation (Svarovsky 1990):

$$u_p = \left(\frac{d^2 (\rho_s - \rho) g}{18 \mu_c} \right)$$

where u_p is the sedimentation velocity of the particle (m/s), d is the particle diameter (m), ρ_s is the particle density (kg/m), ρ is the density of the fluid (kg/m), g is the gravitational acceleration (m/s), and μ_c is the viscosity of the suspension (Ns/m).

Natural sedimentation

Naturally occurring sedimentation would obviously be the most cost effective separation method if it was not for its slow velocity which was measured to be as low as $3 \; 10^{-7}$ to $7 \; 10^{-5}$ m/s (Choi et al. 2006) with the consequence that most of the biomass could deteriorate during the settling time (Greenwell et al. 2010). Griffiths et al. (2012) concluded that gravity sedimentation could be a promising technique for the microalgae *Cylindrotheca fusiformis* and *Tetraselmis suecica* with average biomass recovery of 95% and 89%, respectively, after 24 hr settling. Experience from traditional sedimentation units have shown that efficient sedimentation occurs when the sedimentation velocity is greater than 10^{-4} m/s (Granados et al. 2012). *Arthrospira* was reported to settle spontaneously due to carbohydrate accumulation under nutrient-stressed conditions (Markou et al. 2012). Depraetere et al. (2015) demonstrated that this natural sedimentation was caused by an accumulation of carbohydrates that in turn increased the specific density of the filaments. The settling velocity was 0.64 m/h allowing the biomass to be concentrated 15 times with removal of 94% of the water. However, very seldom is natural sedimentation efficient and therefore alternative techniques are used.

Centrifugation

From the equation above it is clear that there is a direct correlation between gravitational acceleration and speed of sedimentation. Centrifugation up to 5,000 to 10,000 g is currently the norm in large-scale centrifuges used in algae productions. For example, Adam et al. (2012) harvested the microalgae *Nannochloropsis oculata* by centrifugation at around 5,000 rpm, resulting in a 30% dry weight paste for further processing. Similarly, centrifugation of the algal suspension of *Nannochloropsis* sp. was conducted at 3,000 g for 5 min in order to further extract the fatty acid eicosapentaenoic acid (Zou et al. 2000). It is especially important to dewater the biomass to reduce drying costs in large scale production. Most centrifugation setups tend to achieve high capture efficiency (> 90%) with low flow rates through the centrifuge, but this requires high energy. Dassey and Theegala (2013) investigated the use of centrifugation as a cost effective separation of microalgae. By controlling the capture efficiency and the flow rate of the centrifugation process, the final harvesting cost could be reduced by 82%. On the one hand when feeding the centrifuge with a flow of 0.94 L/min, the cell removal efficiency reached 94%, but this required an energy input of 20 kW h/m³ of culture water. On the other hand, a higher feeding flow of 23 L/min only afforded a cell removal efficiency of 17%. However, this could be achieved at a lower energetic cost of 0.80 kW h/m³. Using this lower yielding running conditions, the integration of centrifuges for harvesting microalgae could turn out to be profitable, especially if the energy supply stems from renewable energy such as photovoltaic solar panels or wind power.

Another consideration regarding the use of centrifugation is the physical state of the biomass harvested for further processing. The rheology of the algal cell concentrate becomes non-Newtonian when cell suspension reaches 7–10% (v/v). And when the suspension is concentrated to as much as 15–20% of the volume as cells, the suspension is no longer fluidic and therefore non amenable to pumping (Greenwell et al. 2010), and this is a hindrance for further downstream processing using pumps.

Flocculation

Instead of resorting to costly and energy consuming processes, simple techniques using induced precipitation of microalgae seem to be cheap alternatives. As seen in the above mentioned equation, the sedimentation velocity increases with the size of the particles. Therefore, processes that contribute to an increase of the size of the particles would effectively increase sedimentation speed. Flocculation is the coalescence of individual suspended particles into loosely attached conglomerates termed as flocs which are less buoyant. Flocculation is known to occur naturally, especially in seawater environment, a phenomenon called autoflocculation. Alternatively, flocculation can be induced by adding a flocculant or it can be achieved by co-culture of the microalga of interest with another microorganism (termed as bio-flocculation). Flocculation can also promoted by using ultrasonication or even electro-coagulation-flocculation for batch mode harvesting.

Autoflocculation

Autoflocculation has been observed to take place naturally for the alga *Phaeodactylum tricornutum* when the pH increases above 9 (Spilling et al. 2011). Autoflocculation is the spontaneous formation of flocs due to the precipitation of calcium carbonate and magnesium hydroxide with algal cells at high pH (Sukenik et al. 1985). Hence, autoflocculation often depends on the pH of the culture broth. In algal cultures, high pH is usually the consequence of photosynthesis resulting in CO_2 consumption by the algae. High pH can also be achieved by addition of caustic soda or lime (Nurdogan and Oswald 1996) and even though the flocculation process is induced by addition of chemicals, the process is still referred to as autoflocculation. Şirin et al. (2012, 2013) investigated the autoflocculation of two marine algae, *Nannochloropsis gaditana* and *P. tricornutum*. These two algae did not show significant sedimentation under natural conditions. Effect of pH (from 2 to 11) on flocculation of *N. gaditana*, showed a flocculation efficiency above 90% in the pH range 9.70 to 11, and sedimentation rate as high as 119 cm/h at pH 9.70. Similar results were reported by Spilling et al. (2011) for *P. tricornutum*.

Addition of ammonia has also been reported to be efficient for harvesting marine microalgae (Chen et al. 2012b). The flocculation efficiency was performed on two marine algae, *N. oculata* and a Chinese algal species classified as *Dunaliella*. The cultures were stirred with a magnetic stirrer and added various doses of commercial aqueous ammonia. A removal efficiency of 91.2% after 3 hr settling and an optimum concentration of ammonia of 38.37 mmol/L for *Dunaliella* were achieved. For *N. oculata* the optimum removal efficiency (93%) was performed with 57.31 mmol/L of ammonia measured after 3 hours. This treatment however caused cell damages as observed by scanning electron microscopy (SEM): cell morphology was fuzzy due to membrane disruption.

As autoflocculation is caused by the formation of inorganic precipitates, the removal of these from the biomass has to be envisaged.

Flocculation using flocculants

Where autoflocculation fails, it is possible to induce flocculation of the cultured algae cells. Several types of flocculants are commercially available for that purpose: polyvalent metal salts, synthetic polymers, and natural polymers (usually modified). However salinity levels above 5 g/L tend to inhibit flocculation with polymers used as the sole agent (Bilanovic and Shelef 1988). Flocculation efficiency can be improved by using inorganic coagulants (ferric ion, alum, lime) in conjunction with the polymers (Knuckey et al. 2006).

Flocculation with polyvalent metal salts. Flocculation of microalgae can be achieved by adding a flocculating agent releasing multivalent cationic ions, typically polyelectrolyte salts of aluminium (e.g., alum: aluminium sulfate, $Al_2(SO_4)_3$) and/or iron (e.g., ferric chloride, ferric sulfate), which reduce the inter-particular repulsive forces between the negatively charged surface of the cells yielding aggregates of cells. In the case of aluminium in the range of pH 5.0 to 6.0, flocculation has been explained by a combined action of electrostatic interactions due to positively charged polyaluminium species and entrapment mechanism referred to as sweep flocculation due to amorphous aluminium hydroxide precipitate (Amirtharajah and Mills 1982; Duan and Gregory 2003; Zhang et al. 2004). Aluminium chloride ($AlCl_3$, 30 g/L) was efficient to flocculate the marine algae *Isochrysis galbana* (Sánchez et al. 2013). In this work, the effect of mechanical mixing during harvest was shown to be important in the formation of flocs. An initial vigorous mixing (100–200 rpm) of the aluminium chloride was needed to initiate coagulation of the algae into microflocs, followed by slow mixing (50 rpm), allowing the generation of flocs which would settle within 1 hr. Shen et al. (2013) carried out a similar investigation on *N. oculata* with two cationic salts, aluminium sulfate and ferric chloride. Using single-factor and response-surface-methodology experiments, they developed second-order polynomial models of the final solid concentration of algae (SCA) based upon the initial algal biomass concentration (IABC), pH, and flocculant dose (FD). The optimum flocculation conditions were predicted and validated experimentally at IABC 1.7 g/L, pH 8.3 and FD 383.5 µM for aluminium sulfate, and IABC 2.2 g/L, pH 7.9 and FD 438.1 µM for ferric chloride, resulting in final solid concentrations of algae of 32.98 g/L and 30.10 g/L, respectively. Biomass recovery rate was in the range 82.6 to 100% (on average 94.4%) and 60 to 100% (on average 87.9%) for aluminium sulfate and ferric chloride, respectively. Şirin et al. (2013) performed flocculation experiments on *N. gaditana* with aluminium sulfate (AS) and polyaluminium chloride (PAC), which showed that PAC was 30% more efficient than AS regarding flocculation efficiency. With the optimum concentration of AS (20 ppm), the flocculation efficiency was 70% after 30 min of settling. The optimum concentration of PAC depended on the settling time. Optimum concentrations were 20 and 10 ppm for settling times of 15 and 30 minutes, and flocculation efficiency of 60 and 55%, respectively. Wu et al. (2012a) also investigated the effect of dosage of alum on the recovery of *N. oculata* and found an optimum concentration at 500 ppm and a settling time of 48 h.

Earlier work showed that both alkaline and acidic conditions could induce algal flocculation with polyvalent metal salts. Optimal flocculation of marine microalgae can be achieved efficiently in alkaline conditions (pH > 10.5) due to the precipitation of calcium and/or magnesium salts in the medium (Sukenik and Shelef 1984; Sukenik et al. 1985) while the use of aluminium sulfate and iron chloride as flocculants required acidic conditions (Garzon-Sanabria et al. 2010). Recent studies point to diverging results. On the one hand, it was demonstrated that at high pH, algal flocculation was due to the precipitation of $Ca(OH)_2$ and/or $Mg(OH)_2$ (Wu et al. 2012b). This result was confirmed with the alga *Chlorella vulgaris* grown either in a medium mimicking brackish water (Smith and Davis 2012) or in freshwater (Vandamme et al. 2012). On the other hand, Schlesinger et al. (2012) could not induce algal flocculation of nine different marine and freshwater algae species with $Mg(OH)_2$, but could with $Ca(OH)_2$, NaOH, KOH, and NH_4OH.

Parameters inducing flocculation are different for freshwater and seawater algae. Sukenik et al. (1988) showed that flocculation with aluminium sulfate and ferric chloride of the marine algae *I. galbana* required a 5- to 10-fold greater ionic strength than that for the freshwater algae *C. vulgaris*. Wu et al. (2012b) showed that marine algal cultures use more $Mg(OH)_2$ for flocculation than the freshwater algal cultures as a greater amount of Mg^{2+} was removed from the media during flocculation of the marine algae species.

Various other parameters, such as algal species, cell density, pH, and flocculant concentration, can affect flocculation efficiency. Alum, which has proven to be an efficient flocculant without any pH adjustment for the marine algae *T. suecica* and *Chlorococcum* sp. (> 90% recovery after 5 min settling), turned out to be less effective with the marine algae *Nannochloropsis salina*, *Dunaliella tertiolecta* and *I. galbana* (Elridge et al. 2012). These algae required at least twice as much coagulant to achieve similar recoveries.

Flocculation efficiency for *N. oculata*, was optimal when using 48 mg/L $AlCl_3$ (Garzon-Sanabria et al. 2012). However, this same study also demonstrated that flocculation efficiency also depends upon pH and cell density in the culture as well as of the ionic strength of the growth medium. Optimal flocculation with 48 mg/L $AlCl_3$ was achieved with 3×10^7 cell/mL, a pH of 5.3 and a salt concentration of 15 g/L NaCl. Five times greater $AlCl_3$ dosage was required to obtain the same efficiency with lower cell density culture (10^6 cell/mL). Salt concentration only minimally affected the efficiency of flocculation as removal efficiency was 96% and 97% with 0 and 30 g/L NaCl, respectively, compared to the 15 g/L optimum concentration of NaCl salt which resulted in a 98% removal efficiency.

The efficiency of removal can be affected by the pre-treatment of algae using ozone before polyaluminium chloride flocculation. It was shown to increase floc size probably due to the release of intra-cellular organic matter. The removal of turbidity increased from 61.8% to 80.4% when algae cells

(*Chlorella* sp.) were submitted to ozone for 30 min compared to control cells not treated with ozone (Pranowo et al. 2013).

However, the main obstacle to the use of metal salts as flocculent is that they may end up as a contaminant of the final algal product, thus inhibiting the direct reuse of algae and implying the recovery of the flocculent which in turn would impact the production cost.

Flocculation with Natural Polymers. The most widely used natural polymers are clay, chitosan, and γ-polyglutamic acid. Clay minerals are interesting as flocculants because they are non-toxic and environmentally friendly and at the same time can achieve removal efficiencies of up to 95–99%. Wu et al. (2010) investigated the removal of the marine microalga *Chattonella marina*, responsible of harmful algal blooms along the Chinese coasts, with such mineral clays, Na-montmorillonite, Na-Kaolin, vermiculite, and palygorskite, which they modified with a Gemini surfactant, (2-bromoethyl) tetradecyl dimethylethylammonium. Of these four clays, the Gemini modified vermiculite (6.5 mg/L) exhibited the highest removal rate (100% in 24 hr). The stability of the Gemini modified vermiculite in sea water was good as only 3% of the Gemini surfactant was released in the medium.

Chitosan is a material derived from chitin, the exoskeleton of crustaceans, which is a waste product from the shrimp industry. Chitosan is one of the natural polymers most often investigated for flocculating algae (de Godos et al. 2011; Farid et al. 2013; Divakaran and Pillai 2002; Lee et al. 2012; Beach et al. 2012) mainly due to its low cost. Chitosan is positively charged at neutral and acidic pH due to its amino group (pKa ~ 6.5) and therefore readily binds to the negatively charged surface of algae. The flocculation efficiency is usually compared to multivalent salts. In such a study where chitosan, ferric sulfate, and alum were used for testing the coagulation of *Neochloris oleoabundans*, it was shown that chitosan was the most effective flocculant and the optimal concentration was found to be 100 mg/L, allowing a removal rate of 95% (Beach et al. 2012). Şirin et al. (2013) investigated the effect of chitosan on the flocculation of *N. gaditana* and found 30 ppm as the optimum concentration of chitosan and flocculation efficiency above 50% in the pH range 8.93 to 10.9. In this study, the concentrations of calcium and magnesium ions were monitored and confirm the importance of magnesium hydroxide ($Mg(OH)_2$) precipitation in the flocculation process with chitosan.

As previously mentioned for polyvalent salts, flocculation depends on many parameters. Divakaran and Pillai (2002) demonstrated that effectiveness of chitosan was very sensitive to pH: a maximum removal of 90% was achieved at pH 7.0 for the freshwater algae *Spirulina*, *Oscillatoria* and *Chlorella*. The removal was lower for the marine species *Synechocystis* sp. This same study also showed that the optimal chitosan concentration depended upon the concentration of alga, but a maximum of 15 mg/L of chitosan was found sufficient to clarify all algal concentrations investigated. Ahmad et al. (2011) also found removal efficiency of *Chlorella* sp. (> 99%) for chitosan, but this efficiency was affected by chitosan concentration, mixing time and mixing rate with optimums at 10 ppm, 20 min and 150 rpm, respectively. Studies on mitigating harmful algal blooms also contribute to valuable knowledge to algal cell removal. Chen and Pan (2012c) described the efficiency of xanthan used in combination with clay/soil/sand and calcium hydroxide to remove the marine alga *Amphidinium carterae*. When used alone, xanthan only achieved a maximum cell removal efficiency of 55%. On the other hand, the three minerals (clay, sand, and soil) were ineffective in removing algal cells. However, when xanthan and calcium hydroxide (20 mg/L and 100 mg/L, respectively) were used together with these three minerals the removal efficiency increased to 83–89% within 30 min using 300 mg/L clay, sand, or soil. This approach applied to *Scenedesmus abundans* cultures confirmed that using a combination of chitosan and bentonite clay powder halved the settling time (15.3 vs. 34.2 h) (Moorthy et al. 2017).

The efficiency of removal can be affected by the pre-treatment of algae using ozone before chitosan flocculation. It was shown to increase floc size probably due to the release of intra-cellular organic matter. The removal of turbidity increased from 61.8% to 80.4% when algae cells (*Chlorella* sp.) were submitted to ozone for 30 min compared to control cells not treated with ozone (Pranowo et al. 2013).

Naturally occurring microbial flocculants have been applied to harvest microalgae. The production and application of such flocculants is described in the literature, though not often in connection with microalgae removal: Protein flocculants produced by *Rhodococcus erythropolis* S-1 (Kurane et al. 1986) and *Bacillus* sp. DP-152 (Suh et al. 1997), polyamide flocculants produced by *B. subtilis* DYU1 (Wu and

Ye 2007), and polysaccharide flocculants produced by *Alcaligenes cupidus* KT201 (Toeda and Kurane 1991) and *Bacillus mucilaginosus* (Lian et al. 2008). Among these bioflocculants, poly (γ-glutamic acid) (PGA), is an extracellular product of *Bacillus subtilis* (Yokoi et al. 1996) and *Bacillus licheniformis* (Shih et al. 2001) which can be produced industrially via fermentation (Zheng et al. 2012) and is used commercially as a microbial flocculant in wastewater treatment (Taniguchi et al. 2005). PGA is an anionic water-soluble homo-polyamide consisting of D- and L-glutamic acid monomers connected by amide linkages with α-amino and γ-carboxyl groups (Bajaj and Singhal 2011). The flocculating properties of PGA when applied to the harvest of both marine and freshwater microalgae were recently investigated (Zheng et al. 2012). The flocculating efficiency of PGA was achieved at an optimum concentration of 22.03 mg/L, with a biomass of 0.57 g/L and a salinity of 11.56 g/L for the marine *C. vulgaris*. For the freshwater algae *Chlorella protothecoides* the maximum efficiency was achieved with 19.82 mg/L PGA and 0.60 g/L biomass. The application of the optimized flocculation methods to *Nannochloropsis occulata* LICME 002, *Phaeodactylum tricornutum*, *C. vulgaris* LICME 001, and *Botryococcus braunii* LICME 003 gave no less than 90% flocculation efficiency and a concentration factor of 20.

In another study, the harvesting of *Nannochloropsis* sp. was performed using 20 mL/L mung bean protein extract (MBPE). After 2 hr of settling time, the flocculation efficiency was found to be above 92% (Kandasamy et al. 2017).

Ndikubwimana et al. (2016) performed bioflocculation studies both in pilot scale and *in situ* with the freshwater microalgae *Desmodesmus brasiliensis* and a bacterial broth bioflocculant produced by *Bacillus licheniformis* CGMCC2876 containing γ-PGA. In both cases, the flocculation efficiency was greater than 98%.

Bioflocculation. As can be seen from above, some of the polymers are synthesized by bacteria; hence, the possibility to use concomitant bacterial cultures instead of pure polymers as flocculating agent emerges, as these co-cultured bacteria could produce extracellular polymeric substances (EPS) that would induce flocculation (Toeda and Kurane 1991). Lee et al. (2009) investigated a microbial induced flocculation of the marine microalgae *Pleurochrysis carterae*. Average recovery efficiency over 90% and a concentration factor of 226 were achieved at a low concentration of organic substrate (0.1 g/L) and long mixing time (24 hr). Characterization of flocculating bacteria revealed the presence of *Pseudomonas stutzeri* and *Bacillus cereus*, but could not exclude the presence of other bacteria. The EPS excreted by the bacterial culture was not identified.

Excellent growth of *Nannochloropsis oceanica* IMET1 was observed in Permian groundwater (Wang et al. 2012) at three different temperatures: 15, 25, and 30°C. Interestingly, the algal cells aggregated at the highest temperature, whereas they remained as single cells at the two other temperatures. This led to the isolation of a new bacterium designated as HW001, which showed the ability to aggregate *N. oceanica* IMET1 after three days of incubation. This bacterium was also capable of aggregating the microalgae *N. oceanica* CT-1, *T. suecica*, and *Tetraselmis chuii*. Furthermore, the addition of the bacterium to the algal culture did not affect the lipid content or the lipid composition.

Salim et al. (2011) investigated the sedimentation kinetics and recovery efficiency of *Nannochloropsis oleoabundans*, a non-flocculating microalga with *T. suecica*, a flocculating microalga. It was demonstrated that the combination of both the algae improved the sedimentation of the non-flocculating microalga and that by increasing the amount of flocculating algae, it was possible to double the sedimentation rate of *N. oleoabundans*. The presence of the flocculating microalga in the final biomass concentrate did not impact the further downstream processing of the microalga of interest as the reported lipid contents for *T. suecica* and *N. oleoabundans* were 18–26 and 36–42% DW, respectively. *N. oculata* cultures exposed to a water-soluble extract of senescent cultures of *Skeletonema marinoi* showed flocculation as formation of groups of cells could be observed microscopically (Taylor et al. 2012). The recovery efficiency was 91.2% after 2 hr and increased to 95.3% by 6 hr, which was a significant compared to control cultures. As increasing levels of polyunsaturated aldehydes were detected in the growing media of *S. marinoi* prior to the decline of diatom blooms, it was hypothesized that the diatom-derived polyunsaturated aldehyde decadienal could be responsible for this flocculation event. However, when applying this aldehyde to the *N. oculata* cultures no flocculation was induced, suggesting that other compound(s) in the water-soluble extract of *S. marinoi* was responsible for the flocculation event.

The biomass collected by bioflocculation inherently contains fungi, bacteria, or an additional alga used as a flocculating agent that can interfere with the food or feed it is intended for.

Flocculation with Synthetic Polymers. Synthetic polymers have also been investigated. Knuckey et al. (2006) developed and described a novel technique based on the adjustment of pH of the culture to between 10 and 10.6 using NaOH, followed by the addition of Magnafloc LT-25, a non-ionic polyacrylamide derivative polymer, to a final concentration of 0.5 mg/L. Increasing the pH of seawater to 10 or above resulted in a precipitate of calcium and/or magnesium hydroxides (Knuckey et al. 2006) which flocculated upon addition of LT-25. Algal cells sedimented as they got trapped in this floc. The flocculate was then harvested by siphoning off surface water after the settling period. The method was successfully applied to harvest algal cultures of *Chaetoceros calcitrans, C. muelleri, Thalassiosira pseudomonas, Attheya septentrionalis, Nitzschia closterium, Skeletonema* sp., *T. suecica,* and *Rhodomonas salina,* with efficiencies above 80% and final concentration factor between 200- and 800-fold.

The flocculation efficiency of several cationic polyacrylamide polymers (Zetag 7550, 7570, 8110, 8140, 8180, and 8190) on algal cultures of *T. suecica, Chlorococcum* sp., *N. salina, D. tertiolecta,* and *I. galbana* was found out to be less effective than with Al^{3+} and Fe^{3+} (Elridge et al. 2012). Polyelectrolytic flocculants are very much subject to high salinity suspensions, such as marine culture media, as they shrink and thus cannot act as a bridging molecule capable of initiating flocculation. Addition of an inorganic flocculant alleviates this phenomenon and this can be achieved using lower concentrations of the inorganic flocculant than if it was used alone. Flocculation efficiency close to 100% was achieved (Danquah et al. 2009a) for microalgal culture of *T. suecica* when using the high molecular weight synthetic cationic polyelectrolytic polymer Zetag 7650 in combination with 50 mg/L $Al_2(SO_4)_3$ found to be the optimal dosis of the inorganic flocculant. Flocculation using polyacrylamide polymers is not straightforward as overdosing of the flocculant negatively impacts the flocculation efficiency (Mikulec et al. 2015). Furthermore, this same study also showed that the optimal dose of flocculant not only depends on the algal species where cylindrical shaped species flocculated more efficiently, but it also varies with the life cycle of the algal biomass as the optimum dose was 0.5 mg/L in the linear phase of growth and 2 mg/L in the stationary growth phase for *Scenedesmus obliquus.*

Using 3-aminopropyltriethoxysilane (APTES), Farooq et al. (2013) synthesized aminoclays with a cationic metal center (Mg^{2+} or Fe^{3+}, respectively Mg-APTES and Fe-APTES) presenting functional groups of $(CH_2)_3NH_2$ organic pendants through covalent bonding. In aqueous solutions, these aminoclays formed sheets which presented an amorphous phyllosilicate structure capable of interacting with polysaccharide-based cell walls. When using 1 g/L of Mg-APTES on *N. oculata,* approximately 93 and 99% harvest efficiencies were obtained with the marine media containing 10 g/L and 30 g/L of sea salt, respectively. In absence of the aminoclay Mg-APTES, *N. oculata* did not sediment at all. This result was described by the authors of this work as advantageous as common cationic polymer flocculants only offer limited harvesting efficiencies in high salinity (> 5 g/L) media. Furthermore, the aminoclay showed limited toxicity to microalgae (making possible the recycling of both clay and medium) and thus could offer the possibility of continuous harvesting.

Flocculation using sonication

Besides the use of additives, flocculation can be induced by ultrasounds. Bosma et al. (2003) successfully used ultrasounds to optimize the aggregation efficiency and concentration factor. They achieved 92% separation efficiency and a concentration factor of 20. Ultrasound harvesting requires high energy input and therefore may not be adapted to the industrial scale; however, it may be of interest if used in combination with another flocculating method. Harvest of *Microcystis aeruginosa* was carried out using the combined effect of polyaluminum chloride (PAC) with ultrasounds (Zhang et al. 2009). A short application of sonication (1–5 s) improved the flocculation performance of algal cells and this was even greater at a lower dosage of PAC.

Electrocoagulation flocculation

Electrocoagulation flocculation (ECF) is an alternative technique for the use of metal coagulants and relies on the release of iron or aluminium ions from a sacrificial anode (Vandamme et al. 2013). When using an aluminium anode during ECF, aluminium hydroxides are generated and when using an iron anode, ferric and/or ferrous hydroxides are formed (Vandamme et al. 2011). As the aluminium or iron anodes are oxidized, the main reaction at the cathode is the reduction of water, which generates hydrogen gas. Comparison of the recovery of two marine microalgae species *Chlorococcum* sp. and *Tetraselmis* sp. was carried out (Uduman et al. 2011) by electrocoagulation. High recoveries of up to 98 and 99% for *Chlorococcum* sp. and *Tetraselmis* sp., respectively, were obtained. The technique was also evaluated for harvesting the marine microalga *P. tricornutum* (Vandamme et al. 2011). In this study, the aluminium anode proved to be more efficient than the iron anode and under optimal conditions (pH 4, sedimentation time of 30 min and stirring speed of 150 rpm) the time required to initiate flocculation as well as the final recovery efficiencies were reproducible. Finally, the flocculation obtained was as effective as when using alum, though with the advantage of a limited aluminium content in the harvested biomass (below 1%). Electrocoagulation flocculation has low electricity consumption when used in seawater and therefore could be a promising low-cost technique for harvesting microalgae. Fayad et al. (2017) also found out that aluminium electrodes were best for harvesting *C. vulgaris*. 100% algae recovery was achieved within 50 minutes by using a steering speed of 250 rpm, an inter-electrode distance of 1 cm and a current density of 6.7 mA/cm^2. This same study also optimized the operation parameters to minimize energy consumption down to 1 kWh/kg microalgae: aluminium electrodes were used with a current density of 2.9 mA/cm^2, a stirring speed of 250 rpm and an inter-electrode distance of 1 cm at pH 4 for 60 min electrolysis. ECF harvesting of *C. vulgaris* in this study did not affect significantly the amount of lipids, chlorophyls A and B, and carotenoids.

Flotation and foaming

Flotation

Wastewater treatment sludge removal often resorts to the flotation principle using dissolved air flotation (DAF). The principle of DAF relies on the generation of micro-bubbles (< 10 mm) produced by the decompression of a pressurized fluid. This is a very mature and effective process for removing algae and particles in suspension allowing recovery of up to 90% and a solids concentration of 7 to 10% dry weight after harvesting. Prior to DAF treatment the culture broth is ozonated in order to sensitize the cells and then treated with a flocculating agent, as described in the flocculation section above. The micro-bubbles are then generated and adhere to the flocs thus making them more buoyant. The flocs will rise rapidly to the surface as a cell foam which can be easily removed. Hanotu et al. (2012) performed separation of *Dunaliella salina* through microfiltration with the addition of coagulants and adjustment of pH, combined with dissolved air flotation. A maximum recovery of 99.2% was obtained when using 150 mg/L ferric chloride at pH 5, with average bubble size of 86 μm.

The DAF treatment is applicable to large volumes as demonstrated in pilot to full-scale productions (Christenson and Sims 2011).

Foaming

Foam fractionation has also been investigated. This process is the result of the combined action of a surfactant, which has a polar and a non-polar region allowing hydrophobic interaction to rising gas bubbles (Csordas and Wang 2004). Marine algae are known to secrete surfactants (Žutić et al. 1981), hence the logical hypothesis that foam fractionation could be a suitable harvesting technique. Csordas and Wang (2004) investigated foam fractionation of the marine diatom *Chaetoceros* spp. The diatoms were floated to the surface of the water column by a rising steam of air bubbles. Harvesting efficiencies exceeding 90% were achieved using optimized parameters regarding the bubble size between 0.1–0.5 mm outer diameter, air flow rate of 2.25 L/min, pH 7.5 and run time of 30 min.

This method of separation of algae can only be applied to algae which naturally secrete surfactants or by amphiphilic additives. These additives will then have to be removed from the algal cake, thus making this method less attractive.

Filtration

Filtration of microalgae cultures with filter presses operated under pressure or vacuum were shown to produce a cake with 27% solids when filtering relatively large microalgae such as *Coelastrum proboscideum*, but proved to be inefficient with smaller microalgae such as *Scenedesmus* and *Dunaliella* (Mohn 1980). Harvesting microalgal biomass using membrane filtration technology has also been investigated. Application of submerged microfiltration that applies lower pressure in absence of any cross-flow velocity is commonly used in submerged membrane bioreactors (MBRs) for wastewater treatment. Bilad et al. (2012) investigated the harvesting efficiency of such submerged filtrations fitted with various types of membranes and using the improved flux-step method and batch up-concentration filtrations. The membranes were homemade from 9, 12, and 15% w/w [Polyvinylidene fluoride (PVDF), Mw $\sim$ 534,000/N,N-Dimethylformamide (DMF)] to achieve various pore sizes and casted on a polypropylene non-woven support. For all three polymer concentrations the membrane pore size was far below the size of the diatom *P. tricornutum*. Almost complete algae retention was achieved for all membranes, associated with lower degrees of fouling. Operational costs of such technology shows an energy consumption of 0.91 kW h/m^3 and a final concentration of 22% w/v. This energy consumption is reported as the lowest compared to other harvesting techniques.

Tangential flow filtration (TFF) was evaluated for concentrating *T. suecica* (Danquah et al. 2009a). The unit consisted of a Millipore Pellicon cassette system fitted with a 0.22 μm Pellicon 2 filter. The optimal transmembrane pressure was determined to be 2 bar corresponding to a flow rate of ca. 20 L/m^2 h. Under these optimal conditions, the final retentate concentrations were 4.66 and 8.88% w/v for 45 and 150 minutes run time, respectively. TFF can also be influenced by the microalgal growth phase as demonstrated by Danquah et al. (2009b). Results from TFF (2 bar over 25 min) of low growth rate phase and high growth rate phase of *T. suecica* algal broth showed that the final retentate concentrations were 5.57 and 2.47% w/v for an energy consumption of 0.38 and 0.51 kWh/m^3, respectively.

Two ultrafiltration systems were compared on microalgal suspensions of *C. fusiformis* and *Skeletonema costatum* (Frappart et al. 2011). The first was a cross flow filtration unit equipped with a flat sheet membrane (Rayflow system) and the second was a rotating disk module placed close to the stationary membrane. In both cases, the membrane was an IRIS 3038 membrane made of polyacrylonitrile. This membrane is hydrophilic, with a molecular weight cut-off of 40,000 Da. This study demonstrated a better performance of the rotating disk module which had fluxes twice higher than the Rayflow system. Rossignol et al. (1999) compared the efficiency of cross-flow microfiltration and ultrafiltration using polyvinylidilene difluoride (PVDF), polyethersulfone (PES), and polyacrylnitrile (PAN) membranes applied to cultures of *Haslea ostrearia* and *S. costatum*. For long-term operation and hence reduced running costs, ultrafiltration with PAN membrane turned out to be the most efficient technique when operating at low pressure and low tangential velocity.

Harvesting microalgae was also performed by high gradient magnetic filtration (HGMF) (Cerff et al. 2012). For that purpose, precipitated magnetite (Fe_3O_4) and hydrophilic silica-coated magnetic particles MagSilica 50–85, referred to as precipitated magnetite and MagSilica, respectively, were used in combination with the microalgal cultures of *P. tricornutum* and *N. salina*. For *P. tricornutum*, MagSilica was found to be adsorbed to the cell at pH 8, while at pH 12, the precipitated magnetite was incorporated into large flocules of the algae, probably formed by secreted extracellular polysaccharides. For *N. salina*, low adsorption rates of MagSilica and precipitated silica were observed, but was drastically increased at pH 12, with enhanced flocculation to follow. Separation efficiencies of 90% were obtained for

P. tricornutum with a pH adjusted to 12.2 to 12.5 prior to separation. This separation took place within the first 5 minutes of filtration and that implied that most of the magnetized cells were captured by the filter during their first passage after the external magnetic field had been switched on. Separation efficiencies of 40 and 60% were achieved within 5 min for *N. salina* at pH 8 and 12, respectively.

Whereas filtration can be more cost-effective than centrifugation for small volumes, costs of filtration processes are generally considered too expensive for large scale production and centrifugation may be a more economic method (Molina Grima et al. 2003). However, a recent study suggested a two-step approach where ultrafiltration is carried out prior to centrifugation. This approach resulted in a 45% reduction in energy consumption (Monte et al. 2018).

Drying methods

The biomass moisture content can be responsible for an increase in transportation costs as well as it can impact the downstream processes. For instance, the lipid recovery can be reduced by as much as 50% when the biomass moisture content increases from 4.5 to 85.4% (Balasubramanian et al. 2013). Therefore, drying of the harvested biomass can be an important step in algal processing when the components to be extracted are not thermolabile. A recent study (Balasubramanian et al. 2013) compared three drying methods on the recovery yield of lipids. The harvested biomass of *Nannochloropsis* sp. was subjected to oven drying (60°C, 3 h), freeze drying (16 h), and solar drying (30–34°C, 8 h). The lipid content from the three drying treatments did not show any significant differences in yield. However, when looking at the composition of lipids (free fatty acids, neutral lipids, and polar lipids), it appeared that while freeze and oven drying did not differ significantly, solar drying was responsible for a reduction of neutral and polar lipids and consequently, an increase of free fatty acids. This same alga was also dried under an air flow as well as freeze-dried and the yield of lipids extracted by SC-CO$_2$ was investigated (Crampon et al. 2013). The results showed that the air-dried microalgae required lower CO$_2$/microalgae mass ratio and led to faster extraction kinetic. This phenomenon could be explained by the fact that the freeze-drying preserves the microalgal cells and thus limits the diffusion of the lipids out of the cells.

However, drying of algal biomass is an energy intensive process which can account for as much as 30% of the total production costs (Chen et al. 2009) and consequently, may not be appropriate for industrial applications.

Cell disruption methods

Cell disruption is often required in order to efficiently extract materials from inside the algal cells. Cell disruption is usually performed on concentrated cell preparations (50–200 kg/m^3 dry weight) in order to reduce cost and energy consumption (Greenwell et al. 2013). The most important factor for that processing step is the maximization of the value material to be extracted, which implies a rapid disruption method. Several disruption techniques can be used to perform cell disruption. These are reviewed in a recent article (Lee et al. 2012). However, only techniques which have been assessed on marine microalgae will be described here.

The main disruptive techniques are ultrasounds, hydrodynamic cavitation, and bead milling. However, other less commonly used techniques can be applied such as temperature treatment and laser treatment.

Temperature treatment

N. oculata cells subjected to a water bath treatment (90°C for 20 min) led to 87.72% cell disruption as quantified by direct optical microscopy techniques (McMillan et al. 2013). Cell disruption due to heat treatment is caused by cells bursting as they fail to contain their elevated internal pressure leading to large debris, which is an advantage for further handling and separating steps. However, this technique is not suitable for thermolabile compounds.

Sonication

Sonic waves with a frequency of around 25 kHz can generate cavitation that can implode. Cavitation implosion creates localized shock waves and high temperature which can disrupt algae membranes. Microalgae cultures of *Thalassiosira pseudonana* and *Thalassiosira fluviatilis* were submitted to an ultrasonic ice/water bath during extraction with hexane for 20 min (Neto et al. 2013), yielding higher yields of lipids than traditional extraction technique.

The sonication effect is very much dependent on the energy and duration of sonication treatment as well as on the type of algae treated.

Optimization of the ultrasonic treatment should be considered to maximize both the energy and the time of treatment for cell disruption as shown by Gerde et al. (2012). In that study it was found that for the seawater heterotrophic species *Schizochytrium limacinum*, an energy input of 800 J/10 mL (regardless of cell concentration) achieved optimal lipid extraction. Higher energy levels did not improve lipid yield even though greater cell disruption was observed under microscope.

In another study, Nowotarski et al. (2012) investigated the effect of ultrasonic disruption of two unicellular algal species, *N. oculata* and *D. salina*, with regard to release of lipids for biofuel production. In this study, *D. salina* on the one hand proved to be susceptible to sonication at low algal densities with 90% reduction in cell number after 1 minute and complete cell disruption after 4 minutes of treatment. On the other hand, for *N. oculata* sonication allowed declumping of algal clumps into individual cells (20% increase in cell number), but this species turned out to be more resistant to sonication even after 16 minutes of treatment as a reduction in metabolic activity could be monitored but no cell disruption. McMillan et al. (2013) performed disruption of *N. oculata* with an ultrasonic bath for 20 minutes and evaluated the effect of the treatment by quantifying the damaged cells using direct optical microscopy techniques. This work confirmed that liquid ultrasonic shear treatment was the least effective on this alga as it only achieved 67.66% cell disruption. Halim et al. (2012) investigated the effect of sonication on the microalga *Chlorococcum* sp. and found that ultrasonication of both low-density and high-density stock cultures at low (65 W) and high (130 W) power levels was inefficient to disrupt the cells as observed by microscope with an average disruption of 4.5% of the initial intact cells. However, sonication disintegrated the microalgal colonies as it was observed that the colony diameter was reduced and this effect was directly correlated with the acoustic power input.

Oil extraction from the microalgal biomass of two marine algal species *Nannochloropsis* sp. and *D. tertiolecta* was performed in a Soxhlet apparatus with n-hexane as solvent. Prior to Soxhlet extraction, the biomass was sonicated to achieve cell disruption and treated with propanol (Gouveia and Oliveira 2009) yielding oil contents of 28.7% and 16.7% for *Nannochloropsis* sp. and *D. tertiolecta*, respectively.

Neto et al. (2013) also used sonication as pretreatment of biomasses of *Chlorella minutissima*, *T. fluviatilis*, and *T. pseudonana* for lipid extraction. Lipid content extracted by hexane after sonication was found out to be similar to previous studies for the three species (13.3, 40.3, and 39.3% of DW). However, for the diatoms, it was observed that a vortex-mixing step was necessary to rupture the frustule during the resubmission of cell debris to hexane extraction as the lipid content only reached 8–12% of the dry weight without vortex-mixing. The vortex-mixing step did not improve the lipid recovery from the Chlorophyceae algae and this was explained by the cell fragility when compared to the diatoms. Thus the study concluded that sonication pretreatment improved the yield of lipid recovery when combined with hexane extractions, as well as with vortex-mixing for diatoms.

Efficiency of sonication is also related to the components to be extracted. Ryckebosch et al. (2012) showed that sonication allowed 40% total lipids extraction of dry weight of *P. tricornutum* algae, while only 18% extraction was achieved for non-polar lipids. Balasubramanian et al. (2013) demonstrated that for the biomass of *Nannochloropsis* sp. sonication was not the most efficient method to extract lipids compared to accelerated solvent extraction (ASE), Soxhlet extraction and homogenization as the yield of lipids extracted was 22, 33, 30, and 28%, respectively. Similarly, lipid extraction using sonication from *Nannochloropsis* sp. (Koberg et al. 2011) only afforded 18.9% biodiesel compared to the microwave which afforded 32.8%.

"Solvent-free" ultrasound-assisted extraction of lipids from fresh microalgae cells has been investigated and the sonication parameters optimized (Adam et al. 2012). The optimum conditions identified by this study required an ultrasonic power of 1,000 W, an extraction time of 30 min and a biomass with a dry weight of 5%. Under these conditions the lipid recovery from *N. oculata* was 0.48%, which was 12 times less than the recovery from the Bligh and Dyer conventional method. This study concluded that while the oil recovery was not affected by the power of sonication, it was significantly controlled by the percentage of dry weight of biomass, that is, the medium density.

Homogenizers

The main components extracted from microalgae are lipids, carotenoids, and polymers. Most metabolites of interest produced by microalgae are kept intracellularly. Microalgal lipids are currently very popular as they have the potential to meet the global demand of biodiesel and therefore, many articles are written regarding the perspectives. As most microalgal species do not excrete the lipids they synthesize, cell membrane disruption needs to be performed to free the lipids. Lee et al. (2013) investigated the energy requirement for microalgal cell disruption using atomic force microscope evaluation and found out that the average energy for disruption of an individual cell was 17.4 pJ, which is equivalent to a specific disruption energy of 673 J/kg of dry microalgal biomass, while hydrodynamic cavitation, the most energy efficient mechanical cell disruption process, had a specific disruption energy of 33 MJ/kg of dry biomass. These results demonstrate the inefficiency of existing mechanical cell disruption processes.

Hydrodynamic cavitation

This process is a high pressure homogenization pretreatment which forces the algal fluid through small orifices thus creating a rapid pressure change as well as a high liquid shear that cause cell disruption. High pressure homogenization has been investigated by Samarasinghe et al. (2012). The alga chosen for this study was *N. oculata*. The use of various nozzle sizes did not affect the degree of cell rupture which implies that the pressure differential alone, and not the shear induced by smaller nozzles, had a significant effect on cell walls rupture. The authors therefore note that this will have practical implications as larger nozzles, which will not clog during operation, could be used without compromising the cell lyses. Cell disruption was close to complete after two passes of the cell culture through the high pressure homogenizer operated at 276 MPa.

This process has been used to extract lipids from the marine microalga, *Scenedesmus* sp. (Cho et al. 2012) at 35°C and a pressure of 83 bar for 30 min. The algal lipid yield was about 24.9% through this process while it was only 19.8% when the extraction was carried out with the conventional lipid extraction procedure using the solvent chloroform:methanol (2:1, v/v) at 65°C for 5 h. Hydrodynamic cavitation can therefore be very interesting as the extraction is performed at a low temperature and in a much shorter time. In a comparative study of various disruptive techniques (high pressure homogenization, acid, ultrasonic, and bead beating treatments), Halim et al. (2012) found out that high pressure homogenization was the most efficient technique with an average disruption of 73.8% of the initial intact cells of the microalga *Chlorococcum* sp. stock culture. Increasing operating pressure from 500 to 850 bar was shown to increase the disruption rate of the cells. Furthermore, the efficiency at a given pressure was higher with low-density stock culture as this could be expected as lower density cultures absorb more kinetic energy per individual cell.

Mechanical solid shear with bead milling or mixer

Bead milling is a common process to obtain cell lysis. The algal cells are placed in vessels packed with glass beads that are agitated at great speed. Cell disruption then occurs through physical grinding of the cells against the surface of the beads.

Sheng et al. (2012) showed that extraction of lipids from *Synechocystis* PCC 6803 with bead beating gave the lowest yield of total FAME as percentage of total biomass.

In another study bead beating of lyophilized *P. tricornutum* allowed the recovery of total lipids in similar amount as when performing disruption with sonication (Ryckebosch et al. 2012). In a comparative study of various disruptive techniques (high pressure homogenization, acid, ultrasonic, and bead beating treatments), Halim et al. (2012) found out that bead beating had a moderate efficiency with an average disruption of 17.5% of the initial intact cells of the microalga *Chlorococcum* sp. stock culture as evaluated by microscopic observation. The efficiency of bead beating was improved by a factor 1.5 when bead loading was increased from 1:2 to 1:3 (volumetric ratio of glass beads to microalgal stock culture).

Apart from bead millers, mechanical shear can be achieved by using a simple blender. *N. oculata* cells subjected to a treatment with a hand blender (3,000 rpm for 6 min) led to a 92.95% cell disruption quantified by direct optical microscopy techniques (McMillan et al. 2013).

Pulsed electric field

With this technique, the cell membrane is disrupted by unequal electric charges that accumulate on dipolar molecules, thus creating a pressure on the membrane leading to the formation of irreversible pores in the membrane.

Sheng et al. (2012) investigated this technique for the extraction of lipids from *Synechocystis* PCC 6803. This work showed that pulsed electric field (36 kWh/m^3) and a final temperature of 36°C afforded high yield of suspended lipids (4.4% of total biomass). The authors of this work concluded that pulsed electric field, together with microwave treatment, was the best suited technique for large scale cell disruption because it retained most of the lipids in the suspended phase, thus reducing the use of solvent for downstream extraction.

Microwave

The mechanism by which microwave treatment disrupts the membranes depends on the rotation of molecular dipoles thus disrupting weak hydrogen bonds and resulting in dielectric heating. This increases the solvent penetration into the matrix and facilitates analyte solvation.

Sheng et al. (2012) showed that by using microwaves without temperature control (1.4 kW for a treatment of 1 min and a final temperature of 57°C), lipid extraction as suspend FAME was maximized (4.6% of total biomass) from *Synechocystis* PCC 6803.

N. oculata cells subjected to a microwave treatment with temperature control (90°C, power varying from 210 to 1,025 W) led to a 94.92% cell disruption quantified by direct optical microscopy techniques (McMillan et al. 2013). The efficiency of the microwave treatment could be seen during lipid extraction from *N. oculata* as the lipid recovery was improved from 1.4 to a maximum of 11.3% (weight %) at 120°C (Biller et al. 2013). For this same alga, Koberg et al. (2011) also report that microwave is very efficient for lipid extraction as they obtained a biodiesel yield of 32.8%, a figure which was a factor 1.74 higher than when the extraction was carried out with sonication.

Laser treatment

Algal cells of *N. oculata* subjected to a laser treatment (Nd:YVO4 laser with a transition wavelength of 1,064 nm, 10 W, and 20 kHz) for 60 s at 80% power level led to a 96.53% cell disruption quantified by direct optical microscopy techniques (McMillan et al. 2013) and this treatment was the most efficient to disrupt this algal cell wall when compared to microwave, water bath, blender, and ultrasonic treatments.

Chemical disruption treatments

Enzymatic pretreatment

Microalgae cell walls can be disrupted using an enzymatic approach. Enzymes have to be selected according to the algal species to be treated due to the specificity of their membranes and cell walls. Such a screening was carried out by Horst et al. (2012). Several enzymes with known properties were investigated: viscozyme, driselase, crude papain, lipase from *Rhizomucor miehei*, and proteinase K. These enzymes were selected due to their specific mode of action. Viscozyme contains, among others, a wide range of carbohydratases (Latif and Anwar 2011) acting on branched pectin-like substances found in plant cell walls. Driselase contains (among others) cellulase, pectinase, β-xylanase, and β-mannanase. Crude papain contains among others cysteine endopeptidases papain, chymopapain, glycyl endopeptidase, and caricain. The lipase from *R. miehei* catalyses hydrolysis of TAG molecules while proteinase K is an alkaline serine protease with broad substrate activity. The microalgae investigated were *P. tricornutum*, *T. pseudonana*, and *N. oculata* and the efficiency of the treatment was evaluated on the basis of lipid release. For *P. tricornutum* the most effective enzymes were viscozyme, proteinase K, and papain crude extract. For *N. oculata* the most efficient enzymes were viscozyme and proteinase K and for *T. pseudonana* only driselase was efficient. This study also confirmed that the enzymatic activity is very specific to each algal species and for instance papain had little or no effect on *T. pseudonana* while it had a very promising activity on *P. tricornutum*, despite the fact that both algae are diatoms. This difference lies in the structure of the silica shell. Inhibitory studies of the enzymatic activity showed that the cysteine protease activity of papain was crucial to separate the lower and upper parts of the diatom shell. As enzymatic digestion processes can be pretty costly, this study also investigated the enzyme quantity and residence time in order to minimize them, and found that 2.5 μg enzyme per 200 μg dry weight and a holding time of 2 hr were acceptable.

Similarly, Liang et al. (2012) screened various enzymes for disrupting cell membranes of *Scenedesmus dimorphus*, and *Nannochloropsis* sp. The enzymes tested were cellulase, snailase (a mixture of more than 30 enzymes), neutral protease, alkaline protease, and trypsine. Higher lipid recovery was obtained with a pretreatment by snailase and trypsine. Enzyme dosage and reaction time were also investigated. By the time 4% enzyme dosage was reached, the lipid recovery approached maximum with a duration of treatment of 12 hr.

A wide array of enzymes was tested on the marine microalgae *P. tricornutum* (CCMP 632), *Nannochloropsis* sp., *Franceia* sp., and *Ankistrodesmus falcatus* (Gerken et al. 2013). The two latter strains were resistant to almost all the enzymes tested. *P. tricornutum* was sensitive to the enzymes chitinase, β-glucuronidase, hyaluronidase, and pectinase. *Nannochloropsis* sp. showed significant sensitivity to a number of enzymes: chitinase, chitosanase, β-glucuronidase, β-glucosidase, hyaluronidase, lysozyme, lyticase, pectinase, sulphatase, trypsin, and zymolyase. The enzymes inhibited the growth of the microalgae most probably due to the degradation of cell walls. This could be demonstrated by showing cell permeability to a DNA staining dye, SYTOX Green, which normally cannot pass the cell wall/membrane of living cells. *Nannochloropsis* was sensitive to lysozyme, but the effect was even more drastic when it was combined with one of the following enzymes: sulfatase, trypsin, or lyticase.

Production of bioethanol from biomass requires a preliminary saccharification process step from which fermentable sugars such as glucose and mannose are released (Harun and Danquah 2011). These authors carried out a study using cellulase from *Trichoderma reesel*, ATCC26921, to perform the enzymatic hydrolysis of the marine alga *Chlorococcum* sp. and showed that the highest glucose yield of 64.2% (w/w) was obtained at a temperature of 40°C, pH 4.8 and a substrate concentration of 10 g/L of microalgal biomass. This enzymatic process was evaluated as an effective mechanism to enhance the saccharification process.

Acid pretreatment

Talukder et al. (2012) used sulfuric acid to perform cell lysis. The dried microalgae biomass of *N. salina* (1 g) was treated with sulfuric acid (5% w/v first shaken with the algal biomass at 30°C and 150 rpm for 10 min, followed by hydrolysis at 120°C for 1 h). The algal lipids were then extracted using hexane. For comparison purposes, the same extraction was carried out without acid pretreatment. Lipid yield increased from 48.7% to 85.6% when applying the sulfuric acid lysis pretreatment. This improvement was not only attributed to the disruption or deformation of the microalgae cell wall but also to the acid hydrolysis of fatty acids from polar lipids such as phospholipids and glycolipids. Additionally, in that study, the acidic pretreatment contributed to the release of a maximum of 64.3% of sugar available (as glucose and xylose) in the algal biomass. Similarly, Nguyen et al. (2009) have investigated the pretreatment with sulfuric acid (3% at 110°C for 30 min) of *Chlamydomonas reinhardtii* UTEX 90 biomass for ethanol production. This treatment afforded a maximum glucose release of 58% (w/w) from the biomass polysaccharides and oligosaccharides. The effect of sulfuric acid pretreatment on *Chlorococcum* sp. stock cultures was investigated (Halim et al. 2012) under different conditions: temperature (120 or 160°C), acid concentration (3 or 8 vol%), and treatment duration (15 or 45 min) were evaluated not in terms of yields of a given compound extracted but in terms of cell disruption as observed by microscopy. The disruption efficiency was highest (66%) when using the higher acid concentration and temperature levels. Disruption could be as low as 3% when using the lower temperature and acid concentration with 45 min treatment.

Acid treatment was also used to remove chlorophyll from the astaxanthin SCE-prepared extract of the microalgae *Monoraphidium* sp. GK12. Fujii (2012) showed that the extract still contained chlorophyll and therefore investigated the effect of treatment with H_2SO_4, HCl, H_3PO_4, and CH_3COOH. The two later acids could not remove the chlorophyll. Treatment with 0.1 N H_2SO_4 or HCl was suitable for the removal of more than 80% chlorophyll with no loss of astaxanthin. Miranda et al. (2012) also demonstrated the efficiency of acidic treatment with H_2SO_4 (2N, at 120°C for 30 min) resulting in highest yields of extracted sugars from *S. obliquus* (up to 95.6% compared to harsh quantitative acid hydrolysis).

Surfactant pretreatment

Due to growing concerns regarding the use of organic solvents for lysis and extraction of metabolites from natural sources, non-ionic surfactants have been investigated as potential cell disrupters combined with salting out agents such as sodium inorganic and organic salts in aqueous solutions (Ulloa et al. 2012). The microalgae used for this work was *T. suecica* and the metabolites monitored to evaluate the extraction process were α-tocopherol, β-carotene, and gallic acid. Tween 20, 40, and 80 as well as Triton X-100, X-102, and X-114 were employed as cell lysis agents, and sodium carbonate, sodium citrate, and sodium tartrate as salting out agents. The algal biomass was stirred for two days in a shaker at 150 rpm with the surfactant. Results showed that Tween 20 and Triton-X102 were capable of forming effective phase segregation in the presence of the three salts. Tween 20 and sodium citrate afforded high extraction yields, higher than 99% for α-tocopherol and around 60% for β-carotene and gallic acid. The method using Triton X114 was evaluated against a conventional ultrasound-based method and found out to be more effective as the antioxidant yield was increased by a factor 1.59, 3.69, and 32.5 for gallic acid, α-tocopherol, and β-carotene, respectively.

Extraction, separation, and purification methods

Solvent extraction

Extraction of lipids from microalgae has usually been carried out with a mixture of chloroform-methanol or chloroform-methanol-water, as is the case in the conventional lipids extraction method first described by Bligh and Dyer (1959). This lipid extraction method is quite often used in the literature as a reference method compared to various other described methods. However, the use of chloroform in large scale is precluded by environmental and health risks. The use of other solvents has therefore been investigated. Talukder et al. (2012) used hexane to extract lipids from dried biomass of *N. salina* following an acid hydrolysis pretreatment (5% sulfuric acid at 120°C for 1 h). Incubation of the hydrolysate with hexane (at 40°C and mixing at 200 rpm up to 24 h) was followed by centrifugation (4,000 rpm) allowing separation of the hexane phase from the hydrolysate. The hexane phase afforded a lipid yield of 85.6%. Hexane was also used to extract lipids from the microalga *Chlorococcum* sp. (Halim et al. 2011) with a very low yield (1.5%), though it should be noted that this alga was found to have a maximum lipid yield of 7.1%.

Using various concentrations (50 to 95%) of isopropanol, Yao et al. (2012) performed the extraction of lipids from the microalga *Nannochloropsis* sp. It was concluded that the 70, 88, and 95% IPA concentrations extracted 82–92% of total oil.

Using the marine microalgae *Isochrysis* aff. *galbana*, the extraction of the carotenoid fucoxanthin with different solvents (hexane, acetone, ethyl acetate, methanol, and ethanol) and various extraction times was investigated (Kim et al. 2012a). Results showed that maximum recovery could be achieved with acetone, ethyl acetate, and ethanol. Using ethanol as extraction solvent, it was found out that 5 min extraction time was sufficient to obtain maximum recovery of fucoxanthin. The fucoxanthin degraded gradually with longer extraction times (loss of 14% after 24 h extraction time compared to 5 min). This study also optimized a process for the simultaneous preparation of crude fucoxanthin and lipids from this algae, where the crude ethanolic extract can be partitioned with hexane, yielding crude lipids in the hexane phase and crude fucoxanthin in the hydroalcoholic phase. This same author reports in another work (Kim et al. 2012b) that extraction of fucoxanthin with ethanol from the marine microalga *P. tricornutum* gives the maximum yield (15.71 mg/g DW) compared to acetone and ethyl acetate (4.60 and 2.26 mg/g DW, respectively), and that this maximum yield could only be achieved with an extraction time of 60 min. From these two studies it is clear that the extraction solvent is to be selected and optimized for each type of algae to be extracted. It is widely accepted that carotenoids are best extracted with acetone, for example, Zhu and Jiang (2008) for the extraction of β-carotene from *D. salina*, and while this solvent would have allowed 100% recovery of fucoxanthin from *Isochrysis* aff. *galbana*, this same choice of solvent would have reduced the yield to 29% for *P. tricornutum*.

Solvent extraction is often performed in a Soxhlet extractor as the solvent in contact with the biomass is constantly evaporated and recondensed into the sample container, thus enabling continuous re-establishment of mass transfer equilibria. In a comparative study of static hexane lipid extraction with Soxhlet hexane lipid extraction, Halim et al. (2011) found that the lipid yield increased from 1.5 to 5.7%, respectively. Wu et al. (2012a) investigated Soxhlet extraction of the carotenoid zeaxanthin from *N. oculata* with dichloromethane, hexane, ethanol, and acetone. Highest yields of zeaxanthin were obtained with dichloromethane and ethanol (100 and 98.3% respectively) with an extraction time of 16 h. Increasing the extraction time to 24 h caused a reduction of the yield of extraction of zeaxanthin down to 71.7% probably due to thermal decomposition of zeaxanthin during prolonged heat extraction.

Alternatively, bio-solvents, such as the terpenes d-limonene, α-pinene, and p-cymene, recognized as environmentally safer have also been investigated to extract lipids from the microalgae *N. oculata* and *D. salina* (Dejoye Tanzi et al. 2013) by distillation. Lipid yields from *N. oculata* extracted with p-cymene was highest (21.45%) compared to the two other solvents α-pinene and d-limonene (18.75 and 18.73%, respectively). For *D. salina* the best solvent was α-pinene (3.29%) compared to p-cymene and d-limonene (2.99 and 2.94% respectively). This process also proved to be effective as it allowed elimination of water from the microalgae biomass, extraction of the lipids, and recycling of the terpene solvent (100% recovery).

Supercritical carbon dioxide extraction

Supercritical carbon dioxide fluid extraction (SFE-CO_2) is a technology with low environmental impact due to the absence of harmful residual solvents. Furthermore, supercritical fluids have high diffusivity, high compressibility, low viscosity, and low surface tension, which allow them to diffuse easily through the solid biomass matrix and thus, achieve higher extraction yields.

There are several factors affecting the extraction efficiency of SFE-CO2 among others the type of microalgae and the type of component to be extracted. Furthermore, various operational parameters are also known to significantly affect the extraction efficiency of SFE-CO2, such as load of biomass, flow rate, temperature, pressure, duration of extraction, and addition of modifiers.

Guedes et al. (2013) described how the pressure, temperature, flow rate and the use of a co-solvent could impact the extraction of two types of components, the carotenoids and chlorophylls from *S. obliquus*, and showed that highest yield of chlorophylls was obtained at 250 bar and 40°C while carotenoids required 200 bar and 60°C. The yields of chlorophylls and carotenoids were reduced when the CO_2 flow rate was doubled from 2 to 4 g/min and the addition of 7.7% (v/v) of ethanol afforded a maximum yield for both classes of compounds. Pan et al. (2012) investigated the extraction of astaxanthin from *Haematococcus pluvialis* by supercritical carbon dioxide fluid with ethanol modifier and found that the conditions for optimum extraction of astaxanthin were a loading of biomass of 21.67 g/L, a CO_2 flow rate of 6 NL/min (where NL/min is the flow rate in L/min measured under normal air conditions), an extraction time of 20 min, an extraction pressure of 4,500 psi, a volume of ethanol modifier of 9.23 mL/g, an extraction temperature of 50°C, and a modifier composition of 99.5%. Astaxanthin was also extracted using supercritical carbon dioxide (60°C and 20 MPa for 1 h) from the vegetative green microalga *Monoraphidium* sp. GK12 (Fujii 2012) affording 83% recovery of the carotenoid. Addition of ethanol to the algal biomass (20/1 v/w) prior to SFE could increase the recovery of astaxanthin to 101%.

Supercritical fluid extraction of lipids from *Crypthecodinium cohnii* was also investigated (Couto et al. 2010). This work found out that 50% of total oil contained in the raw material was extracted after 3 h at optimum extraction conditions 30.0 MPa and 323 K, and that the docosahexaenoic acid (DHA) attained 72% w/w of total fatty acids. Assessment of bio-oil extraction from *Tetraselmis chui* with SFE-CO$_2$ (Grierson et al. 2012) without and with methanol or ethanol as co-solvents resulted in low recoveries (0.01 to 4.3% wt) of natural oil compared to the amount of lipids as recorded for this species in the literature (17% wt). Similarly, extraction of carotenoids from the marine strain *Synechococcus* sp. with supercritical CO$_2$ could not compare with the extraction of carotenoids with dimethylformamide used as a reference, as it only afforded 50% of the total carotenoids at a working pressure of 500 bar (Montero et al. 2005).

Pressurized fluid extraction

Pressurized fluid extraction (PFE) allows efficient extraction of compounds with solvents mainly due to the use of high temperatures which increase compounds solubility. PFE could be compared to a Soxhlet extraction but with the difference that the extraction is performed at high pressure, thus keeping the solvent in the liquid state, even if the temperature applied is above the boiling point of the extracting solvent. The high pressure also favors the penetration of the solvent into the biological matrix. Pieber et al. (2012) investigated the use of pressurized fluid extraction on *N. oculata* for the recovery of lipids, and more specifically the PUFA yield. For his extraction solvents applicable in the food and pharmaceutical industries, hexane, hexane/isopropanol (2:1 vol.%) and ethanol (96 vol.%) were tested. The highest extraction yield was obtained from ethanol extraction (36 mass%), while hexane afforded the lowest yield (6.1 mass%). The use of ethanol also resulted in the highest PUFA yield (5.7 mass%), total FA yield (ca. 17 mass%), and EPA yield (3.7 mass%).

Direct transesterification

When the microalgae biomass is mixed with alcohol and a catalyst at high temperature, lipid extraction, and transesterification occur simultaneously, thus reducing the operational cost of biodiesel production. Patil et al. (2011) carried out a response surface methodology (RSM) of the direct conversion of wet algae (*Nannochloropsis* sp.) to biodiesel under supercritical methanol conditions. The optimal conditions were then reported as wet algae to methanol ratio of around 1:9 (w/v), reaction temperature, and time of about 255°C and 25 min, respectively. During this process, fatty acid methyl esters (FAME) were produced from polar phospholipids, free fatty acids, and triglycerides. In their subsequent investigation (Patil et al. 2012), a comparison of direct transesterification of algal biomass with supercritical methanol (SCM) and microwave irradiation (MW) conditions was performed. Maximum yields of FAME were obtained with optimum parameters as 25 and 8, and 6 and 9 for time of reaction (min) and biomass to methanol ratio (w/v), respectively, as well as a temperature of 250°C for the SCM process and 2% KOH (wt.%) for the MW process. Both processes resulted comparable extraction yields of FAME (84.15 and 80.13% for SCM and MW, respectively), though with higher energy requirements for the supercritical methanol process. Koberg et al. (2011) also investigated the direct transesterification of the crude dried solid of *Nannochloropsis* sp. microalgae. To perform the transesterification, the dried algal biomass was suspended in methanol:chloroform (1:2 v/v) with a SrO catalyst and heated using either sonication (50°C) or microwave irradiation (60°C). For comparison purposes, the reaction was also carried out by reflux using the conventional protocol. Both the yield and conversion of biodiesel (conversion of triglycerides to biodiesel) were 37.1 and 99.9%, and 20.9 and close to 95% for microwave and sonication heating, respectively. Compared to these results the reflux technique yielded 6.9% biodiesel with only 74.6% conversion.

Milking

Milking is a method for simultaneous production and extraction of target compounds directly from live cells without harvesting or killing them. The extraction relies on an aqueous-organic biphasic system where the organic solvent is in direct contact with the cells and therefore must be biocompatible. Zhang et al. (2011) performed a screening of biocompatible solvents for enhancement of lipid milking from *Nannochloropsis* sp. The solvents tested were hexanol, heptanol, octanol, hexane, heptane, octane, nonane, decane, dodecane, tetradecane, and hexadecane, with partition coefficient (Log P_{oct}), ranging from 2.03 for hexanol to 8.80 for hexadecane. This study concluded that on the one hand hydrophilic solvents with Log $P_{oct} > 5.5$ (the alcohols and the alkanes with C-6 to C-8) were not biocompatible with the microalgae *Nannochloropsis* sp. due to deactivated dehydrogenase and increased cell membrane permeability. On the other hand, dodecane, tetradecane, and hexadecane (10% v/v) were biocompatible with the microalgae investigated, with hexadecane allowing extraction (22%) of the biomass total lipids. Furthermore, it was possible to recover 89% of the solvent used in the milking process.

Similarly, milking of *D. salina* cultures with various solvents to extract β-carotene showed that solvents having Log $P_{oct} > 5$ or having a molecular weight above 150 g/mol were considered biocompatible for this microalga (Mojaat et al. 2008). This study showed that mixing a toxic polar solvent (such as dichloromethane or methyl ethyl ketone) to the biocompatible solvent decane in appropriate ratio could achieve efficient extraction of β-carotene while preserving the microalgal cell membrane integrity. Applying this result to centrifugal partition chromatography, this group investigated the use of ethyl oleate with 5% dichloromethane (v/v) and compared it with the solvent mixture decane:dichloromethane (95:5 and 90:10 v/v) for the extraction of β-carotene from *D. salina* cultures (Marchal et al. 2013). With the addition of 10% dichloromethane, the cell viability was below 5%. With the addition of 5% dichloromethane in decane and ethyl oleate, the cell viability reached 80 and 65% while extraction yields of β-carotene were 37 and 65%, respectively.

Recently the use of nonionic surfactants (among others Triton X-114) has been reported (Glembin et al. 2013) to extract *in situ* the fatty acids from *S. obliquus*. Cell viability was performed showing that cell growth remained unaffected after exposure to Triton X-114 and that relative photosynthetic activity recovered from 82 (start) to 100% (after 10 days). This study showed that the algae cells were concentrated in the aqueous phase, whereas the fatty acids were enriched in the micellar phase.

The milking process seems to be very promising as it can be carried out *in situ* directly from the culture medium without affecting the algal culture which can then be cultivated continuously. This technological concept is so promising that it has been patented by OriginOil (USA) (Guedes et al. 2011).

Precipitation

Extraction of phycobilliproteins from microalgae usually resorts to ammonium sulfate precipitation (Duerring et al. 1991; Ficner et al. 1992; Minkova et al. 2003). This can be exemplified by the extraction of β-phycoerythrin from the red alga *Rhodosorus marinus* which was performed by successive precipitation steps with ammonium sulfate (Básaca-Loya et al. 2009). The algal cells were first saturated with 40% ammonium sulfate and the precipitate discarded. The supernatant was then saturated with 60% ammonium sulfate to precipitate the β-phycoerythrin. Further purification was performed by chromatography.

Chromatography

Purification of a specific metabolite may require techniques with higher selectivity. Chromatography is a well-known and established technique used to separate individual components from a complex mixture. Belarbi et al. (2000) employed column chromatography to extract selectively eicosapentaenoic acid (EPA) methyl ester, which is a ω-3 C20-polyunsaturated fatty acid methyl ester, from cultures of *P. tricornutum*. For that purpose, the fatty acid esters were first chromatographed on an argentated silica gel chromatography yielding 70% of the EPA present in the biomass with a purity of 83%, followed by a second chromatographic step on silica gel to remove pigments from the EPA ester fraction and thus leading to highly purified EPA ester (purity up to 96%).

Zeaxanthin was extracted from *N. oculata* by acetone using an ultrasonic extractor (Chen et al. 2012a). The extract obtained was then chromatographed on a reverse phase polystyrene-based resin, PS100. Elution was carried out isocratically with the mobile phase methanol:acetone (90:10 v/v) at a flow rate of 65 mL/min. This chromatography step increased the concentration of zeaxanthin from 36.2 mg/g of the ultrasonic extract to 425.6 mg/g of the collected column fraction.

Phycobilliproteins obtained by precipitation usually require further purification. Básaca-Loya et al. (2009) for instance first chromatographed the protein fraction on Sephadex G-25 to remove the ammonium sulfate, then on an anionic HiPrep 16/10 Q XL chromatographic column and finally on a Zorbax GF-250 HPLC column to get purified β-phycoerythrin.

Soluble proteins from the alga *Tetraselmis* sp. were isolated under mild, non-denaturing conditions (Schwenzfeier et al. 2011). Chromatography was one of the purification steps of this process. The stationary phase was an EBA ion exchange adsorbent Streamline DEAE and after application of the alga extract the bound proteins were eluted using a 35 mM potassium phosphate buffer containing 2M NaCl.

Crystallization

A zeaxanthin extract of *N. oculata* was further purified by using supercritical anti-solvent recrystallization (Chen et al. 2012a) after a response surface methodology was systematically applied to optimize the process. Zeaxanthin-rich particles with a purity of 84.2% and a recovery of 85.3% were obtained from the recrystallization process using the optimized feed concentration of 1.5 mg/mL, CO_2 flow rate of 48.6 g/min and a pressure of 135 bar.

Conclusion

On the one hand, most work carried out on culture of microalgae to date mainly focuses on the extraction of lipids from the algal biomass for the production of biodiesel, as microalgae could be a good alternative to terrestrial crops. Chisti (2008) estimated that microalgae biodiesel production would only need 3% of the United States cultivable land to replace all of US petroleum based transport fuels. However, the cost of production of microalgae biodiesel is still not competitive with fossil fuels and consequently, the commercialization of microalgae biodiesel is still far from reality. On the other hand, some high value compounds, such as carotenoids, have been extracted and commercialized from microalgae. Due to their market prices, these compounds justify the use of separation and extraction processes which are more energy demanding. However, the current harvesting, dewatering, and cell disrupting techniques are still too energy and time consuming, and are therefore a bottleneck hindering economic industrial-scale production of microalgae. Most of these techniques have to be optimized for each specific alga and for the compounds to be extracted from the algal biomass, and this contributes to the overall production costs. Further research is clearly needed to streamline these novel technologies to an industrial scale so that the algal components can be exploited by the industry.

References

Adam, F., M. Abert-Vian, G. Peltier and F. Chemat. 2012. "Solvent-free" ultrasound-assisted extraction of lipids from fresh microalgae cells: a green, clean and scalable process. Bioresour. Technol. 114: 457–465.

Ahmad, A.L., N.H. Mat Yasin, C.J.C. Derek and J.K. Lim. 2011. Optimization of microalgae coagulation process using chitosan. Chem. Eng. J. 173: 879–882.

Amirtharajah, A. and K.M. Mills. 1982. Rapid-mix design for mechanism of alum coagulation. J. Am. Water Works Ass. 74: 210–216.

Bajaj, I.B. and R.S. Singhal. 2011. Flocculation properties of poly(γ-glutamic acid) produced from *Bacillus subtilis* isolate. Food Bioprocess. Technol. 4: 745–752.

Balasubramanian, R.K., T.T.Y. Doan and J.P. Obbard. 2013. Factors affecting cellular lipid extraction from marine microalgae. Chem. Eng. J. 215-216: 929–936.

Básaca-Loya, G.A., M.A. Valdez, E.A. Enríquez-Guevara, L.E. Gutiérrez-Millán and M.G. Burboa. 2009. Extraction and purification of B-phycoerythrin from the red microalga *Rhodosorus marinus*. Cienc. Mar. 35: 359–368.

Beach, E.S., M.J. Eckelman, Z. Cui, L. Brentner and J.B. Zimmerman. 2012. Preferential technological and life cycle environmental performance of chitosan flocculation for harvesting of the green algae *Neochloris oleoabundans*. Bioresour. Technol. 121: 445–449.

Belarbi, E.H., E. Molina and Y. Chisti. 2000. A process for high yield and scaleable recovery of high purity eicosapentaenoic acid esters from microalgae and fish oil. Enzyme Microb. Tech. 26: 516–529.

Bilad, M.R., D. Vandamme, I. Foubert, K. Muylaert and I.F.J. Vankelecom. 2012. Harvesting microalgal biomass using submerged microfiltration membranes. Bioresour. Technol. 111: 343–352.

Bilanovic, D. and G. Shelef. 1988. Flocculation of microalgae with cationic polymers—Effects of medium salinity. Biomass 17: 65–76.

Biller, P., C. Friedman and A.B. Ross. 2013. Hydrothermal microwave processing of microalgae as a pre-treatment and extraction technique for bio-fuels and bio-products. Bioresour. Technol. 136: 188–195.

Bligh, E.G. and W.J. Dyer. 1959. A rapid method for total lipid extraction and purification. Can. J. Biochem. Physiol. 37: 911–917.

Bosma, R., W.A. van Spronsen, J. Tramper and R.H. Wijffels. 2003. Ultrasound, a new separation technique to harvest microalgae. J. Appl. Phycol. 15: 143–153.

Cerff, M., M. Morweiser, R. Dillschneider, A. Michel, K. Menzel and C. Posten. 2012. Harvesting fresh water and marine algae by magnetic separation: screening of separation parameters and high gradient magnetic filtration. Bioresour. Technol. 118: 289–295.

Chen, P., M. Min, Y. Chen, L. Wang, Y. Li, Q. Chen, C. Wang, Y. Wan, Y. Cheng, S. Deng, K. Hennessy, X. Lin, Y. Liu, Y. Wang, B. Martinez and R. Ruan. 2009. Review of the biological and engineering aspects of algae to fuels approach. Int. J. Agric. & Biol. Eng. 2: 1–30.

Chen, C.-R., S.-E. Hong, Y.-C. Wang, S.-L. Hsu, D. Hsiang and C.-M.J. Chang. 2012a. Preparation of highly pure zeaxanthin particles from sea water-cultivated microalgae using supercritical anti-solvent recrystallization. Bioresour. Technol. 104: 828–831.

Chen, F., Z. Liu, D. Li, C. Liu, P. Zheng and S. Chen. 2012b. Using ammonia for algae harvesting and as nutrient in subsequent cultures. Bioresour. Technol. 121: 298–303.

Chen, J. and G. Pan. 2012c. Harmful algal blooms mitigation using clay/soil/sand modified with xanthan and calcium hydroxide. J. Appl. Phycol. 24: 1183–1189.

Chisti, Y. 2008. Biodiesel from microalgae beats bioethanol. Trends Biotechnol. 26: 126–131.

Cho, S.-C., W.-Y. Choi, S.-H. Oh, C.-G. Lee, Y.-C. Seo, J.-S. Kim, C.-H. Song, S.-Y. Lee, D.-H. Kang and H.-Y. Lee. 2012. Enhancement of lipid extraction from marine microalga, *Scenedesmus* associated with high-pressure homogenization process. J. Biomed. Biotechnol. Article ID 359432, doi: 10.1155/2012/359432.

Choi, S.K., J.Y. Lee, D.Y. Kwon and K.J. Cho. 2006. Settling characteristics of problem algae in the water treatment process. Water Sci. Technol. 53: 113–119.

Christenson, L. and R. Sims. 2011. Production and harvesting of microalgae for wastewater treatment, biofuels and bioproducts. Biotechnol. Adv. 29: 686–702.

Couto, R.M., P.C. Simões, A. Reis, T.L. Da Silva, V.H. Martins and Y. Sánchez-Vicente. 2010. Supercritical fluid extraction of lipids from the heterotrophic microalga *Crypthecodinium cohnii*. Eng. Life Sci. 10: 158–164.

Crampon, C., A. Mouahid, S.A. Toudji, O. Lépine and E. Badens. 2013. Influence of pretreatment on supercritical CO_2 extraction from Nannochloropsis oculata. J. Supercrit. Fluid. 79: 337–344.

Csordas, A. and J.-K. Wang. 2004. An integrated photobioreactor and foam fractionation unit for the growth and harvest of *Chaetoceros* spp. In open systems. Aquacult. Eng. 30: 15–30.

Danquah, M.K., L. Ang, N. Uduman, N. Moheimani and G.M. Forde. 2009a. Dewatering of microalgal culture for biodiesel production: exploring polymer flocculation and tangential flow filtration. J. Chem. Technol. Biotechnol. 84: 1078–1083.

Danquah, M.K., B. Gladman, N. Moheimani and G.M. Forde. 2009b. Microalgal growth characteristics and subsequent influence on dewatering efficiency. Chem. Eng. J. 151: 73–78.

Dassey, A.J. and C.S. Theegala. 2013. Harvesting economics and strategies using centrifugation for cost effective separation of microalgae cells for biodiesel applications. Bioresour. Technol. 128: 241–245.

Dejoye Tanzi, C., M. Abert Vian and F. Chemat. 2013. New procedure for extraction of algal lipids from wet biomass: A green clean and scalable process. Bioresour. Technol. 134: 271–275.

Depraetere, O., G. Pierre, F. Deschoenmaeker, H. Badri, I. Foubert, N. Leys, G. Markou, R. Wattiez, P. Michaud and K. Muylaert. 2015. Harvesting carbohydrate-rich *Arthrospira platensis* by spontaneous settling. Bioresour. Technol. 180: 16–21.

Divakaran, R. and V.N.S. Pillai. 2002. Flocculation of algae using chitosan. J. Appl. Phycol. 14: 419–422.

Duan, J. and J. Gregory. 2003. Coagulation by hydrolyzing metals salts. Adv. Colloid Interfac. 100–102: 475–502.

Duerring, M., G.B. Schmidt and R. Huber. 1991. Isolation, crystallization, crystal structure analysis and refinement of constitutive C-phycocyanin from the chromatically adapting cyanobacterium *Fremyella diplosiphon* at 1.66 Å resolution. J. Mol. Biol. 217: 577–592.

Elridge, R.J., D.R.A. Hill and B.R. Gladman. 2012. A comparative study of the coagulation behaviour of marine microalgae. J. Appl. Phycol. 24: 1667–1679.

Farid, M.S., A. Shariati, A. Badakhshan and B. Anvaripour. 2013. Using nano-chitosan for harvesting microalga *Nannochloropsis* sp. Bioresour. Technol. 131: 555–559.

Farooq, W., Y.-C. Lee, J.-I. Han, C.H. Darpito, M. Choi and J.-W. Yang. 2013. Efficient microalgae harvesting by organo-building blocks of nanoclays. Green Chem. 15: 749–755.

Fayad, N., T. Yehya, F. Audonnet and C. Vial. 2017. Harvesting of microalgae *Chlorella vulgaris* using electro-coagulation-flocculation in the batch mode. Algal Res. 25: 1–11.

Ficner, R., K. Lobeck, G. Scmidt and R. Huber. 1992. Isolation, crystallization, crystal structure analysis and refinement of B-phycoerythrin from the red alga *Porphyridium sordidum* at 2.2 Å resolution. J. Mol. Biol. 228: 935–950.

Frappart, M., A. Massé, M.Y. Jaffrin, J. Pruvost and P. Jaouen. 2011. Influence of hydrodynamics in tangential and dynamic ultrafiltration systems for microalgae separation. Desalination 265: 279–283.

Fujii, K. 2012. Process integration of supercritical carbon dioxide extraction and acid treatment for astaxanthin extraction from a vegetative microalga. Food Bioprod. Process 90: 762–766.

Garzon-Sanabria, A.J., R.T. Davis and Z.L. Nikolov. 2012. Harvesting *Nannochloris oculata* by inorganic electrolyte flocculation: effect of initial cell density, ionic strength, coagulant dosage and media pH. Bioresour. Technol. 118: 418–424.

Gerde, J.A., M. Montalbo-Lomboy, L. Yao, D. Grewell and T. Wong. 2012. Evaluation of microalgae cell disruption by ultrasonic treatment. Bioresour. Technol. 125: 175–181.

Gerken, H.G., B. Donohoe and E.P. Knoshaug. 2013. Enzymatic cell wall degradation of *Chlorella vulgaris* and other microalgae for biofuels production. Planta 237: 239–253.

Glembin, P., M. Kerner and I. Smirnova. 2013. Cloud point extraction of microalgae cultures. Sep. Purif. Technol. 103: 21–27.

Godos (de), I., H.O. Guzman, R. Soto, P.A. García-Encina, E. Becares, R. Muñoz and V.A. Vargas. 2011. Coagulation/flocculation-based removal of algal-bacterial biomass from piggery wastewater treatment. Bioresour. Technol. 102: 923–927.

Goiris, K., K. Muylaert, I. Fraeye, I. Foubert, J. De Brabanter and L. De Cooman. 2012. Antioxidant potential of microalgae in relation to their phenolic and carotenoid content. J. Appl. Phycol. 24: 1477–1486.

Gouveia, L. and A.C. Oliveira. 2009. Microalgae as a raw material for biofuels production. J. Ind. Microbiol. Biotechnol. 36: 269–274.

Granados, M.R., F.G. Acién, C. Gómez, J.M. Fernández-Sevilla and E. Molina Grima. 2012. Evaluation of flocculants for the recovery of freshwater microalgae. Bioresour. Technol. 118: 102–110.

Greenwell, H.C., L.M.L. Laurens, R.J. Shields, R.W. Lovitt and K.J. Flynn. 2010. Placing microalgae on the biofuels priority list: a review of the technological challenges. J. R. Soc. Interface 7: 703–726.

Greenwell, H.C., L.M.L. Laurens, R.J. Shields, R.W. Lowitt and K.J. Flynn. 2013. Placing microalgae on the biofuels priority list: a review of the technological challenges. J. R. Soc. Interface 7: 703–726.

Grierson, S., V. Strezov, S. Bray, R. Mummacari, L.T. Danh and N. Foster. 2012. Assessment of bio-oil extraction from *Tetraselmis chui* microalgae comparing supercritical CO_2, solvent extraction, and thermal processing. Energy Fuels 26: 248–255.

Griffiths, M.J., R.P. van Hille and S.T.L. Harrison. 2012. Lipid productivity, settling potential and fatty acid profile of 11 microalgal species grown under nitrogen replete and limited conditions. J. Appl. Phycol. 24: 989–1001.

Gudin, C. and C. Thepenier. 1986. Bioconversion of solar energy into organic chemicals by microalgae. Adv. Biotechnol. Process 6: 73–110.

Guedes, A.M., H.M. Amaro and F.X. Malcata. 2011. Microalgae as sources of high added-value compounds—A brief review of recent work. Biotechnol. Prog. 27: 597–613.

Guedes, A.C., M.S. Gião, A.A. Matias, A.V.M. Nunes, M.E. Pintado, C.M.M. Duarte and F.X. Malcata. 2013. Supercritical fluid extraction of carotenoids and chlorophylls *a*, *b* and *c*, from a wild strain of *Scenedesmus obliquus* for use in food processing. J. Food Eng. 116: 478–482.

Halim, R., B. Gladman, M.K. Danquah and P.A. Webley. 2011. Oil extraction from microalgae for biodiesel production. Bioresour. Technol. 102: 178–185.

Halim, R., R. Harun, M.K. Danquah and P.A. Webley. 2012. Microalgal cell disruption for biofuel development. Appl. Energy 91: 116–121.

Hanotu, J., H.C. Hemaka Bandulasena and W.B. Zimmerman. 2012. Microflotation performance for algal separation. Biotechnol. Bioeng. 109: 1663–1673.

Harun, R. and M.K. Danquah. 2011. Enzymatic hydrolysis of microalgal biomass for bioethanol production. Chem. Eng. J. 168: 1079–1084.

Horst, I., B.M. Parker, J.S. Dennis, C.J. Howe, S.A. Scott and A.G. Smith. 2012. Treatment of *Phaeodactylum tricornutum* cells with papain facilitates lipid extraction. J. Biotechnol. 162: 40–49.

Kandasamy, G. and S.R.M. Saleh. 2017. Harvesting of microalga *Nannochloropsis* sp. by bioflocculation with mung bean protein extract. Appl. Biochem. Biotechnol. 182: 586–597.

Kim, S.M., S.-W. Kang, O.-N. Kwon, D. Chung and C.-H. Pan. 2012a. Fucoxanthin as a major carotenoid in *Isochrysis* aff. *galbana*: characterization of extraction for commercial application. J. Korean Soc. Appl. Biol. Chem. 55: 477–483.

Kim, S.M., Y.-J. Yung, O.-N. Kwon, K.H. Cha, B.-H. Um, D. Chung and C.-H. Pan. 2012b. A potential commercial source of fucoxanthin extracted from the microalga *Phaeodactylum tricornutum*. Appl. Biochem. Biotechnol. 166: 1843–1855.

Koberg, M., M. Cohen, A. Ben-Amotz and A. Gedanken. 2011. Bio-diesel production directly from the microalgae biomass of *Nannochloropsis* by microwave and ultrasound radiation. Bioresour. Technol. 102: 4265–4269.

Knuckey, R.M., M.R. Brown, R. Robert and D.M.F. Frampton. 2006. Production of microalgal concentrates by flocculation and their assessment as aquaculture feeds. Aquacult. Eng. 35: 300–313.

Kurane, R., K. Toeeda, K. Takeda and T. Suzuki. 1986. Microbial flocculant. 2. Culture conditions for production of microbial flocculant by *Rhodococcus erythropolis*. Agric. Biol. Chem. 50: 2309–2313.

Latif, S. and F. Anwar. 2011. Aqueous enzymatic sesame oil and protein extraction. Food Chem. 125: 679–684.

Lee, A.K., D.M. Lewis and P.J. Ashman. 2009. Microbial flocculation, a potentially low-cost harvesting technique for marine microalgae for the production of biodiesel. J. Appl. Phycol. 21: 559–567.

Lee, A.K., D.M. Lewis and P.J. Ashman. 2012. Disruption of microalgal cells for the extraction of lipids for biofuels: processes and specific energy requirements. Biomass Bioenerg. 46: 89–101.

Lee, A.K., D.M. Lewis and P.J. Ashman. 2013. Force and energy requirement for microalgal cell disruption: an atomic force microscope evaluation. Bioresour. Technol. 128: 199–206.

Lian, B., Y. Chen, J. Zhao, H.H. Teng, L. Zhu and S. Yuan. 2008. Microbial flocculation by *Bacillus mucilaginous*: Applications and mechanisms. Bioresour. Technol. 99: 4825–4831.

Liang, K., Q. Zhang and W. Cong. 2012. Enzyme-assisted aqueous extraction of lipid from microalgae. J. Agric. Food Chem. 60: 11771–11776.

Marchal, L., M. Mojaat-Guemir, A. Foucault and J. Pruvost. 2013. Centrifugal partition extraction of β-carotene from *Dunaliella salina* for efficient and biocompatible recovery of metabolites. Bioresour. Technol. 134: 396–400.

Markou G., I. Chatzipavlidis and D. Georgakakis. 2012. Carbohydrates production and bioflocculation characteristics in cultures of *Arthrospira* (*Spirulina*) *platensis*: Improvements through phosphorus limitation process. Bioenerg. Res. 5: 915–925.

McMillan, J.R., I.A. Watson, M. Ali and W. Jaafar. 2013. Evaluation and comparison of algal cell disruption methods: microwave, waterbath, blender, ultrasonic and laser treatment. Appl. Energ. 103: 128–134.

Mikulec J., G. Polakovičová and J. Cvengroš. 2015. Flocculation using polyacrylamide polymers for fresh microalgae. Chem. Eng. Technol. 38(4): 595–601.

Minkova, K.M., A.A. Tchernov, M.I. Tchorbadjieva, S.T. Fournadjieva, R.E. Antova and M.Ch. Busheva. 2003. Purification of C-phycocyanin from *Spirullina (Arthrospira) fusiformis*. J. Biotechnol. 102: 55–59.

Miranda, J.R., P.C. Passarinho and L. Gouveia. 2012. Pre-treatment optimization of *Scenedesmus obliquus* microalga for bioethanol production. Bioresour. Technol. 104: 342–348.

Mohn, F.H. 1980. Experiences and strategies in the recovery of biomass from mass cultures of microalgae. pp. 547–571. *In*: G. Shelef and C.J. Soeder (eds.). Algae Biomass. Elsevier, Amsterdam.

Mojaat, M., A. Foucault, J. Pruvost and J. Legrand. 2008. Optimal selection of organic solvents for biocompatible extraction of β-carotene from *Dunaliella salina*. J. Biotechnol. 133: 433–441.

Molina Grima, E., E.-H. Belarbi, F.G.A. Fernández, A.R. Medina and Y. Chisti. 2003. Recovery of microalgal biomass and metabolites: process options and economics. Biotechnol. Adv. 20: 491–515.

Monte, J., M. Sá, C.F. Galinha, L. Costa, H. Hoekstra, C. Brazinha and J.G. Crespo. 2018. Harvesting of *Dunaliella salina* by membrane filtration at pilot scale. Sp. Purif. Technol. 190: 252–260.

Montero, O., M.D. Macías-Sánchez, C.M. Lama, L.M. Lubián, C. Mantell, M. Rodríguez and E.M. de la Ossa. 2005. Supercritical CO_2 extraction of b-carotene from a marine strain of the Cyanobacterium *Synechococcus* species. J. Agric. Food Chem. 53: 9701–9707.

Moorthy, R.K., M. Premalatha and M. Arumugam. 2017. Batch sedimentation studies for freshwater green alga *Scenedesmus abundans* using combination of flocculants. Front. Chem. 5: 37.

Ndikubwimana, T., X. Zeng, T. Murwarnashyaka, E. Manirafasha, N. He, W. Shao and Y. Lu. 2016. Harvesting of freshwater microalgae with microbial bioflocculant: a pilot-scale study. Biotechnol. Biofuels 9: 47.

Neto, A.M.P., R.A.S. de Souza, A.D. Leon-Nino, J.A.A. da Costa, R.S. Tiburcio, T.A. Nunes, T.C.S. de Mello, F.T. Kanemoto, F.M.P. Saldanha-Corrêa and S.M.F. Gianesella. 2013. Improvement in microalgae lipid extraction using a sonication-assisted method. Renew. Energ. 55: 525–531.

Nguyen, M.T., S.P. Choi, J. Lee, J.H. Lee and S.J. Sim. 2009. Hydrothermal acid pretreatment of *Chlamydomonas reinhardtii* biomass for ethanol production. J. Microbiol. Biotechnol. 19: 161–166.

Nowotarski, K.K.P.M., E.M. Joyce and T.J. Mason. 2012. Ultrasonic disruption of algae cells. AIP Conf. Proc. 1433: 237–240. (doi: 10.1063/1.3703179).

Nurdogan, Y. and W.J. Oswald. 1996. Tube settling of high-rate pond algae. Wat. Sci. Tech. 33: 229–241.

Pan, J.-L., H.-M. Wang, C.-Y. Chen and J.-S. Chang. 2012. Extraction of astaxanthin from *Haematococcus pluvialis* by supercritical carbon dioxide fluid with ethanol modifier. Eng. Life Sci. 12: 638–647.

Patil, P.D., V.G. Gude, A. Mannarswamy, S. Deng, P. Cooke, S. Munson-McGee, I. Rhodes, P. Lammers and N. Nirmalakhandan. 2011. Optimization of direct conversion of wet algae to biodiesel under supercritical methanol conditions. Bioresour. Technol. 102: 118–122.

Patil, P.D., V.G. Gude, A. Mannarswamy, P. Cooke, N. Nirmalakhandan, P. Lammers and S. Deng. 2012. Comparison of direct transesterification of algal biomass under supercritical methanol and microwave irradiation conditions. Fuel 97: 822–831.

Pieber, S., S. Schober and M. Mittelbach. 2012. Pressurized fluid extraction of polyunsaturated fatty acids from the microalga *Nannochloropsis oculata*. Biomass Bioenerg. 47: 474–482.

Pranowo, R., D.J. Lee, J.C. Liu and J.S. Chang. 2013. Effect of O_3 and O_3/H_2O_2 on algae harvesting using chitosan. Wat. Sci. Technol. 67: 1294–1301.

Rossignol, N., L. Vandanjon, P. Jaouen and F. Quéméneur. 1999. Membrane technology for the continuous separation microalgae/culture medium : compared performances of cross-flow microfiltration and ultrafiltration. Aquacult. Eng. 20: 191–208.

Ryckebosch, E., K. Muylaert and I. Foubert. 2012. Optimization of an analytical procedure for extraction of lipids from microalgae. J. Am. Oil Chem. Soc. 89: 189–198.

Salim, S., R. Bosma, M.H. Vermuë and R.H. Wijffels. 2011. Harvesting of microalgae by bio-flocculation. J. Appl. Phycol. 23: 849–855.

Samarasinghe, N., S. Fernando, R. Lacey and W.B. Faulkner. 2012. Algal cell rupture using high pressure homogenization as a prelude to oil extraction. Renew. Energ. 48: 300–308.

Sánchez, Á., R. Maceiras, Á. Cancela and A. Pérez. 2013. Culture aspects of *Isochrysis galbana* for biodiesel production. Appl. Energy 101: 192–197.

Schlesinger, A., D. Eisenstadt, A. Bar-Gil, H. Carmely, S. Einbinder and J. Gressel. 2012. Inexpensive non-toxic flocculation of microalgae contradicts theories; overcoming a major hurdle to bulk algal production. Biotechnol. Adv. 30: 1023–1030.

Schwenzfeier, A., P.A. Wierenga and H. Gruppen. 2011. Isolation and characterization of soluble protein from the green microalgae *Tetraselmis* sp. Bioresour. Technol. 102: 9121–9127.

Shen, Y., Y. Cui and W. Yuan. 2013. Flocculation optimization of microalga *Nannochloropsis oculata*. Appl. Biochem. Biotechnol. 169: 2049–2063.

Sheng, J., R. Vannela and B.E. Rittmann. 2012. Disruption of *Synechocystis* PCC 6803 for lipid extraction. Water Sci. Technol. 65: 567–573.

Shih, I.L., Y.T. Van, L.C. Yeh, H.G. Lin and Y.N. Chang. 2001. Production of a biopolymer flocculant from *Bacillus licheniformis* and its flocculation properties. Bioresour. Technol. 78: 267–272.

Şirin, S., R. Trobajo, C. Ibanez and J. Salvadó. 2012. Harvesting the microalgae *Phaeodactylum tricornutum* with polyaluminium chloride, aluminium sulphate, chitosan and alkalinity-induced flocculation. J. Appl. Phycol. 24: 1067–1080.

Şirin, S., E. Clavero and J. Salvadó. 2013. Potential pre-concentration methods for *Nannochloropsis gaditana* and a comparative study of pre-concentrated sample properties. Bioresour. Technol. 132: 293–304.

Smith, B.T. and R.H. Davies. 2012. Sedimentation of algae flocculated using naturally-available magnesium-based flocculants. Algal Res. 1: 32–39.

Spilling, K., J. Seppälä and T. Tamminen. 2011. Inducing autoflocculation in the diatom *Phaeodactylum tricornutum* through CO_2 regulation. J. Appl. Phycol. 23: 959–966.

Suh, H.-H., G.-S. Kwon, C.-H. Lee, H.-S. Kim, H.-M. Oh and B.-D. Yoon. 1997. Characterization of bioflocculant produced by *Bacillus* sp. DP–152. J. Ferment. Bioeng. 84: 108–112.

Sukenik, A. and G. Shelef. 1984. Algal autoflocculation—Verification and proposed mechanism. Biotechnol. Bioeng. 26: 142–147.

Sukenik, A., W. Schröder, J. Lauer, G. Shelef and C.J. Soeder. 1985. Coprecipitation of microalgal biomass with calcium and phosphate ions. Water Res. 19: 127–129.

Sukenik, A., D. Bilanovic and G. Shelef. 1988. Flocculation of microalgae in brackish and sea waters. Biomass 15: 187–199.

Svarovsky, L.D. 1990. Solid Liquid Separation. 3rd Ed., Butterworth & Co., London, UK.

Talukder, Md.M.R., P. Das and J.C. Wu. 2012. Microalgae (*Nannochloropsis salina*) biomass to lactic acid and lipid. Biochem. Eng. J. 68: 109–113.

Taniguchi, M., K. Kato, A. Shimauchi, P. Xu, H. Nakayama, K.I. Fujita, T. Tanaka, Y. Tarui and E. Hirasawa. 2005. Proposals for wastewater treatment by applying flocculating activity of cross-linked poly-gamma-glutamic acid. J. Biosci. Bioeng. 99: 245–251.

Taylor, R.L., J.D. Rand and G.S. Caldwell. 2012. Treatment with algae extracts promotes flocculation, and enhances growth and neutrl lipid content in *Nannochloropsis oculata*—a candidate for biofuel production. Mar. Biotechnol. 14: 774–781.

Toeda, K. and R. Kurane. 1991. Microbial flocculant from *Alcaligenes cupidus* KT201. Agr. Biol. Chem. 55: 2793–2799.

Uduman, N., V. Bourniquel, M.K. Danquah and A.F.A. Hoadley. 2011. A parametric study of electrocoagulation as a recovery process of marine microalgae for biodiesel production. Chem. Eng. J. 174: 249–257.

Ulloa, G., C. Coutens, M. Sánchez, J. Sineiro, J. Fábregas, F.J. Deive, A. Rodríguez and M.J. Núñez. 2012. On the double role of surfactants as microalga cell lysis agents and antioxidants extractants. Green Chem. 14: 1044–1051.

Vandamme, D., I. Foubert, B. Meeschaert and K. Muylaert. 2010. Flocculation of microalgae using cationic starch. J. Appl. Phycol. 22: 525–530.

Vandamme, D., S.C.V. Pontes, K. Goiris, I. Foubert, L.J.J. Pinoy and K. Muylaert. 2011. Evaluation of electro-coagulation-flocculation for harvesting marine and freshwater microalgae. Biotechnol. Bioeng. 108: 2320–2329.

Vandamme, D., I. Foubert, I. Fraeye, B. Meesschaert and K. Muylaert. 2012. Flocculation of *Chlorella vulgaris* induced by high pH: Role of magnesium and calcium and practical implications. Bioresour. Technol. 105: 114–119.

Vandamme, D., I. Foubert and K. Muylaert. 2013. Flocculation as a low cost method for harvesting microalgae for bulk biomass production. Trends Biotechnol. 31: 233–239.

Wang, H., H.D. Laughinghouse IV, M.A. Anderson, F. Chen, E. Williams, A.R. Place, O. Zmora, Y. Zohar, T. Zheng and R.T. Hill. 2012. Novel bacterial isolate from Permian groundwater, capable of aggregating potential biofuel-producing microalga *Nannochloropsis oceanica* IMET1. Appl. Environ. Microbiol. 78: 1445–1453.

Wu, J.-Y. and H.-F. Ye. 2007. Characterization and flocculating properties of an extracellular biopolymer produced from a *Bacillus subtilis* DYU1 isolate. Process Biochem. 42: 1114–1123.

Wu, T., X. Yan, X. Cai, S. Tan, H. Li, J. Liu and W. Yang. 2010. Removal of *Chattonella marina* with clay minerals modified with a gemini surfactant. Appl. Clay Sci. 50: 604–607.

Wu, J.J., S.-E. Hong, Y.-C. Wang, S.-L. Hsu and C.-M.J. Chang. 2012a. Microalgae cultivation and purification of carotenoids using supercritical anti-solvent recrystallization of CO_2 + acetone solution. J. Supercrit. Fluid. 66: 333–341.

Wu, Z., Y. Zhu, W. Huang, C. Zhang, T. Li, Y. Zhang and A. Li. 2012b. Evaluation of flocculation induced by pH increase for harvesting microalgae and reuse of flocculated medium. Bioresour. Technol. 110: 496–502.

Yao, L., J.A. Gerde and T. Wang. 2012. Oil extraction from microalga *Nannochloropsis* sp. with isopropyl alcohol. J. Am. Oil Chem. Soc. 89: 2279–2287.

Yokoi, H., T. Arima, J. Hirose, S. Hayashi and Y. Takasaki. 1996. Flocculation properties of poly(γ-glutamic acid) produced by *Bacillus subtilis*. J. Ferment. Bioeng. 82: 84–87.

Zhang, P., H.H. Hahn, E. Hoffman and G. Zeng. 2004. Influence of some additives to aluminium species distribution in aluminium coagulants. Chemosphere. 57: 1489–1494.

Zhang, G., P. Zhang and M. Fan. 2009. Ultrasound-enhanced coagulation for *Microcystis aeruginosa* removal. Ultrason. Sonochem. 16: 334–338.

Zhang, F., L.-H. Cheng, X.-H. Xu, L. Zhang and H.-L. Chen. 2011. Screening of biocompatible organic solvents for enhancement of lipid milking from *Nannochloropsis* sp. Process Biochem. 46: 1934–1941.

Zheng, H., Z. Gao, J. Yin, X. Tang, X. Ji and H. Huang. 2012. Harvesting of microalgae by flocculation with poly(γ-glutamic acid). Bioresour. Technol. 112: 212–220.

Zhu, Y.-H. and J.-G. Yiang. 2008. Continuous cultivation of *Dunaliella salina* in photobioreactor for the production of β-carotene. Eur. Food Res. Technol. 227: 953–959.

Zou, N., C. Zhang, Z. Cohen and A. Richmond. 2000. Production of cell mass and eicosapentaenoic acid (EPA) in ultrahigh cell density cultures of *Nannochloropsis* sp. (Eustigmatophyceae). Eur. J. Phycol. 35: 127–133.

Žutić, V., B. Ćosović, E. Marčenko and N. Bihari. 1981. Surfactant production by marine phytoplankton. Marine Chem. 10: 505–520.

Marine Algal Bioactivities

Catarina Vizetto-Duarte, Carolina Bruno de Sousa, Maria Rodrigues,
Luísa Custódio, Luísa Barreira and *João Varela**

Introduction

Biological activity can be defined as the specific effect of, or a reaction to, exposure of a living organism/tissue/cell to a given compound or mix of compounds (e.g., extracts). Humans have explored different terrestrial plants as sources of bio-compounds for centuries. However, historically, marine organisms have had a limited number of reported applications in traditional medicine as compared to their terrestrial counterparts. Even so, the exploration of the marine environment has resulted in the isolation of thousands of structurally distinctive bioactive natural products (Haefner 2003).

Algae are promising sources of novel natural compounds with biological activities. Because of their exclusive features during evolution, algae spread to every known habitat including aquatic, terrestrial, and sub-aerial environments (Metting 1996). Since biodiversity is vital in the prospection for new chemical entities in drug discovery research, algae can be seen as a crucial field for the exploration of new biologically active compounds. Apart from being potential sources of natural products, algae present unique features such as easy cultivation and rapid growth for many species with commercial value, as well as the possibility of controlling the production of some bioactive compounds by manipulating the cultivation conditions (Kim et al. 2007; Sánchez et al. 2008). In this way, macro- and microalgae can be considered as genuine natural reactors being, in some cases, a good alternative to chemical synthesis for certain compounds. Several species live in complex habitats submitted to extreme abiotic conditions (e.g., salinity, temperature, nutrients, UV-Vis irradiation, and pressure) and their adaptation to these extreme environmental conditions includes the production of a great variety of secondary metabolites, which cannot be found in other organisms (Carlucci et al. 1999).

Formerly called blue-green algae, cyanobacteria are a diverse group of Gram-negative photosynthetic bacteria that produce an array of secondary metabolites with relevant bioactivities such as antifungal, antiviral, antibiotic, and cytotoxic, which make them also interesting candidates of potential pharmaceutical importance (Dittmann et al. 2001). However, this chapter will not focus on bioactivities displayed by compounds from cyanobacteria since nowadays most authors do not consider them as algae.

Secondary metabolites are generally produced either as a result of the acclimation and/or adaptation of an organism to its surrounding environment, or as a defence mechanism against predators (Dewick 2002). The processes by which an organism produces these compounds are found in specific groups of

Centre of Marine Sciences, Universidade do Algarve, Ed. 7, Campus de Gambelas, 8005-139 Faro, Portugal.
* Corresponding author: jvarela@ualg.pt

organisms and are an expression of the individuality of a species, which is defined as secondary metabolism (Maplestone et al. 1992; Dewick 2002). This metabolism provides the majority of pharmacologically active natural products, in contrast to the synthesis and breakdown of carbohydrates, proteins, lipids, and nucleic acids, which are vital for the sustainability of all living organisms, and is known as primary metabolism. Therefore, compounds involved in these pathways are known as 'primary metabolites'. It is the biosynthesis of secondary metabolites that provides algae with novel chemical structures possessing unique biological activities.

Today, there are several marine-derived drugs available commercially. For example, Ecteinascidin 743 (ET743; Yondelis™; trabectedin) was isolated in very low yields from the ascidian *Ecteinascidia turbinata* (Wright et al. 1990). The amount of ET743 required for advanced preclinical and clinical studies was only achieved via large-scale aquaculture of *E. turbinata* in open ponds, followed by compound isolation, as its synthesis was too complex (Manzanares et al. 2001). The production of ET743 was carried out later by a semi-synthetic, simpler method. In 2007, ET743 was the first marine anticancer drug to be approved in the European Union (Molinski et al. 2009) for ovarian, soft tissue sarcoma, breast, endometrial, prostate, non-small cell lung, and paediatric cancers, and the first of a novel class of DNA-binding agents (Henríquez et al. 2005). Other examples are ziconotide (Prialt, Elan Pharmaceuticals) and aplidine (dehydrodidemnin B). The former is a peptide first discovered in a tropical cone snail (*Conus magus*) and that was approved for the treatment of pain, whereas aplidine was isolated from the Mediterranean tunicate *Aplidium albicans* (Urdiales et al. 1996). Aplidine is used to treat various cancers, including melanoma, pancreatic, head and neck, small and non-small cell lung, bladder and prostate cancers, as well as non-Hodgkin lymphoma and acute lymphoblastic leukaemia (Henríquez et al. 2005).

Although marine algae are among the richest sources of chemically diverse natural products (Liu et al. 2011), their potential in drug discovery has remained largely unexplored. Nevertheless, great effort has been done in the last decades in order to investigate this resource which have led to important discoveries that are reviewed below.

Antioxidant activity

In aerobic life, oxygen metabolism is essential for energy production but also produces toxic metabolites such as reactive oxygen species (ROS), namely superoxide anion (O_2^-), hydroxyl radical (OH·), and hydrogen peroxide (H_2O_2) (Chew et al. 2008; Ko et al. 2012). In regular physiological conditions, organisms are able to repair or reduce the damage caused by ROS through an antioxidant defence system, including non-enzymatic and enzymatic factors (Ko et al. 2012). If these mechanisms are insufficient, the generation of ROS can overburden the cell, which might lead to oxidative damage to macromolecules (e.g., DNA, lipids, and proteins) and to the onset of different ailments (Kuda et al. 2005; Souza et al. 2012).

Antioxidants belong to the non-enzymatic protection defence system and can alleviate the adverse effects of oxidative stress (Chew et al. 2008; Ko et al. 2012). Since they act as a protection mechanism against oxidative stress, it has been put forward that antioxidant compounds may play a key role in preventing mutations and several diseases comprising cancer, cardiovascular, inflammatory, and neurodegenerative disorders (Kuda et al. 2005; Costa et al. 2010). Several synthetic antioxidants— butylated hydroxyanisole (BHA; E320), butylated hydroxytoluene (BHT; E321), propyl gallate (PG; E310), and tert-butylhydroquinone (TBHQ; E319)—have been used to avoid oxidative damage in food, but they have also been associated with non-desirable side effects such as toxicity and carcinogenicity (Cho et al. 2011; Souza et al. 2012). Therefore, safety issues have underlined the need for safer compounds from natural origin to replace artificial antioxidants (Kumar et al. 2011; Cho et al. 2011). Natural antioxidants may protect organisms against free radicals and delay the development of several chronic diseases (Heo and Jeon 2009). For example, astaxanthin is a pigment found in natural sources (e.g., algae, yeast, and crustacean by-products), which has been associated to a reduced risk of developing chronic diseases, including cardiovascular disorders, cancer, and *Helicobacter pylori* infection (Higuera-Ciapara et al. 2006). Moreover, astaxanthin is a powerful antioxidant with the capacity to modulate and stimulate the immune system (Higuera-Ciapara et al. 2006).

To evaluate antioxidant activity in natural extracts, there are several methods differing in their reaction mechanism, oxidant, and target species as well as reaction conditions (Karadag et al. 2009). Several *in vitro* methods generally used to evaluate the antioxidant potential of natural products comprise assays related with lipid peroxidation, in which peroxidation levels are assessed using ferric thiocyanate, the conjugated diene assay, β-carotene bleaching test, thiobarbituric acid reactive substances (TBARS), aldehyde/carboxylic acid assay, and formic acid measurements (Miguel 2010). Other procedures determine the free radical scavenging activity: 2,2-diphenyl-1-picrylhydrazyl (DPPH) assay, Trolox® equivalent antioxidant capacity (TEAC) or 2,2'-azino-bis(3-ethylbenzothiazoline-6-sulphonic acid) (ABTS$^{•+}$), ferric reducing/antioxidant power (FRAP) assay, reducing power (RP), chelating activity, hydroxyl (OH$^•$) radical scavenging and superoxide anion (O$_2$) scavenging activity (Miguel 2010).

With the purpose of identifying novel sources of natural and non-toxic antioxidants from marine organisms, various species of marine algae have already revealed their antioxidant capacity, and many compounds have been identified as responsible for these bioactivities (Nahas et al. 2007; El Gamal 2010; Kumar et al. 2011). Table 1 lists several compounds identified in marine algae with antioxidant capacity. Nine groups of compounds with proven antioxidant potential were identified, which are present in 24 species of marine algae belonging to 11 families (Heo and Jeon 2009; Li et al. 2009; Ananthi et al. 2010; Chakraborty and Paulraj 2010; Costa et al. 2010; El Gamal 2010; Cho et al. 2011; Ko et al. 2012; Souza et al. 2012). Bioactive polysaccharides are the most represented group, including the crude and sulphated polysaccharides, which are present in eight species belonging to four different families, namely Sargassaceae, Caulerpaceae, Codiaceae, and Gracilariaceae. Sulphated polysaccharides from *Gracilaria caudata* showed the best total antioxidant capacity with an IC$_{50}$ of 53.9 mg/g of acid ascorbic equivalents (Costa et al. 2010). In addition, eight terpenoids displaying antioxidant ability (IC$_{50}$ ranging between 0.07 [isoepitaondiol] and 0.21 mg algae extract/g DPPH [stypodiol] in DPPH assay) were found in six species of Sargassaceae, Dictyotaceae, Ishigeaceae, and Ulvaceae (Nahas et al. 2007; El Gamal 2010). Quinones, as sargaquinone (IC$_{50}$ of 0.20 mg algae extract/g DPPH) and stypoldione (IC$_{50}$ of 0.18 mg algae extract/g DPPH), have also demonstrated to possess this bioactivity, being present in three species, namely *Cystoseira crinita, Sargassum micracanthum*, and *Taonia atomaria* (Nahas et al. 2007; El Gamal 2010).

Anti-inflammatory activity

Inflammation is the first response of the immune system, an essential step to fight infection and heal wounds. Persistent activation of the immune system may, however, cause chronic inflammation giving rise to chronic diseases, such as arthritis, hepatitis, diabetes, several heart ailments, irritable bowel syndrome, Alzheimer's and Parkinson's diseases, allergies, asthma, and cancer (D'Orazio et al. 2012).

There are several types of cells involved in the inflammatory process as, for example, monocytes that differentiate into macrophages (Chatter et al. 2011). These are key cells in inflammatory disorders involving the increased production of cytokines (interleukin-1 beta [IL-1β], interleukin-6 [IL-6], tumour necrosis factor-alpha [TNF-α]), and other inflammatory mediators (ROS, nitric oxide [NO], prostaglandin E$_2$ [PGE$_2$], inducible nitric oxide synthase [iNOS] and ciclooxygenase-2 [COX-2]) (Heo et al. 2010). Diverse signalling molecules, such as nuclear factor-kappa B (NF-κB) and arachidonic acid (AA), participate and regulate inflammation through the production of pro- and anti-inflammatory mediators (Chatter et al. 2011). NF-κB is an important molecular player due to its fast activation and its capacity to activate the transcription of target genes involved in inflammation, such as those encoding pro-inflammatory cytokines, adhesion molecules, chemokines, and inducible enzymes (COX-2 and iNOS; Folmer et al. 2008; Kim et al. 2010). Conversely, AA is released by phospholipase A$_2$ (PLA$_2$), being converted into inflammatory mediators by COX, which catalyses the conversion from AA to prostaglandins (PGs), and by 5-lipooxygenase (5-LOX), which plays a role in the biosynthesis of leukotrienes (Lts) from AA (Funk 2001).

Hence, modulation of the production of the inflammatory mediators is a central goal in the treatment of inflammatory medical conditions (Heo et al. 2010). Several natural anti-inflammatory products exert their function by targeting and modulating the NF-κB signalling pathway (Chatter et al. 2011).

Table 1. Antioxidant compounds from marine algae.

Fraction/Compound Structure/Name	Marine Algae			Assay(s)[1]	Reference(s)
	Phylum	Family	Species		
Bromophenols					
Bromophenols	Rhodophyta	Rhodomelaceae	*Polysiphonia urceolata*	DPPH	El Gamal 2010; Li et al. 2007
(2R)-2-(2,3,6-tribromo-4,5-dihydroxybenzyl)cyclohexanone	Rhodophyta	Rhodomelaceae	*Symphyocladia latissula*	DPPH	El Gamal 2010
Glycerides					
Diacylglycerols	Ochrophyta	Sargassaceae	*Sargassum thunbergii*	DPPH and ONOO⁻	El Gamal 2010
Monogalactosyl	Ochrophyta	Sargassaceae	*Sargassum thunbergii*	DPPH and ONOO⁻	El Gamal 2010
Peptides					
Hexapeptide Leu-Asn-Gly-Asp-Val-Trp	Chlorophyta	Chlorellaceae	*Chlorella ellipsoidea*	DPPH, OH· and ROO·	Ko et al. 2012
Polysaccharides					
Crude polysaccharides	Ochrophyta	Sargassaceae	*Turbinaria ornata*	DPPH, ABTS·⁺, NO and LPO	Ananthi et al. 2010
Sulphated polysaccharides	Chlorophyta	Caulerpaceae	*Caulerpa cupressoides*	TAC	Costa et al. 2010
	Chlorophyta	Caulerpaceae	*Caulerpa prolifera*	TAC and ferrous-chelating capacity	Costa et al. 2010
	Chlorophyta	Caulerpaceae	*Caulerpa sertularioides*	TAC, OH·, O₂·⁻, ferrous-chelating activity and reducing power	Costa et al. 2010
	Chlorophyta	Codiaceae	*Codium isthmocladum*	TAC and reducing power	Costa et al. 2010
	Ochrophyta	Sargassaceae	*Sargassum filipendula*	TAC and reducing power	Costa et al. 2010
	Rhodophyta	Gracilariaceae	*Gracilaria caudata*	TAC, OH·, ferrous-chelating activity and reducing power	Costa et al. 2010
			Gracilaria birdiae	DPPH and OH·	Souza et al. 2012
Protein hydrolysates					
Hydrolysate	Rhodophyta	Bangiaceae	*Porphyra columbina*	DPPH, ABTS·⁺ and copper-chelating activity	Cian et al. 2012

Quinones					
Phenyl toluquinones	Ochrophyta	Sargassaceae	*Cystoseira crinita*		El Gamal 2010
Plastiquinones	Ochrophyta	Sargassaceae	*Sargassum micracanthum*		El Gamal 2010
Sargaquinone	Ochrophyta	Dictyotaceae	*Taonia atomaria*	DPPH and OH·	Nahas et al. 2007
Stypoldione	Ochrophyta	Dictyotaceae	*Taonia atomaria*	DPPH and OH·	Nahas et al. 2007
Tannins					
Eckstolonol	Ochrophyta	Lessoniaceae	*Ecklonia stolonifera*	DEPP	El Gamal 2010
Fucodiphlorethol	Ochrophyta	Lessoniaceae	*Ecklonia cava*	DPPH	El Gamal 2010
Phlorotannins	Ochrophyta	Lessoniaceae	*Ecklonia cava*	TAC, ESR, DPPH, OH·, O_2^{-}, ROO·, ROS determination by DCFH-DA, cellular membrane protein oxidation, intracellular GSH and MPO	Li et al. 2009
Tetrapyrroles					
Pheophorbide	Chlorophyta	Ulvaceae	*Enteromorpha prolifera*	DPPH, OH· and reducing power	Cho et al. 2011
Terpenoids					
Diphlorethohydroxycarmalol	Ochrophyta	Ishigeaceae	*Ishige okamurae*	ABTS·⁺, H_2O_2-induced cell damage, apoptosis and intracellular ROS RSA	Heo and Jeon 2009
Isoepitaondiol	Ochrophyta	Dictyotaceae	*Taonia atomaria*	DPPH and OH·	Nahas et al. 2007
Sargachromanols	Ochrophyta	Sargassaceae	*Sargassum siliquastrum*	DPPH and BchE	El Gamal 2010
Sargaol	Ochrophyta	Dictyotaceae	*Taonia atomaria*	DPPH and OH·	Nahas et al. 2007
Sesquiterpenoids	Chlorophyta	Ulvaceae	*Ulva fasciata*	DPPH and ABTS·⁺	Chakraborty and Paulraj 2010
Stypodiol	Ochrophyta	Dictyotaceae	*Taonia atomaria*	DPPH and OH·	Nahas et al. 2008
Taondiol	Ochrophyta	Dictyotaceae	*Taonia atomaria*	DPPH and OH·	Nahas et al. 2008
Thunbergols	Ochrophyta	Sargassaceae	*Sargassum thunbergii*	DPPH and ONOO⁻	El Gamal 2010

[1]ABTS·⁺ – 2,2'-azino-bis(3-ethylbenzothiazoline-6-sulphonic acid; BchE – butyrylcholinesterase; GSH – cellular glutathione; LPO – lipid peroxidation; MPO – myeloperoxidase; NO – nitric oxide; O_2^{-} – superoxide anion; OH· – hydroxyl radical; DCFH-DA – dichlorofluorescin diacetate; DPPH – 2,2-diphenyl-1-picrylhydrazyl; ESR – electron spin resonance spectrometry; ONOO⁻ – peroxynitrite; ROO· – peroxyl radical; TAC – total antioxidant capacity.

The most frequently used assays to evaluate the anti-inflammatory activity of natural compounds can be divided into *in vitro* and *in vivo* tests. The *in vitro* tests involve the (i) assessment of effects on the AA metabolism (COX and lipoxygenase [LOX] pathways), and/or the (ii) presence of different eicosanoids (prostaglandins [PGs] and leukotrienes [LTs]) in various cells; (iii) the effects on cytokines production (lipopolysaccharide [LPS]-induced IL-1β and TNF-α); (iv) the modulation of pro-inflammatory gene expression of NO, and isoforms of the NO synthase (NOS): constitutive (cNOS), endothelial (eNOS), the inducible (iNOS) NOS; and also (v) signalling pathways where the NF-κB transcription factor and mitogen-activated protein kinases (MAPKs) play an important role (Miguel 2010). The *in vivo* models include, among others, granuloma models and oedema induction via, for example, carrageenan (Souto et al. 2011).

Natural products and folk medicine are an important field in the search for bioactive compounds and development of drugs for the treatment of inflammation (Gautam and Jachak 2009). Algal natural products have already been shown to inhibit pro-inflammatory mediators, suggesting their potential in the inflammation treatment (Pangestuti and Kim 2011). As algae are normally exposed to high light and oxygen concentrations, which can trigger the accumulation of inflammatory molecules (NO and ROS) in vertebrate cells, it is likely that they have evolved biochemical mechanisms for their own protection against external stressors. This fact makes them a potential source of antioxidant and anti-inflammatory compounds (Heo et al. 2010). Table 2 provides an overview of compounds displaying anti-inflammatory properties identified in different marine algae. A set of 9 groups of compounds was identified in 20 species belonging to 9 different families. Among these biochemicals, pigments are one of the most representative group of bioactive compounds namely pheophytin, fucoxanthin, and a fucoxanthin derivative, which have been shown to reduce the production of NO, PGE$_2$, and pro-inflammatory cytokines. These pigments were isolated from five species belonging to four different families (Table 2). Crude sulphated polysaccharides, such as fucan and fucoidan, isolated from the Sargassaceae and Lessoniaceae families, also displayed anti-inflammatory activity measured by the carregenan permeability oedema (CPO) and vascular permeability tests, NO production in LPS-induced macrophages and transcriptional analysis of inflammatory mediator assays (Ananthi et al. 2010; Dore et al. 2012; Lee et al. 2012). Furthermore, the sesquiterpenes pacifenol and prepacifenol found in three species of the *Laurencia* genus were able to inhibit leukotriene B4 (LTB4) and thromboxane B2 (TXB2) production, modulate the COX pathway, and inhibit phospholipase A$_2$ (D'Orazio et al. 2012). In addition, the anti-inflammatory flavonoids catechol, hesperidin, rutin, and stypotriol triacetate were found in the rhodophyte *Porphyra dentata* and the ochrophyte *Stypopodium flabelliforme* (Kazłowska et al. 2010; Jaswir and Monsur 2011; D'Orazio et al. 2012).

Anti-proliferative activity

The balance between cell division and cell death is a basic feature in the development and maintenance of the body homeostasis. Disturbances in this balance can cause disease: too much cell death can cause injury, but too little of it is a prerequisite for cancer development (Elmore 2007). Thus, a tight control of the equilibrium between cell survival and death is necessary. Under typical conditions, this balance is maintained by tightly regulating both processes. However, when one or both processes are deregulated, cancer may ensue.

Andreeff et al. (2003) state that cancer is primarily the accumulation of clonal cells, leading to therapies that consist in (1) trying to reduce the number of tumour cells and (2) preventing their accumulation, either by stimulating cancer cell death (preferentially by apoptosis) or via cytostatic effects. There is a need for improving existing therapies as well as searching for new drugs that provide higher survival rates and lower the impact of side effects.

Several biomolecules with proved anti-proliferative activity are presented in Table 3. These molecules act via different cell mechanisms, described below, and present a diverse array of chemical structures, including phenols, alkaloids, terpenoids, polyesters and other secondary metabolites (Cabrita et al. 2010; El Gamal 2010; Güven et al. 2010; Liu et al. 2011; Wijesekara et al. 2011).

Table 2. Anti-inflammatory compounds from marine algae.

Fraction/Compound Structure/Name	Phylum	Family	Specie	Assay(s)[1]	Reference(s)
		Marine Algae			
Bromophenols					
Vidalols	Rhodophyta	Rhodomelaceae	*Vidalia obtusiloba*	Inhibition of phospholipase A_2	El Gamal 2010
Dienes					
3-0-b-D-glucopyranosy-lstigmasta-5,25-diene	Chlorophyta	Ulvaceae	*Ulva lactuca*		El Gamal 2010
Diphenyl ethers					
2-(20,40-dibromophenoxy)-4,6-dibromoanisol	Chlorophyta	Cladophoraceae	*Cladophora fascicularis*		El Gamal 2010
Flavonoids					
Catechol	Rhodophyta	Bangiaceae	*Porphyra dentata*	LPS-induced macrophages	Jaswir and Monsur 2011; Kazłowska et al. 2010
Hesperidin	Rhodophyta	Bangiaceae	*Porphyra dentata*	LPS-induced macrophages	
Rutin	Rhodophyta	Bangiaceae	*Porphyra dentata*	LPS-induced macrophages	
Stypotriol triacetate	Ochrophyta	Dictyotaceae	*Stypopodium flabelliforme*	Inhibition of LTB4 and TXB2, modulation of COX pathway and inhibition of phospholipase A_2	D' Orazio et al. 2012
Peptides					
cis- and trans-ceratospongamide	Rhodophyta	Lomentariaceae	*Ceratodictyon spongiosum*	Inhibition of phospholipase A_2	El Gamal 2010
Pigments					
Pheophytin	Chlorophyta	Ulvaceae	*Enteromorpha prolifera*	Production of O_2^- in macrophages, chemotaxis of human PMNs to FMLP and ear oedema	Okai and Higashi-Okai 1997
Fucoxanthin	Chlorophyta	Ishigeaceae	*Ishige okamurae*	NO, PGE_2, iNOS, COX-2, TNF-α, IL-1β, and IL-6 levels, degradation of IκB-α and Phosphorylation of p50, p60, and MAPKs in LPS-induced macrophages	Kim et al. 2010
	Ochrophyta	Sargassaceae	*Myagropsis myagroides*	LPS-induced NO production, PGE_2, iNOS, COX-2, IL-1β, IL-6, and TNF-α	Heo et al. 2010

Table 2 contd. ...

...Table 2 contd.

Fraction/Compound Structure/Name	Marine Algae			Assay(s)[1]	Reference(s)
	Phylum	Family	Specie		
Fucoxanthin derivative	Ochrophyta	Sargassaceae	Sargassum siliquastrum	LPS-induced NO production, PGE_2, iNOS, COX-2, IL-1β, IL-6, and TNF-α	Heo et al. 2010
	Ochrophyta	Dictyotaceae	Dictyota coriacea	LPS-induced NO production, PGE_2, iNOS, COX-2, IL-1β, IL-6, and TNF-α	Heo et al. 2010
Polysaccharides					
Crude polysaccharides	Ochrophyta	Sargassaceae	Turbinaria ornata	CPO and vascular permeability	Ananthi et al. 2010
Fucan sulphated polysaccharides	Ochrophyta	Sargassaceae	Sargassum vulgare	CPO	Dore et al. 2012
Fucoidan	Ochrophyta	Lessoniaceae	Ecklonia cava	NO production in LPS-induced macrophages, transcriptional analysis of iNOS, COX-2, TNF-α, IL-6, and IL-1β	Lee et al. 2012
Protein hydrolysates					
Hydrolysate	Rhodophyta	Bangiaceae	Porphyra columbina	Inhibition of TNF-α in LPS-stimulated macrophages and splenocytes, and inhibition of IL-6 and IFNγ in LPS- and ConA-stimulated macrophages	Cian et al. 2012
Terpenoids					
Epitaondiol	Ochrophyta	Dictyotaceae	Stypopodium flabelliforme	Inhibition of LTB4 and TXB2, modulation of COX pathway and inhibition of phospholipase A_2	D' Orazio et al. 2012
Neorogioltriol	Rhodophyta	Rhodomelaceae	Laurencia glandulifera	CPO, NF-κB transactivation, COX-2 expression and TNF-α and NO release	Chatter et al. 2011

Pacifenol	Rhodophyta	Rhodomelaceae	*Laurencia claviformis*	Inhibition of LTB4 and TXB2, modulation of COX pathway and inhibition of phospholipase A$_2$	D' Orazio et al. 2012
	Rhodophyta	Rhodomelaceae	*Laurencia filiformis*	Inhibition of LTB4 and TXB2, modulation of COX pathway and inhibition of phospholipase A$_2$	D' Orazio et al. 2012
Prepacifenol	Rhodophyta	Rhodomelaceae	*Laurencia filiformis*	Inhibition of LTB4 and TXB2, modulation of COX pathway and inhibition of phospholipase A$_2$	D' Orazio et al. 2012

[1]ConA – concanavalin A; COX – cyclooxygenase; CPO – carrageenan permeability oedema; FMLP – formyl-Met-Leu-Phe; IFN-γ – interferon-gamma; IL-1β – interleukin-1-β; IL-6 – interleukin-6; iNOS – inducible nitric oxide synthase; IκB – inhibitory protein B; LPS – lipopolysaccharide; LTB4 – leukotriene B4; MAPKs – mitogen activated protein kinases; NF-κB – nuclear factor-kappaB; NO – nitric oxide; PGE2 – prostaglandin E2; PMNS – human polymorphonuclear leukocytes; TNF-α – tumour necrosis factor – alpha; TXB2 – thromboxane B2.

Table 3. Anti-proliferative compounds from marine algae.[1]

Fraction/Compound	Marine algae			Reference(s)
Structure/Name	Phylum	Family	Species	Reference(s)
Alkaloids				
Lophocladine B	Rhodophyta	Rhodomelaceae	*Lophocladia* sp.	Gross et al. 2006
Bromoethers				
Thyrsiferyl 23-acetate	Rhodophyta	Rhodomelaceae	*Laurencia obtusa*	Suzuki et al. 1985
Carboxylic acid				
Turbinaric acid	Ochrophyta	Sargassaceae	*Turbinaria ornata*	Asari et al. 1989
Hydrocarbons				
Pyrene	Rhodophyta	Plocamiaceae	*Plocamium cartilagineum*	Argandoña et al. 2002; de Inés et al. 2004
Organic halides				
Isorawsonol	Chlorophyta		*Avrainvillea rawsonii*	Chen et al. 1994
Tetrachlorinated cyclohexane	Rhodophyta	Plocamiaceae	*Plocamium cartilagineum*	Argandoña et al. 2002; de Inés et al. 2004
Peptides				
Kahalalide F	Chlorophyta	Bryopsidaceae	*Bryopsis* sp.	Hamman and Scheuer 1993; Dmitrenok et al. 2006
Kahalalide P	Chlorophyta	Bryopsidaceae	*Bryopsis* sp.	Hamman and Scheuer 1993; Dmitrenok et al. 2006
Kahalalide Q	Chlorophyta	Bryopsidaceae	*Bryopsis* sp.	Hamman and Scheuer 1993; Dmitrenok et al. 2006
Polysaccharides				
Fucose-containing sulfated polysaccharide	Ochrophyta	Sargassaceae	*Sargassum henslowianum*	Ale et al. 2011
Fucose-containing sulfated polysaccharide	Ochrophyta	Fucaceae	*Fucus vesiculosus*	Ale et al. 2011
Quinones				
Bis-prenylated quinones	Ochrophyta	Sporochnaceae	*Perithalia capillaris*	Blackman et al. 1979
7-hydroxycymopolone	Chlorophyta	Dasycladaceae	*Cymopolia barbata*	Badal et al. 2012
Sterols				
Cycloartenol disulfates	Chlorophyta		*Tydemania expeditionis*	Govindan et al. 1994
6-hydroxy-24-ethylcholesta-4,24(28)-dien-3-one (8)	Ochrophyta	Sargassaceae	*Turbinaria conoides*	Sheu et al. 1999

Compound	Phylum	Species	Family	Reference
24-hydroperoxy-6-hydroxy-24-ethylcholesta-4,28(29)-dien-3-one (9)	Ochrophyta	*Turbinaria conoides*	Sargassaceae	Sheu et al. 1999
24-ethyl-cholesta-4,24(28)-dien-3-one (4)	Ochrophyta	*Turbinaria conoides*	Sargassaceae	Sheu et al. 1999
24-hydroperoxy-24-ethylcholesta-4,28(29)-dien-3-one (5)	Ochrophyta	*Turbinaria conoides*	Sargassaceae	Sheu et al. 1999
24-ethylcholesta-4,24(28)-dien-3,6-dione (6)	Ochrophyta	*Turbinaria conoides*	Sargassaceae	Sheu et al. 1999
24-hydroperoxy-24-ethylcholesta-4,28(29)-dien-3,6-dione (7)	Ochrophyta	*Turbinaria conoides*	Sargassaceae	Sheu et al. 1999
Sterols	Ochrophyta	*Sargassum carpophyllum*	Sargassaceae	Tang et al. 2002
Sulfur-containing polybromoindoles	Rhodophyta	*Laurencia brongniartii*	Rhodomelaceae	El Gamal et al. 2005
Terpenoids				
4-acetoxydictyolactone	Ochrophyta	*Dictyota dichotoma*	Dictyotaceae	Ishitsuka et al. 1988
Atomarianones A and B	Ochrophyta	*Taonia atomaria*	Dictyotaceae	Abatis et al. 2005
Bifurcadiol	Ochrophyta	*Bifurcaria bifurcata*	Sargassaceae	Guardia et al. 1999
Bromophycolides C-I	Rhodophyta	*Callophycus serratus*	Rhodophyta incertae sedis	Kubanek et al. 2006
Chromazonarol	Ochrophyta	*Dictyopteris undulata*	Dictyotaceae	Ochi et al. 1979; Kurata et al. 1996; Laube et al. 2005
Cuparene sesquiterpenes	Rhodophyta	*Laurencia microcladia*	Rhodomelaceae	Kladia et al. 2005
Cyclozonarone	Ochrophyta	*Dictyopteris undulata*	Dictyotaceae	Ochi et al. 1979; Kurata et al. 1996; Laube et al. 2005
Cystoseirol monoacetate	Ochrophyta	*Cystoseira myrica*	Sargassaceae	Ayyad et al. 2003
Dictyol F monoacetate	Ochrophyta	*Cystoseira myrica*	Sargassaceae	Ayyad et al. 2003
Dictyone acetate	Ochrophyta	*Cystoseira myrica*	Sargassaceae	Ayyad et al. 2003
Dictyotalide A	Ochrophyta	*Dictyota dichotoma*	Dictyotaceae	Ishitsuka et al. 1988
Dictyotalide B	Ochrophyta	*Dictyota dichotoma*	Dictyotaceae	Ishitsuka et al. 1988
Dictyotins A, B and C	Ochrophyta	*Dictyota dichotoma*	Dictyotaceae	Wu et al. 1990
4,18-dihydroxydictyolactone	Ochrophyta	*Dictyota* sp.	Dictyotaceae	Jongaramruong and Kongkam 2007

Table 3 contd....

...Table 3 contd.

Structure/Name	Marine algae			Reference(s)
Fraction/Compound	**Phylum**	**Family**	**Species**	
Dilopholide	Ochrophyta	Dictyotaceae	*Dilophus ligulatus*	Bouaicha et al. 1993
Dolabellane	Ochrophyta	Dictyotaceae	*Dictyota* sp.	Tringali et al. 1984
Furoplocamioid C	Rhodophyta	Plocamiaceae	*Plocamium cartilagineum*	Argandoña et al. 2002; de Inés et al. 2004
Halimedalactone	Chlorophyta	Udoteaceae	*Halimeda* sp.	Paul and Fenical 1983
Halimediatrial	Chlorophyta	Udoteaceae	*Halimeda* sp.	Paul and Fenical 1984
Halimedatrial	Chlorophyta	Udoteaceae	*Halmida lamouroux*	Paul and Fenical 1983
Halmon	Rhodophyta	Rhizophyllidaceae	*Portieria hornemannii*	Fuller et al. 1992, 1994
12-hydroxygeranylgeraniol	Ochrophyta	Sargassaceae	*Bifurcaria bifurcata*	Gulioli et al. 2004
Isodictytriol monoacetate	Ochrophyta	Sargassaceae	*Cystoseira myrica*	Ayyad et al. 2003
Isozonarol	Ochrophyta	Dictyotaceae	*Dictyopteris undulata*	Ochi et al. 1979; Kurata et al. 1996; Laube et al. 2005
Isozonarone	Ochrophyta	Dictyotaceae	*Dictyopteris zonarioides*	Fenical et al. 1973
Laurenditerpenol	Rhodophyta	Rhodomelaceae	*Laurencia intricata*	Mohammed et al. 2004
Laurinterol	Rhodophyta	Rhodomelaceae	*Laurencia okamurae*	Kim et al. 2008
Nordictyotalide	Ochrophyta	Dictyotaceae	*Dictyota dichotoma*	Ishitsuka et al. 1988
Perfuroplocamioid	Rhodophyta	Plocamiaceae	*Plocamium cartilagineum*	Argandoña et al. 2002; de Inés et al. 2004
Sargol, Sargol I and II	Ochrophyta	Sargassaceae	*Sargassum tortile*	Numata et al. 1991
Terpenoid C	Ochrophyta	Dictyotaceae	*Stypopodium zonale*	Dorta et al. 2002
Yahazunol	Ochrophyta	Dictyotaceae	*Dictyopteris undulata*	Ochi et al. 1979; Kurata et al. 1996; Laube et al. 2005
Zonarol	Ochrophyta	Dictyotaceae	*Dictyopteris undulata*	Ochi et al. 1979; Kurata et al. 1996; Laube et al. 2005
Zonarone	Ochrophyta	Dictyotaceae	*Dictyopteris undulata*	Ochi et al. 1979; Kurata et al. 1996; Laube et al. 2005

[1]This list of anti-proliferative compounds is not comprehensive.

Hypoxia-inducible factor 1 (HIF-1) is an important molecule as it regulates the transcription of many of the genes involved in key aspects of cancer biology, including immortalization, maintenance of stem cell pools, cellular dedifferentiation, genetic instability, vascularization, metabolic reprogramming, autocrine growth factor signalling, invasion/metastasis, and treatment failure. In general, HIF-1 promotes tumour cell adaptation and survival under hypoxic conditions (Mohammed et al. 2004). Thus, specific HIF-1 inhibitors are validated as an important class of potential tumour-selective therapeutic agents. In further support of this conclusion, immunohistochemical detection of HIF-1 overexpression in biopsy sections is a prognostic factor in many cancers. Gaining insight into the cell and molecular biology of cancer cells and how they survive and spread has led to additional lines of research involving bioactive compounds from algae. For example, using a T47D human breast tumour cell-based luciferase reporter assay to monitor HIF-1 activity, extracts from the red alga *Laurencia intricata* (Rhodomelaceae) were evaluated for HIF-1 inhibitory activity. The bioassay-guided fractionation of the lipid extract of a Jamaican collection of the red alga *Laurencia intricata* yielded the first marine natural product inhibiting HIF-1 activation (Mohammed et al. 2004). The active compound was a structurally novel bicyclic diterpene called laurenditerpenol that inhibited hypoxia (1% O_2)-induced HIF-1 activation in T47D cells (IC_{50} = 0.4 µM). Laurenditerpenol was shown to inhibit HIF-1 activation by blocking hypoxia-induced HIF-1α protein accumulation. Respiration studies established that laurenditerpenol suppresses mitochondrial oxygen consumption at the electron transport complex I (IC_{50} = 0.8 µM). Further studies with laurenditerpenol were hindered by a lack of compound supply. However, total synthesis has resolved the absolute configuration of laurenditerpenol (Chittiboyina et al. 2007) and together with other synthetic efforts (e.g., Jung and Im 2008) may provide sufficient compound to allow further evaluation as a potential chemotherapeutic agent.

The preclinical pharmacology of dehydrothyrsiferol (DT), a polyether triterpenoid isolated from a Canary island collection of the red alga *Laurencia viridis*, was evaluated by the biochemical nature of the cytotoxic effect of DT on the human oestrogen receptor+ (ER+) and oestrogen receptor⁻ (ER⁻) breast cancer cell lines (Pec et al. 2003). Although they were able to exclude the possibility that DT functions as a mitotic inhibitor, they noted that induction of apoptosis was induced more efficiently and with distinct cell cycle-related patterns in the more aggressive ER⁻ cells' while being less complete in ER+ breast cancer cell lines. Moreover, other polyether triterpenoids such as iubol, 22-hydroxy-15(28)-dehydrovenustatriol and secodehydrothyrsiferol isolated from the latter alga showed effectiveness against Jurkat leukemic cells (IC_{50} = 2.0–3.5 µM). The capacity of the above compounds to inhibit cell proliferation was likely due to their ability to induce apoptosis, as assessed by the appearance of a sub-G1/G0 subpopulation in cell cycle analysis, indicative of DNA breakdown (Pacheco et al. 2011). Two cyclized meroditerpenoids, atomarianones A and B, were isolated from the organic extract of the brown alga *Taonia atomaria* collected at Serifos island in the Central Aegean Sea. The cytotoxicity of atomarianones A and B was assayed against NSCLC-N6 and A549 lung cancer cell lines. Both metabolites showed significant cytotoxicity in the two cell lines with IC_{50} values < 7.35 µM (Abatis et al. 2005). Halogenated monoterpenes isolated from the red alga *Plocamium cartilagineum* were evaluated for their cytotoxic effects on murine colon adenocarcinoma CT26, human colon adenocarcinoma SW480, human cervical adenocarcinoma HeLa, and human malignant melanoma SkMel28 cells as well as on the mammalian non-tumoural cell line CHO (Chinese hamster ovary cells). Four of the nine isolated monoterpenes exhibited selective cytotoxicity against colon and cervical adenocarcinoma cells. Interestingly, the effect of one of the compounds was specific and irreversible to human colon adenocarcinoma SW480 cells, which overexpress the transmembrane P-glycoprotein, a drug efflux transporter often related to chemoresistance. None of the anti-tumoural doses of these compounds was cytotoxic against CHO cells (de Inés et al. 2004).

Certain polysaccharides are known to inhibit proliferation of different cancer cell lines. The extracellular acidic polysaccharide GA3P, a D-galactan sulphate associated with L-(+)-lactic acid produced by the marine microalga *Gymnodinium* sp., has been shown to be a potent inhibitor of the DNA topoisomerases I and II (Umemura et al. 2003). Furthermore, GA3P exhibited moderate *in vitro* cytotoxicity against 38 human tumours (IC_{50} values ranged from 0.67 to 11 µg/mL; Umemura et al. 2003). Fucose-containing sulphated polysaccharides extracted from *Sargassum henslowianum* and *Fucus vesiculosus* decreased the proliferation of melanoma cells in a dose-response manner via induction of apoptosis dependent on regulatory cascades involved in the activation of caspase-3 (Ale et al. 2011). Polysaccharides from the algae *Sargassum latifolium* showed selective cytotoxicity against lymphoblastic leukaemia (1301 cells), leading to a major disturbance of the cell cycle, including arrest in S-phase and apoptotic (rather than necrotic) induced-cell death (Gamal-Eldeen et al. 2009).

Polyunsaturated fatty acids (PUFA) have also been shown to possess anti-proliferative potential. Chiu et al. (2004) investigated whether docosahexaenoic acid (DHA) from the cultured microalga *Crypthecodinium cohnii* was able to inhibit growth of human breast carcinoma MCF-7 cells. Indeed, DHA produced a dose-dependent growth inhibition of breast cancer cells upon a 72-h incubation with 40–160 μM of the fatty acid. DNA flow cytometry showed that DHA induced sub-G1 cells, or apoptotic cells, by 64.4% to 171.3% of control levels upon incubations with 80 μM of the fatty acid. Western blot studies further suggested that DHA did not modulate the expression of pro-apoptotic Bax protein but induced a time-dependent downregulation of anti-apoptotic Bcl-2 expression. Therefore, Chiu et al. (2004) suggested that DHA from the cultured *Crypthecodinium cohnii* is effective in controlling cancer cell growth and in inducing apoptosis by downregulation of Bcl-2.

Carotenoids are well known antioxidant molecules that may display a strong anti-proliferative effect. Cha et al. (2008) isolated carotenoids from the marine microalga *Chlorella ellipsoidea* displaying antiproliferative activity. Partially purified extracts of *C. ellipsoidea* inhibited HCT116 cell growth in a dose-dependent manner, yielding IC_{50} values of 40.73 μg/mL. In addition, treatment with *Chlorella* extracts enhanced the fluorescence intensity typical of early apoptotic cell population in HCT-116 cells. Astaxanthin, a red carotenoid accumulated by *Haematoccocus pluvialis* cysts, showed a potent anti-tumoural activity against oral and colon cancers. A few *Dunaliella* species also accumulate considerable amounts of isoforms of β-carotene, which displayed a strong anti-proliferative effect against MCF-7 cell line, along with astaxanthin (Guedes et al. 2011).

Degradation of extracellular matrix is crucial for malignant tumour growth, invasion, metastasis, and angiogenesis. Matrix metalloproteinases (MMP) are a family of zinc-dependent neutral endopeptidases collectively capable of degrading essentially all matrix components. Elevated levels of distinct MMP can be detected in tumour tissue or in the serum of patients with advanced cancer and their role as prognostic indicators in cancer has been assessed. Also, MMP inhibitors are in clinical trials with cancer patients (Vihinen and Kähäri 2002). 6,6'-bieckol is a phloroglucinol derivative isolated from the marine alga *Ecklonia cava* that inhibited MMP-2 and -9 gene expression by down-regulation of NF-κB. Furthermore, 6,6'-bieckol suppressed the migration of HT1080 cell line (Zhang et al. 2010). An *Ecklonia cava* phloroglucinol derivative, dioxinodehydroeckol, not only inhibited the proliferation of human breast cancer cells (Kong et al. 2009), but also exerted a higher anti-proliferative activity in MCF-7 human cancer cells as well as apoptosis induction in a dose-dependent manner. Treatment with dioxinodehydroeckol induced caspase-3 and -9 activities, DNA repair enzyme poly-(ADP-ribose) polymerase (PARP) cleavage, and pro-apoptotic (p53 and Bax) gene expression. It also led to a down-regulation in Bcl-2 anti-apoptotic gene. In addition, the NF-κB family and NF-κB-dependent activated genes were down-regulated (Kong et al. 2009). Apart from the inflammatory response, the inducible transcription factor NF-κB plays an important role in the regulation of carcinogenic responses. Although normal, NF-κB activation is necessary for cell survival and immunity, deregulated NF-κB expression is common in cancer development (Folmer et al. 2008).

Prenylated bromohydroquinones are cymopol-related metabolites that are known to accumulate in the marine chlorophyte *Cymopolia barbata*. One of these compounds, 7-hydroxycymopolone, was investigated for cytotoxicity against three cancerous cell lines and one non-tumoural cell line. 7-hydroxycymopolone selectively impacted the viability of HT29 colon cancer cells (IC_{50} = 19.82 μM) with similar potency to the known chemotherapeutic drug, fluorouracil (IC_{50} = 23.50 μM). In comparison, the effect on non-tumoural colon cells resulted in an IC_{50} value of 55.65 and 55.51 μM for 7-hydroxycymopolone and fluorouracil, respectively (Badal et al. 2012). The endemic New Zealand brown alga *Perithalia capillaris* afforded a bis-prenylated quinone through bioactivity-directed isolation that was potent at inhibiting the proliferation of HL60 cells (IC_{50} 0.34 μM; Sansom et al. 2007).

Fucosterol, identified as the most abundant phytosterol from the hexane fraction of *Sargassum angustifolium*, was considered to be responsible for the cytotoxic effect of this extract against human ductal breast epithelial tumour (T47D) and human colon carcinoma (HT29) cell lines (IC$_{50}$ of 27.94 and 70.41 µg/mL; Khanavi et al. 2012).

Algae extracts have also showed promising preliminary results, although the final goal would always be to identify the bioactive compound(s). Ethyl acetate extracts from *Colpomenia sinuosa*, *Halimeda discoidea*, and *Galaxaura oblongata* inhibited the growth of human hepatoma HuH-7 cells and leukaemia U937 and HL-60 cells in a time- and dose-dependent manner (Huang et al. 2005). The extracts induced apoptosis of U937 and HL-60 cells as evaluated by detection of hypodiploid cells using flow cytometry and observation of condensed and fragmented nuclei in algae extract-treated cells. However, the antioxidant N-acetylcysteine effectively blocked algal extract-induced apoptosis, suggesting that the extracts induced apoptosis in human leukaemia cells through the generation of ROS (Huang et al. 2005). The hexane fraction of *Sargassum swartzii* and *Cystoseira myrica* showed selective cytotoxicity against proliferation of Caco-2 cells (IC$_{50}$ < 100 µg/mL) and T47D cell line (IC$_{50}$ < 100 µg/mL), respectively, increasing apoptosis in these cells (Khanavi et al. 2010).

Marine algae are thus prolific producers of biologically active secondary metabolites with cytotoxic chemicals. Global research towards the discovery of novel and clinically useful antitumoural agents derived from marine organisms continues at a remarkably active pace.

In vivo anti-cancer activity

Suffness and Douros (1982) posited that the terminology with reference to *in vitro* and *in vivo* anticancer activity had been used too loosely. Many compounds cited as anticancer or antitumoural agents are in fact only cytotoxic to tumour cells *in vitro* and may not display particular selectivity towards tumour cells as opposed to non-tumoural cells. For that reason, the authors have suggested that the term "cytotoxicity" means toxicity to tumour cells in *in vitro* cell cultures. Thus, the terms "antitumoural" or "antineoplastic" should not be used to express results of *in vitro* assays; rather it should refer to data of *in vivo* trials with animal models exclusively. In addition, the term "anticancer" should be reserved for reporting clinical trials data in humans.

The Food and Drug Administration (FDA) is the federal agency responsible for the approval of all drugs commercialized in the United States. The FDA has established an approval protocol that implies several drug trials, grouped in stages. The first stage involves *in vitro* and animal testing. Only if adverse side effects are not observed should the testing proceed to the second stage, where human subjects are subjected to clinical trials. In the clinical trials stage, compounds are typically submitted to four phases. In Phase I, researchers gather preliminary information on the chemical action and safety to find a safe testing dose in a small group of people. In Phase II trials, the experimental treatment is given to a larger group of subjects to provide knowledge on the efficacy of the drug and additional information on its safety. In Phase III trials, the treatment is given to large groups of people (1000–3000) to confirm its effectiveness, monitor side effects, compare it to commonly used treatments, assess dosage effects, and collect further information that will allow it to be safely used. In Phase IV trials, post-marketing studies delineate additional information, including the risks and benefits of the treatment as well as the optimal use of the novel drug (Meinert 2006). However, one has to take into account that going through all the steps of this protocol can take several years in order to verify whether treatment generates long-term side effects.

Kahalalide F is a marine natural product and an anticancer drug candidate in clinical development at PharmaMar (Hamann and Scheuer 1993). This cyclic depsipeptide (of the family of dehydroaminobutyric) was isolated from the sea slug *Elysia rufescens* but is most probably derived from *Bryopsis* sp., its green algal diet (reviewed by Varela et al. 2013). *E. rufescens* is able to sequester algal chloroplasts, which synthesize secondary metabolites. The compound has shown antitumoural activity, probably by interfering with lysosome function in prostate, colorectal, and lung cancer cell lines as well as in animal models of lung and breast cancer (García-Rocha et al. 1996; Suárez et al. 2003). The evidence of *in vivo* activity in experimental human cancer models of androgen independent prostate cancer and other solid tumours established a rationale to implement a clinical program with this innovative compound. A phase I trial investigating the feasibility of a weekly schedule with kahalalide F given as a 1-hour intravenous infusion in patients with advanced pre-treated solid tumours is now complete (Pardo et al. 2008). A dose limit was set due to the occurrence of toxicity symptoms, mainly acute transaminitis, which precluded the administration of the compound in a weekly fashion; however, administration of this compound led to the remarkable absence of bone marrow suppression, alopecia, and other organ toxicities. Such early data suggests lack of cumulative toxicities that may allow chronic therapy. The pharmacokinetic profile demonstrates a short terminal half-life, a finding supporting additional studies with longer infusion schedules. Evidence of activity in pre-treated patients with melanoma (Martín-Algarra et al. 2009), colorectal cancer, and hepatocellular carcinoma has also been reported (Jimeno et al. 2004). A phase II trial in patients with advanced liver cancer is on-going and further studies in different tumour types has been planned (Jimeno et al. 2004). A second phase I trial with a schedule of five drug administrations per day in patients with androgen independent prostate cancer is also on-going and expanding the cohort of patients at the recommended dose (Rademaker-Lakhai et al. 2005).

Neuroprotective activities

Dementia (from the Latin *demens*, meaning "without mind") is a group of symptoms that may accompany several neurological disorders. It occurs mainly in the elderly population and is characterized by the progressive and permanent deterioration of neuronal processes and the loss of synaptic connections (Selkoe 2002; Honer 2003; Parihar and Hemnani 2004; Tanzi 2005; Blennow et al. 2006; Scheff et al. 2006), which results in the decline of multiple cognitive functions, such as memory, language, thinking, comprehension, and calculation. In 2030 it is estimated that 63 million people will suffer from dementia, 65% of which will be in less developed countries (Wimo et al. 2003). Dementia can be caused by several degenerative neurological diseases, namely Alzheimer's (AD), dementia with Lewy bodies, Parkinson's, and Huntington's.

AD is the most common cause of senile dementia, and is estimated to account for 50–60% of dementia cases among people age 65 or older (Filho et al. 2006). It has no cure and is the fourth leading cause of death in developed nations, after heart disease, cancer, and stroke (Natarajan et al. 2009). Pathologically, AD is characterized by extracellular deposits of plasma amyloid beta peptide (Aβ) in senile plaques, intracellular formation of neurofibrillary tangles, and the loss of neuronal synapses and pyramidal neurons (Weinreb et al. 2011). The molecular mechanism of AD can be explained by two major hypotheses: the cholinergic hypothesis and the amyloid cascade hypothesis (Parihar and Hemnani 2004). The amyloid cascade theory suggests that the sequence of events leading to AD is initiated by the deposition of the amyloid-β peptide in the brain parenchyma (Karran et al. 2011). Conversely, the cholinergic premise postulates that the renewal of acetylcholine (ACh) levels, the major neurotransmitter in the central nervous system (CNS), which are progressively lost during the progression of AD, delays the loss of cognitive function (Filho et al. 2006). Therefore, one of the main strategies for symptomatic relief of AD is the enhancement of the levels of ACh, which is hydrolized primarily by acetylcholinesterase (AChE) and secondly by butyrylcholinesterase (BChE). The inhibition of AChE through the use of AChE inhibitors (AChEi) has so far been considered the main approach for the symptomatic treatment of several neurological disorders (e.g., AD, Parkinson disease, and myasthenia gravis), associated with reduced ACh levels in the synapse (Whitehouse et al. 1983; Mukherjee et al. 2007). In fact, several AChE inhibitors such as tacrine, donzepil, and galanthamine have been approved for the treatment of AD (Racchi et al. 2004).

The search for new cholinesterase inhibitors (ChEi) among natural resources, especially plants, is still however a growing area of investigation since the compounds mentioned above can display several side effects. There are more than 180 natural compounds namely alkaloids, terpenoids, and flavonoids with a direct relevance to AD therapeutics, mostly from plant origin (reviewed in Williams et al. 2011). However, knowledge of AChE inhibitors from marine photosynthetic organisms is particularly scarce. Recently, effort has been made to find AChEi from marine origin, namely from macroalgae. The AChE inhibitory activity was evaluated on methanol extracts of seven macroalgae species (*Caulerpa racemosa* var. *laetevirens*, *Codium capitatum*, *Halimeda cuneata*, *Ulva fasciata*, *Amphiroa bowerbankii*, *A. ephedraea*, and *Dictyota humifusa*) collected from the east coast of South Africa (Stirk et al. 2007).

The species *D. humifusa* and *U. fasciata* had the highest capacity to inhibit the activity of AChE, with IC_{50} values of 4.8 mg/mL. The species *Hypnea valentiae* (IC_{50} = 2.6 mg/mL), and *Gracilaria edulis* (IC_{50} = 3 mg/mL) collected from Hare Island, Gulf of Mannar (India) also exhibited promising AChE inhibitory potential. Strong AChE inhibitory activity was detected on methanol extracts of the species *Gracilaria gracilis*, *Sargassum*, and *Cladophora fasicularis* at the concentrations of 1.5, 1, and 2 mg/mL respectively (Suganthy et al. 2010). In addition, *Hypnea valentiae* and *Ulva reticulate* exhibited dual anticholinesterase activity, that is, they are active against both AChE and BChE, with IC_{50} values on BChE of 3.9 mg/mL and 10 mg/mL, respectively (Suganthy et al. 2010). Compounds with dual anti-ChE activity may be appropriate to patients at a moderate stage of AD if ACh levels have not yet significantly declined and BChE could hydrolyse ACh (Mesulam et al. 2002). Indeed, tacrine and physostigmine, which are drugs used in the clinical treatment of AD, exhibit mixed AChE-BChE inhibition.

Myung et al. (2005) reported that dieckol and phlorofucofluoroeckol, two phlorotannins found in the brown algae *Ecklonia cava* have memory enhancing and AChE inhibitory activities. Working with ethanolic extracts of 27 Korean marine algae, Yoon et al. (2008) found that *Ecklonia stolonifera* has significant inhibitory activity against AChE. The bioassay-guided fractionation of the active hexane and ethyl acetate fractions, obtained from the ethanolic extract of *E. stolonifera* allowed the isolation of eight phlorotannins (phloroglucinol, eckstolonol, eckol, phlorofucofuroeckol-A, dieckol, triphlorethol-A, 2-phloroeckol, and 7-phloroeckol), and two sterols (fucosterol and 24-hydroperoxy 24-vinylcholesterol). Eckol, dieckol, 2-phloroeckol, and 7-phloroeckol exhibited selective dose dependent inhibitory activities toward AChE; while eckstolonol and phlorofucofuroeckol-A had inhibitory activities against both AChE and BChE. Plastoquinones isolated from *Sargassum sagamianum* have also been found to be potent ChEi (Choi et al. 2007).

Microalgae have a distinct biochemistry and are considered as one of the most promising sources of functional molecules (Pulz and Gross 2004; Chacón-Lee and González-Maríño 2010). In spite of this, little is known of its neuroprotective potential. Working with methanol and hexane extracts of different microalgae species, Custódio et al. (2012) found a significant AChE inhibitory activity in the hexane extracts of *Mychonastes homosphaera* (formerly known as *Chlorella minutissima*), *Tetraselmis chuii*, and *Rhodomonas salina*, which resulted in an AChE inhibition ranging between 79 and 86% at a concentration of 10 mg/mL. A high AChE inhibitory capacity was also detected by the same authors in water and diethyl ether extracts of *Scenedesmus* sp. (Custódio et al. 2013). Taken together, these results point to a possible therapeutic value of bioactive molecules present in macro- and microalgae as ChEi, with application in the management of AD and other neurological disorders. Moreover, some compounds isolated from macroalgae exhibit dual anti-ChE activity, which is considered to be more effective in the treatment of AD.

Anti-protozoal activities

Vector-borne parasitic diseases—for example, malaria, Chagas disease, African trypanosomiasis, and leishmaniasis—caused by unicellular flagellates such as *Plasmodium* sp., *Trypanosoma* sp., and *Leishmania* sp. constitute one of the major challenges for global health. The human importance of these infectious diseases is annually confirmed by WHO reports and initiatives (WHO 2010a,b, 2011, 2012a,b). Besides being a major cause of mortality in various tropical and subtropical regions (WHO 2008), these diseases generate an enormous impact in terms of disease burden, quality of life, loss of productivity, and the aggravation of poverty as well as the high cost of long-term care, being a serious obstacle to the socio-economic development, especially in developing countries (WHO 2010b).

Terrestrial plants have been used commonly as natural sources of antiprotozoal compounds (Wright and Phillipson 1990). Although to a less extent, marine algae have also been used by coastal Asia and Caribbean people in traditional medicine (Moo-Puc et al. 2008), and there are ancient Chinese records of its use upon boiling in antiparasitic treatments (Tseng and Chang 1984).

As in land plants, the interest in the chemical wealth of marine organisms as a potential source of antiprotozoal pharmacological agents has increased over the last few years (Fattorusso and Taglialatela-Scafati 2009; Mayer et al. 2011; Tempone et al. 2011). Since the early 2000s there has been a focus on screening macroalgae for antiprotozoal activity (Tempone et al. 2011). Despite the promising availability of novel bioactive compounds, very little research has, until now, been directed towards marine algae and their antiprotozoal potential (Moo-Puc et al. 2008; Vonthron-Sénécheau et al. 2011).

The search for novel antiprotozoal compounds from algae is reported in a few screening papers that highlight the inhibition of protozoan parasites activity by extracts from different algae belonging to the Chlorophyta, Ochrophyta, and Rhodophyta phyla (Nara et al. 2005; Sabina et al. 2005; Lakshmi et al. 2006; Orhan et al. 2006; Freire-Pelegrin et al. 2008; Moo-Puc et al. 2008; Chen et al. 2009; Genovese et al. 2009; Léon-Deniz et al. 2009; Allmendinger et al. 2010; Felício et al. 2010; Spavieri et al. 2010a,b; Süzgeç-Selçuk et al. 2010; Dos Santos et al. 2010, 2011; Fouladvand et al. 2011; Vonthron-Sénécheau et al. 2011; Table 4).

In spite of the interest on this subject, very few published articles describe the identification of bioactive compounds from marine algae, in contrast with the literature on terrestrial plants (Schmidt et al. 2012a,b).

Data reported by these authors reflects the well-known phenomenon that marine organisms collected from different environments have different chemistries, which affects their biological activities (Spavieri et al. 2010b). For example, antileishmanial activity of three samples of either *Ulva lactuca* or *Dictyota dichotoma* from different locations against *L. donovani* axenic amastigote forms, evaluated by the resazurin method, resulted in IC$_{50}$ values between 5.9 and 12 µg/mL and 8.8 and 52 µg/mL, respectively (Orhan et al. 2006; Spavieri et al. 2010b; Vonthron-Sénécheau et al. 2011). The observed discrepancies may stem from several factors ranging from abiotic (e.g., salinity) and biotic (e.g., predation) components to the use of different extraction methods and solvents resulting in extracts of diverse chemical composition.

Also different trypanosomatid parasite species have different responses and sensibilities to the same species extracts. For example, both Orhan et al. (2006) and Spavieri et al. (2010b) evaluated antiprotozoal activity of *U. lactuca* on *L. donovani, T. cruzi,* and *T. brucei rhodesiensis* by the resazurin method. Although the Turkish sample was more potent against *L. donovani* (IC$_{50}$ = 5.9 µg/mL (Orhan et al. 2006), it had weaker anti-*T. brucei rhodesiense* activity (IC$_{50}$ = 22.3 µg/mL) and had no activity against *T. cruzi* (Spavieri et al. 2010b). Also Sabina et al. (2005) and Lakshimi (2006) reported non-concordant anti-leishmanial activities using *Codium elongatum* (Chlorophyta) and *Scinaia indica* (Rhodophyta) extracts during surveys developed for the screening of extracts of Pakistani and Indian marine samples.

Different bioactivity results were also registered using the same algal species but tested in specific forms of the parasite. Dos Santos et al. (2011) observed a higher sensibility of the promastigote forms (IC$_{50}$ = 2.0 µg/mL) as compared with intracellular (IC$_{50}$ = 4.0 µg/mL) and especially when compared with axenic amastigote forms (IC$_{50}$ = 12.0 µg/mL) of *L. amazonensis* to the (4R,9S,14S)-4α-acetoxy-9β,14α-dihydroxydolast-1(15),7-diene isolated from *Canistrocarpus cervicornis*.

Since 2005, few bioactive compounds with antiprotozoal activity have been isolated from marine macroalgae. Most data report the effects of crude extracts, obtained by different sequential extractions usually using solvents (hexane, dichloromethane, ether, ethyl acetate, chloroform, water, ethanol, and methanol) of various polarities. However, Marcolino (2010), Veiga-Santos et al. (2010), and Dos Santos et al. (2010, 2011) studied the effect of purified compounds isolated from macroalgae from Brazil on protozoan parasites, *L. amazonensis* and *T. cruzi*. The identified compounds are sesquiterpene elatol, isolated from *Laurencia dendroidea* (Veiga-Santos et al. 2010; Dos Santos et al. 2010), 4-acetoxy-dolastane diterpene obtained from *Canistrocarpus cervicornis* (Dos Santos et al. 2011) and sulphated polysaccharides obtained from *Gayralia oxysperma, Gymnogongrus griffithsiae,* and *Eucheuma denticulatum*. All these compounds had already been recognized as secondary metabolites with important roles in ecological interactions, such as anti-herbivore activity and potential defence against infection by microorganisms (Marcolino 2010).

Regarding the type of solvents used for extraction, Freire-Pelegrin et al. (2008) mentioned that organic extracts of tropical marine algae from Gulf of Mexico and Caribbean coast showed activity against *L. mexicana* promastigote forms. On a screening study of French macroalgae, Vonthron-Sénécheau et al. (2011) observed that the majority of the active extracts were obtained using ethyl acetate as extraction solvent, while the hydroalcoholic extracts were mainly inactive, suggesting that, in general, the active anti-protozoal compounds were relatively non polar (Genovese et al. 2009). However, Fouladvand et

Table 4. Marine algae with antileishmanial, antitrypanosomal, and anti-plasmodial activities with an IC$_{50}$ <25 µg/mL.

Marine algae		Parasite species			Reference(s)
Phylum/Family	**Species**	*Leishmania*	*Trypanosoma*	*Plasmodium*	
Chlorophyta					
Caulerpaceae	*Caulerpa racemosa*	*L. donovani*[a]		*P. falciparum*[es]	Süzgeç-Selçuk et al. 2010
Cladophoraceae	*Cladophora glomerata*		*T.b. rhodesiense*[ia]	*P. falciparum*[es]	Orhan et al. 2006
Cladophoraceae	*Cladophora rupestris*	*L. donovani*[a]	*T.b. rhodesiense*[ia]		Spavieri et al. 2010b
Codiaceae	*Codium fragile* ssp. *Tomentosoize*	*L. donovani*[a]	*T.b. rhodesiense*[ia]		Spavieri et al. 2010b
Codiaceae	*Codium bursa*		*T.b. rhodesiense*[ia]	*P. falciparum*[es]	Süzgeç-Selçuk et al. 2010
Ulvaceae	*Ulva intestinalis*	*L. donovani*[a]	*T.b. rhodesiense*[ia]		Spavieri et al. 2010b
Ulvaceae	*Ulva lactuca*	*L. donovani*[a]	*T.b. rhodesiense*[ia]		Orhan et al. 2006
Ochrophyta					
Chordaceae	*Chorda filum*	*L. donovani*[a]	*T.b. rhodesiense*[ia]		Spavieri et al. 2010a
Dictyotaceae	*Dictyopteris polypodioides*	*L. donovani*[a]			Vonthron-Sénécheau et al. 2011
Dictyotaceae	*Dictyota dichotoma*	*L. donovani*[a]	*T.b. rhodesiense*[ia]	*P. falciparum*[es]	Orhan et al. 2006
Dictyotaceae	*Dictyota caribaea*	*L. mexicana*[p]	*T. cruzi*[t]		Léon-Deniz et al. 2009
Dictyotaceae	*Lobophora variegata*		*T. cruzi*[t]		Léon-Deniz et al. 2009
Fucaceae	*Fucus ceranoides*		*T.b. rhodesiense*[ia]		Spavieri et al. 2010a
Fucaceae	*Fucus serratus*		*T.b. rhodesiense*[ia]		Spavieri et al. 2010a
Fucaceae	*Fucus spiralis*		*T.b. rhodesiense*[ia]		Spavieri et al. 2010a
Fucaceae	*Fucus vesiculosus*		*T.b. rhodesiense*[ia]		Spavieri et al. 2010a
Fucaceae	*Pelvetia canaliculata*		*T.b. rhodesiense*[ia]		Spavieri et al. 2010a
Himanthaliaceae	*Himanthalia elongata*			*P. falciparum*[es]	Vonthron-Sénécheau et al. 2011
Laminariaceae	*Laminaria digitata*		*T.b. rhodesiense*[ia]		Spavieri et al. 2010a
Phaeophyceae	*Turbinaria turbinata*	*L. mexicana*[p]	*T. cruzi*[t]		Léon-Deniz et al. 2009
Phyllariaceae	*Saccorhiza polyschides*		*T.b. rhodesiense*[ia]		Spavieri et al. 2010a

Table 4 contd....

...Table 4 contd.

Phylum/Family	Marine algae Species	Parasite species Leishmania	Trypanosoma	Plasmodium	Reference(s)
Sargassaceae	*Sargassum muticum*		*T.b. rhodesiense*[ia]	*P. falciparum*[es]	Spavieri et al. 2010a; Vonthron-Sénécheau et al. 2011
Sargassaceae	*Sargassum natans*		*T.b. rhodesiense*[ia]	*P. falciparum*[es]	Orhan et al. 2006
Sargassaceae	*Bifurcaria bifurcata*	*L. donovani*[a]	*T.b. rhodesiense*[ia]		Spavieri et al. 2010a; Vonthron-Sénécheau et al. 2011
Sargassaceae	*Cystoseira baccata*	*L. donovani*[a]	*T.b. rhodesiense*[ia]		Spavieri et al. 2010a
Sargassaceae	*Cystoseira barbata*	*L. donovani*[a]		*P. falciparum*[es]	Süzgeç-Selçuk et al. 2010
Sargassaceae	*Cystoseira tamariscifolia*	*L. donovani*[a]	*T.b. rhodesiense*[ia]		Spavieri et al. 2010a
Sargassaceae	*Halidrys siliquosa*	*L. donovani*[a]	*T.b. rhodesiense*[ia]		Spavieri et al. 2010a
Stypocaulaceae	*Stypocaulon scoparium*		*T.b. rhodesiense*[ia]		Spavieri et al. 2010a
Rhodophyta					
Acrotylaceae	*Claviclonium ovatum*		*T.b. rhodesiense*[ia]		Allmendinger et al. 2010
Bangiaceae	*Porphyra linearis*		*T.b. rhodesiense*[ia]		Allmendinger et al. 2010
Bonnemaisoniaceae	*Asparagopsis armata*	*L. donovani*[a]			Genovese et al. 2009
Bonnemaisoniaceae	*Asparagopsis taxiformis*	*L. donovani*[a]			Genovese et al. 2009
Ceramiaceae	*Ceramium virgatum*		*T.b. rhodesiense*[ia]		Allmendinger et al. 2010
Ceramiaceae	*Halurus flosculosus*			*P. falciparum*[es]	Vonthron-Sénécheau et al. 2011
Ceramiaceae	*Ceramium rubrum*	*L. donovani*[a]		*P. falciparum*[es]	Süzgeç-Selçuk et al. 2010
Champiaceae	*Chylocladia verticillata*		*T.b. rhodesiense*[ia]		Allmendinger et al. 2010
Corallinaceae	*Corallina granifera*	*L. donovani*[a]		*P. falciparum*[es]	Süzgeç-Selçuk et al. 2010
Corallinaceae	*Corallina officinalis*	*L. donovani*[a]	*T.b. rhodesiense*[ia]		Allmendinger et al. 2010
Corallinaceae	*Jania rubens*		*T.b. rhodesiense*[ia]		Allmendinger et al. 2010
Cystocloniaceae	*Calliblepharis jubata*			*P. falciparum*[es]	Vonthron-Sénécheau et al. 2011
Dasyaceae	*Dasya pedicellata*	*L. donovani*[a]	*T.b. rhodesiense*[ia]	*P. falciparum*[es]	Süzgeç-Selçuk et al. 2010
Delesseriaceae	*Cryptopleura ramosa*		*T.b. rhodesiense*[ia]		Allmendinger et al. 2010

Family	Species	Leishmania/Trypanosoma	*T.b. rhodesiense*[ia]	*P. falciparum*[es]	Reference
Dumontiaceae	*Dilsea carnosa*	*L. donovani*[a]		*P. falciparum*[es]	Vonthron-Sénécheau et al. 2011
Furcellariaceae	*Furcellaria lumbricalis*		*T.b. rhodesiense*[ia]		Allmendinger et al. 2010
Galaxauraceae	*Scinaia furcellata*		*T.b. rhodesiense*[ia]		Orhan et al. 2006
Galaxauraceae	*Scinaia hatei Børgesen*	*L. major*[p]			Sabina et al. 2005
Gelidiaceae	*Gelidium crinale*	*L. donovani*[a]		*P. falciparum*[es]	Süzgeç-Selçuk et al. 2010
Gelidiaceae	*Gelidium latifolium*				Vonthron-Sénécheau et al. 2011
Gelidiaceae	*Gelidium pulchellum*		*T.b. rhodesiense*[ia]		Allmendinger et al. 2010
Gigartinaceae	*Chondrus crispus*			*P. falciparum*[es]	Vonthron-Sénécheau et al. 2011
Gracilariaceae	*Gracilaria gracilis*		*T.b. rhodesiense*[ia]	*P. falciparum*[es]	Allmendinger et al. 2010; Vonthron-Sénécheau et al. 2011
Halymeniaceae	*Grateloupia turuturu*				Vonthron-Sénécheau et al. 2011
Lomentariaceae	*Lomentaria articulata*		*T.b. rhodesiense*[ia]		Allmendinger et al. 2010
Phyllophoraceae	*Mastocarpus stellatus*		*T.b. rhodesiense*[ia]	*P. falciparum*[es]	Allmendinger et al. 2010; Vonthron-Sénécheau et al. 2011
Plocamiaceae	*Plocamium cartilagineum*	*L. donovani*[a]	*T.b. rhodesiense*[ia]		Allmendinger et al. 2010
Polyidaceae	*Polyides rotundus*		*T.b. rhodesiense*[ia]		Allmendinger et al. 2010
Rhodomelaceae	*Bostrychia tenella*	*L. amazonensis*[p]			Felício et al. 2010
Rhodomelaceae	*Halopitys incurvus*	*L. donovani*[a]	*T.b. rhodesiense*[ia]		Allmendinger et al. 2010
Rhodomelaceae	*Laurencia microcladia*	*L. mexicana*[p]	*T. cruzi*[t]		Freire-Pelegrin et al. 2008; Léon-Deniz et al. 2009
Rhodomelaceae	*Laurencia pinnatifida*	*L. major*[p]			Sabina et al. 2005
Rhodomelaceae	*Osmundea pinnatifida*		*T.b. rhodesiense*[ia]		Allmendinger et al. 2010

[a]axenic amastigotes, [p]promastigotes, [ia]intracellular amastigotes, [t]Trypomastigotes, [es]Erythrocytic stages.

al. (2011) have more recently reported that interesting activities can also be found with water extracts of Persian Gulf rhodophytes from the Gracilariaceae family (*Gracilaria corticata, G. salicornia, G. corticata*), displaying IC_{50} values ranging from 38 to 74 μg/mL.

Antileishmanial activity

Concerning the algae activity against *Leishmania* sp., 122 species from 47 families of Chlorophyta (26 species/9 families), Ochrophyta (33 species/13 families), and Rhodophyta (63 species/25 families) phyla from European, Asian, Middle-East, and South American countries were screened by various authors (Sabina et al. 2005; Lakshmi et al. 2006; Orhan et al. 2006; Freire-Pelegrin et al. 2008; Genovese et al. 2009; Allmendinger et al. 2010; Felício et al. 2010; Spavieri et al. 2010a,b; Süzgeç-Selçuk et al. 2010; Dos Santos et al. 2010, 2011; Vonthron-Sénécheau et al. 2011; Fouladvand et al. 2011) towards different *Leishmania* species (*L. donovani, L. major, L. amazonensis*, and *L. mexicana*).

Regarding the algae extracts activity against *Leishmania* promastigote forms, the IC_{50} values ranged from 10.9 to 105.0 μg/mL, 6.3 to 74.0 μg/mL and 34 to 125.0 μg/mL for the Ochrophyta, Rhodophyta, and Chlorophyta phyla, respectively. The most potent activities against this parasite form were found in rhodophytes (Table 4) with six species displaying $IC_{50} < 25$ μg/mL, namely *Laurencia pinnatifida* (Sabina et al. 2005), *Laurencia microcladia* (Freire-Pelegrin et al. 2008), *Bostrychia tenella* (Felício et al. 2010), *Asparagopsis armata* and *A. taxiformis* (Genovese et al. 2009), and *Scinaia hatei* (Sabina et al. 2005).

As red algae are known to contain a wide range of secondary metabolites, such as halogenated mono- and diterpenes, sterols, alkaloids, polyphenols, and sulphated sugars (Blunt et al. 2009) related with their antifungal and antibacterial activity, Genovese et al. (2009) suggested that the inhibitory properties of *Asparagopsis* could also be due to its contents in halogenated compounds.

From the Ochrophyta phylum, six species from three families were screened. The most relevant results were observed with *Turbinaria turbinata, Dictyota caribaea*, and *Lobophora variegata*, displaying antileishmanial activity with $IC_{50} < 50$ μg/mL (Freire-Pelegrin et al. 2008).

Chlorophyta was the phylum with less potent activities against *Leishmania* promastigote forms, with all reported 17 species from 6 families not revealing activities under 25 μg/mL. The lowest IC_{50} values were obtained with the ethanol extracts of *Caulerpa faridii* ($IC_{50} = 34$ μg/mL), *C. racemosa* ($IC_{50} = 37.5$ μg/mL), and *Codium flabellatum* ($IC_{50} = 34$ μg/mL) (Sabina et al. 2005).

Regarding the amastigote form of the parasites, most studies used *L. donovani*. The inhibitory effect of the tested marine macroalgae extracts ranged from 3.8 to 90.9 μg/mL, 9.5 to 85.6 μg/mL and 5.9 to 39.2 μg/mL for the Ochrophyta, Rhodophyta, and Chlorophyta phyla, respectively.

However, in contrast with the results observed with promastigote forms described above, higher activities were found in species from the Ochrophyta phylum, namely in *Bifurcaria bifurcata* (Spavieri et al. 2010a; Vonthron-Sénécheau et al. 2011), *Halidrys siliquosa* (Spavieri et al. 2010a), *Dictyota dichotoma*, and *Dictyopteris polypodioides* (Vonthron-Sénécheau et al. 2011). All these species presented extracts with $IC_{50} < 11$ μg/mL. From the reviewed data, the Ulvaceae appears as one of the main sources, among chlorophytes, of antileishmanial compounds for the axenic amastigote form (Orhan et al. 2006; Spavieri et al. 2010b).

Within the 19 Rhodophyta families screened for antileishmanial on axenic amastigote forms, only seven (Dumontiaceae, Rhodomelaceae, Ceramiaceae, Gelidiaceae, Plocamiaceae, Corallinaceae, Dasyaceae) evidenced activities with $IC_{50} < 25$ μg/mL (Table 4).

Considering the data obtained with pure compounds isolated from marine seaweeds, Marcolino (2010) demonstrated that the sulphated acid polysaccharides obtained from *Gayralia oxysperma, Gymnogongrus griffithsiae*, and *Eucheuma denticulatum* have leishmanicidal activity towards *in vitro* intracellular amastigote forms of *L. amazonensis*, inhibiting parasite activity above 50% with a concentration of 10 μg/mL. Likewise, Dos Santos et al. (2010) obtained promising results using the sesquiterpene elatol isolated from the red alga *Laurencia dendroidea* against *L. amazonensis* promastigote ($IC_{50} = 4.0$ μM) and intracellular amastigote ($IC_{50} = 0.45$ μM) forms.

Felício et al. (2010) have tested leishmanicidal activity of the hexane and dichloromethane extracts of the marine Brazilian red alga *Bostrychia tenella* ($IC_{50} = 17.4$ μg/mL) against *L. amazonensis* promastigotes, and proceeded with the identification of the bioactive compounds by GC/MS. In this work, IC_{50} values between 1.5 and 4.3 µg/mL were observed using subfractions of the hexane and dichloromethane extracts. Although the bioactive compounds have yet to be identified, the volatile compounds found in *B. tenella* seem to be fatty acids, low molecular mass hydrocarbons, esters and steroids, and also some less common compounds as neophytadiene.

Cruz et al. (2009) have also reported leishmanicidal activity for kahalalide F, a cyclic depsipeptide that can be found in the green alga *Bryopsis* sp. (Smit 2004), and its synthetic analogues. This peptide seems to alter and depolarize the plasma membrane of the parasite with consequently fatal biochemical imbalances. The authors refer to this lethal mechanism as considerably less prone to the development of resistance than classical drugs.

Antitrypanosomal activity

Nara et al. (2005), Orhan et al. (2006), Allmendinger et al. (2010), Felício et al. (2010), Spavieri et al. (2010a,b), Süzgeç-Selçuk et al. (2010), and Vonthron-Sénécheau et al. (2011) screened different marine macroalgae extracts for antitrypanosomal activity against *Trypanosoma brucei rhodesiense* and *T. cruzi*, encompassing 147 species from 56 families of Chlorophyta (15 species/5 families), Ochrophyta (47 species/16 families) and Rhodophyta (85 species/35 families) phyla from Japan, Turkey, France, United Kingdom, and Ireland. The species displaying antitrypanosomal activity below 25 µg/mL are summarized in Table 4.

The data shown in Table 4 was obtained towards *T. cruzi* trypomastigote using resazurin staining and intracellular *T. brucei rhodesiense* amastigote forms by means of the MTT or the β-galactosidase (chlorophenolred-ß-D-galactopyranoside) assays.

Concerning the activities against *T. brucei rhodesiense*, the majority of the species with IC_{50} values under 25 µg/mL belong to the Ochrophyta and Rhodophyta phyla. From this group of macroalgae, the chloroform:methanol extracts of the ochrophytes such as *Halidrys siliquosa*, *Bifurcaria bifurcata*, *Cystoseira tamariscifolia*, *Cystoseira baccata*, *Sargassum natans*, *Sargassum muticum*, *Fucus spiralis*, *Fucus vesiculosus*, *Fucus ceranoides*, *Pelvetia canaliculata*, *Chorda filum*, and *Saccorhiza polyschides* displayed IC_{50} < 11 µg/mL (Orhan et al. 2006; Spavieri et al. 2010a).

Although the Ochrophyta phylum displayed the most prominent antiparasitic activities, algae of the Rhodophyta phylum were also highly active against *T. brucei rhodesiense*, being the methanolic extract of *Dasya pedicellata*, the most potent of the species screened (IC_{50} = 0.37 µg/mL; Süzgeç-Selcuk et al. 2010). The chloroform:methanol extracts of other algae from different families as, for example, *Corallina officinalis*, *Ceramium virgatum*, *Gelidium pulchellum*, *Halopitys incurvus*, *Osmundea pinnatifida*, and *Porphyra linearis*, exhibited higher IC_{50} values (IC_{50} < 10 µg/mL, Allmendinger et al. 2010).

The rhodophyte *Dasya pedicellata* showed the lowest IC_{50} value (62.02 µg/mL) against *T. cruzi* trypomastigote forms (Süzgeç-Selcuk et al. 2010).

The search for inhibitors of *T. cruzi* dihydroorotate dehydrogenase (DHOD), a key enzyme for pyrimidine biosynthesis of the parasite, in methanol extracts of Japanese algae showed that the brown algae *Fucus evanescens* and *Pelvetia babingtonii* were able to halve the recombinant DHOD activity at a concentration of 50 µg/mL, and were also able to inhibit the growth of *T. cruzi* cells (Nara et al. 2005).

As observed for the antileishmanial activity of macroalgae, so far, only a few works have explored the antitrypanosomal activity of isolated compounds from algae. As described by Dos Santos et al. (2010) for *L. donovani* parasites, Veiga-Santos et al. (2010) showed that the sesquiterpene elatol isolated from *Laurencia dendroidea* is active against the *T. cruzi* trypomastigote (IC_{50} = 1.38 µg/mL) and the intracellular amastigote (IC_{50} = 1.01 µg/mL) forms, being also active against the epimastigote form (IC_{50} = 45.4 µg/mL).

Antiplasmodial activity

Antiplasmodial activity has been screened on different algae from France, Turkey, India, Brazil and Philippines, namely Chlorophyta (11 species/5 families), Ochrophyta (12 species/6 families), and

Rhodophyta (19 species/13 families) algae (Wright and König 1996; Topcu et al. 2003; Mendiola-Martínez et al. 2005; Lakshimi et al. 2006; Orhan et al. 2006; Süzgeç-Selçuk et al. 2010; Vonthron-Sénécheau et al. 2011). Algae with IC_{50} for antiplasmodial activity, under 25 µg/mL are summarized in Table 4.

The data reported for the screening of antiplasmodial activity of marine algae, were obtained towards *P. falciparum* erythrocytic stages using [3H]-hypoxanthine incorporation assay. Only Lakshimi et al. 2006 have tested chlorophyta algae species against *P. berghei* without antiplasmodial results.

As for the *L. donovani* and *T. brucei rhodesiense*, the methanolic extracts of the Rhodophyta *Dasya pedicellata* and the Chlorophyta *Codium bursa* were the most potent against *P. falciparum* evidencing IC_{50}'s of 0.38 µg/mL and 1.38 µg/mL, respectively (Süzgeç-Selçuk et al. 2010). The ethyl acetate extracts of the rhodophytes *Mastocarpus stellatus*, *Chondrus crispus*, *Grateloupia turuturu*, *Gracilaria gracilis*, *Gelidium latifolium*, *Dilsea carnosa*, *Halurus flosculosus* and Ochrophytas *Sargassum muticum*, *Dictyota dichotoma*, *Himanthalia elongata*, and the methanolic extract of the Chlorophyta *Caulerpa racemosa*, revealed inhibitory activities between 2.8 and 4.6 µg/mL (Süzgeç-Selçuk et al. 2010; Vonthron-Sénécheau et al. 2011).

Concerning the works involving isolated compounds, once more the sesquiterpenes ((8R)-8-bromo-10-epi-beta-snyderol) and aromatic compounds (*p*-hydroxybenzaldehyde and p-methoxy-benzyl) isolated from *Laurencia* sp. show antimalarial activity (Wright et al. 1996; Topcu et al. 2003). Chen et al. (2009) indicates that fucoidan, a sulfated polysaccharide, isolated from the Korean brown algae *Undaria pinnatifida* inhibits the invasion of *Plasmodium falciparum* merozoites into erythrocytes *in vitro* and in *in vivo Plasmodium berghei*-infected mice, and had no toxic effect on RAW 264.7 cells.

Orhan et al. (2006) had evaluated the ability of five ethanolic extracts of turkish seaweeds to inhibit the recombinant key enzyme (FabI) of the fatty acid biosynthesis of *P. falciparum in vitro*. Despite its moderate antiplasmodial activity, the green algae *C. glomerata* (IC_{50} = 33.7 µg/mL) and *U. lactuca* (IC_{50} = 48.8 µg/mL) efficiently inhibited the FabI enzyme with IC_{50} values of 1.0 and 4.0 µg/mL, respectively, suggesting that this species might represent alternative sources in the search of new antiprotozoal agents.

Concerning the extraction procedures for antiplasmodial screenings of marine algae, only ethyl acetate, ethanol, and methanol extracts were surveyed. In agreement with previously described for antileishmanial and antitrypanossomal activities, ethanolic extracts were less active or even inactive against *P. falciparum* parasites (Vonthron-Sénécheau et al. 2011).

The search for bioactive compounds originating from the sea is recent. However, based on the enormous amount of data collected so far, it is possible to note that marine algal extracts are promising sources of novel antiparasitic chemotherapeutic compounds and that further research is needed to identify them.

Concluding remarks

Marine algae are novel and important sources of bioactive compounds, which can be used as pharmaceutical agents in the near future. Often, however, the structure of natural bioactive compounds is complex, which hinders their identification, isolation, and synthesis, as well as a better understanding of the molecular mechanisms involved. Moreover, insufficient yields can also become a limitation. Nonetheless, the wide spectrum of bioactivities found in marine algae underlines the important potential application of algal compounds in the pharmaceutical industry, which can complement and inspire the synthesis of novel synthetic drugs that can improve the quality of life for all humankind.

Acknowledgements

This work was supported by Portuguese FCT – Fundação para a Ciência e a Tecnologia through the PTDC/MAR/103957/2008 and CCMAR/Multi/04326/2013 projects. CVD and CBS were supported by FCT doctoral grants (SFRH/BD/81425/2011 and SFRH/BD/78062/2011, respectively) and LC by the FCT Investigator Programme (IF/00049/2012).

References

Abatis, D., C. Vagias, D. Galanakis, J.N. Norris, D. Moreau, C. Roussakis and V. Roussis. 2005. Atomarianones A and B: two cytotoxic meroditerpenes from the brown alga *Taonia atomaria*. Tetrahedron Lett. 46: 8525–8529.

Ale, M.T., H. Maruyama, H. Tamauchi, J.D. Mikkelsen and A.S. Meyer. 2011. Fucose-containing sulfated polysaccharides from brown seaweeds inhibit proliferation of melanoma cells and induce apoptosis by activation of caspase-3 *in vitro*. Mar. Drugs 9: 2605–2621.

Allmendinger, A., J. Spavieri, M. Kaiser, R. Casey, S. Hingley-Wilson, A. Lalvani, M. Guiry, G. Blunden and D. Tasdemir. 2010. Antiprotozoal, antimycobacterial and cytotoxic potential of twenty-three British and Irish red algae. Phytother. Res. 24: 1099–1103.

Ananthi, S., H.R.B. Raghavendran, A.G. Sunil, V. Gayathri, G. Ramakrishnan and H.R. Vasanthi. 2010. *In vitro* antioxidant and *in vivo* anti-inflammatory potential of crude polysaccharide from *Turbinaria ornata* (Marine Brown Alga). Food Chem. Toxicol. 48: 187–192.

Andreeff, M., D.W. Goodrich and A.B. Pardee. 2003. Cell proliferation and differentiation. *In*: D.W. Kufe, R.E. Pollock, R.R. Weichselbaum, R.C. Bast, Jr, T.S. Gansler, J.F. Holland and E. Frei III (eds.). Holland-Frei Cancer Medicine. BC Decker, Hamilton.

Argandoña, V.H., J. Rovirosa, A. San-Martin, A. Riquelme, A.R. Diaz-Marrero, M. Cueto, J. Darias, O. Santana, A. Guadano and A. Gonzalez-Coloma. 2002. Antifeedant effects of marine halogenated monoterpenes. J. Agric. Food Chem. 50: 7029–7033.

Asari, F., T. Kusumi and H. Kakisawa. 1989. Turbinaric acid, a cytotoxic secosqualene carboxylic acid from the brown alga *Turbinaria ornata*. J. Nat. Prod. 52: 1167–1169.

Ayyad, S.-E.N., O.B. Abdel-Halim, W.T. Shier and T.R. Hoye. 2003. Cytotoxic hydroazulene diterpenes from the brown alga *Cystoseira myrica*. Z. Natuforsch. C. Biosci. 58: 33–38.

Badal, S., W. Gallimore, G. Huang, T.R. Tzeng and R. Delgoda. 2012. Cytotoxic and potent CYP1 inhibitors from the marine algae *Cymopolia barbata*. Org. Med. Chem. Lett. 2: 21.

Blackman, A.J., C. Dragar and R.J. Wells. 1979. A new phenol from the brown alga *Perithalia caudata* containing a "reverse" isoprene unit at the 4-position. J. Aust. J. Chem. 32: 2783–2786.

Blennow, K., M.J. de Leon and H. Zetterberg. 2006. Alzheimer's disease. Lancet 368: 387–403.

Blunt, J.W., B.R. Copp, W.P. Hu, M.H. Munro, P.T. Northcote and M.R. Prinsep. 2009. Marine natural products. Nat. Prod. Rep. 26: 170–244.

Bouaicha, N., D. Pesando, D. Puel and C. Tringali. 1993. Cytotoxic diterpenoids from the brown alga *Dilophus ligulatus*. J. Nat. Prod. 56: 1747–1752.

Cabrita, M.T., C. Vale and A.P. Rauter. 2010. Halogenated compounds from marine algae. Mar. Drugs 8: 2301–2317.

Carlucci, M.J., L.A. Scolaro and E.B. Damonte. 1999. Inhibitory action of natural carrageenans on *Herpes simplex* virus infection of mouse astrocytes. Chemotherapy 45: 429–436.

Cha, K.H., S.Y. Koo and D.U. Lee. 2008. Antiproliferative effects of carotenoids extracted from *Chlorella ellipsoidea* and *Chlorella vulgaris* on human colon cancer cells. J. Agric. Food Chem. 56: 10521–10526.

Chacón-Lee, T.L. and G.E. González-Mariño. 2010. Microalgae for "healthy" foods-possibilities and challenges. Compr. Rev. Food Sci. F. 9: 655–675.

Chakraborty, K. and R. Paulraj. 2010. Sesquiterpenoids with free-radical-scavenging properties from marine macroalga *Ulva fasciata* Delile. Food Chem. 122: 31–41.

Chatter, R., R.B. Othman, S. Rabhi, M. Kladi, S. Tarhouni, C. Vagias, V. Roussis, L. Guizani-Tabbane and R. Kharrat. 2011. *In vivo* and *in vitro* anti-inflammatory activity of neorogioltriol, a new diterpene extracted from the red algae *Laurencia glandulifera*. Mar. Drugs 9: 1293–1306.

Chen, I.L., W.H. Gerwick, R. Schatzman and M. Laney. 1994. Isorawsonol and related IMO dehydrogenase inhibitors from the tropical alga *Avrainvillea rawsoni*. J. Nat. Prod. 57: 947–952.

Chen, J.-H., J.-D. Lim, E.-H. Sohn, Y.-S. Choi and E.-T. Han. 2009. Growth-inhibitory effect of a fucoidan from brown seaweed *Undaria pinnatifida* on *Plasmodium* parasites. Parasitol. Res. 104: 245–250.

Chew, Y.L., Y.Y. Lim, M. Omar and K.S Khoo. 2008. Antioxidant activity of three edible seaweeds from two areas in South East Asia. LWT 41: 1067–1072.

Chi, J.T., Z. Wang, D.S.A. Nuyten, E.H. Rodriguez, M.E. Schaner, A. Salim, Y. Wang, G.B. Kristensen, Å. Helland, A.-L. Børresen-Dale, A. Giaccia, M.T. Longaker, T. Hastie, G.P. Yang, M.J. van de Vijver and P.O. Brown. 2006. Gene expression programs in response to hypoxia: cell type specificity and prognostic significance in human cancers. PLoS Med. 3, e47.

Chittiboyina, A.G., G.M. Kumar, P.B. Carvalho, Y. Liu, Y.-D. Zhou, D.G. Nagle and M.A. Avery. 2007. Total synthesis and absolute configuration of laurenditerpenol: a HIF-1 activation inhibitor. J. Med. Chem. 50: 6299–6302.

Chiu, L.C., E.Y. Wong and V.E. Ooi. 2004. Docosahexaenoic acid from a cultured microalga inhibits cell growth and induces apoptosis by upregulating Bax/Bcl-2 ratio in human breast carcinoma MCF-7 cells. Ann. N.Y. Acad. Sci. 1030: 361–368.

Cho, M., H.-S. Lee, I.-J. Kang, M.-H. Won and S. You. 2011. Antioxidant properties of extract and fractions from *Enteromorpha prolifera*, a type of green seaweed. Food Chem. 127: 999–1006.

Choi, B.W., G. Ryu, S.H. Park, E.S. Kim, J. Shin, S.S. Roh, H.C. Shin and B.H. Lee. 2007. Anticholinesterase activity of plastoquinones from *Sargassum sagamianum*: Lead compounds for Alzheimer's disease therapy. Phytother. Res. 21: 423–442.

Cian, R.E., R. López-Posadas, S.R. Drago, F. Sánchez de Medina and O. Martínez-Augustin. 2012. A *Porphyra columbina* hydrolysate upregulates IL-10 production in rat macrophages and lymphocytes through an NF-κB, and p38 and JNK dependent mechanism. Food Chem. 134: 1982–1990.

Costa, L., G.P. Fidelis, S.L. Cordeiro, R.M. Oliveira, D.A. Sabry, R.B.G. Câmara, L.T.D.B. Nobre, M.S.S.P. Costa, J. Almeida-Lima, E.H.C. Farias, E.L. Leite and H.A.O. Rocha. 2010. Biological activities of sulfated polysaccharides from tropical seaweeds. Biomed. Pharmacother. 64: 21–28.

Cruz, L.J., J.R. Luque-Ortega, L. Rivas and F. Albericio. 2009. Kahalalide F, an antitumor depsipeptide in clinical trials, and its analogues as effective antileishmanial agents. Mol. Pharm. 6: 813–824.

Custódio, L., T. Justo, L. Silvestre, A. Barradas, C. Vizetto Duarte, H. Pereira, L. Barreira, A.P. Rauter, F. Alberício and J. Varela. 2012. Microalgae of different phyla display antioxidant, metal chelating and acetylcholinesterase inhibitory activities. Food Chem. 131: 134–140.

de Inés, C., V.H. Argandoña, J. Rovirosa, A. San-Martín, A.R. Díaz-Marrero, M. Cueto and A. González-Coloma. 2004. Cytotoxic activity of halogenated monoterpenes from *Plocamium cartilagineum*. Z. Naturforsch. 59: 339–344.

Dewick, P.M. 2002. Medicinal natural products: A biosynthetic approach. 2nd Ed. John Wiley and Son. Great Britain. 520 p.

Dittmann, E., B.A. Neilan and T. Börner. 2001. Molecular biology of peptide and polyketide biosynthesis in cyanobacteria. Appl. Microbiol. Biotechnol. 57: 467–473.

Dmitrenok, A., T. Iwashita, T. Nakajima, B. Sakamoto, M. Namikoshi and H. Nagai. 2006. New cyclic depsipeptides from the green alga *Bryopsis* species; application of a carboxypeptidase hydrolysis reaction to the structure determination. Tetrahedron 62: 1301–1308.

D'Orazio, N., M.A. Gammone, E. Gemello, M. De Girolamo, S. Cusenza and G. Riccioni. 2012. Marine bioactives: pharmacological properties and potential applications against inflammatory diseases. Mar. Drugs 10: 812–833.

Dore, C.M.P.G., M.G.D.C.F. Alves, L.S.E.P. Will, T.G. Costa, D.A. Sabry, L.A.R.D.S. Rêgo, C.M. Accardo, H.A.O. Rocha, L.G.A. Filgueira and E.L. Leite. 2012. A sulfated polysaccharide, fucans, isolated from brown algae *Sargassum vulgare* with anticoagulant, antithrombotic, antioxidant and anti-inflamatory effects. Carbohyd. Polym. (in press).

Dorta, E., M. Cueto, I. Bito and J. Darias. 2002. New terpenoids from the brown alga *Stypopodium zonale*. J. Nat. Prod. 65: 1727–1730.

Dos Santos, A.O., P. Veiga-Santos, T. Ueda-Nakamura, B.P. Dias-Filho, D.B. Sudatti, E.M. Bianco, R.C. Pereira and C.V. Nakamura. 2010. Effect of elatol, isolated from red seaweed *Laurencia dendroidea*, on *Leishmania amazonensis*. Mar. Drugs 8: 2733–2743.

Dos Santos, A.O., E. Britta, E.M. Bianco, T. Ueda-Nakamura, B.P. Dias-Filho, R.C. Pereira and C.V. Nakamura. 2011. 4-Acetoxydolastane diterpene from the brazilian brown alga *Canistrocarpus cervicornis* as antileishmanial agent. Mar. Drugs 9: 2369–2383.

El Gamal, A.A., W.-L. Wang and C.-Y. Duh. 2005. Sulfur-containing polybromoindoles from the Formosan red alga *Laurencia brongniartii* . J. Nat. Prod. 68: 815–817.

El Gamal, A.A. 2010. Biological importance of marine algae. Saudi Pharmaceut. J. 18: 1–25.

Elmore, S. 2007. Apoptosis: a review of programmed cell death. Toxicol. Pathol. 35: 495–516.

Fattorusso, E. and O. Taglialatela-Scafati. 2009. Marine antimalarials. Mar Drugs. 7: 130–152.

Felício, R., S. Albuquerque, M.C. Marx Younge, N.S. Yokoyad and H.M. Debonsia. 2011. Trypanocidal, leishmanicidal and antifungal potential from marine red alga *Bostrychia tenella* J. Agardh (Rhodomelaceae, Ceramiales). J. Pharm. Biomed. Anal. 52: 763–769.

Fenical, W., J.J. Sims, D. Squatrito, R.M. Wing and P. Radlick. 1973. Marine natural products. VII. Zonarol and isozonarol, fungitoxic hydroquinones from the brown seaweeds Dictyopteris zonarioides. J. Org. Chem. 38: 2383–2386.

Filho, J., K. Medeiros, M. Dinin, L. Batista, R. Athayde Filho, M. Silva, F. da Cunha, J. Almeida and J. Quintans-Júnior. 2006. Natural products inhibitors of the enzyme acetylcholinesterase. Braz. J. Pharmacogn. 16: 258–285.

Folmer, F., M. Jaspars, M. Dicato and M. Diederich. 2008. Marine natural products as targeted modulators of the transcription factor NF-κB. Biochem. Pharmacol. 75: 603–617.

Fouladvand, M., H. Malekizadeh, A. Barazesh, K. Sartavi and F. Farolkhza. 2011. Evaluation of *in vitro* anti-leishmanial activity of some brown, green and red algae from the Persian Gulf. Eur. Rev. Med. Pharmacol. Sci. 15: 597–600.

Freire-Pelegrin, Y., D. Robledo, M.J. Chan-Bacab and B. Ortega-Morales. 2008. Antileishmanial properties of tropical marine algae extracts. Fitoterapia 79: 374–377.

Fuller, R.W., J.H. Cardellina, Y. Kato, L.S. Brinen, J. Clardy, K.M. Sander and M.R. Boyad. 1992. A pentahalogenated monoterpene frm the red alga *Portieria hornemannii* produced a novel cytotoxicity profile against a diverse panel of human tumor celllin. J. Med. Chem. 35: 3007–3011.

Fuller, R.W., J.H. Cardellina, J. Jurek, P.J. Scheuer, B. Alvarado-Linder, M. McGuier, G.N. Gray, I.R. Steiner, J. Clardy, E. Menez, R.H. Shoemaker, D.I. Newman, K.M. Sander and M.R. Boyad. 1994. Isolation and structure/activity features of halomon-related antitumor monoterpenes from the red alga *Portieria hornemannii*. J. Med. Chem. 37: 4407–4411.

Funk, C.D. 2001. Prostaglandins and Leukotrienes: Advances in Eicosanoid Biology. Sci. 294: 1871–1875.

Gamal-Eldeen, A.M., E.F. Ahmed and M.A. Abo-Zeid. 2009. *In vitro* cancer chemopreventive properties of polysaccharide extract from the brown alga, *Sargassum latifolium*. Food Chem. Toxicol. 47: 1378–1384.

García-Rocha, M., P. Bonay and J. Avila. 1996. The antitumoral compound Kahalalide F acts on cell lysosomes. Cancer Lett. 99: 43–50.

Gautam, R. and S.M. Jachak. 2009. Recent developments in anti-inflammatory natural products. Med. Res. Rev. 5: 767–820.

Genovese, G., L. Tedone, M. Hamann and M. Morabito. 2009. The mediterranean red alga *Asparagopsis*: A source of compounds against *Leishmania*. Mar. Drugs 7: 361–366.

Govindan, M., S.A. Abbas, E.I. Schmitz, R.H. Lee, I.S. Papkoff and D.L. Slate. 1994. New cycloartanol sulfates from the alga *Tydemania expeditionis*: inhibitor of the protein tyrosin kinase pp60. J. Nat. Prod. 57: 74–78.

Gross, H., D.E. Goeger, P. Hills, S.L. Mooberry, D.L. Ballantine, T.F. Murray, F.A. Valeriote and W.H. Gerwick. 2006. Lophocladines, bioactive alkaloids from the red alga *Lophocladia* sp. J. Nat. Prod. 69: 640–644.

Guardia, S.D., R. Valls, V. Mesguiche, J.-M. Brunei and G. Gulioli. 1999. Enantioselective synthesis of (–)-bifuracadiol: a natural antitumor marine product. Tetrahedron Lett. 40: 8359–8360.

Guedes, A.C., H.M. Amaro and F.X. Malcata. 2011. Microalgae as sources of high added-value compounds—a brief review of recent work. Biotechnol. Prog. 27: 597–613.

Gulioli, G., A. Oratalo-Magne, M. Daoudi, H. Thomas Guyon, R. Vallis and L. Piovetti. 2004. Trihydroxylated linear diterpenes from the brown alga *Bifurcaria bifurcata*. Phytochemistry 65: 2063–2069.

Güven, K.C., A. Percot and E. Sezik. 2010. Alkaloids in marine algae. Mar. Drugs 8: 269–284.

Haefner, B. 2003. Drugs from the deep: marine natural products as drug candidates. Drug Discov. Today 8: 536–544.

Hamann, M.T. and P.J. Scheuer. 1993. Kahalalide F: a bioactive depsipeptide from the sacoglossan mollusk *Elysia rufescens* and the green alga *Bryopsis* sp. J. Am. Chem. Soc. 115: 5825–5826.

Henríquez, R., G. Faircloth and C. Cuevas. 2005. Ecteinascidin 743 (ET-743, Yondelis), aplidin and kahalalide F. *In*: G.M. Cragg, D.G.I. Kingston and D.J. Newman (eds.). Anticancer Agents from Natural Products. Taylor and Francis, Boca Raton, Florida.

Heo, S.-J. and Y.-J. Jeon. 2009. Evaluation of diphlorethohydroxycarmalol isolated from *Ishige okamurae* for radical scavenging activity and its protective effect against H_2O_2-induced cell damage. Process Biochem. 44: 412–418.

Heo, S.-J., W.-J. Yoon, K.-N. Kim, G.-N. Ahn, S.-M. Kang, D.-H. Kang, A. Affan, C. Oh, W.-K. Jung and Y.-J. Jeon. 2010. Evaluation of anti-inflammatory effect of fucoxanthin isolated from brown algae in lipopolysaccharide-stimulated RAW 264.7 macrophages. Food Chem. Toxicol. 48: 2045–2051.

Higuera-Ciapara, I., L. Félix-Valenzuela and M. Goycoolea. 2006. Astaxanthin: a review of its chemistry and applications. Crit. Rev. Food Sci. 46: 185–196.

Honer, W.G. 2003. Pathology of presynaptic proteins in Alzheimer's disease: more than simple loss of terminals. Neurobiol. Aging 24: 1047–1062.

Huang, H.L., S.L. Wu, H.F. Liao, C.M. Jiang, R.L. Huang, Y.Y. Chen, Y.C. Yang and Y.J. Chen. 2005. Induction of apoptosis by three marine algae through generation of reactive oxygen species in human leukemic cell lines. J. Agric. Food. Chem. 53: 1776–1781.

Ishitsuka, O.M., T. Kusumi and H. Kakisawa H. 1988. Antitumor xenicane and norxenicane lactones from the brown algae *Dictyota dichotoma*. J. Org. Chem. 53: 5010–5013.

Jaswir, I. and H.A. Monsur. 2011. Anti-inflammatory compounds of macro algae origin: A review. J. Med. Plants Res. 5: 7146–7154.

Jimeno, J., G. Faircloth, J.M. Fernández Sousa-Faro, P. Scheuer and K. Rinehart. 2004. New marine derived anticancer therapeutics—a journey from the sea to clinical trials. Mar. Drugs 2: 14–29.

Jongaramruong, J. and N. Kongkam. 2007. Novel diterpenes with cytotoxic, anti-malarial and anti-tuberculosis activities from a brown alga *Dictyota* sp. J. Asian Nat. Prod. Res. 9: 743–751.

Jung, M.E. and G.-Y.J. Im. 2008. Convergent total synthesis of the racemic HIF-1 inhibitor laurenditerpenol. Tetrahedron Lett. 49: 4962–4964.

Karadag, A., B. Özcelik and Ş. Saner. 2009. Review of methods to determine antioxidant capacities. Food Anal. Methods 2: 41–60.

Karran, E., M. Mercken and B. Strooper. 2011. The amyloid cascade hypothesis for Alzheimer's disease: an appraisal for the development of therapeutics. Nat. Rev. Drug. Discov. 10: 698–712.

Kazłowska, K., T. Hsu, C.-C. Hou, W.-C. Yang and G.-J Tsai. 2010. Anti-inflammatory properties of phenolic compounds and crude extract from *Porphyra dentata*. J. Ethnopharmacol. 128: 123–130.

Khanavi, M., M. Nabavi, N. Sadati, M. Shams Ardekani, J. Sohrabipour, S.M. Nabavi, P. Ghaeli and S.N. Ostad. 2010. Cytotoxic activity of some marine brown algae against cancer cell lines. Biol. Res. 43: 31–37.

Khanavi, M., R. Gheidarloo, N. Sadati, M.R. Ardekani, S.M. Nabavi, S. Tavajohi and S.N. Ostad. 2012. Cytotoxicity of fucosterol containing fraction of marine algae against breast and colon carcinoma cell line. Pharmacogn. Mag. 8: 60–64.

Kim, K.-N., S.-J. Heo, W.-J. Yoon, S.-M. Kang, G. Ahn, T.-H. Yi and Y.-J. Jeon. 2010. Fucoxanthin inhibits the inflammatory response by suppressing the activation of NF-κB and MAPKs in lipopolysaccharide-induced RAW 264.7 macrophages. Eur. J. Pharmacol. 649: 369–375.

Kim, M.K., J.W. Park, C.S. Park, S.J. Kim, K.H. Jeune, M.U. Chang and J. Acreman. 2007. Enhanced production of *Scenedesmus* spp. (green microalgae) using a new medium containing fermented swine wastewater. Biores. Technol. 98: 2220–2228.

Kim, M.M., E. Mendis and S.K. Kim. 2008. *Laurencia okamurai* extract containing laurinterol induces apoptosis in melanoma cells. J. Med. Food 11: 260–266.

Kladia, M., C. Vagias, G. Furnari, D. Morreau, C. Roussakis and V. Roussis. 2005. Cytotoxic cuparene sesquiterpenes from *Laurencia microcladia*. Tetrahedron Lett. 46: 5723–5726.

Ko, S.-C., D. Kim and Y.-J. Jeon. 2012. Protective effect of a novel antioxidative peptide purified from a marine *Chlorella ellipsoidea* protein against free radical-induced oxidative stress. Food Chem. Toxicol. 50: 2294–2302.

Kong, C.S., J.A. Kim, N.Y. Yoon and S.K. Kim. 2009. Induction of apoptosis by phloroglucinol derivative from *Ecklonia cava* in MCF-7 human breast cancer cells. Food Chem. Toxicol. 47: 1653–1658.

Kubanek, J., A.C. Prusak, T.W. Snell, R.A. Giese, C.R. Fairchild, W. Aalbersberg and M.E. Hay. 2006. Bromophycolides C-I from the Fijian red alga *Callophycus serratus*. J. Nat. Prod. 69: 731–735.

Kuda, T., M. Tsunekawa, T. Hishi and Y. Araki. 2005. Antioxidant properties of dried 'kayamo-nori', a brown alga *Scytosiphon lomentaria* (Scytosiphonales, Phaeophyceae). Food Chem. 89: 617–622.

Kumar, M., V. Gupta, P. Kumari, C.R.K. Reddy and B. Jha. 2011. Assessment of nutrient composition and antioxidant potential of Caulerpaceae seaweeds. J. Food Comp. Anal. 24: 270–278.

Kurata, K., K. Tanguchi and M. Suzuki. 1996. Cyclozonarone, a sesquiterpene-substituted benzoquinone derivative from the brown alga *Dictyopteris undulata*. Phytochemistry 41: 749–752.

Lakshmi, V., A.K. Goel, M.N. Srivastava, D.K. Kulshreshtha and R. Raghubir. 2006. Bioactive of marine organism - Part IX: Screen of some marine flora from the Indian coasts. Indian J. Exp. Biol. 44: 137–141.

Laube, T., W. Beil and K. Seifert. 2005. Total synthesis of two 12-nordrimanes and the pharmacological active sesquiterpene hydroquinone yahazunol. Tetrahedron 61: 1141–1148.

Lee, S.-H., C.-I. Ko, G. Ahn, S.G. You, J.-S. Kim, M.S. Heu, J. Kim, Y. Jee and Y.-J. Jeon. 2012. Molecular characteristics and anti-inflammatory activity of the fucoidan extracted from *Ecklonia cava*. Carbohyd. Polym. 89: 599–606.

León-Deniz, L.V., E. Dumonteil, R. Moo-Puc and Y. Freile-Pelegrin. 2009. Antitrypanosomal *in vitro* activity of tropical marine algae extracts. Pharm. Biol. 47: 864–871.

Li, K., X.-M. Li, N.-Y. Ji and B.-G. Wang. 2007. Natural bromophenols from the marine red alga *Polysiphonia urceolata* (Rhodomelaceae): structural elucidation and DPPH radical-scavenging activity. Bioorg. Med. Chem. 15: 6627–6631.

Li, Y., Z.-J. Qian, B. Ryu, S.-H. Lee, M.-M. Kim and S.-K. Kim. 2009. Chemical components and its antioxidant properties *in vitro*: an edible marine brown alga, *Ecklonia cava*. Bioorg. Med. Chem. 7: 1963–1973.

Liu, M., P.E. Hansen and X. Lin. 2011. Bromophenols in marine algae and their bioactivities. Mar. Drugs 9: 1273–1292.

Manalo, D.J., A. Rowan, T. Lavoie, L. Natarajan, B.D. Kelly, S.Q. Ye, J.G. Garcia and G.L. Semenza. 2005. Transcriptional regulation of vascular endothelial cell responses to hypoxia by HIF-1. Blood 105: 659–669.

Manzanares, I., C. Cuevas, R. Garcia-Nieto, E. Marco and F. Gago. 2001. Advances in the chemistry and pharmacology of ecteinascidins, a promising new class of anticancer agents. Curr. Med. Chem. Anticancer Agents 1: 257–276.

Maplestone, R.A., M.J. Stone and D.H. Williams. 1992. The evolutionary role of secondary metabolites—a review. Gene 115: 151–157.

Marcolino, M. 2010. Avaliação da atividade leishmanicida *in vitro* de heteropolissacarídeos ácidos: não sulfatados e naturalmente sulfatados. M.S. Thesis, Universidade Federal do Paraná, Brasil.

Martín-Algarra, S., E. Espinosa, J. Rubió, J.J. López López, J.L. Manzano, L.A. Carrión, A. Plazaola, A. Tanovic and L. Paz-Ares. 2009. Phase II study of weekly Kahalalide F in patients with advanced malignant melanoma. Eur. J. Cancer. 45: 732–735.

Mayer, A., A. Rodríguez, R. Berlinck and N. Fusetani. 2011. Marine pharmacology in 2007–8: Marine compounds with antibacterial, anticoagulant, antifungal, anti-inflammatory, antimalarial, antiprotozoal, antituberculosis, and antiviral activities; affecting the immune and nervous system, and other miscellaneous mechanisms of action. Comp. Biochem. Physiol. C Toxicol. Pharmacol. 153: 191–222.

Meinert, C.L. 2006. Clinical trials, overview. pp. 24–47. *In*: Sean Pidgeon (ed.). Wiley Handbook of Current and Emerging Drug Therapies, Volumes 1–4. John Wiley & Sons, Hoboken, NJ.

Mendiola-Martínez, J., H. Hernández, D. Acuña, M. Esquivel, R.S. Lizama and J.A. Payrol. 2005. Inhibiting activity of the *in vitro growth of Plasmodium falciparum of extracts from algae of genus Laurencia*. Rev. Cubana Med. Trop. 57: 192 f.

Mesulam, M., A. Guillozet, P. Shaw and B. Quinn. 2002. Widely spread butyrylcholinesterase can hydrolyse acetylcholine in the normal and Alzheimer brain. Neurobiol. Dis. 9: 88–93.

Metting, F.B. 1996. Biodiversity and application of microalgae. J. Ind. Microbiol. 17: 477–489.

Miguel, M.G. 2010. Antioxidant and anti-inflammatory activities of essential oils: A short review. Molecules 15: 9252–9287.

Mohammed, K.A., C.F. Hossain, L. Zhang, R.K. Bruick, Y.D. Zhou and D.G. Nagle. 2004. Laurenditerpenol, a new diterpene from the tropical marine alga *Laurencia intricata* that potently inhibits HIF-1 mediated hypoxic signaling in breast tumor cells. J. Nat. Prod. 67: 2002–2007.

Molinski, T.F., D.S. Dalisay, S.L. Lievens and J.P. Saludes. 2009. Drug development from marine natural products. Nat. Rev. Drug Discov. 8: 69–85.

Moo-Puc, R., D. Robledo and Y. Freile-Pelegrin. 2008. Evaluation of selected tropical seaweeds for *in vitro* anti-trichomonal activity. J. Ethnopharmacol. 120: 92–97.

Mukherjee, P.K., V. Kumar, M. Mal and P.J. Houghton. 2007. Acetylcholinesterase inhibitors from plants. Phytomedicine 14: 289–300.

Myung, C.-S., H.-C. Shin, H.Y. Bao, S.J. Yeo, B.H. Lee and J.S. Kang. 2005. Improvement of memory by dieckol and phlorofucofuroeckol in ethanol-treated mice: possible involvement of the inhibition of acetylcholinesterase. Arch. Pharm. Res. 28: 691–698.

Nahas, R., D. Abatis, M. Anagnostopoulou, P. Kefalas, C. Vagias and V. Roussis. 2007. Radical-scavenging activity of Aegean Sea marine algae. Food Chem. 102: 577–581.

Nara, T., Y. Kameib, A. Tsubouchia, T. Annouraa, K. Hirotaa, K. Iizumia, Y. Dohmotob, T. Onob and T. Aokia. 2005. Inhibitory action of marine algae extracts on the *Trypanosoma cruzi* dihydroorotate dehydrogenase activity and on the protozoan growth in mammalian cells. Parasitol. Int. 54: 59–64.

Natarajan, S., K.P. Shanmugiahthevar and P.D. Kasi. 2009. Cholinesterase inhibitors from *Sargassum* and *Gracilaria gracilis*: seaweeds inhabiting South Indian coastal areas (Hare Island, Gulf of Mannar). Nat. Prod. Res. 23: 355–369.

Numata, A., S. Kambara, C. Takahashi, R. Fujiki, M. Yoneda, E. Fujita and Y. Nabeshima. 1991. Cytotoxic activity of marine algae and a cytotoxic principle of the brown alga *Sargassum tortile*. Chem. Pharm. Bull. 39: 2129–2131.

Ochi, M., H. Kotsuki, K. Muraoka and T. Tokoroyama. 1979. The structure of yahazunol, a new sesquiterpene-substituted hydroquinone from the brown seaweed *Dictyopteris undulata* Okamura Bull. Chem. Soc. Jpn. 52: 629–630.

Okai, Y. and K. Higashi-Okai. 1997. Potent anti-inflammatory activity of pheophytin *a* derived from edible green alga, *Enteromorpha prolifera* (Sujiao-nori). Int. J. Immunopharmacol. 19: 355–358.

Orhan, I., B. Senera, T. Atıcıb, R. Brunc, R. Perozzod and D. Tasdemir. 2006. Turkish freshwater and marine macrophyte extracts show *in vitro* antiprotozoal activity and inhibit FabI, a key enzyme of *Plasmodium falciparum* fatty acid biosynthesis. Phytomedicine 13: 388–393.

Pacheco, F.C., J.A. Villa-Pulgarin, F. Mollinedo, M.N. Martín, J.J. Fernández and A.H. Daranas. 2011. New polyether triterpenoids from *Laurencia viridis* and their biological evaluation. Mar. Drugs 9: 2220–2235.

Pangestuti, R. and S.-K. Kim. 2011. Biological activities and health benefit effects of natural pigments derived from marine algae. J. Funct. Foods 3: 255–266.

Pardo, B., L. Paz-Ares, J. Tabernero, E. Ciruelos, M. García, R. Salazar, A. López, M. Blanco, A. Nieto, J. Jimeno, M.A. Izquierdo and J.M. Trigo. 2008. Phase I clinical and pharmacokinetic study of kahalalide F administered weekly as a 1-hour infusion to patients with advanced solid tumors. Clin. Cancer Res. 14: 1116–1123.

Parihar, M.S. and T. Hemnani. 2004. Alzheimer's disease pathogenesis and therapeutic interventions. J. Clin. Neurosci. 11: 456–467.

Paul, V.J. and W. Fenical. 1983. Isolation of halimedtrial: chemical defense adaptation in the calcareous reef-building alga *Halimeda*. Science 221: 747–749.

Paul, V.J. and W. Fenical. 1984. Novel bioactive diterpenoid metabolites from tropical marine algae of the genus *Halimeda* (Chlorophyta). Tetrahedron 40: 3053–3062.

Pec, M.K., A. Aguirre, K. Moser-Thier, J.J. Fernández, M.L. Souto, J. Dorta, F. Díaz-González and J. Villar. 2003. Induction of apoptosis in estrogen dependent and independent breast cancer cells by the marine terpenoid dehydrothyrsiferol. Biochem. Pharmacol. 65: 1451–1461.

Pulz, O. and W. Gross. 2004. Valuable products from biotechnology of microalgae. Appl. Microbiol. Biotechnol. 65: 635–648.

Racchi, M., M. Mazzucchelli, E. Porrello, C. Lanni and S. Govoni. 2004. Acetylcholinesterase inhibitors: novel activities of old molecules. Pharmacol. Res. 50: 441–451.

Rademaker-Lakhai, J.M., S. Horenblas, W. Meinhardt, E. Stokvis, T.M. de Reijke, J.M. Jimeno, L. Lopez-Lazaro, J.A. Lopez Martin, J.H. Beijnen and J.H. Schellens. 2005. Phase I clinical and pharmacokinetic study of kahalalide F in patients with advanced androgen refractory prostate cancer. Clin. Cancer Res. 11: 1854–1862.

Sabina, H., S. Tasneem, Y. Samreen, M. Kausar, M.I. Choudhary and R. Aliya. 2005. Antileishmanial activity in the crude extract of various seaweed from coast of Karachi, Pakistan. Pak. J. Bot. 37: 163–168.

Sánchez, J.F., J.M. Fernández, F.G Acién, A. Rueda, J. Pérez-Parra and E. Molina. 2008. Influence of culture conditions on the productivity and lutein content of the new strain *Scenedesmus almeriensis*. Proc. Biochem. 43: 398–405.

Sansom, C.E., L. Larsen, N.B. Perry, M.V. Berridge, E.W. Chia, J.L. Harper and V.L. Webb. 2007. An antiproliferative bis-prenylated quinone from the New Zealand brown alga *Perithalia capillaris*. J. Nat. Prod. 70: 2042–2044.

Scheff, S.W., D.A. Price, F.A. Schmitt and E.J. Mufson. 2006. Hippocampal synaptic loss in early Alzheimer's disease and mild cognitive impairment. Neurobiol. Aging 27: 1372–1384.

Schmidt,T.J., S.A. Khalid, A.J. Romanha, T.M.A. Alves, M.W. Biavatti, R. Brun, F.B. da Costa, S.L. de Castro, V.F. Ferreira, M.V.G. de Lacerda, J.H.G. Lago, L.L. Leon, N.P. Lopes, R.C. das Neves Amorim, M. Niehues, I.V. Ogungbe, A.M. Pohlit, M.T. Scotti, W.N. Setzer, M. de NC Soeiro, M. Steindel and A.G. Tempone. 2012a. The potential of secondary metabolites from plants as drugs or leads against protozoan neglected diseases - Part I. Curr. Med. Chem. 19: 2128–2175.

Schmidt,T.J., S.A. Khalid, A.J. Romanha, T.M.A. Alves, M.W. Biavatti, R. Brun, F.B. da Costa, S.L. de Castro, V.F. Ferreira, M.V.G. de Lacerda, J.H.G. Lago, L.L. Leon, N.P. Lopes, R.C. das Neves Amorim, M. Niehues, I.V. Ogungbe, A.M. Pohlit, M.T. Scotti, W.N. Setzer, M. de NC Soeiro, M. Steindel and A.G. Tempone. 2012b. The potential of secondary metabolites from plants as drugs or leads against protozoan neglected diseases - Part II. Curr. Med. Chem. 19: 2176–2228.

Selkoe, D.J. 2002. Alzheimer's disease is a synaptic failure. Science 298: 789–791.

Sheu, J.H., G.H. Wang, P.J. Sung and C.Y. Duh. 1999. New cytotoxic oxygenated Fucosterols from the brown alga, *Turbinaria conoides*. J. Nat. Prods. 62: 224–227.

Smit, A.J. 2004. Medicinal and pharmaceutical uses of seaweed natural products: A review. J. Appl. Phycol. 16: 245–262.

Souto, A.L., J.F. Tavares, M.S. da Silva, M.F.F.M. Diniz, P.F. de Athayde-Filho and J.M. Barbosa Filho. 2011. Anti-inflammatory activity of alkaloids: an update from 2000 to 2010. Molecules 16: 8515–8534.

Souza, B.W.S., M.A. Cerqueira, A.I. Bourbon, A.C. Pinheiro, J.T. Martins, J.A. Teixeira, M.A. Coimbra and A.A. Vicente. 2012. Chemical characterization and antioxidant activity of sulfated polysaccharide from the red seaweed *Gracilaria birdiae*. Food Hydrocolloids 27: 287–292.

Spavieri, J., A. Allmendinger, M. Kaiser, R. Casey, S. Hingley-Wilson, A. Lalvani, M.D. Guiry, G. Blunden and D. Tasdemir. 2010a. Antimycobacterial, antiprotozoal and cytotoxic potential of twenty-one brown algae (Phaeophyceae) from British and Irish waters. Phytother. Res. 24(11): 1724–9.

Spavieri, J., M. Kaiser, R. Casey, S. Hingley-Wilson, A. Lalvani, G. Blunden and D. Tasdemir. 2010b. Antiprotozoal, antimycobacterial and cytotoxic potential of some British green algae. Phytother. Res. 24: 1095–1098.

Stirk, W., D. Reinecke and J. van Staden. 2007. Seasonal variation in antifungal, antibacterial and acetylcholinesterase activity in seven South African seaweeds. J. Appl. Phycol. 19: 271–276.

Suárez, Y., L. González, A. Cuadrado, M. Berciano, M. Lafarga and A. Muñoz. 2003. Kahalalide F, a new marine-derived compound, induces oncosis in human prostate and breast cancer cells. Mol. Cancer Ther. 2: 863–872.

Suffness, M. and J. Douros. 1982. Current status of the NCI plant and animal product program J. Nat. Prod. 45: 1–14.

Suganthy, N., S. Karutha Pandian and K. Pandima Devi. 2010. Neuroprotective effect of seaweeds inhabiting South Indian coastal area (Hare Island, Gulf of Mannar marine biosphere reserve): Cholinesterase inhibitory effect of *Hypnea valentiae* and *Ulva reticulata*. Neurosci. Lett. 468: 216–219.

Süzgeç-Selçuk, S., A.H. Meriçli, K.C. Güven, M. Kaiser, R. Casey, S. Hingley-Wilson, A. Lalvani and D. Tasdemir. 2010. Evaluation of Turkish seaweeds for antiprotozoal, antimycobacterial and cytotoxic activities. Phytother. Res. 25: 778–83.

Suzuki, T., A. Furusaki, T. Matsumoto, A. Kato, Y. Lmanaka and E. Kurosawa. 1985. Teurilene and thyrsiferyl 23 acetate, meso and remarkably cytotoxic compounds from the marine red alga *Laurencia obtuse*. Tetrahedron Lett. 26: 1329–1332.

Tang, H.-F., Y.-H. Yi, X.-S. Yao, Q.-Z. Xu, S.-Y. Zhang and H.-W. Lin. 2002. Bioactive steroids from the brown alga *Sargassum carpophyllum*. J. Asian Nat. Prod. Res. 4: 95–101.

Tanzi, R.E. 2005. The synaptic Abeta hypothesis of Alzheimer disease. Nat. Neurosci. 8: 977–979.

Tempone, A.G., C. Martins de Oliveira and R.G. Berlinck. 2011. Current approaches to discover marine antileishmanial natural products. Planta Med. 77: 572–85.

Topcu, G., Z. Aydogmus, S. Imre, A.C. Gören, J.M. Pezzuto, J.A. Clement and D.G. Kingston. 2003. Brominated sesquiterpenes from the red alga *Laurencia obtusa*. J. Nat. Prod. 66: 1505–8.

Tringali, C., M. Prattellia and G. Nicols. 1984. Structure and conformation of new diterpenes based on the dolabellane skeleton from *Dictyota* species. Tetrahedron 40: 703–799.

Tseng, Z.C. and Z.J. Chang. 1984. Chinese seaweeds herbal medicine. Hydrobiologia. 116/117: 152–154.

Umemura, K., K. Yanase, M. Suzuki, K. Okutani, T. Yamori and T. Andoh. 2003. Inhibition of DNA topoisomerases I and II, and growth inhibition of human cancer cell lines by a marine microalgal polysaccharide. Biochem. Pharmacol. 66: 481–487.

Urdiales, J.L., P. Morata, I.N. De Castro and F. Sanchez-Jimenez. 1996. Anti-proliferative effect of dehydrodidemnin B (DDB), a depsipeptide isolated from Mediterranean tunicates. Cancer Lett. 102: 31–37.

Varela, J., C. Vizetto-Duarte, L. Custódio, L. Barreira and F. Albericio. 2013. Marine peptides and proteins with cytotoxic and antitumoral properties. pp. 407–430. *In*: S.-K. Kim (ed.). Marine Proteins and Peptides: Development, Biological Activities, and Industrial Perspectives. Wiley-Blackwell, Chichester, UK.

Veiga-Santos, P., K.J. Pelizzaro-Rocha, A.O. Santos, T. Ueda-Nakamura, B.P. Dias-Filho, S.O. Silva, D.B. Sudatti, E.M. Bianco, R.C. Pereira and C.V. Nakamura. 2010. *In vitro* anti-trypanosomal activity of elatol isolated from red seaweed *Laurencia dendroidea*. Parasitology 14: 1–10.

Vihinen, P. and V.M. Kähäri. 2002. Matrix metalloproteinases in cancer: prognostic markers and therapeutic targets. Int. J. Cancer 99: 157–166.

Vonthron-Sénécheau, C., M. Kaiser, I. Devambez, A. Vastel, I. Mussio and A M. Rusig. 2011. Antiprotozoal activities of organic extracts from french marine seaweeds. Mar. Drugs. 9: 922–933

Weinreb, O., S. Mandel, O. Bar-Am and T. Amit. 2011. Iron-chelating backbone coupled with monoamine oxidase inhibitory moiety as novel pluripotential therapeutic agents for Alzheimer's disease: A tribute to Moussa Youdim. J. Neural Transm. 118: 479–492.

Whitehouse, P.J., J.C. Hedreen, C.L. White and D.L. Price. 1983. Basal forebrain neurons in the dementia of Parkinson disease. Ann. Neurol. 13: 243–248.

Wijesekara, I., R. Pangestuti and S. Kim. 2011. Biological activities and potential health benefits of sulfated polysaccharides derived from marine algae. Carbohyd. Polym. 84: 14–21.

Williams, P., A. Sorribas and M.J. Howes. 2011. Natural products as a source of Alzheimer's drug leads. Nat. Prod. Rep. 28: 48–77.

Wimo, A., B. Winblad, H. Aguero-Torres and E. von Strauss. 2003. The magnitude of dementia occurrence in the world. Alzheimer Dis. Assoc. Disord. 17: 63–67.

[WHO] World Health Organization. 2008. The global burden of disease: 2004 update. WHO press, Geneva: 160pp.

[WHO] World Health Organization. 2010a. Control of the leishmaniases. Report of a meeting of the WHO Expert Committee on the Control of Leishmaniases. WHO Technical Report Series n.º 949. WHO Press, Geneva: 186pp.

[WHO] World Health Organization. 2010b. Working to overcome the global impact of neglected tropical diseases. First WHO report on neglected tropical diseases. WHO Press, Geneva: 184pp.

[WHO] World Health Organization. 2011. World malaria report: 2011. WHO Press, Geneva: 278pp.

[WHO] World Health Organization. 2012a. Fact sheet on Trypanosomiasis, Human African (sleeping sickness) - N.º 259.

[WHO] World Health Organization. 2012b. Fact sheet on Chagas disease (American trypanosomiasis) - N.º 340.

Wright, A.D. and G.M. König. 1996. Antimalarial activity: the search for marine-derived natural products with selective antimalarial activity? J. Nat. Prod. 59: 710–716.

Wright, A.E., D.A. Forleo, G.P. Gunawardana, S.P. Gunasekera, F.E. Koehn and O.J. McConnell. 1990. Antitumor tetrahydroisoquinoline alkaloids from the colonial ascidian *Ecteinascidia turbinata*. J. Org. Chem. 55: 4508–4512.

Wright, C.W. and J.D. Phillipson. 1990. Natural products and the development of selective antiprotozoal drugs. Phytotherapy Res. 4: 127–139.

Wu, C.X., Z.G. Li and H.W. Li. 1990. Effect of berbamine on action potential in isolated human atrial tissues. Asia Pac. J. Pharmacol. 5: 191–193.

Yoon, N., H. Chung, H. Kim and J. Choi. 2008. Acetyl and butyrylcholinesterase inhibitory activities of sterols and phlorotannins from *Ecklonia stolonifera*. Fish. Sci. 74: 200–207.

Zhang, C., Y. Li, X. Shi and S.K. Kim. 2010. Inhibition of the expression on MMP-2, 9 and morphological changes via human fibrosarcoma cell line by 6,6'-bieckol from marine alga *Ecklonia cava*. BMB Rep. 43: 62–68.

Macroalgal Phycocolloids

Leonel Pereira

Introduction

Seaweeds have been utilized by mankind for several hundreds of years, directly for food, for medicinal purposes, and for agriculture fertilizers. Today, seaweed is used in many countries for many different purposes: directly as food, phycocolloids extraction, extraction of compounds with antiviral, antibacterial or antitumor activity, and as biofertilizers (Rudolph 2000; Pereira 2010a,b).

About four million tonnes of seaweed are harvested annually worldwide. The major producers are China and Japan, followed by America and Norway. France used to import Japanese seaweed in the 70s but ten years later, went on to produce algae for food and for biological products users. Unlike what happens in East Asia, the West is more interested in thickeners and gelling properties of some polysaccharides extracted from seaweeds. The use of seaweeds for fertilizer and soil improvement was also well-known in Europe, and both large brown algae and calcified red algae have been collected for this purpose (Guiry and Bluden 1991; Pereira 2010b).

Besides its vegetable nature, algae are sought due to the conjunction of variety of colors (and shapes) and the blue ocean—a great selling point, both for food and cosmetics, especially after certain substances of animal origin have become suspicious of transmitting the spongiform encephalopathy (BSE), commonly known as mad-cow disease (Pereira 2004, 2010a,b).

The marine algae are a rich contributor to health—vitamins and trace elements, and also offer a wide variety of flavors, fragrances, and textures. In fact, algae, as well as being a vitamin and mineral treasure, are low in fat, an essential feature in weight loss diets. In addition, the algae are rich in dietary fiber, which may facilitate intestinal transit, lower the rate of blood cholesterol, and reduce certain diseases such as colon cancer (Pereira 2010a,b, 2011).

Phycocolloids

Phycocolloids refer to those polysaccharides extracted from both freshwater and marine algae. Until now, only the polysaccharides extracted from marine red and brown algae, such as agar, carrageenan, and algin are of economic and commercial significance, since these polysaccharides exhibit high molecular weights, high viscosity, and excellent gelling, stabilizing and emulsifying features. They are also extracted in fairly

Department of Life Sciences, Faculty of Sciences and Technology, and MARE - Marine and Environmental Sciences Centre University of Coimbra, Portugal.
Email: leonel.pereira@uc.pt, Webpage: www.uc.pt/seaweeds

high amount from algae. All these polysaccharides are water-soluble, and can be extracted with hot water or alkaline solutions (Minghou 1990).

Commercial hydrocolloids

Colloids are compounds that form colloidal solutions, an intermediate state between a solution and a suspension, and are used as thickeners, gelling agents, and stabilizers for suspensions and emulsions (see Table 1). Hydrocolloids are carbohydrates that, when dissolved in water, form viscous solutions. The phycocolloids are hydrocolloids extracted from algae and represent a growing industry, with more than one million tons of seaweeds extracted annually for hydrocolloid production (Ioannou and Roussis 2009; Pereira et al. 2009a; Pereira 2010b).

Many seaweeds produce hydrocolloids, associated with the cell wall and intercellular spaces. Members of the red algae (Rhodophyta) produce sulfated galactans (e.g., carrageenans and agars) and the brown algae (Heterokontophyta, Phaeophyceae) produce uronates (alginates) (Pereira 2010b; Bixler and Porse 2011; Jiao et al. 2011; Pereira and van de Velde 2011).

The various phycocolloids used in food industry as natural additives are (European codes of phycocolloids):

Alginic acid – E400
Sodium alginate – E401
Potassium alginate – E402
Ammonium alginate – E403
Calcium alginate – E404
Propylene glycol alginate – E405
Agar – E406
Carrageenan – E407
Semi-refined carrageenan or processed *Eucheuma* seaweed – E407A

Sulfated galactans

The commonest and most abundant cell wall constitutes encountered to date in Rhodophyta are families of galactans bearing the trivial names agars and carrageenans. The pioneering studies conducted in Japan and Canada established that the backbone structure of both agars and carrageenans was based on repeating galactose and 3,6-anhydrogalactose residues, β-(1→4) and α-(1→3) linked, respectively. The main feature distinguishing the highly sulfated carrageenans from the less sulfated agars was the presence of D galactose and anhydro-D-galactose in the former and D-galactose, L-galactose, or anhydro-L-galactose in the later. Classification of these polysaccharides based on their solubility and gelling properties proved unsatisfactory, so attention was focused on the common underlying structural patterns (Anderson et al. 1965; Craigie 1990). The seminal concept of the masked repeating structure first reported for the agar-like porphyran (Anderson and Rees 1965) is now widely accepted for both agars and carrageenans (Craigie 1990).

Agar. Agar is the phycocolloid of most ancient origin. In Japan, agar is considered to have been discovered by Minoya Tarozaemon in 1658, and a monument is Shimizu-mura commemorates the first time it was manufactured. Originally, and even at present times, it was made and sold as an extract in solution (hot) or in gel form (cold), to be used promptly in areas near the factories; the product was then known as "tokoroten". Its industrialization as a dry and stable product started at the beginning of the 18th century, and it has since been called "kanten". The word "agar-agar", however, has a Malayan origin and agar is the most commonly accepted term, although in French- and Portuguese-speaking countries it is also called "gelosa" (Minghou 1990; McHugh 2003).

Agar production by modern techniques of industrial freezing was initiated in California by Matsuoka who registered patents in 1921 and 1922 in the United States. The present manufacturing method by freezing is the classic one, and derives from the American one developed in California prior to World War

Table 1. Applications of macroalgae phycocolloids (adapted from van de Velde and de Ruiter 2002; Dhargalkar and Pereira 2005; Pereira 2004; 2008; Pereira et al. 2017).

Use	Phycocolloid	Function
Food additives		
Baked food	Agar Kappa, Iota, Lambda	Improve quality and control moisture
Beer and wine	Alginate Kappa	Promote flocculation and sedimentation of suspended solids
Canned and processed meat	Alginate Kappa	Hold liquid inside the meat for texturing
Cheese	Kappa	Texturing
Chocolate milk	Kappa, lambda	Keep cocoa in suspension
Cold preparation puddings	Kappa, Iota, Lambda	Thicken and gel
Condensed milk	Iota, lambda	Emulsify
Dairy Creams	Kappa, iota	Stabilize emulsion
Fillings for pies and cakes	Kappa	Give body and texture
Frozen fish	Alginate	Give adhesion and moisture retention
Gelled water-based desserts	Kappa + Iota Kappa + Iota + CF	Gel
Gums and sweets	Agar Iota	Gel, texturize
Hot preparation flans	Kappa, Kappa + Iota	Gel and improve mouth-feel
Jelly tarts	Kappa	Gel
Juices	Agar Kappa, Lambda	Provide viscosity, emulsify
Low calorie gelatins	Kappa + Iota	Gel
Milk ice-cream	Kappa + GG, CF, X	Stabilize emulsion and prevent ice crystal formation
Milkshakes	Lambda	Stabilize emulsion
Salad dressings	Iota	Stabilize suspension
Sauces and condiments	Agar Kappa	Thicken
Soymilk	Kappa + iota	Stabilize emulsion and improve mouth-feel
Cosmetics		
Shampoos	Alginate	Vitalize interface
Toothpaste	Carrageenan	Increase viscosity
Lotions	Alginate	Emulsify, provide elasticity, and skin firmness
Lipstick	Alginate	Provide elasticity, viscosity
Medicinal and Pharmaceutical uses		
Dental mold	Alginate	Aid in form retention
Laxatives	Alginate Carrageenan	Circumvent indigestibility and provide lubrication
Tablets	Alginate Carrageenan	Encapsulate
Metal poisoning	Carrageenan	Bind metal
HSV	Alginate	Inhibit virus

Table 1 contd. ...

...Table 1 contd.

Use	Phycocolloid	Function
Industrial and Lab Uses		
Paints	Alginate	Provide viscosity, glazing
Textiles	Agar, Carrageenan	Provide sizing and glazing
Paper making	Alginate, Agar, Carrageenan	Provide viscosity and thickening
Analytical separation	Alginate, Carrageenan	Gel
Bacteriological media	Agar	Gel
Electrophoresis gel	Agar, Carrageenan	Gel

Non-seaweed colloids: CF – Carob flour; GG – Guar gum; X – Xanthan.

II by H.H. Selby and C.K. Tseng. This work was supported by the American Government that wanted the country to be self-sufficient in strategic needs, especially with regard to bacteriological culture media (Armisen and Galatas 2000).

Apart from the above American production, practically the only producer of this phycocolloid until World War II was the Japanese industry, which held a very traditional industrial structure based on numerous small factories (about 400 factories operated simultaneously). These factories were family-operated, producing a non-standardized quality, and had a high employment rate as production was not mechanized. For this reason, and in spite of the later installation of some factories of a medium to small size, only in recent times has Japan operated modern industrial plants especially for the agar plants (McHugh 2003).

During the Second World War, the shortage of available agar acted as an incentive for those countries with coastal resources of *Gelidum corneum* (formerly *Gelidium sesquipedale*), which is very similar to the *Gelidium pacificum* used by the Japanese industry. Hence, Loureiro started the agar industry in Oporto (Portugal), while at the same time J. Mejias and F. Cabrero, in Spain, commenced studies that led to establishment of the important Iberian agar industry. Other European countries that did not have agarophytes tried to prepare agar substitutes from other seaweed extracts (Minghou 1990; Sousa-Pinto 1998; McHugh 2003; Pereira 2008).

Agar is a phycocolloid named by Malaysians that means "red alga" in general, and has traditionally been applied to what we now know taxonomically as—*Eucheuma*. Ironically, we now know this to be the commercial source of iota carrageenan. Agar is composed of two polysaccharides: agarose and agaropectin. The first is responsible for gelling, while the latter has thickening properties (Armisen and Galatas 2000; Pereira 2011).

Most agar is extracted from species of *Gelidium* and *Gracilaria*. Closely related to *Gelidium* are species of *Pterocladiella*, and small quantities of these are collected mainly in Azores (Portugal) and New Zealand. *Gelidiella acerosa* is the main source of agar in India. *Ahnfeltia* species have been used in both Russia and Japan, one source being the island of Sakhalin (Russia) (Cardoso et al. 2014) *Gelidium* spp. and *Gracilaria* spp. are collected in Portugal, Morocco, Tunisia, and Chile for agar production (Mouradi-Givernaud et al. 1999; Givernaud et al. 2005; Givernaud and Mouradi 2006; Krisler 2006; Ortiz et al. 2009).

Agar is a relatively mature industry in terms of manufacturing methods and applications. Nowadays, most processors are using press/syneresis technology; although some still favor freeze/thaw technology or a mixture of these technologies. While the basic processes may not have changed, improvements in presses and freezing equipment should be noted. High-pressure membrane presses have greatly improved dewatering of agar, and thereby reduced energy requirements for final drying before powder milling.

The origin of agar as a food ingredient lies in Asia, where it has been consumed for several centuries. Its extraordinary qualities as a thickening, stabilizing, and gelling agent make it an essential ingredient for preparing processed food products. Furthermore, its satiating and gut regulating characteristics make it an ideal fiber ingredient in the preparation of low calorie food products. The principal applications of agar food grade are (see Table 2): fruit jellies, milk products, fruit pastilles, caramels, chewing gum, canned meat, soups, confectionery and baked goods, icing, and frozen and salted fish (Armisen and Galatas 2000).

Table 2. Agar grades (adapted from Armisen 1995).

	Agar type	Seaweed source
Natural Agar	Strip Square	Only *Gelidium* by old traditional methods
Industrial Agar	Food grade	*Gelidium, Gracilaria, Pterocladiella, Gelidiella, Ahnfeltia*
	Pharmacological grade	Only *Gelidium*
	Clonic plants production grade	*Gelidium, Pterocladiella*
	Bacteriological agar	Only *Gelidium, Pterocladiella*
	Purified agar	*Gelidium*

About 80 percent of the agar produced globally is for food applications (see Tables 1 and 2), the remaining 10 percent being used for bacteriological plates and other biotechnology uses (in particular agarose electrophoresis). Agar has been classified as GRAS (Generally Recognized as Safe) by the United States of America Food and Drug Administration, which has set maximum usage levels depending on particular applications. In the baked goods industry, the ability of agar gels to withstand high temperatures allows for its use as a stabilizer and thickener in pie fillings, icings, and meringues. Cakes, buns, etc., are often pre-packed in various kinds of modern wrapping materials and often stick to them, especially in hot weather; by reducing the quantity of water and adding some agar, a more stable, smoother, non-stick icing may be obtained (McHugh 2003; Pereira et al. 2009a). Some agars, especially those extracted from *Gracilaria chilensis*, can be used in confectionery with very high sugar content, such as fruit candies. These agars are said to be "sugar reactive" because the sugar (sucrose) increases the strength of the gel. Since agar is tasteless, it does not interfere with the flavors of foodstuffs; this contrasts with some of its competitive gums that require the addition of calcium or potassium salts to form gels. In Asian countries, it is a popular component of jellies; this has its origin in the early practice of boiling seaweed, straining it, and adding flavors to the liquid before cooling to form a jelly (McHugh 2003). The remaining 20 percent is accounted for biotechnological applications (Armisen and Galatas 2000). A list of different uses and the corresponding type of algae required can be found in Table 2 (Armisen 1995).

Agar is fundamental in biotechnology studies, and is used in the preparation of inert, solidified culture media for bacteria, microalgae, fungi, and tissue culture. It is also used to obtain monoclonal antibodies, interferons, steroids, and alkaloids. The biotechnological applications of agar are increasing —as being essential for the separation of macromolecules by electrophoresis, chromatography and DNA sequencing (Pereira 2008; 2010b).

Carrageenans. Carrageenans represent one of the major texturising ingredients used by the food industry; they are natural ingredients, which have been used for decades in food applications and are generally regarded as safe (GRAS). The phycocolloid "carrageen *in*", as it was first called, was discovered by the British pharmacist Stanford in 1862, who extracted it from Irish moss (*Chondrus crispus*). The name was later changed to "carrageen *an*" so as to comply with the "-an" suffix for the names of polysaccharides. The modern carrageenan industry dates from the 1940s, receiving its impetus from the dairy applications (see carrageenan applications in Table 1), where carrageenan was found to be the ideal stabilizer for the suspension of cocoa in milk chocolate (van de Velde and de Ruiter 2002; van de Velde et al. 2004; Pereira et al. 2009a).

The commercial carrageenans are normally divided into three main types: kappa (κ), iota (ι), and lambda (λ) carrageenan. The idealized disaccharide repeating units of these carrageenans are given in Fig. 1. Generally, seaweeds do not produce these idealized and pure carrageenans, but more likely a range of hybrid structures (Fig. 2) and/or precursors (see Fig. 1). Several other carrageenan repeating units exist, e.g., xi (ξ), theta (θ), beta (β), mu (μ), and nu (ν) (Fig. 1). The precursors (mu and nu), when exposed to alkaline conditions, are modified into kappa and iota, respectively, via formation of the 3,6-anhydrogalactose bridge (Myslabodski 1990; Rudolph 2000; van de Velde et al. 2005; Pereira et al. 2009a; Pereira and van de Velde 2011). This is a feature used extensively in extraction and industrial modification.

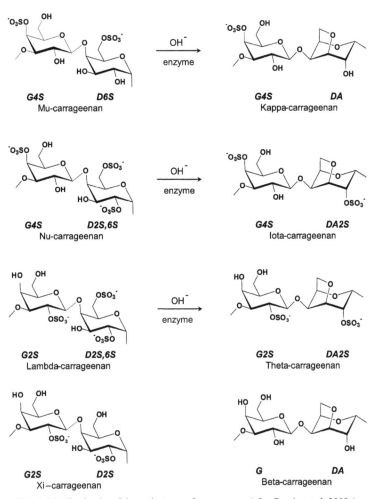

Fig. 1. Idealized units of the main types of carrageenan (after Pereira et al. 2009a).

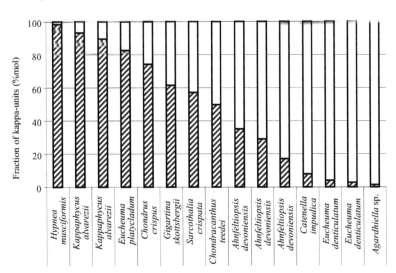

Fig. 2. κ-content of purified κ/ι-hybrid carrageenans - Stancioff diagram (adapted from Pereira 2004; van de Velde et al. 2005).

Carrageenans are the third most important hydrocolloid in the food industry, after gelatin (animal origin) and starch (plant origin) (van de Velde and de Ruiter 2002). The most commonly used, commercial carrageenans are extracted from *Kappaphycus alvarezii* and *Eucheuma denticulatum* (McHugh 2003).

Primarily, wild-harvested genera such as *Chondrus*, *Furcellaria*, *Gigartina*, *Chondracanthus*, *Sarcothalia*, *Mazzaella*, *Iridaea*, *Mastocarpus*, and *Tichocarpus* are also mainly cultivated as carrageenan raw materials and producing countries include Argentina, Canada, Chile, Denmark, France, Japan, Mexico, Morocco, Portugal, North Korea, South Korea, Spain, Russia, and USA (Pereira et al. 2009c; Bixler and Porse 2011).

The original source of carrageenans was from the red seaweed *Chondrus crispus*, which continues to be used but in limited quantities. *Betaphycus gelatinum* is used for the extraction of beta (β) carrageenan. Some South American red algae (used previously only in minor quantities) have been receiving attention from carrageenan producers, as they seek to increase diversification of raw materials in order to provide for the extraction of new carrageenan types with different physical functionalities, and therefore promote product development, which in turn stimulates demand (McHugh 2003). *Gigartina skottsbergii*, *Sarcothalia crispata*, and *Mazzaella laminaroides* are currently the most valuable species, and all are harvested from natural populations in Chile and Peru. The recent earthquake in Chile (27 February 2010), which caused the elevation of intertidal areas and the consequent large losses of harvestable biomass (Pereira 2011) should be mentioned in this regard. Small quantities of *Gigartina canaliculata* are harvested in Mexico and *Hypnea musciformis* has been used in Brazil (Furtado 1999). The use of high value carrageenophytes as a dissolved organic nutrient sink to boost economic viability of integrated multitrophic aquaculture (IMTA) operations has also been considered (Pereira 2004; Chopin 2006; Sousa-Pinto and Abreu 2011).

Large carrageenan processors have fuelled development of *Kappaphycus alvarezii* (which goes by the name "cottonii" to the trade) and *Eucheuma denticulatum* (commonly referred to as "spinosum") farming in several countries including Philippines, Indonesia, Malaysia, Tanzania, Kiribati, Fiji, Kenya, and Madagascar (McHugh 2003). Indonesia has recently overtaken the Philippines as the world's largest producer of dried carrageenophytes biomass (Pereira 2011).

Shortages of carrageenan-producing seaweeds suddenly appeared in mid-2007, resulting in doubling of its price; some of this price increase was due to increased fuel costs and a weak US dollar (most seaweed polysaccharides are traded in US dollars). The reasons for shortages of the raw materials for processing are less certain: perhaps it is a combination of environmental factors, sudden increases in demand, particularly from China, and some market manipulation by farmers and traders. Most hydrocolloids are indeed experiencing severe price changes.

However, the monocultures of some carrageenophytes (namely *Kappaphycus alvarezii*) have several problems due to environmental change and also diseases. The problems with ice-ice and epiphytes have resulted in large scale crop losses (Vairappan et al. 2008; Hayashi et al. 2010; Hurtado et al. 2016).

Uronates

Alginates. "Alginate" is the term usually used for the salts of alginic acid, but it can also refer to all the derivatives of alginic acid and alginic acid itself; in some publications, the term "algin" is used instead of alginate. Alginate is a linear copolymer of β-D-mannuronic acid (M) and α-L-guluronic acid (G) (1→4)-linked residues, arranged either in heteropolymeric (MG) and/or homopolymeric (M or G) blocks (Larsen et al. 2003; Pereira et al. 2003; Leal et al. 2008).

Alginic acid was discovered in 1883 by Stanford, a British pharmacist who called it "algin". In seaweeds, algin is extracted as a mixed salt of sodium and/or potassium, calcium, and magnesium. The exact composition varies with algal species. Since its discovery, the name has been applied to a number of substances, for example, alginic acid and all alginates, derived from alginic acid. The extraction process is based on the conversion of an insoluble mixture of alginic acid salts of the cell wall in a soluble salt (alginate), which is appropriate for the water extraction (Lobban et al. 1988; Lahaye 2001).

Alginic acid is present in the cell walls of brown seaweeds, and it is partly responsible for the flexibility of the seaweed. Consequently, brown seaweeds that grow in more turbulent conditions usually have higher alginate content than those in calmer waters. While any brown seaweed could be used as a source of alginate, the actual chemical structure of the alginate varies from one genus to another, and similar variability is found in the properties of the alginate that is extracted from the seaweed. Since the main applications of alginate are in thickening aqueous solutions and forming gels, its quality is judged on how well it performs in these uses (McHugh 2003).

Twenty-five to 30 years ago, almost all extraction of alginates took place in Europe, USA, and Japan. The major change in the alginates industry over the last decade has been the emergence of producers in China in the 1980s. Initially, production was limited to low cost, low quality alginate for the internal industrial markets produced from the locally cultivated *Saccharina japonica*. By the 1990s, Chinese producers were competing in western industrial markets to sell alginates, primarily based on low cost. A high-quality alginate forms strong gels and gives thick, aqueous solutions. A good raw material for alginate extraction should also give a high yield of alginate. Brown seaweeds that fulfill the above criteria are species of *Ascophyllum*, *Durvillaea*, *Ecklonia*, *Fucus*, *Laminaria*, *Lessonia*, *Macrocystis*, and *Sargassum*. However, *Sargassum*, is only used when nothing else is available: its alginate is usually borderline quality and the yield is usually low (Draget et al. 2004; Pereira 2008).

The goal of the extraction process is to obtain dry, powdered, sodium alginate. The calcium and magnesium salts do not dissolve in water; the sodium salt conversely does. The rationale behind the extraction of alginate from the seaweed is to convert all the alginate salts to the sodium salt, dissolve this in water, and remove the seaweed residue by filtration (McHugh 2003).

Water-in-oil emulsions, such as mayonnaise and salad dressings, are less likely to separate into their original oil and water phases if thickened with alginate. Sodium alginate is not useful when the emulsion is acidic, because insoluble alginic forms acid; for these applications, propylene glycol alginate (PGA) is used since this is stable in mild acid conditions. Alginate improves the texture, body, and sheen of yoghurt, but PGA is also used in the stabilization of milk proteins under acidic conditions, as found in some yoghurts. Some fruit drinks have fruit pulp added and it is preferable to keep it in suspension; addition of sodium alginate, or PGA under acidic conditions, can prevent sedimentation of the pulp and create foams. In chocolate milk, the cocoa can be kept in suspension by an alginate/phosphate mixture, although in this application it faces strong competition from carrageenan (see Table 1). Small amounts of alginate can thicken and stabilize whipped cream (Nussinovitch 1997; Onsoyen 1997).

Methods and techniques for analyzing phycocolloids

Preparation of ground seaweed samples for FTIR-ATR and FT-Raman

The seaweed samples were rinsed in distilled freshwater to eliminate salt and debris from the thallus surface and dried to constant weight at 60°C. The dried seaweeds were finely ground in order to render the samples uniform. The samples do not need additional treatment for FTIR analysis. The analysis by FT-Raman requires that these are devoid of pigmentation. The lack of pigmentation can be achieved by sun drying (process used by collectors/producers of commercial seaweeds) or by pigment elimination in the laboratory by the addition of acetone/methanol moisture (V/V) or by the addition of calcium hypochlorite solution (4%, for 30 to 60s, 4°C) (Pereira 2004, 2006).

Phycocolloid extraction

Before phycocolloid extraction, the ground dry material is rehydrated and pre-treated in acetone followed by ethanol to eliminate the organosoluble fraction (Zinoun and Cosson 1996).

For extraction of the native phycocolloid, the seaweed samples are placed in distilled water (50 mL/g), pH 7 at 85°C for 3 h. For an alkaline-extraction (resembling the industrial method), the samples are placed in a solution (150 mL/g) of NaOH (1 M) at 80–85°C for 3–4 h (according to Pereira and Mesquita 2004), and neutralized to pH 6–8 with HCl (0.3 M).

The solutions are hot filtered, twice, under vacuum, through cloth and glass fiber filter. The extract is evaporated under vacuum to one-third of the initial volume. The carrageenan is precipitated by adding the warm solution to twice its volume of ethanol (96 percent).

Vibrational spectroscopy

In order to identify the seaweed carrageenan, agar, and alginate nature, vibrational spectroscopy can reveal detailed information concerning the properties and structure of materials at a molecular level. Until now, this type of analysis required the extraction of colloids, through lengthy and complicated procedures. With the development of FTIR diffuse-reflectance spectroscopy (DRIFTS), it became possible to directly analyze ground, dried seaweed material (Chopin and Whalen 1993).

Recently Pereira and coworkers (2003, 2009a,b) developed an analysis technique based on FTIR-ATR (attenuated total reflectance) and FT-Raman spectroscopy, which allowed for the accurate identification of diverse phycocolloids (see Tables 3 and 4).

Table 3. Identification of carrageenan types by infrared spectroscopy (adapted from Pereira 2006; Pereira et al. 2009a).

Wavenumbers (cm⁻¹)	Bond(s)/Group(s)	Letter Code	Type of carrageenan							
			Kappa (κ)	Mu (μ)	Iota (ι)	Nu (ν)	Beta (β)	Theta (θ)	Lambda (λ)	Xi (ξ)
1240–1260	S=O of sulphate esters		+	++	++	+++	-	++	+++	++
1070	C-O of 3,6-anhydrogalactose	DA	+	-	+	-	+	+	-	-
970–975	Galactose	G/D	+	s	+	s	+	+	-	-
930	C-O of 3,6-anhydrogalactose	DA	+	-	+	-	+	+	-	-
905	C-O-SO3 on C2 of 3,6-anhydrogalactose	DA2S	-	-	+	-	-	+	-	-
890–900	Unsulphated β-D-galactose	G/D	-	-	-	-	+	-	-	-
867	C-O-SO3 on C6 of galactose	G/D6S	-	+	-	+	-	-	+	-
845	C-O-SO3 on C4 of galactose	G4S	+	+	+	+	-	-	-	-
825–830	C-O-SO3 on C2 of galactose	G/D2S	-	-	-	+	-	+	+	n
815–820	C-O-SO3 on C6 of galactose	G/D6S	-	+	-	+	-	-	+	-
805	C-O-SO3 on C2 of 3,6-anhydrogalactose	DA2S	-	-	+	-	-	+	-	-

-, absent; +, medium; ++, strong; +++, very strong; s, shoulder peak; n, narrow peak.

Table 4. Identification of carrageenan types by Raman spectroscopy (adapted from Pereira 2006; Pereira et al. 2009a).

Wavenumbers (cm⁻¹)	Bond(s)/Group(s)	Letter code	Type of carrageenan							
			Kappa (κ)	Mu (μ)	Iota (ι)	Nu (ν)	Beta (β)	Theta (θ)	Lambda (λ)	Xi (ξ)
1240–1260	S=O of sulphate esters		++	++	++	+++	-	++	++	++
1075–1085	C-O of 3,6-anydrogalactose	DA	+++	-	+++	-	+	+	-	-
970–975	Galactose	G/D	+	+	s	s	+	+	-	-
925–935	C-O of 3,6-anydrogalactose	DA	+	-	+	-	+	+	-	-
905–907	C-O-SO₄ on C₂ of 3,6-anydrogalactose	DA2S	-	-	+	-	-	+	+	+
890–900	Un-sulphated β-D-galactose	G/D	-	-	-	-	+	-	-	-
867–871	C-O-SO₄ on C₆ of galactose	G/D6S	-	s	-	+	-	-	+	-
845–850	C-O-SO₄ on C₄ of galactose	G4S	++	+	++	+	-	-	-	+
825–830	C-O-SO₄ on C₂ of galactose	G/D2S	-	-	-	+	-	+	+	-
815–825	C-O-SO₄ on C₆ of galactose	G/D6S	-	s	-	s	-	-	+	+
804–808	C-O-SO₄ on C₂ of 3,6-anhydrogalactose	DA2S	-	-	++	-	-	+	-	-

-, absent; +, medium; ++, strong; +++, very strong; s, shoulder peak.

FTIR spectroscopy. Infrared (IR) spectroscopy was, until recently the most frequently used vibrational technique for the study of the chemical composition of phycocolloids. This technique presents two main advantages: it requires minute amounts of sample (milligrams), and it is non-aggressive method with reliable accuracy (Pereira et al. 2003). However, conventional IR spectroscopy requires laborious procedures to obtain spectra with a good signal/noise ratio (Chopin and Whalen 1993). This limitation was overcome with the development of interferometric IR techniques (associated with the Fourier transform algorithm), known as FTIR spectroscopy (Fourier Transform IR). More recently, Pereira and collaborators used a technique of analysis on the basis of FTIR-ATR (from Attenuated Total Reflectance) spectroscopy, allowing for determination of the composition of the different phycocolloids from dried ground seaweed, without having to prepare tablets of KBr (Pereira and Mesquita 2004; Pereira 2006; Pereira et al. 2009a).

Raman spectroscopy. In contrast with FTIR, the application of traditional Raman spectroscopy was limited until recently, due to the laser-induced fluorescence (strong background signal that is detected when some samples, such as biochemical compounds, are excited with visible lasers) and risk of sample destruction by light energy. The use of Nd:YAG lasers operating at 1064 nm has been generalized to decrease the fluorescence level. Opto-electronic devices have progressed dramatically in the past decade, because of major achievements in solid-state technology. As a result, compact, efficient and reliable diode lasers are now available from visible to infrared light and have been shown to work correctly on Raman instruments in combination with suitable filter sets (Pereira 2006; Pereira et al. 2009b).

Raman spectroscopy comprises the family of spectral measurements made on molecular media based on inelastic scattering of monochromatic radiation. During this process, energy is exchanged between the photon and the molecule such that the scattered photon is of higher or lower energy than the incident photon. The difference in energy is made up by a change in the rotational and vibrational energy of the molecule, and gives information on its energy levels. The modern FT-Raman spectrometers have been used to produce good quality Raman spectra from seaweed samples (Matsuhiro 1996; Pereira et al. 2003; Dyrby et al. 2004; Pereira 2006).

Four different FT-Raman spectra are presented in Fig. 3 (*Chondrus crispus,* female gametophytes), corresponding to the different tests of depigmentation to reduce the fluorescence caused by the laser beam in Raman spectroscopy. The spectrum "a" corresponds to the ground seaweed treated with a mixture of acetone and methanol; this presents some fluorescence, particularly in the spectral area 600–875 cm^{-1} and the peaks are ill-defined. The spectrum "b" corresponds to the fresh seaweed treated with calcium hypochlorite 4% (30 s), then dried and milled. The spectrum "c" concerns the ground seaweed (obtained from a herbarium sample) treated with calcium hypochlorite 4% (30 s). Finally, the spectrum "d" was obtained from the native carrageenan (*C. crispus* water-extracted) analysis. The last three spectra (b, c, d) so not exhibit fluorescence, with peaks well-defined and without background noise (after Pereira et al 2009b).

NMR *spectroscopy*

Since natural carrageenans are mixtures of different sulfated polysaccharides, their composition differs from batch to batch. Therefore, the quantitative analysis of carrageenan batches is of greatest importance for both ingredient suppliers and food industries to ensure ingredient quality. From the pioneering work of Usov and coworkers (Yarotsky et al. 1977; Usov 1984), NMR-spectroscopy is nowadays one of the preferred techniques to determine and quantify the composition of carrageenan batches (van de Velde et al. 2002). Starting with their early work, chemical shifts of carrageenan resonances are generally converted to values relative to tetrametylsilane (TMS) via an internal dimethylsulphoxide (DMSO) or methanol (MeOH) standard (Usov et al. 1980; Usov 1984; Knutsen et al. 1994). The use of DMSO or MeOH as internal standard resulted in a generally accepted set of chemical shifts for different types of carrageenans as summarized by van de Velde et al. (2002). However, to convert chemical shifts from aqueous internal DMSO or MeOH to values relative to TMS is not obvious as TMS is only sparingly soluble in highly polar solvents, such as water or D$_2$O. Therefore, the IUPAC commission for molecular structure and spectroscopy recently recommended the use of 2,2-dimethyl-2-silapentane-

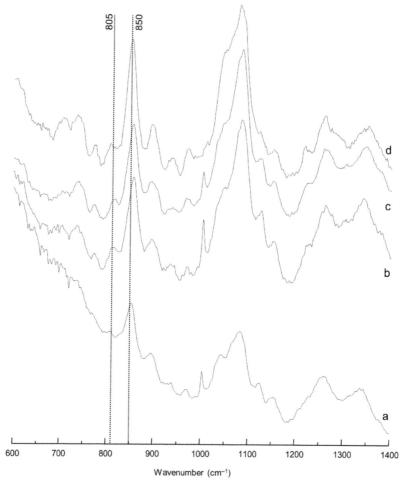

Fig. 3. (a) FT-Raman spectrum of ground seaweed (*Chondrus crispus* female gametophyte) treated with a mixture of acetone and methanol. (b) FT-Raman spectrum of fresh seaweed treated with calcium hypochlorite 4% (30 s), then dried and ground. (c) FT-Raman spectrum of ground seaweed (obtained from a herbarium sample) treated with calcium hypochlorite 4% (30 s). (d) FT-Raman spectrum of *C. crispus* extracted carrageenan (after Pereira et al. 2009b).

3,3,4,4,5,5-d_6-5-sulfonate sodium salt (DSS) as the primary reference for both ^{1}H and ^{13}C-NMR spectroscopy in highly polar solvents. For most purposes, the difference between DSS and TMS when dissolved in the same solvent is negligible and, therefore, the data from DSS and TMS scales may be validly compared without correction (see Table 5) (Harris et al. 2001).

Table 5. Chemical shifts for common internal standards for NMR-spectroscopy in aqueous systems (adapted from van de Velde et al. 2004; Pereira 2004; Pereira 2016).

Compound Abbr.	Chemical shift (ppm)		Compound chemical name
	^{13}C	^{1}H	
DSS	0.000	0.000	2,2-Dimethyl-2-silapentane-3,3,4,4,5,5-d_6-5-sulfonate sodium salt
TSP	−0.18	−0.017	3-(Trimethylsilyl) propionic-2,2,3,3-d_4 acid sodium salt
MeOH	51.43	3.337	Methanol
DMSO	41.53	2.696	Dimethyl sulfoxide
Acetone	32.69	2.208	Acetone

Recorded in D_2O containing Na_2HPO_2 (20 mM) at 65°C.

However, application of the IUPAC recommendations in NMR-spectroscopy to carrageenans results in changed chemical shifts for all common carrageenan types. Pereira et al. (2003) reported chemical shift values relative to DSS for κ- and ι-carrageenan that are 2.5 ppm larger than those reported by Usov (1984). Therefore, the detailed applications of DSS as internal standard for NMR spectroscopy of carrageenans are published by Pereira (2004) and van de Velde et al. (2004).

^{13}C- and ^{1}H-NMR shift data of representative carrageenan samples containing the most common repeating units, e.g., κ-, ι-, λ-, μ-, ν-, θ-, β-, and ξ-carrageenan are presented in Tables 5 and 6

Table 6. ^{13}C-NMR chemical shifts for the most common carrageenan structural units[a] (adapted from van de Velde et al. 2004; Pereira 2004; Pereira 2016).

Carrageenan	Unit[b]	Chemical shift (ppm) relative to DSS as internal standard					
		C-1	C-2	C-3	C-4	C-5	C-6
β (beta)	G DA	104.81	71.72	82.58	68.56	77.55	63.49
		96.81	72.40	81.64	80.33	79.26	71.72
ι (iota)	G4S DA2S	104.43	71.53	79.06	74.34	77.04	63.52
		94.29	77.15	80.04	80.55	79.29	72.02
κ (kappa)	G4S DA	104.70	71.72	80.98	76.25	77.00	63.49
		97.34	72.11	81.41	80.54	79.07	71.72
λ (lambda)	G2S D2S,6S	105.61	79.61	77.99	66.35	76.51	63.45
		93.85	77.04	71.76	82.61	70.89	70.25
μ (mu)[c]	G4S D6S	107.00	72.69	80.54	76.25	77.1	63.48
		100.26	70.7	72.8	81.40	70.5	69.89
ν (nu)	G4S D2S,6S	106.96	72.40	82.42	73.36[d]	77.15	63.58
		100.53	78.58	70.37	82.13	70.37	70.02
θ (theta)[e]	G2S DA2S	102.57	79.8	79.4	69.97	77.05	63.38
		97.81	77.05	79.6	81.75	79.2	72.35
ξ (xi)[f]	G2S D2S	105.44			66.92		
		94.94					

[a] Carrageenan (30 mg ml^{-1}), DSS (10 mM) and Na$_2$HPO$_4$ (20 mM) in D$_2$O recorded at 65°C.
[b] Codes refer to the nomenclature developed by Knutsen et al. 1994.
[c] Chemical shifts given with one decimal place are obtained from literature and corrected with the offset calculated for the chemical shifts of the anomeric carbon atoms (Chiovitti et al. 1998).
[d] Chemical shift differs from the value given in literature (Chiovitti et al. 1998).
[e] Chemical shifts given with one decimal place are recalculated from literature (van de Velde et al. 2004); however they may be interchanged. TSP is used as the internal standard; however chemical shifts are given relative to DSS.
[f] Only a few resonances of the ξ-carrageenan spectrum are assigned in literature (Falshaw and Furneaux 1995). TSP is used as the internal standard; however chemical shifts are given relative to DSS.

Table 7. Chemical shifts (ppm) of the α-anomeric protons of carrageenans referred to DSS as an internal standard at 0 ppm[a] (adapted from van de Velde et al. 2004; Pereira 2004).

Carrageenan	Monosaccharide[b]	Chemical shift (ppm)
β (beta)	DA	5.074
ι (iota)	DA2S	5.292
κ (kappa)	DA	5.093
λ (lambda)	D2S,6S	5.548
ν (nu)	D2S,6S	5.501
μ (mu)	D6S	5.238
θ (theta)	DA2S	5.30
ξ (xi)	D2S	5.49

[a] Carrageenan (30 mg ml^{-1}), DSS (10 mM) and Na$_2$HPO$_4$ (20 mM) in D$_2$O recorded at 65°C.
[b] Codes refer to the nomenclature developed by Knutsen et al. (1994).

respectively. The chemical shifts of all carbon atoms, as well as the anomeric protons are reported relative to DSS internal standard ($\delta = 0.000$ ppm) according to IUPAC recommendations (Table 7) (Harris et al. 2001).

Acknowledgements

This work had the support of "Fundação para a Ciência e Tecnologia" (FCT), through the strategic project UID/MAR/04292/2013 granted to MARE.

References

Anderson, N.S. and D.A. Rees. 1965. Porphyran: a polysaccharide with a masked repeating structure. J. Chem. Soc. 0: 5880–5887.

Anderson, N.S., T.C.S. Dolan and D.A. Rees. 1965. Evidence for a common structural pattern in polysaccharide sulphates of Rhodophyceae. Nature 205(4976): 1060–1068.

Armisen, R. 1995. World-wide use and importance of *Gracilaria*. J. Appl. Phycol. 7(3): 231–243.

Armisen, R. and F. Galatas. 2000. Agar. pp. 21–40. *In*: G. Phillips and P. Williams (eds.). Handbook of Hydrocolloids. CRC Press, Boca Raton, FL.

Bixler, H.J. and H. Porse. 2011. A decade of change in the seaweed hydrocolloids industry. J. Appl. Phycol. 3(23): 321–335.

Cardoso, S.M., L.G. Carvalho, P.J. Silva, M.S. Rodrigues, R.O. Pereira and L. Pereira. 2014. Bioproducts from seaweeds: a review with special focus on the Iberian Peninsula. Current Organic Chemistry 18(7): 896–917.

Chiovitti, A., A. Bacic, D.J. Craik, G.T. Kraft, M.-L. Liao, R.H. Falshaw and R.H. Furneaux. 1998. Carbohydr. Res. 310: 77–83.

Chopin, T. and E. Whalen. 1993. A new and rapid method for carrageenan identification by FT-IR diffuse-reflectance spectroscopy directly on dried, ground algal material. Carbohydr. Res. 246: 51–59.

Chopin, T. 2006. Integrated multi-trophic aquaculture. What it is, and why you should care... and don't confuse it with polyculture. Northern Aquaculture 12: 4.

Craigie, J.S. 1990. Cell walls. pp. 221–257. *In*: K.M. Cole and R.G. Sheath (eds.). Biology of the Red Algae. Cambridge University Press, Cambridge.

Dhargalkar, V.K. and N. Pereira. 2005. Seaweeds: promising plants of the millennium. Science and Culture 71: 60–66.

Draget, K.I., O. Smidsrød and S. Skjåk-Broek. 2004. Alginates from algae. pp. 215–224. *In*: S.D. Baets, E. Vandamme and A. Steinbüchel (eds.). Polysaccharides II: Polysaccharides from Eukaryotes (Biopolymers, Vol. 6). Wiley-Blackwell, Weinheim.

Dyrby, M., R.V. Petersen, B. Larsen, B. Rudolph, L. Norgaard and S.B. Engelsen. 2004. Towards on-line monitoring of the composition of commercial carrageenan powders. Carbohydrate Polymers 57(3): 337–348.

Falshaw, R.H. and R.H. Furneaux. 1995. Carbohydr. Res. 276: 155–165.

Furtado, M.R. 1999. Alta Lucratividade Atrai Investidores em Hidrocolóides. Química e Derivados 377: 20–29.

Givernaud, T., N. Sqali, O. Barbaroux, A. Orbi, Y. Semmaoui, N.E. Rezzoum, A. Mouradi and R. Kaas. 2005. Mapping and biomass estimation for a harvested population of *Gelidium sesquipedale* (Rhodophyta, Gelidiales) along the Atlantic coast of Morocco. Phycologia 44: 66–71.

Givernaud, T. and A. Mouradi. 2006. Seaweed resources of Morocco. *In*: A.T. Critchley, M. Ohno and D.B. Largo (eds.). World Seaweed Resources—An Authoritative Reference System V1.0, ETI Information Services Ltd. Hybrid Windows and Mac DVD-ROM.

Guiry, M.D. and G. Blunden. 1991. Seaweed Resources in Europe: Uses and Potential. John Wiley & Sons, Chichester.

Harris, R.K., E.D. Becker, S.M. Cabral de Menezes, R. Goodfellow and P. Granger. 2001. NMR nomenclature. Nuclear spin properties and conventions for chemical shifts (IUPAC recommendations 2001). Pure Appl. Chem. 73(11): 1795–1818.

Hayashi, L., A.G. Hurtado, F.E. Msuya, G. Bleicher-Lhonneur and A.T. Critchley. 2010. A review of *Kappaphycus* farming: Prospects and constraints. pp. 251–283. *In*: J. Seckbach (ed.). Seaweeds and Their Role in Globally Changing Environments (Cellular Origin, Life in Extreme Habitats and Astrobiology). Springer, Netherlands.

Hurtado, A.Q., P.-E. Lim, J. Tan, S.-M. Phang, I.C. Neish and A.T. Critchley. 2016. Biodiversity and biogeography of commercial tropical carrageenophytes in the Southeast Asian region. pp. 51–74. *In*: L. Pereira (ed.). Carrageenans: Sources and Extraction Methods, Molecular Structure, Bioactive Properties and Health Effects. Nova Science Publisher, N.Y.

Ioannou, E. and V. Roussis. 2009. Natural products from seaweeds. pp. 51–81. *In*: A.E. Osbourn and V. Lanzotti (eds.). Plant Derived Natural Products: Synthesis, Function, and Application. Springer US, NY.

Jiao, G., G. Yu, J. Zhang and H.S. Ewart. 2011. Chemical structures and bioactivities of sulfated polysaccharides from marine algae. Marine Drugs 9: 196–223.

Knutsen, S.H., D.E. Myslabodski, B. Larsen and A.I. Usov. 1994. A modified system of nomenclature for red algal galactans. Bot. Mar. 37(2): 163–169.

Krisler, A.V. 2006. Seaweed resources of Chile. *In*: A.T. Critchley, M. Ohno and D.B. Largo (eds.). World Seaweed Resources —An Authoritative Reference System V1.0, ETI Information Services Ltd. Hybrid Windows and Mac DVD-ROM.

Lahaye, M. 2001. Chemistry and physico-chemistry of phycocolloids. Cahiers de Biologie Marine 42(1-2): 137–157.

Larsen, B., D.M.S.A. Salem, M.A.E. Sallam, M.M. Mishrikey and A.I. Beltagy. 2003. Characterization of the alginates from algae harvested at the Egyptian Red Sea coast. Carbohydr. Res. 338: 2325–2336.

Leal, D., B. Matsuhiro, M. Rossi and F. Caruso. 2008. FT-IR spectra of alginic acid block fractions in three species of brown seaweeds. Carbohydr. Res. 343: 308–316.

Lobban, C.S., J.D. Chapman and B.P. Kremer. 1988. Experimental Phycology: A Laboratory Manual. Phycological Society of America. Cambridge University Press, Cambridge, 320 p.

Matsuhiro, B. 1996. Vibrational Spectroscopy of Seaweed Galactans. Hydrobiologia 327: 481–489.

McHugh, D.J. 2003. A Guide to the Seaweed Industry. FAO, Fisheries Technical Paper 441: 73–90.

Minghou, Ji. 1990. Processing and extraction of phycocolloids. *In*: Regional Workshop on the Culture and Utilization of Seaweeds, Volume II. FAO Technical Resource Papers. Cebu City, Philippines.

Mouradi-Givernaud, A., L. Amina Hassani, T. Givernaud, Y. Lemoine and O. Benharbet. 1999. Biology and Agar Composition of *Gelidium sesquipedale* Harvested Along the Atlantic Coast of Morocco. Hydrobiologia 398: 391–395.

Myslabodski, D.E. 1990. Red-Algae Galactans: Isolation and Recovery Procedures—Effects on the Structure and Rheology. Ph.D. Thesis. Norwegian Institute of Technology, Trondheim.

Nussinovitch, A. 1997. Hydrocolloid Applications: Gum Technology in the Food and Other Industries. Chapman and Hall, London.

Onsoyen, E. 1997. Alginates. pp. 22–44. *In*: A. Imeson (ed.). Thickening and Gelling Agents for Food. Blackie Academic and Professional, London.

Ortiz, J., E. Uquiche, P. Robert, N. Romero, V. Quitral and C. Llantén. 2009. Functional and nutritional value of the Chilean seaweeds *Codium fragile*, *Gracilaria chilensis* and *Macrocystis pyrifera*. European Journal of Lipid Science and Technology 111: 320–327.

Pereira, L. 2004. Estudos em Macroalgas Carragenófitas (Gigartinales, Rhodophyceae) da Costa Portuguesa—Aspectos Ecológicos, Bioquímicos e Citológicos. Ph.D. Thesis, Departamento de Botânica, FCTUC, Universidade de Coimbra, Coimbra.

Pereira, L. and J.F. Mesquita. 2004. Population Studies and Carrageenan Properties of *Chondracanthus teedei* var. *lusitanicus* (Gigartinaceae, Rhodophyta). J. Appl. Phycol. 16(5): 369–383.

Pereira, L. 2006. Identification of phycocolloids by vibrational spectroscopy. *In*: A.T. Critchley, M. Ohno and D.B. Largo (eds.). World Seaweed Resources—An Authoritative Reference System V1.0, ETI Information Services Ltd. Hybrid Windows and Mac DVD-ROM.

Pereira, L. 2008. As algas marinhas e respectivas utilidades. Monografias. PDF URL: http://br.monografias.com/trabalhos913/algas-marinhas-utilidades/algas-marinhas-utilidades.pdf.

Pereira, L. 2010a. Seaweed: An Unsuspected Gastronomic Treasury. Chaîne de Rôtisseurs Magazine 2: 50.

Pereira, L. 2010b. Littoral of Viana do Castelo – ALGAE. Uses in Agriculture, Gastronomy and Food Industry (Bilingual). Município de Viana do Castelo, Viana do Castelo.

Pereira, L. 2011. A review of the nutrient composition of selected edible seaweeds. pp. 15–47. *In*: V.H. Pomin (ed.). Seaweed: Ecology, Nutrient Composition and Medicinal Uses. Nova Science Publishers Inc., New York.

Pereira, L. and F. van de Velde. 2011. Portuguese Carrageenophytes: Carrageenan Composition and Geographic Distribution of Eight Species (Gigartinales, Rhodophyta). Carbohydrate Polymers 84: 614–623.

Pereira, L., A. Sousa, H. Coelho, A.M. Amado and P.J.A. Ribeiro-Claro. 2003. Use of FTIR, alFT-Raman and ¹³C-NMR spectroscopy for identification of some seaweed phycocolloids. Biomol. Eng. 20: 223–228.

Pereira, L., A.M. Amado, A.T. Critchley, F. van de Velde and P.J.A. Ribeiro-Claro. 2009a. Identification of selected seaweed polysaccharides (Phycocolloids) by vibrational spectroscopy (FTIR-ATR and FT-Raman). Food Hydrocolloids 23: 1903–1909.

Pereira, L., A.M. Amado, P.J.A. Ribeiro-Claro and F. van de Velde. 2009b. Vibrational spectroscopy (FTIR-ATR AND FT-RAMAN)—a rapid and useful tool for phycocolloid analysis. Biodevices. Porto: 131–136.

Pereira, L., A.T. Critchley, A.M. Amado and P.J.A. Ribeiro-Claro. 2009c. A comparative analysis of phycocolloids produced by underutilized *versus* industrially utilized carrageenophytes (Gigartinales, Rhodophyta). J. Appl. Phycol. 21: 599–605.

Pereira, L. 2016. Chapter 12—Analysis of carrageenan molecular composition by NMR: Principles and applications. pp. 219–241. *In*: S.-K. Kim (ed.). Marine OMICS—Principles and Applications. CRC-Press, Taylor & Francis Group, Boca Raton, FL.

Pereira, L., F. Soares, A.C. Freitas, A.C. Duarte and P. Ribeiro-Claro. 2017. Extraction, characterization and use of carrageenans. pp. 37–90. *In*: P.N. Sudha (ed.). Industrial Applications of Marine Biopolymers. CRC Press, Taylor & Francis Group, Boca Raton, FL.

Rudolph, B. 2000. Seaweed products: red algae of economic significance. pp. 515–529. *In*: R.E. Martin, E.P. Carter, L.M. Davis and G.J. Flich (eds.). Marine and Freshwater Products Handbook. Technomic Publishing Company Inc. Lancaster, USA.

Sousa-Pinto, I. 1998. The seaweed resources of Portugal. pp. 176–184. *In*: A.T. Critchley and M. Ohno (eds.). Seaweed Resources of the World. Japan International Cooperation Agency, Yokosuka, Japan.

Sousa-Pinto, I. and H. Abreu. 2011. Aquacultura Multitrófica Integrada – O que é? pp. 10–27 *In*: U.V. Ferreiro, M.I. Filgueira, R.F. Otero and J.M. Leal (eds.). Macroalgas na Aquacultura Multitrófica Integrada Peninsular. Valorização da Biomassa. CETMAR, Iberomare.

Usov, A.I., S.V. Yarotsky and A.S. Shashkov. 1980. C-13-NMR spectroscopy of red algal galactans. Biopolymers 19(5): 977–990.

Usov, A.I. 1984. NMR-spectroscopy of red seaweed polysaccharides—agars, carrageenans, and xylans. Botanica Marina 27(5): 189–202.

Vairappan, C., C. Chung, A. Hurtado, F. Soya, G. Lhonneur and A. Critchley. 2008. Distribution and symptoms of epiphyte infection in major carrageenophyte-producing farms. J. Appl. Phycol. 20: 477–483.

van de Velde, F. and G.A. de Ruiter. 2002. Carrageenan. pp. 245–274. *In*: E.J. Vandamme, S.D. Baets and A. Steinbèuchel (eds.). Biopolymers V6, Polysaccharides II, Polysaccharides from Eukaryotes. Wiley, Weinheim.

van de Velde, F., S.H. Knutsen, A.I. Usov, H.S. Rollema and A.S. Cerezo. 2002. H-1 and C-13 high resolution NMR spectroscopy of carrageenans: application in research and industry. Trends in Food Science & Technology 13(3): 73–92.

van de Velde, F., L. Pereira and H.S. Rollema. 2004. The Revised NMR Chemical Shift Data of Carrageenans. Carbohyd. Res. 339: 2309–2313.

van de Velde F., A.S. Antipova, H.S. Rollema, T.V. Burova, N.V. Grinberg, L. Pereira, P.M. Gilsenan, R.H. Tromp, B. Rudolph and V.Y. Grinberg. 2005. The structure of kappa/iota-hybrid carrageenans II. Coil-helix transition as a function of chain composition. Carbohydr. Res. 340(6): 1113–1129.

Yarotsky, S.V., A.S. Shashkov et al. 1977. Analysis of C-13-NMR spectra of some red seaweed galactans. Bioorganicheskaya Khimiya 3(8): 1135–1137.

Zinoun, M. and J. Cosson. 1996. Seasonal variation in growth and carrageenan content of *Calliblepharis jubata* (Rhodophyceae, Gigartinales) from the normandy coast, France. Journal of Applied Phycology 8(1): 29–34.

Microalgae for Feed

Vítor Verdelho,[1,2,*] *João Navalho*[1,2] and *Ana P. Carvalho*[1,3]

Introduction

World population growth and the concomitant increasing average standard of living have created an exponential demand for animal food products; therefore, it is vital that the feed industry can respond to this challenge in a sustainable and safe way.

Microalgae can play an important role in feed industry as several species present unique and interesting biochemical properties, when compared with higher plants, which place microalgae as priority ingredient candidates. Furthermore, microalgal biomass can be daily harvested all year around, and cultivated in controlled conditions, thus providing a raw material free of organic or chemical contaminants, that is, safe. The actual challenge is to "domesticate" these organisms, as has been done with higher plants, in order to manage their large-scale production for feeds.

This chapter aims to bring together the most relevant information on microalgae for feed production and thus promote the management of knowledge necessary to accelerate the development of microalgal feeds.

Using microalgae as feed ingredients

The survival, growth, productivity, and fertility of animals are a reflection of their health. Feed quality is the most important exogenous factor influencing animal health, especially in connection with intensive breeding conditions and the trend to avoid chemical products such as antibiotics (Pulz and Gross 2004).

Microalgae possess various properties that may play an important role in many aspects of animal daily lives, from simple nutrition to health maintenance (Sweetman 2009). Due to their metabolic flexibility they are able to, until a certain limit, modulate their chemical composition, thus providing a specifically targeted source of unique compounds (Carvalho et al. 2009). Due to this enlarged portfolio of possible compositions, microalgae products may have a role as functional ingredients (products incorporated in

---
[1] A4F_S.A. • design, build, operate, and transfer microalgae plants • www.a4f.pt; Campus do Lumiar, Edifício E - R/C, Estrada do Paço do Lumiar, 1649-038 Lisboa, PORTUGAL.

[2] NECTON, Companhia Portuguesa de Culturas Marinhas, S.A., P-8700-152 Olhão, PORTUGAL.
Email: jnavalho@necton.pt

[3] REQUIMTE/LAQV–ISEP, Porto Polytechnic Institute, Rua Dr, António Bernardino de Almeida, 431, P-4249-015 Porto, PORTUGAL.
Email: ana.p.santos.carvalho@gmail.com

* Corresponding author: vvv@a4f.pt

the feed that present a benefit above and beyond fulfilling the basic nutritional needs). Such capacity is extremely important, since the feed industry is constantly searching for ingredients that could add value to their traditional feeds during sensitive periods of the production cycle, or specific situations resulting from farming practices. Sensitive periods of the aquaculture production cycle where functional ingredients could be highly valued are early feeding stages, specific physiological stages (e.g., high stressful conditions), and finishing stages (for enhancement of quality criteria). In land-based animals, critical periods of development include piglets and calves weaning phase, egg productivity/quality in laying hens, milk yield and productivity in dairy cows, and recovery after stressing conditions for athletic animals (e.g., race horses) (Dias and Sendão unpublished data). Due to their composition, digestibility, and lack of toxicity, the introduction of microalgae meal can thus target specialty feed applications, apart from the nutritional commodities market.

Besides the nutritional improvement that microalgae incorporation in feeds may bring to animals health, they are the only biomass material that allows daily harvest production during all the year, an advantage that ensures a constant supply of raw materials to the feed market. In fact, microalgae biomass can be produced in large-scale bioreactors, under straight controlled conditions, so that desired, customer-specific, active components are produced. Thus, biomass can be manufactured to a consistent, standardized quality, reproducible during the whole year, irrespective of location, weather and climatic conditions, and without the seasonal variation typically associated with cultivated plants. Furthermore, microalgae also possess advantages in terms of productivity, when compared with higher plants.

Animal nutritional requirements

Feeds are a blend from various raw materials and additives, formulated according to the specific requirements of the target animal, and manufactured as meal type, pellets, or crumbles. The nutritive value of a diet is directly related with its ingredients constitution (Wondra 1995; McDonald 2002).

Animals need a variety of nutrients to meet their basic needs. These nutrients generally include lipids and carbohydrates to supply energy, proteins that provide amino acids, vitamins used as co-factors for enzymes and other functions, and specific compounds to fulfil specific needs. Depending on the amounts needed, they can be categorized as either macronutrients (needed in relatively large amounts) or micronutrients (needed in smaller quantities). The former includes carbohydrates, fats, fibre, proteins, and water, whereas the later comprise minerals and vitamins (Freeman 2003).

Macronutrients

Carbohydrates

Carbohydrates in plants can be divided into those that serve as storage and energy reserves, available for metabolism (e.g., sugars, starch, pectin, and some cellulose in barley grain) and those that are structural (e.g., fibrous cellulose, hemi-cellulose, and lignin in straw). Carbohydrates are a major source of energy in livestock feeds (Anonymous 2006). The main functions of this nutrient mainly comprise energy supply (to power muscular movement), body heat, and building block for other nutrients. Carbohydrates utilization by animals depend on their digestion system, although its dietary excess is always stored as fat. Simple stomached animals cannot digest large amounts of fibre, and their ration must be made up of mostly cereal grains. Ruminant animals can eat large amounts of fibre, and a high percentage of their ration is roughage.

Lipids

Lipids (fats or oils), like carbohydrates, are neutral chemical compounds essentially composed of carbon, hydrogen, and oxygen, although they contain more carbon and hydrogen atoms than carbohydrates. Concomitantly, fats provide 2.25 times more energy per gram than carbohydrates. Their main functions include energy supply, heat maintenance, body protection, fat-soluble vitamins carrying, and immune functions through essential fatty acids. Animals easily digest lipids and common feed sources are mostly soybean oil, corn oil, fish oil, and by-products.

Proteins

Proteins are present in animal cells and tissues, and are continuously used to replace dying body cells and supply building body tissues (ligaments, hair, hooves, skin, organs, and muscle are partially formed by protein). Apart from their important role as a basic structural unit, they are also needed for metabolism, hormone, antibody, and DNA production. When proteins are fed in excess, they are converted into energy and fat.

Proteins are complex organic compounds of high molecular weight. As with carbohydrates and fats, they contain carbon, hydrogen, and oxygen, but in addition they all contain nitrogen and generally sulphur. Proteins are composed of amino acids, arranged in a linear chain and folded into a globular form, which are released when the former are hydrolysed by enzymes, acids, or alkalis. Although over 200 amino acids have been isolated from biological materials, only 20 of these are commonly found as components of proteins. Within all known amino acids, 10 are classified as essential (EAA), as animals are not able to produce them, and they must be therefore supplied in the diet. Although non-essential amino acids (NEAA) are not classified as dietary essential nutrients, they are necessary because they perform many essential functions at the cellular or metabolic level. In fact, they are termed dietary non-essential nutrients only because the body tissues can synthesize them on demand (McDonald 2002). From a feed formulation viewpoint, it is important to know that the NEAA's cystine and tyrosine can be synthesized within the body from the EAA's methionine and phenylalanine respectively, and consequently the dietary requirement for these EAA is dependent on the concentration of the corresponding NEAA within the diet (FAO 2012).

The requirements of amino acids in animals are well defined in various sets of recommendations such as those of National Research Council in US. Requirements vary depending on the species and age of animals. For instance, the dietary EAA for fish and shrimp are Threonine, Valine, Leucine, Isoleucine, Methionine, Tryptophan, Lysine, Histidine, Arginine, and Phenylalanine (FAO 2012).

Crude protein contains both "true" protein and other nitrogenous products (non-protein nitrogen), although monogastric animals can only digest the former. However, ruminants are able to convert non-protein nitrogen into true protein through rumen bacteria. Therefore, this class of animals is theoretically independent of the dietary protein sources provided, once the rumen microorganisms have become established (MacDonald 2002).

Animal source proteins are considered as good-quality proteins since they contain a balanced level of essential amino acids. Plant proteins are thought to be poor-quality proteins because they lack some amino acids

Micronutrients

Vitamins

Vitamins are organic compounds required as a nutrient in minor amounts. They are classified as water-soluble and fat-soluble vitamins. The former include vitamins B and C; because they are not readily stored, a continuous daily intake is important. Vitamin B complex is necessary for body chemical reactions, and also helps improving appetite, growth, and reproduction, whereas vitamin C helps teeth and bone formation and also prevents infections. Lipid-soluble vitamins A, D, E, and K, are absorbed through the intestinal tract. Because they are more likely to accumulate in the body, they may lead to hipervitaminosis conditions. Vitamin A is associated with healthy eyes, good conception rate, and disease resistance; vitamin D is related with bone development and mineral balance of the blood; vitamin E is associated with normal reproduction and muscle development, and can also improve immune system; finally, vitamin K has a role in blood clotting and prevents excessive bleeding from injuries. Most vitamins have multiple functions in body, and both deficiencies and excesses lead to diseases.

Minerals

Minerals are chemical elements required by living organisms, other than the four elements carbon, hydrogen, nitrogen, and oxygen, which are present in nearly all the organic molecules. They are needed in small amounts and may be classified, regarding its nutrition requirement rate, as major or trace minerals. Some dieticians recommend that these should be supplied from foods in which they naturally occur, as complex compounds, or from natural inorganic sources, as their bioavailability is higher under these ionic forms. Minerals provide material for bone and teeth growth, regulate chemical processes, improve muscular activities, release energy for body heat, and have a role in protein synthesis, oxygen transport, and many other processes.

Macro-minerals, also called major minerals, are required in gram quantities; the following are essential to animals: calcium and phosphorus (major constituents of the skeleton), sodium, magnesium (participates in muscle contractions), potassium (plays an important role in heart functioning and helps to promote neuromuscular development), sulphur, and chlorine.

Micro-minerals, also called trace minerals, are required in milligram or microgram amounts. They often serve as components of enzyme cofactors or hormones, and examples are cobalt, iodine (activates the synthesis of thyroid hormones), zinc (essential for skin healing and renewal; acts in synergy with linoleic acid to enhance brightness of the fur), copper (participates in the synthesis of melanin that give colour to the coat), manganese, and selenium (protects the integrity of cell membranes) (Anonymous 2006).

Antioxidants

As cellular metabolism requires oxygen, potentially damaging (e.g., mutation causing) compounds, known as free radicals, can be formed. Most of these are oxidizers and must be sufficiently neutralized by antioxidant compounds in order to maintain normal cellular processes. Human body produces some antioxidants with adequate precursors, such as vitamin D, which is synthesized from cholesterol by sunlight, or beta-carotene, which is converted to vitamin A by the body; those that body cannot produce may only be obtained in the diet via direct sources.

Phytochemicals

Phytochemicals and especially their subgroup of polyphenols represent a growing area of interest. These nutrients are typically found in edible plants, especially colour fruits and vegetables, but also in other organisms including seafood, algae, and fungi.

Both antioxidants and phytochemicals are supposed to influence (or protect) body systems, although their mechanism of action and necessities are not as well established as in the case of other nutrients. Although phytochemicals may act as antioxidants, not all the phytochemicals are antioxidants.

Probiotics

A probiotic can be a live (viable) culture of microbial species, a dead (non-viable) product of microbial fermentation, or an extract of plant origin. It improves growth and development of the desirable microbial population in an animal's gut, facilitating their domination over undesirable organisms. Probiotics are not essential in feed diets, although their inclusion has proven beneficial effects.

Microalgae chemical composition

Microalgae spontaneously respond to physico-chemical variations in the surrounding environment, concomitantly modifying their chemical composition in order to adapt to those new environmental conditions. Therefore, depending on growth conditions (including nutrient supplies, temperature, and sunlight, among others), microalgae, even from the same strain, can exhibit large variations in composition. This plasticity in composition is an important attribute in their use as animal feeds, as they can be modulated to overproduce the desired nutritional compounds.

As for higher plants, the chemical composition of microalgae is constituted mainly by proteins, carbohydrates, lipids, and trace compounds, including also vitamins, antioxidants, and trace minerals; a brief comparison of nutritional composition in conventional foods and several microalgae is presented in Table 1.

Several factors can contribute to the nutritional value of a microalga, including its size and shape, digestibility (related to cell wall structure and composition), biochemical composition (e.g., nutrients, enzymes, and toxins if present), and the requirements of the animal feeding on the alga (Barsanti and Gualtieri 2006).

Table 1. General composition of some animal feeds and foods, and microalgae species (% of dry matter of maximum values achieved in each commodity).

Commodity	Crude protein	Carbohydrates	Lipids
Meat	43	1	34
Fish	55	-	38
Egg	49	3	45
Milk	26	38	28
Soybean	37	30	20
Corn	10	85	4
Fish-meal	60–72	-	6–10
Chlorella	51–58	12–17	14–22
Dunaliella salina	57	32	6
Porphyridium	28–39	40–57	9–14
Scenedesmus	50–56	10–17	12–14
Arthrospira maxima	60–71	13–16	6–7
Arthrospira platensis	46–63	8–14	4–9
Isochrysis	29	13	23
Tetraselmis	31	12	17
Haematococcus	17–27	37–40	7–21
Nannochloropsis	35	8	18
Porphyridium	28–39	40–57	9–14

Adapted from Aaronson and Dubinsky 1982; Fabregas and Herrero 1985; Becker 1994; Miles and Chapman 2009; unpublished data from Necton and A4F.

Proteins and essential amino acids

When animals utilize amino acids for body protein synthesis, their utilization is limited to the amount of the amino acid most deficient in the feed (the so-called limiting amino acid), regardless of the amount of the other amino acids. Consequently, the surplus portion of the other amino acids is wasted. Therefore, the nutritional value of any protein is directly related to the amino acid composition of that protein. A protein that does not contain the proper amount of required (essential) amino acids would be an imbalanced protein and would have a lower nutritional value.

Generally, livestock feeds consist of a combination of energy sources such as corn and wheat, and protein sources such as soybean meal. Soybean meal is rich in lysine, an amino acid deficient in corn and wheat. However, due to the high price of soybean meal relative to grains such as corn and wheat, using more soybean meal to meet lysine requirements is generally regarded as uneconomical. Therefore, feed formulators are inclined to decrease the cost of feeds by slightly increasing the proportion of corn and wheat. This tends to create an insufficiency of lysine and an excess of the other nutrients, particularly other amino acids, resulting in adverse effects on livestock. Generally, proteins of cereal grains and most other plant protein concentrates fail to supply the complete amino acid needs of poultry, due to a shortage of methionine and/or lysine. Soybean meal, which is widely used in poultry diets, is a good source of lysine and tryptophan, but it is low in the sulfur-containing amino acids methionine and cysteine. Fishmeal is an excellent source of all of these amino acids.

Standard fishmeal typically contains 64% to 67% crude protein, although it may reach 72% in special fish products. High quality fishmeal provides a balanced amount of all essential amino acids, minerals, phospholipids, and fatty acids (e.g., docosahexaenoic and eicosapentaenoic acids) for optimum development, growth, and reproduction. Nutrients in fishmeal also aid in disease resistance by boosting and helping to maintain a healthy functional immune system.

Interest in algae as protein source arose with the realization that many species are extremely rich in such nutrient. In fact, reported protein values vary with algal species, from 28–39% protein such as in *Porphyridium cruentum*, to 60–70% protein as in *Spirulina maxima* (Becker 2004), with an interesting amino acid profile.

The amino acid profile of various algae are presented in Table 2 and compared with some basic conventional food items.

The diversity of microalgal relatively rich amino acid profiles and high mineral composition make them a promising candidate for incorporation into animal feed.

Table 2. Amino acid profile of different algae as compared with conventional protein sources (g per 100 protein).

	Arg	His	Ile	Leu	Lys	Met	Phe	Thr	Trp	Val
Meat	6.6	3.2	5.1	7.8	8.2	2.4	4.2	4.5	-	5.3
Egg	6.2	2.4	6.6	8.8	5.3	3.2	5.8	5	1.7	7.2
Milk	3.3	2.6	4.3	9.2	7.8	2.5	5.6	4.5	-	5.7
Soybean	7.4	2.6	5.3	7.7	6.4	1.3	5	4	1.4	5.3
Fishmeal from Peruvian anchovy	3.8	1.6	3.2	5.0	5.0	2.0	2.8	2.8	0.7	3.5
Chlorella vulgaris	6.4	2	3.8	8.8	8.4	2.2	5	4.8	2.1	5.5
Scenedesmus obliquus	7.1	2.1	3.6	7.3	5.6	1.5	4.8	5.1	0.3	6
Arthrospira maxima	6.5	1.8	6	8	4.6	1.4	4.9	4.6	1.4	6.5
Arthrospira platensis	7.3	2.2	6.7	9.8	4.8	2.5	5.3	6.2	0.3	7.1
Isochrysis galbana		2.5	4.9	10.5	12.1	0.7	6.1	6.1	-	6
Tetraselmis suecica	7.6	2.5	4.1	9.3	9.8	1.5	5.9	5.3	-	5.6

Adapted from Fabregas and Herrero 1985; Becker 2007; Miloo and Chapman 2009, IFFO.

Carbohydrates

Carbohydrates in algae can be present in the form of starch, cellulose, sugars, and other polysaccharides, which will improve the metabolizable energy of the feed with an algae supplement. The cellulose content will affect the digestibility of the algae by non-ruminant animals, but the reported algae tests performed so far indicate that their overall digestibility is good (Becker 2004).

Lipids

The main fraction of the non-polar (neutral) lipids of microalgae is composed of triglycerides, which may account for up to 80% of the total lipid content, whereas polar lipids are mainly constituted by phospholipids and glycolipids. Algae average lipid content varies between 1 and 40% dry weight, although it may reach 85% under certain conditions such as nitrogen limitation (Becker 2004). Furthermore, nutritional and environmental parameters can affect the relative proportion of the component fatty acids, as well as their total amount. Some microalgae contain large amounts of polyunsaturated fatty acids (PUFA) from $\omega 3$ and $\omega 6$ families, known to be important in animal's health. Long chain $\omega 3$ fatty acids (eicosapentaenoic and docosahexaenoic acids) are known to possess anti-inflammatory action, whereas the quality of ω6 fatty acids influences the production of sebum responsible for the shiny coat.

Vitamins, minerals, and pigments

A similar comparison can be made for vitamins, demonstrating that microalgae contain high levels of essential vitamins, similar to the best food sources such as bakers yeast and liver, and are largely superior to all commodity feeds (such as soybean, corn, and fishmeal, among others) which possess little if any vitamin content (Table 3). Microalgae represent a valuable source of nearly all the important vitamins that improve the nutritional value of algal biomass, and they are generally highly bioavailable (Becker 2004).

The nutritionally necessary trace elements have similarly high levels in microalgae, compared to most feed sources. However, data on trace elements is more limited, as contents are variable between and even within species, and strongly depending on growth conditions. Indeed, microalgae with desired high concentrations of trace elements could be produced on demand by adjusting the trace elements in their growth medium. An example of application is the growth in crustaceans, which requires the periodic shedding of the shell, a process called molting. Adequate iodine levels in the water or feeds have been closely associated to molting success. As some microalgae are naturally, or could be produced to be, rich in iodine and serve as a natural iodine source/supplement in shrimp feeds, they could be used as functional ingredients in feeds developed for this specific physiological stage.

In the case of both vitamins and trace elements, the bioavailability of these trace nutrients is often more important and decisive in terms of feed quality than just their bulk constituents. An also important property of microalgae is that their biomass usually has little ash content (less than 10% dry weight) (Aaronson and Dubinsky 1982).

There is a wide range of pigments usually present in microalgae, from chlorophylls to phycocyanin, beta-carotene, and other carotenoids, just to mention a few. Apart from their important role within improving and intensifying coloration in fish, pigments possess antioxidant and probiotic properties, already validated by scientific studies (Latcscha 1990; Mortensen and Skibsted 1997; Bennedsen et al. 1999; Wang et al. 2000).

Table 3. Vitamin content of different microalgae in comparison with common foodstuffs (mg Kg^{-1} dry matter).

	Vit A	Vit B1	Vit B2	Vit B6	Vit B12	Vit C	Vit E	Nicotinate	Biotin	Folic Acid	Pantothenic Acid
Liver	60	3	29	7	0 7	310	10	126	1	0	7.3
Spinach	130	1	2	2	-	470	-	6	Trace	Trace	3
Baker's yeast	Trace	7	17	21	-	Trace	112	4	5	53	
Arthrospira platensis	840	44	37	3	7	80	120	-	trace	trace	13
Chlorella pyrenoidoisa	480	10	36	23		-	-	240	Trace	-	20
Scenedesmus quadricauda	554	12	27	-	1	396	-	108	-	-	46

Adapted from Becker 2004.

Case study: Chlorella

Chlorella is one of the most produced microalgae species in the world. Due to its component constituents, it can be an interesting ingredient in feed products: according to reported nutritional values, its composition varies within the following ranges: 50–60% of proteins, with all the essential amino acids, 10–20% of carbohydrates, and 10–20% of lipids. It contains complex vitamins B and P, folic and pantothenic acids, and choline, which contribute at different levels to promote the skin barrier function and preserve its hydration. *Chlorella* also contains the highest percentage of chlorophyll known in the plant world; chlorophyll is known to detoxify and purify pet's internal system and to control breath, body odour, and promote the growth of red cells. Lutein and zeaxanthin exist both in *Chlorella* biomass, and play an important role on the eye vision system's maintenance, preventing macular degeneration. Beta-carotene and astaxanthin are powerful antioxidants also present, and are used in competition animals to prevent oxidation in the body caused by reactive oxygen species.

Effect of processing on the digestibility of microalgae

The cell walls accounts for ca. 10% of the algal dry matter, and although its composition is strain-specific, it contains cellulose in most of the species. This cellulosic cell wall is not digestible for non-ruminants, and therefore presents a problem in digesting, or more broadly, utilizing algal biomass. In fact, effective disrupting treatments are necessary in order to allow the access of digestive enzymes to the algal protein and remaining components. Apart from *Spirulina*, all the commercially important algae (Chlorophyceae, Rodophyceae) present an indigestible cell wall, which obliges a disruption process to be performed. Such disruptive process can be performed by physical (e.g., boiling, drying) or chemical methods (Becker 2004).

Toxicological issues

In order to prove their safety, unconventional food sources have to fulfil a series of toxicological tests for the presence of biogenic (toxic compounds synthesized by the algae or formed by decomposition of metabolic products) or non-biogenic toxins (environmental contaminants, which can usually be avoided by proper cultivation techniques and selection of cultivation places free of pollution).

Biogenic toxins include nucleic acids and other algal toxins. One of the very few constituents present in all living organisms and hence in algae, and which under certain circumstances may be counted as toxin, are nucleic acids (RNA and DNA). They are the sources of purines, which may lead to an increase of plasma uric acid concentration in humans. Consequently, the recommended daily intake of nucleic acids from unconventional sources should not exceed 2.0 g per day, which corresponds to 20 g of algae per day. Nevertheless, these possible problems were reported for humans, and there are no studies of this possible toxic effect on animals (Becker 2004).

Although poisoning of livestock and other animals due to toxic blooms of algae frequently occurs, algal strains associated with this phenomenon are predominantly cyanobacteria. There are no reports of toxicity cases in connection with mass-cultured algae, and numerous chemical and toxicity studies, as well as feeding trials carried on along the years never revealed any pathological symptoms due to algal toxins.

Concerning non-biogenic toxins, most problematic issues are related with the possible presence of heavy metals and polycyclic aromatic compounds in the algal biomass. There are no characteristic levels of these compounds in microalgae, they simply depend on the environmental conditions prevailing but can be completely eliminated through microalgal cultivation in closed photobioreactors.

Microalgae as ingredients for animal feeds

About 30% of the current world annual production of *Spirulina* is sold as ingredient for animal feed applications. Although most of its positive effects rely in nutritional field (e.g., increased growth rate, colour enhancement, and general tissue quality), the fact that growth rates are improved, even at 0.1% *Spirulina* supplementation, suggest the presence of substances that may mimic the effects of or stimulate production of growth hormones. Within these non-nutritional effects, the most promising one may be an immune enhancement effect due to the antiviral and anti-bacterial properties presented by *Spirulina* or its extracts, especially during the early stages of animal life (Belay et al. 1996).

The use of microalgae as general animal feed ingredient is more recent when compared with their utilization in human food. Within this new market, a growing sector is the utilization of microalgae in aquaculture; although aquaculture already successfully uses several microalgal species as an essential component of the live food chain in the early development phases of fish production, the potential of microalgal biomass to partially replace fish oils in adult feeding, thus decreasing the levels of environmental contamination generated by these industries, and concomitantly providing higher quality fish flesh, offers an outstanding opportunity for the expansion of global aquaculture production (Sweetman 2009).

Main feed market areas were microalgae may have a valuable application can be classified into four groups: (i) pets (dogs, cats, birds, tropical fish, rabbits and rodents); (ii) aquaculture (e.g., salmonidae, sparidae, and bivalves); (iii) husbandry (e.g., poultry, cattle, swine); (iv) and others (e.g., race horses). Apart from responding to nutritional daily demands, microalgae can be tailored for specialty feed applications in critical growth stages of the animal.

Microalgae in pet feeds

A promising application for microalgal biomass is the pet food market, where effects on the external appearance of the pet (shiny hair, beautiful feathers) are as important as the health-promoting ones (Pulz and Gross 2004).

Within this sector, consumers are noticed to be increasingly aware and demanding while purchasing feed, paying close attention to labels, packaging and marketing claims, and directing their preferences to natural products and/or supplements with multiple benefits. Pet humanization has thus led to specialized formulas, with foods addressing weight loss and management, arthritis relief, sensitive skin, diabetes, digestive/urinary tract health, heart health, immune system support, and mental development and health. In Japan, pet foods focused on the aging and senior pet population are popular (Dias and Sendão unpublished data report; Anonymous 1985). Main active compounds and effects of microalgae species on pet feeds are presented in Table 4.

The pet food market is a substantial one, worth almost US$80 billion in 2010, and representing a growth of 4% over 2009 (Dias and Sendão unpublished data). However, the market is mature offering little prospect of exciting growth over the next decade. As a result, pet food manufacturers looking for sales growth are prospecting for product segmentation (specialty feeds and treats).

Table 4. Selected examples of microalgae species used in farm animals.

Market sector	Microalgae species	Active compounds	Process
PETS	*Arthrospira* sp. and *Chlorella* sp.	Proteins, Fatty acids, Carotenoids	Proteins and fatty acids provide health-promoting and external appearance effects (shinny hair, beautiful feathers). Benefits associated to microalgae are associated to an enhanced resistance and recovery of muscle electrolyte balance, its overall antioxidant role, prevention of myocardial (heart) lesions, and joint health.

(Hu 2004; Pulz and Gross 2004)

Microalgae in aquaculture

The largest current application of microalgae feeds is in aquaculture. Microalgae are commonly used in aquaculture as live feeds for all growth stages of bivalve molluscs, for the larval/early juvenile stages of abalone, crustaceans, and some fish species, and for zooplankton used in aquaculture food chains (Carvalho and Malcata 2000). In order to be used in aquaculture, microalgae must possess a number of key attributes: they must be of an appropriate size for ingestion, must have rapid growth rates, be amenable to mass culture, and also be stable in culture to any fluctuations in temperature, light, and nutrients as may occur in hatchery systems; finally, they must have a good nutrient composition, including an absence of toxins that might be transferred up the food chain (Muller-Feuga 2004).

Apart from feeding larvae and zooplankton, the addition of microalgal biomass to adult fish feed compositions seems to be a promising market. Initially, the colour-enhancing effects of phycocyanin-containing *Spirulina* biomass or carotenoids from *Dunaliella* were exploited in ornamental fish. Furthermore, the addition of astaxanthin from *Haematococcus pluvialis* to feed formulations is usually employed to provide the characteristic colour of the muscles of salmons. However, in recent years, the use of microalgae for adult fish feed utilization became increasingly important not only because of colour but also due to enhancing health effects.

Fish feeding represents over 50% of operating costs in intensive aquaculture, with protein being the most expensive dietary source. Therefore, continuous efforts have been made over the past decades to find alternative protein sources for aqua feeds, with special emphasis on terrestrial plants. There are limitations to the use of plant protein sources in fish diets, attributed to their deficiency in certain essential amino acids, their content of anti-nutritional compounds and to palatability problems. Additionally, recent interest in several crops as biodiesel raw materials has increased the world's demand for these crops and consequently, their price. Besides, its production requires agricultural land and freshwater, which are becoming very limited resources, and therefore, there is a need to search for alternatives (Pereira et al. 2012). Several improvements on fish health status by feeding with microalgal biomass are detailed in Table 5.

Apart from their general use in the growth phases previously described, and irrespective of the fish, shrimp, or mollusc species under farming, sensitive periods of the production cycle where microalgae as functional ingredients could be highly valued are:

i) Early feeding stages: microalgae are successfully used for larval nutrition, either for direct consumption, in the case of molluscs and penaeid shrimp, or indirectly, as food for the live prey (rotifers, *Artemia* or copepods) fed to small fish larvae. They are also been used during rearing larvae and juveniles stages of commercially important molluscs, penaeid prawn larvae, crustaceans, and fish. Although nutritional requirements for some target species have been defined, no set of nutritional criteria have yet been advanced. Generally, the microalgae must be non-toxic, in an acceptable size for ingestion, the cell wall should be digestible, and with sufficient biochemical constituents. Protein content and lipid quality are major factors determining the nutritional value of microalgae. In marine microalgae strains, its vitamin and mineral contents are also of importance.

ii) Specific physiological stages: during salmon production cycle animal's undergo a process of physiological changes (smoltification), which allow them to adapt from fresh water to salt water environment. It is a critical stage for the industry; in some cases it is still associated to high mortality rates. Among a vast array of metabolic changes during smoltification, animals present a rise on thyroid hormones and increased salinity tolerance. As some microalgae are naturally, or could be produced to be, rich in minerals, they could be of major interest to adequately modulate the salinity tolerance process in fish.

Table 5. Selected examples of microalgae species used in aquaculture.

Market sector	Microalgae species	Active compounds	Process
AQUACULTURE	**Bivalve Molluscs**. *Isochrysis* sp. *Chaetoceros gracilis* *Tetraselmis suecica* *Pavlova lutheri* *Skeletonema costatum* **Penaid Shrimp:** *Tetraselmis chui* *C. gracilis* *S. costatum* **Fish Feed Composition:** *Hypnea cervicornis* *Cryptonemia crenulata* **Marine fish larvae "Green Water":** *Chlorella* sp. *Isochrysis galbana* *Tetraselmis chui* **Adults:** *Chlorella* sp. *Spirulina* sp. **Salmonids:** *Haematococcus*	Aminoacids Lipids (PUFA) Vitamins Minerals	• Direct enhancement in the immune system • Partial replacement of animal sourced proteins in adult diets • Pigmentation in salmon's muscles
ORNAMENTAL FISH	*Spirulina* sp., *Dunaliella* sp. *Chlorella vulgaris*	Pigments (phycocyanin, beta-carotene and other carotenoids, chlorophyll)	Carotenoids improve and intensify the coloration in fish

Schreckenbach et al. 2001; Becker 2004; da Silva and Barbosa 2008; Azaza et al. 2008.

iii) High stress periods: stress is a condition in which an organism is unable to maintain its normal physiologic state. It results from physical (handling, capture, confinement, and transport) and/or abiotic challenges (contaminants, temperature and salinity fluctuation, low oxygen, high ammonia), which modify animal's natural or homeostatic state. The above stressors can be categorized as either acute or chronic, depending on their duration and frequency, and have severe negative effects on an animal's growth and health. Microalga-based diets have been shown to bring positive effects on fish immune system, mainly attributed to their antioxidant potential and polyphenol contents. They may also be used following fish vaccination, as their bioactive constituents could act as vaccine adjuvants and potentiate its antigenic effect (Dias and Sendão unpublished data).

Microalgae in husbandry

Studies with incorporation of 5–10% microalgae in poultry feeds revealed their positive influence in terms of (i) visual enhancements on skin's colour and egg yolk, as well as weight gain; (ii) healthier final products (egg yolk with reduced cholesterol levels and increased linoleic and arachidonic acid contents); and (iii) probiotic effects (algal meal reduced caecal colonization of Clostridium perfringens) (Ginzberg et al. 2000; Waldenstedt et al. 2003; Spolaore et al. 2006). Other poultry feeding studies with Spirulina (up to 30%) showed that both protein and energy efficiency of this alga were similar to other conventional protein carriers (Becker 2004). A few more examples are detailed in Table 6.

Because of the pressure for decreasing costs, the introduction of microalgae in land based animal feeds may not be target to the commodities market, but to specialty feeds applications. The main targets for specialty nutritional applications in which microalgae protein products could be commercially interesting comprise: (a) Piglet and calves weaning phase, a sensitive growth phase because of environmental, disease, and nutrition stress factors; in terms of nutrition, these young animals require highly digestible ingredients, adapted to the gut maturation. The current protein sources used (skim milk powder, died egg protein, vegetable protein concentrates, yeast, and fish protein hydrolisates) reach 4000€/ton, therefore microalgae should either be cost competitive or demonstrate clear additional benefits to justify a higher price, for example, through reinforcement of the immune system of the animals and the palatability of feeds (Dias and Sendão unpublished data); (b) Productivity and egg quality traits of laying hens, due to the increase in calcium content and strength of the shell, and increase in both the number of eggs per year and egg laying period promoted by algal diets. Additionally, there is also the potential for tailoring the nutritional composition of eggs, in terms of PUFA and cholesterol levels, as well as shell and yolk colour; (c) Milk yield and productivity in dairy cows: use of algae has been reported to increase milk yield and prolong milking periods. Depending on the microalgae composition, its use has also been tested as a tool to enhance PUFA levels in milk and increase butter fat percentage. Morcover, diets incorporating microalgae have been associated with having beneficial effects on the prevention of osteoporosis and mastitis, two major problems in the dairy industry.

Table 6. Selected examples of microalgae species used in farm animals.

Market sector	Microalgae species	Active compounds	Process
FARM ANIMALS	**Poultry:** *Arthrospira* sp. *Chlorella* sp. *Porphyridium* sp. *Haematococcus* sp.	Proteins Astaxanthin	Partial (5–10%) replacement for conventional proteins, resulting in 10% less consumed food and lower cholesterol levels. Egg yolk has higher carotenoid content.
	Swine: *Schizochytrium* sp.	DHA	DHA-rich microalgae administered in the diet result in higher weight gain and feed conversion efficiency.

Firman 1984; Ginzberg et al. 2000; Abril et al. 2003; Cysewski and Lorenz 2004; Spolaore et al. 2006; Coffey 2008

High performance/Athletic animals

Nutrition is critical for optimal performance in athletic animals, either for a racehorse, equestrian show jumping or greyhound race dog, among others. Animals in training and following competition are submitted to considerable stress, which can upset their body system. Besides the enormous energy expenses and recovery needs, managing the balance of dietary electrolytes is a key factor in maximizing performance and recovery.

The dietary use of microalgae in these particular niche markets has already been reported, although in a few cases. Benefits associated to microalgae are associated to an enhanced resistance and recovery of muscle electrolyte balance, its overall antioxidant role, prevention of myocardial (heart) lesions, and joint health. High levels of chlorophyll in blue green algae have also been associated with an enhanced blood detoxifying capacity after strenuous exercise (Dias and Sendão unpublished data).

Table 7. Selected examples of microalgae species used in competition animals.

Market sector	Microalgae species	Active compounds	Process
Athletic/Competition animals	*Arthrospira* sp. *Haematococcus* sp.	Proteins Fatty acids Carotenoids	Benefits associated to an enhanced resistance and recovery of muscle electrolyte balance, its overall antioxidant role, prevention of myocardial (heart) lesions, and joint health. High levels of chlorophyll in blue green algae have also been associated with an enhanced blood detoxifying capacity after strenuous exercise.

Hu 2004; da Silva et al. 2008.

Market statistics and regulations

Statistical data on feed market

The most relevant entity establishing feed safety international standards is IFIF—International Feed Industry Federation. According to data from this organization, in a survey among 132 countries, China was the main feed producer in 2011, with 175.400 million metric tons; US and Brazil followed, with 164.920 million metric tons and 59.629 million metric tons, respectively, and EU feed production was 200 million metric tons. Several new countries, Japan, India, and Canada, are emerging among the top ten as strong players in the global feed scene (Anonymous 2012).

When analysed by species and according to Table 8, poultry (broilers, layers, turkey, and other fowl) had the dominant share of the tonnage market, followed by ruminant feed; such high consumption of poultry feeds reflects costs, health, and religious preferences for this white meat. Aquaculture was the fastest growing feed sector, whereas pig, equine, and pet feed numbers did not indicate a significant change when compared with previous studies (Anonymous 2012).

Table 8. Global feed tonnage by species in 2011 (million metric tons).

Region	Pig	Poultry	Ruminant	Aqua	Other**
Asia	81	116	80.12	24.4	4.03
Europe*	63.09	70.25	57.11	1.33	8
North America	31.23	91.07	45.5	0.286	17.09
Middle East/Africa	0.87	27.71	17.04	0.60	0.72
Latin America	24.80	71.26	22.34	1.88	4.46
Other**	2	4.60	3.49	0.20	0.86
Total	202.99	380.89	225.6	28.696	35.16

*EU27 & Non-EU Europe and former Soviet Union; ** include horse (9.24M) and pets (25.6M) (Source: Anonymous 2012).

Regulations

Safe animal feed is important for the animal's health, the environment, and the safety of foods from animal origin. There are many examples of the close link between the safety of animal feed and human foods; a known example succeeded with the inclusion of mammalian and bone meal in farm animals feed, the consequent spread of Bovine Spongiform Encephalopathy (BSE) in cattle, and the association of its infected meat with the variant Creutzfeldt-Jakob Disease (vCJD) in humans (http://www.efsa.europa.eu/en/topics/topic/feed.htm).

Although microalgae are able to enhance the nutritional content of conventional foods and hence, positively affect the health of humans and animals, algal material must also be analysed for the presence of toxic compounds to prove their harmlessness, prior to commercialization. There is always a long way for a new microalgae potential supplement from the Petri dish to the market place, and a crucial step is to get approval for incorporation as dietary supplements or additives. The most relevant entities in food supplement regulation are European Food Safety Authority (EFSA), in Europe, and Food and Drug Administration (FDA) in US.

The standing Committee on the Feed Chain and Animal Health has adopted on July 2012 a revised version of the EU Catalogue of Feed Materials. This Catalogue is intended to be a non-exhaustive list of feed materials, aimed to improve market transparency, and facilitate the exchange of information on feed materials properties between the parties, and therefore, providing a common system within the EU for the description and labeling of feed materials and compound feeds.

Regarding algae, the Part C of the Annex (7. other plants, algae, and products derived thereof) on the Catalogue (Commission Regulation (EU) No. 575/2011) details the specifications algal ingredients must fulfil in order to be approved as feed constituents (Table 9).

Although the use of this Catalogue by the feed business operators is voluntary, the name of a feed material listed in Part C may be used only for a feed material complying with the requirements of the entry concerned. Nevertheless, all entries in the list of feed materials shall comply with the restrictions on the use of feed materials in accordance with the relevant legislation of the Union. Feed business operators using algae and microalgae as feed material entered in the Catalogue shall ensure that it complies with Article 4 of Regulation (EC) No. 767/2009. Finally, and in accordance with good practice as referred to in Article 4 of Regulation (EC) No. 183/2005, feed materials shall be free from chemical impurities resulting from their manufacturing process and from processing aids, unless a specific maximum content is fixed in the Catalogue. Concerning algae entries, only antifoaming agents (with a general maximum level of 0,1%) were inserted in the revised version of the Catalogue.

Table 9. Excerpt of Section 7 "Other plants, algae and products derived thereof" on Part C of the Catalogue of Feed Materials.

Number	Name	Description	Compulsory declarations
7.1.1	Algae[21]	Algae, live or processed, regardless of their presentation, including fresh, chilled, or frozen algae. May contain up to 0.1% of antifoaming agents.	Crude protein Crude fat Crude ash
7.1.2	Dried algae[21]	Product obtained by drying algae. This product may have been washed to reduce the iodine content. May contain up to 0.1% of antifoaming agents.	Crude protein Crude fat Crude ash
7.1.3	Algae meal[21]	Product of algae oil manufacture, obtained by extraction of algae. May contain up to 0.1% of antifoaming agents.	Crude protein Crude fat Crude ash
7.1.4	Algal oil[21]	Product of the oil manufacture from algae obtained by extraction. May contain up to 0.1% of antifoaming agents.	Crude fat Moisture if > 1%
7.1.5	Algae extract[21]; [Algae fraction][21]	Watery or alcoholic extract of algae that principally contains carbohydrates. May contain up to 0.1% of antifoaming agents.	
7.2.6	Seaweed meal	Product obtained by drying and crushing macro-algae, in particular brown seaweed. This product may have been washed to reduce the iodine content. May contain up to 0.1% of antifoaming agents.	Crude ash

[21]—the name shall be supplemented by the species.

If a company wants to enter into the positive list, it has to file an application at the Commission. The primary criteria for the inclusion of a straight feeding stuff in the positive list comprise (i) a substantiated feed value (the product must be consumed orally in an effective quantity and thereby make a relevant contribution to the supply of energy and/or nutrition, or make a contribution to satiety and maintaining the function of intestinal tract, be safe with regard to the health of animals and humans, not negatively affect the quality of animal products, and do not present a hazard to the ecological balance due to undesirable substances it may contain); (ii) a recognisable importance in the market (it must be currently traded and used regardless of its market share); (iii) the legal admissible use as a straight feeding stuff (Anonymous 2011).

An example of an approved ingredient is astaxanthin [(3S,3'S)-3,3'-dihydroxy-b,b-carotene-4,4'-dione), a natural carotenoid with red-pigmenting properties occurring in yeasts, algae, crustaceans, and predator fish like salmons. It is approved at EU level as feed additive for salmon and trout (at 100 mg kg^{-1} complete feed, from 6 months of age onwards without time limit) and for ornamental fish (http://www.efsa.europa.eu/en/efsajournal/pub/291.htm).

Commercially important microalgae species

Autotrophic microalgae production is approx. 12,000 ton/year, divided by the species described in Table 10. Heterotrophic microalgae production is approx. 6,000 ton/year with the species *Schizochytrium*, *Crypthecodinium*, and *Ulkenia*. From these amounts, ca. 2,000 ton/year is used in aquaculture hatcheries.

Table 10. Main current commercial autotrophic microalgal species (sold as dried biomass) and corresponding market values.

Species name	Average price of dried biomass (€/kg)	Annual production (ton)	Market value (M€)
Arthrospira	5–50	6,500–8,000	110–140
Chlorella	10–60	2,600–2,900	80–100
Dunaliella	60–100	1,000–1,600	70–110
Haematococcus	150–340	400–500	80–100
Total		10,000–12,000	340–450

Source: Unpublished data from Necton, S.A. and A4F.

Conclusions

World feed production in 2011 presented an estimated value of 870 million ton, whereas the turnover of global commercial feed manufacturing generated annual sales values equivalent to US$350 billion worldwide (source: IFIF).

An increase in world population plays an important role on increasing feed demand. There are currently 6.5 billion people in the world, and it is estimated that by 2030 there will be two billion more. This growth is driving meat consumption, and more meat means more feed, therefore feed demand and animal feed supplements are rising. The ability to produce increasing amounts of feed is one of the greatest challenges facing mankind, perhaps even more critical than the environmental, energy, global warming, and resource crisis.

As feed quality is the most important exogenous factor influencing animal health, the pressure for higher quality feed ingredients increases, thus providing space in the market for high quality microalgal biomass as nutritional/functional ingredient in feeds. Microalgae present unique and interesting biochemical properties when compared to higher plants: they contain a much higher level of vitamins and trace elements than commodity feeds, and these are generally highly bio-available. Besides, they can be harvested on a daily basis all year round, and they can be cultured under controlled conditions, thus avoiding the presence of non-biogenic toxins. Microalgae are thus able to enhance the nutritional content of conventional feed preparations and hence, to positively affect the health of animals through their biochemical composition. Besides, feeds can be designed for specific growth conditions.

Besides their high nutritional value, microalgae protein products benefiting from a nature-like image are well positioned to enter the functional ingredients market and gain an important role in the raising trend for the development of specialty feeds (Dias and Sendão unpublished data report).

References

Aaronson, S. and Z. Dubinsky. 1982. Mass production of microalgae. Experientia 38: 36–40.

Anonymous. 1985. NRC – National Research Council. Nutrient Requirements of Dogs. National Academy Press, Washington.

Anonymous. 2006. Alberta Agriculture and Rural Development (www.agriculture.alberta.ca, accessed on September 2012).

Anonymous. 2011. Standards Commission for straight feeding stuffs at the central committee of the German agriculture – positive list for straight feeding stuffs 9th edition, March 2011.

Anonymous. 2012. Alltech Global Feed Survey, USA (www.alltech.com, accessed on September 2012).

Abril, R., J. Garrett, S.G. Zeller, W.J. Sander and R.W. Mast. 2003. Safety Assessment of DHA-rich microalgae from *Schizochytrium* sp. Part V: target animal safety/toxicity study in growing swine. Regulatory Toxicology and Pharmacology 37: 73–82.

Azaza, M.S., F. Mensi, J. Ksouri, M.N. Dhraief, B. Brini, A. Abdelmouleh and M.M. Kraiem. 2008. Growth of Nile tilapia (*Oreochromis niloticus* L.) fed with diets containing graded levels of green algae ulva meal (*Ulva rigida*) reared in geothermal waters of southern Tunisia. J. Appl. Ichthyol. 24: 202–207.

Barsanti, L. and P. Gualtieri. 2006. Algae – Anatomy, Biochemistry and Biotechnology. CRC Press, Taylor & Francis Group, USA.

Becker, E.W. 1986. Nutritional properties of microalgae: potentials and constrains. pp. 339–420. *In*: A. Richmond (ed.). CRC Handbook of Microalgae Mass Culture. CRC Press, Florida.

Becker, E.W. 1994. Microalgae Biotechnology and Microbiology. Cambridge University Press, Cambridge.

Becker, W. 2004. Microalgae in human and animal nutrition. pp. 312–351. *In*: A. Richmond (ed.). Handbook of Microalgae Culture - Biotechnology and Applied Phycology. Blackwell Science, Oxford.

Becker, E.W. 2007. Microalgae as a source of protein. Biotechnol. Adv. 25: 207–210.

Belay, A., T. Kato and Y. Ota. 1996. *Spirulina* (*Arthrospira*): potential application as an animal feed supplement. J. Appl. Phycol. 24: 202–207.

Bennedsen, M., X. Wang, R. Willén, T. Wadström and L.P. Andersen. 1999. Treatment of *H. pylori* infected mice with antioxidant astaxanthin reduces gastric inflammation, bacterial load and modulates cytokine release by splenocytes. Immunol. Lett. 70: 185–189.

Carvalho, A.P. and F.X. Malcata. 2000. Effect of culture media on production of polyunsaturated fatty acids by *Pavlova lutheri*. Cryptogamie Algologie 21: 59–71.

Carvalho, A.P., C.M. Monteiro and F.X. Malcata. 2009. Simultaneous effect of irradiance and temperature on biochemical composition of the microalga *Pavlova lutheri*. J. Appl. Phycol. 21: 543–552.

Coffey, R. 2008. Digestive physiology of farm animals. Introduction to Animals and Food Sciences. University of Kentucky.

Cysewski, G.R. and R.T. Lorenz. 2004. Industrial production of microalgal cell-mass and secondary products – species of high potential: *Haematococcus*. pp. 281–288. *In*: A. Richmond (ed.). Handbook of Microalgae Culture - Biotechnology and Applied Phycology. Blackwell Science, Oxford.

da Silva, R.L. and J.M. Barbosa. 2008. Seaweed meal as a protein source for the white shrimp *Lipopenaeus vannamei*. J. Appl. Phycol. 21: 193–197.

Dias, J. and J. Sendão. 2011. (Unpublished data) Report on Microalgae Biomass coproducts: Prospective Analysis of Nutritional Applications. Sparos Lda. and CRIA – University of Algarve.

EFSA http://www.efsa.europa.eu/en/topics/topic/feed.htm (accessed on September 2012).

EFSA http://www.efsa.europa.eu/en/efsajournal/pub/291.htm (accessed on September 2012).

Fabregas, J and C. Herrero. 1985. Marine microalgae as a potential source of single cell protein (SCP). Appl. Microbiol. Biotechnol. 23: 110–113.

FAO: The nutrition and feeding of farmed fish and shrimp; a training manual. http://www.fao.org/docrep/field/003/ab470e/AB470E02.htm (accessed on November 2012).

Firman, J. 1984. Nutrient Requirements of Chickens and Turkeys. University of Missouri. Department of Animal Sciences.

Freeman, S. 2003. Biological Science. Prentice Hall has offices worldwide and therefore it is not possible to indicate the place of publication.

Ginzberg, A., M. Cohen, U.A. Sod-Mariah, S. Shany, A. Rosenshtrauch and S. Arad. 2000. Chickens fed with biomass of the red microalga *Porphyridium* sp. have reduced blood cholesterol level and modified fatty acid composition in egg yolk. J. Appl. Phycol. 12: 325–330.

Hu, Q. 2004. Industrial production of microalgal cell-mass and secondary products – major industrial species: *Arthrospira* (*Spirulina*) *platensis*. pp. 264–272. *In*: A. Richmond (ed.). Handbook of Microalgae Culture - Biotechnology and Applied Phycology. Blackwell Science, Oxford.

IFIF - http://www.ifif.org/pages/t/The+global+feed+industry (accessed on September 2012).

IFFO – International Fishmeal and Fishoil Organization http://www.iffo.net (accessed on October 2012).

Latcscha, T. 1990. Carotenoids in Animal Nutrition. Hoffman-La Roche Ltd., Basel, Switzerland.

McDonald, P., R. Edwards, J. Grenhalgh and C. Morgan. 2002. Animal Nutrition. Prentice Hall.

Merchen, N. and E. Titgemeyer. 1992. Manipulation of amino acid supply to the growing ruminant. J. Animal Sci. 70: 3238–3240.

Miles, R. and R. Chapman. 2009. The benefits of fishmeal in aquaculture diets. Institute of Food and Agricultural Sciences (IFAS), University of Florida.

Mortensen, A. and L.H. Skibsted. 1997. Importance of carotenoid structure in radical scavenging reactions. J. Agric. Food Chem. 45: 2970–2977.

Muller-Feuga, A. 2004. Microalgae for aquaculture—the current global situation and future trends. pp. 352–364. *In*: A. Richmond (ed.). Handbook of Microalgae Culture. Blackwell, Oxford.

Pereira, R., L.M.P. Valente, I. Sousa-Pinto and P. Rema. 2012. Apparent nutrient digestibility of seaweeds by rainbow trout (*Oncorhynchus mykiss*) and *Nile tilapia* (*Oreochromis niloticus*) Algal Res. 1: 77–82.

Pulz, O. and W. Gross. 2004. Valuable products from biotechnology of microalgae. Appl. Microbiol. Biotechnol. 65: 635–648.

Schreckenbach, K., C. Thürmer, K. Loest, G. Träger and R. Hahlweg. 2001 Der Einfluss von mikroalgen (*Spirulina platensis*) in trockenmischfutter auf karpfen (*Cyprinus carpio*). Fischer. Teichwirt. 1: 10–13.

Spolaore, P., C. Joannis-Cassan, E. Duran and A. Isambert. 2006. Commercial applications of microalgae. J. Biosc. Bioeng. 101: 86–96.

Sweetman, E. 2009. Microalgae: its applications and potential. International Aqua Feed. Perendale Publishers Ltd., UK.

Waldenstedt, L., J. Inborr, I. Hansson and K. Elwinger. 2003. Effects of astaxanthin-rich algal meal (*Haematococcus pluvalis*) on growth performance, caecal campylobacter and clostridial counts and tissue astaxanthin concentration of broiler chickens. Animal Feed Sci. Technol. 108: 119–132.

Wang, X., R. Willén and T. Wadström. 2000. Astaxanthin-rich algal meal and vitamin C inhibit *Helicobacter pylori* infection in BALB/cA mice. Antimicrobial Agents and Chemotherapy 44: 2452–2457.

Wondra, K., J. Hancock, K. Behnke and C. Stark. 1995. Effects of mill type and particle size uniformity on growth performance, nutrient digestibility, and stomach morphology in finishing pigs. Department of Animal Sciences and Industry. Kansas State University.

10

Potential Use of Extracts of Seaweeds Against Plant Pathogens

Jatinder Singh Sangha,[1,2] *Robin E. Ross,*[3] *Sowmyalakshmi Subramanian,*[1,4] *Alan T. Critchley*[5] *and Balakrishnan Prithiviraj*[1,*]

Introduction

Seaweeds are marine macroalgae, and classified into three phyla based on their pigmentation: Chlorophyta (green algae), Phaeophyta (brown algae), or Rhodophyta (red algae) (Guiry 2012; Guiry and Guiry 2018). Seaweeds are exploited for numerous uses, as food, in industry, and in agriculture. The use of seaweeds as an organic amendment in soil has been documented as common practice by farmers in many parts of the world for many centuries. Some of these whole seaweeds such as kelps (*Laminaria digitata*) and rockweed (*Ascophyllum nodosum*) are still used in agriculture as soil amendments, the application of extracts in plant protection has been an emerging practice. This is due to the fact that macroalgae are rich sources of nutrients and also an exceptional source of several biologically active compounds, including complex polysaccharides, a great variety of secondary metabolites which exhibit a broad spectrum of bioactivity (Kulik 1995; Lordan et al. 2011). In fact, various algal species have the potential to be used in innumerable fields such as food, industrial raw materials, therapeutic uses (Gupta and Abu-Ghannam 2011; Marfaing 2017; Srikong et al. 2017) and agriculture and especially in plant production as biostimulants (Yakhin et al. 2017).

Seaweeds, in general, are suggested to have inherent bioactive properties due to the presence of various components and or metabolites, which are produced under normal growth conditions, as well as during periods when they are challenged by stressors (i.e., desiccation and or pressure of herbivores) (Contreras-

---

[1] Department of Plant, Food and Environmental Sciences, Faculty of Agriculture, Dalhousie University, PO Box 550, Truro B2N 5E3, NS, Canada.

[2] Swift Current Research and Development Centre, Agriculture and Agri-Food Canada, 1 Airport, Road, Swift Current, Saskatchewan, Canada S9H 3X2.

[3] Acadian Seaplants Limited, 30 Brown Avenue, Dartmouth, B3B 1X8, NS, Canada.

[4] Department of Plant Sciences, 21111, Lakeshore road, Macdonald Campus, McGill University, Montreal, Canada H9X 3V9.

[5] Verschuren Centre, Cape Breton University, 1250 Grand Lake Rd, Sydney, Nova Scotia, Canada B1P 6L2.

Emails: jatinder.sangha2@canada.ca; rross@acadian.ca; sowmyalakshmi.subramanian@mail.mcgill.ca; alan.critchley2016@gmail.com

* Corresponding author: bprithiviraj@dal.ca

Porcia et al. 2010). The major bioactive substances in seaweeds can be broadly identified as aromatics, sterols, dibutanoids, proteins, peptides, and sulphated polysaccharides. Some of these are unique compounds, only found in macroalgae, for example, carrageenans, fucoidans (Khan et al. 2009; Sangha et al. 2010; Seth and Shanmugam 2016). Many of the secondary metabolites are abundant in marine seaweeds and are probably synthesized as chemical defences to cope with extreme environments such as, salinity, extreme temperature variations, high and low light intensities, and UV rays amongst others. In addition, these compounds also protect and mitigate against damage which may be caused by colonization by epiphytes, microbial organisms, or grazing herbivores (Kubanek et al. 2003). Taken together, these attributes may be part of the reasons why certain seaweeds and their extracts have biological activity, especially against a variety of microbes, including plant pathogens. Some of the algal-derived compounds within seaweed extracts have direct or indirect effects on the suppression of pathogens infecting plants and humans and thus have attracted scientific interest in the utilization of macroalgal extracts for plant disease management. In fact, a growing body of literature suggests that the bioactive compounds present and as extracted from certain seaweeds may be an economically viable resource of agricultural importance that can influence not only the plant growth, but also plant-pathogen interactions, thereby assisting crop production through the relief of stress from plant pathogens (Cox et al. 2010; Craigie 2010; de Freitas et al. 2015; Esserti et al. 2017). Identification of the active components in seaweeds will facilitate the commercialization of seaweed products, further the success for their industrial uses, and lead to very specific and surgical applications with recommended rates and timings (Patier et al. 1993; Jiao et al. 2011).

A large number of seaweed extracts have been tested against various microbes infecting humans, animals and plants (Reichelt and Borowitzka 1984; Gonzalez del Val et al. 2001; Paulert et al. 2007; Khan et al. 2009; Craigie 2010; Bahar et al. 2016; Shi et al. 2017). With information on the anti-microbial activity of seaweed extracts against human pathogens and the benefits of seaweed extracts to the human gut microbiome being of great interest and discussed elsewhere (Wang et al. 2009; Charoensiddhi et al. 2017), the documentation of many marine algal extracts and their effects against plant pathogens is also a fast emerging area of phycology (Jayaraj et al. 2008; Paulert et al. 2010; Sangha et al. 2010; Subramanian et al. 2011; Stadnik and de Freitas 2014; Bhattacharyya et al. 2015; Abouraïcha et al. 2017; Esserti et al. 2017). Extracts of red, green, and brown seaweeds, including examples from the Phaeophyceae such as: *Ecklonia maxima, Saccharina (Laminaria) saccharina, Fucus serratus, F. vesiculosus, Sargassum* spp. and *Ascophyllum nodosum,* Chlorophyceae *Ulva lactuca, Codium* sp., and Rhodophyceae such as *Kappaphycus alvarezii* are potential crop biostimulants and anti-microbial agents in agriculture (Khan et al. 2009; Craigie 2010; Burketova et al. 2015; Esserti et al. 2017; Abouraïcha et al. 2017). Information on seaweed extracts against plant microbes generated by *in vitro* and *in vivo* bioassays, in conjunction with greenhouse and field trials, reflect the potential for the application of seaweed extracts in plant disease management. The widespread application of seaweed extracts for the suppression of various diseases is possible for selective pathogen control (McLachlan 1985; Craigie 2010). Furthermore, the natural origin and renewable, sustainable nature of the seaweed resources, from which many of the extracts are responsibly and commercially extracted, so too along with cost effectiveness and safety to the environment, make these seaweed extracts compatible with an integrated approach to biocontrol and plant disease management, that may find multiple and widespread applications in agriculture (Aziz et al. 2003; Chandia and Matsuhiro 2008; Courtois 2009).

Herewith, we review the successful use of selected seaweed extracts to suppress plant diseases through direct or indirect activity in agricultural systems. This information will help to identify certain bioactive seaweeds and their extracts and encourage new ways for their applications in an integrated approach to plant disease management.

Anti-microbial bioactive compounds in seaweeds

Seaweeds possess several metabolites, some of which exhibit antimicrobial activities. For example, the brown seaweed *A. nodosum* is a major source of polyphenols, sulphated galactans and fucans, and volatile halogenated compounds, for example, brominated vanadium peroxides that have been shown to have bioactivity against microorganisms (Cardozo et al. 2007). Similarly, compounds such as aromatic esters, sterols, phenols, fatty acids, terpenoids, and polysaccharides present in different seaweed species, either exert direct anti-microbial action or may act indirectly to elicit plant responses to pathogen infection (Kamenarska et al. 2002; Khan et al. 2009; Sangha et al. 2011; Subramanian et al. 2011). The number of compounds isolated from various seaweeds is indeed large, but by no means have all been tested against plant pathogens. Some of the common and important components which are thought to have roles against plant pathogens are considered in further detail here:

Polysaccharides: Polysaccharides are major components in seaweeds that are present to a varying extent in brown, red, and green seaweeds. Cell wall polysaccharides in brown seaweeds, for example, alginates, sulphated galactans and fucans, as well as the storage polysaccharide laminarin (Klarzynski et al. 2003; Khan et al. 2009), can elicit defense responses in plants against pathogens (Klarzynski et al. 2000; Klarzynsci et al. 2003). The polysaccharide fucoidan (obtained from *Lessonia vadosa*) is another example of a bioactive polysaccharide that has demonstrated significant activation of defence responses when applied to plants (Chandia and Matsuhiro 2008). Similarly, sulphated polysaccharides, obtained from red and green algae, have also been reported to elicit anti-microbial activities. For example, λ-carrageenans from red seaweeds, and ulvan the principle cell wall polysaccharide obtained from green seaweeds, have been shown to induce resistance in several plant species after their application (Mercier et al. 2001; Araújo et al. 2008; Cluzet et al. 2004; Paulert et al. 2011; Jaulneau et al. 2010; Hernández-Herrera et al. 2014; Esserti et al. 2017; Abouraïcha et al. 2017; Van Oosten et al. 2017). Interestingly, the levels of bioactivity of polysaccharides seems to vary as per the level of sulphation of the molecules (Mercier et al. 2001; Menard et al. 2005; Sangha et al. 2010) and should be given prior consideration when being used to elicit plant defense responses.

Terpenes: Terpenes are highly volatile organic compounds produced as secondary metabolites in plants, seaweeds, insects, and many other organisms. Seaweed terpenes are thought to be produced as a chemical defense strategy to ward off herbivores and for competitive advantage over reef corals (Bianco et al. 2010; Rasher et al. 2012). Terpenes from brown seaweeds; mostly halogenated, have been tested for microbial suppression (Katayama 1962; Fenical et al. 1973; Venkatesh et al. 2011; Peres et al. 2012). *Staphylococcus aureus* and *Eschericia coli* were inhibited by seaweed terpenes (Katayama 1962). Terpenes have been shown to exhibit anti-microbial activity against plant pathogens (Bassolé and Juliani 2012). Sesquiterpenes, such as, zonarol and isozonarol from the brown alga, *Dictyopteris zonarioides,* were active against 10 species of plant fungi (Fenical et al. 1973). Similarly, the meroditerpenoid metabolite, methoxybifurcarenone, from the brown alga *Cystoseira tamariscifolia,* showed anti-fungal activity against three tomato pathogenic fungi, for example, *Botrytis cinerea, Fusarium oxysporum* f.sp. *lycopersici,* and *Verticillium alboatrum* whereas *in vitro* anti-bacterial activity was also observed against *Agrobacterium tumefaciens* and *Escherichia coli* (Bennanmara et al. 1999). Five meroditerpenes were isolated from the brown alga *Cystoseira* spp. (Navarro et al. 2004), that also showed anti-microbial activity. Peres et al. (2012) reported anti-microbial terpenes present in ethanolic extracts of various seaweeds viz., the browns *Stypopodium zonale, Ascophyllum nodosum, Pelvetia canaliculata, Fucus spiralis, Sargassum muticum, S. filipendula, S. stenophyllum,* and *Laminaria hyperborea* and the reds *Laurencia dendroidea* and *Gracilaria edulis* were seen to be effective against two plant pathogens. These reports suggested that seaweed terpenes could be used potentially against plant pathogens.

Fatty acids: Fatty acids have been shown to have strong anti-microbial activity against various microorganisms (Desbois and Smith 2010). Seaweeds are a well-known source of various fatty acids, and comprise mainly of polyunsaturated fatty acids (PUFAs) (Khotimchenko et al. 2002; Dawczynski et al. 2007; Schmid et al. 2017). For some time, it has been known that lipid extracts, from various brown, green and red seaweeds have been effective against different pathogens under *in vitro* conditions (Caccamese et al. 1981) indicating that fatty acids can play a role in the suppression of plant pathogens. Downstream fractionation of lipophilic extracts from the red *Gracilaria edulis,* the brown *Sargassum wightii,* and the green *Enteromorpha* flexuosa* (**Enteromorpha* are properly named *Ulva*), and testing for anti-bacterial activity revealed that their fatty acids, predominantly palmitic acids, were anti-bacterial against *Xanthomonas oryzae* pv. *oryzae* (Arunkumar and Rengasamy 2000). Fatty acids such as palmitic acid, followed by oleic and myristic acids from the brown alga *Padina pavonia* have also shown anti-microbial activity (Kamenarska et al. 2002). Similarly, chloroform and methanol fractions of an ethanolic extract of the brown seaweed *Spatoglossum asperum,* which was rich in various fatty acid esters, exhibited strong anti-fungal activities against *Macrophomina phaseolina,* whereas the n-hexane fraction was active against *Rhizoctonia solani* and *Fusarium solani* (Ara et al. 2005). Similarly, a lipophilic compound, primarily composed of pyrrole-2-carboxylic acid, pentadecanoic acid, and octadecanoic acid, from the green coenocyte *Asparagopsis taxiformis,* was active against various pathogenic microorganisms (Manilal et al. 2009). Sulphoglycerolipid 1-0-palmitoyl-3-0 (6′-sulpho-α-quinovopyranosyl)-glycerol, isolated from the methanolic extract of the brown seaweed *Sargassum wightii,* showed activity against *Xanthomonas oryzae* pv. *oryzae* (*Xoo*)—the causative agent of bacterial blight in rice (Arunkumar et al. 2005).

Sterols: Sterols are an important family of lipids that are present in eukaryotic cells (Nabil and Cassan 1996). Sterols are also known to have bioactivity in biological systems that can increase the innate immunity of plants against pathogens (Wang et al. 2012). Biologically active sterols such as cholesterol, fucosterol, stigmasterol, and ergosterol are predominant in certain red, green, and brown seaweeds (Kamenarska et al. 2002; Sánchez-Machado et al. 2004). Ghazala et al. (2004) reported biologically active sterols, the majority of which was β-Sitosterol isolated from the methanol extracts of freshwater, stonewort algae, *Chara contraria,* and *Nitella flexilis* that were highly effective against fungal pathogens. Sterols were demonstrated to induce morphological abnormalities in some plant pathogens, for example, the effect of an extract of *Sargassum carpophyllum* on the fungus *Pyricularia oryza* (Blunt et al. 2007), thus exhibiting the potential to suppress fungal pathogens.

Phenolic compounds: Macroalgae are also known sources of phenolic compounds which possess anti-microbial properties. These compounds are also known antioxidants and they can elicit anti-microbial activities. Phenols, polyphenols, tannins, and tannic acid are abundant in certain brown, green, and red seaweeds (see Sieburth and Conover 1965; Zhang et al. 2006; Liu and Gu 2012; Eom et al. 2012). A particular type of polyphenol called phlorotannin, is unique to the brown algae; it has been demonstrated to have strong anti-microbial activity and potential for applications against plant pathogens (Wang et al. 2009; Gupta and Abu-Ghannam 2011). Seaweed extracts, rich in phenolic compounds, along with sulphated galactans and fucans, as well as other halogenated compounds (Cardozo et al. 2007), together may have the ability to reduce pathogenic microbes, including those causing plant diseases.

Other secondary metabolites: The strong anti-microbial activity of some seaweed extracts may be linked with other metabolites, as found in various seaweeds. A variety of halogenated compounds, as well as other primary and secondary metabolites may be anti-microbial, for example, acrylic acid from seaweeds such as: *Gracilaria corticata* and *Ulva lactuca* (Katoyama 1961; Dandara et al. 1988) was demonstrated to have anti-microbial activity. Ethanol soluble extracts from *Gracilaria chilensis* and *Lessonia trabeculata* contained unspecified active compound(s) against the mycelial growth of *P. cinnamomi* and also activated plant defense mechanisms to protect the challenged plant tissue against *B. cinerea* infection (Jiménez et al. 2011). These authors suggested that the active fraction was neither polysaccharide nor proteinaceous in nature, which indicated that unknown secondary metabolites or some of the above-mentioned compounds, singly or in combination, could have been present in the bioactive extracts. Venkatesh et al. (2011) reported 2,3 dihydro-7-methy-1,4-benzoxazine-3-one was one of the bioactive components in the carragenophyte *Kappaphycus alvarezii* that has shown anti-microbial potential. Apart from terpenes, various anti-microbial compounds such as neophytadiene, cartilagineol, obtusol elatol, and the ester ethyl hexadecanoate were identified in the *L. dendroidea* extract (Peres et al. 2012). The results were promising and demonstrated the ability of selected seaweed extracts against different types of plant pathogens, including fungi and virus.

Seaweed extracts against plant pathogens

In their natural environment, some seaweeds maybe exposed to a number of seemingly stressful conditions including microbial attack. Thus, it is natural that a variety of compounds produced during interactions with the environment might be anti-microbial in nature. In order to understand the real potential of seaweed extracts, as applied towards the priming or stimulation of plant responses against pathogens, we present the following information from *in vitro* and *in vivo* experiments on the biostimulatory activity of certain seaweed extracts; including their direct and indirect effects on microbes.

Direct activity of seaweed extracts related to plant/microbial activity

Though whole seaweeds directly, or as various extracts, have been used in agriculture since ancient times, documented anti-microbial activity attributed to the algal component, has only been presented since the mid-20th century (Pratt et al. 1951; Ross 1957; Welch 1962). The number of reports of *in vitro* anti-microbial activities attributed to crude seaweed extracts increased tremendously after including anti-bacterial (Fenical and Paul 1984; Gonzalez et al. 2001; Yi et al. 2001; Esserti et al. 2017), anti-fungal (Fenical et al. 1973; Sultana et al. 2009; Esserti et al. 2017) and anti-viral activities (Klarzynsci et al. 2003; Jiménez et al. 2011). This was probably due to increasing interests in new sources of ecologically, safe anti-microbial compounds. Although many of these studies targeted human pathogens, some plant pathogens, were also equally susceptible to certain seaweed extracts. Nevertheless, marine algal extracts demonstrated activity against a wide range of potential plant pathogenic bacteria, fungi, and viruses.

Bacteria: A wide range of extracts from red, brown, and green algae have been shown to suppress activity of Gram positive and Gram negative bacterial pathogens *in vitro*. It is known that whole marine algae produce their own defensive compounds enabling them to reduce the effects of colonization of their thalli by epiphytes including marine bacteria, thereby helping to protect the host. The same, or similar, compounds in hydrolysed seaweed extracts may subsequently protect plants from microbial attack (Weinberger 2007; Sangha et al. 2010; Jiao et al. 2011). Amongst the various seaweeds, extracts of the browns appeared to have the highest levels of anti-bacterial activity (Subramanian et al. 2011). The phenolic and halogenated compounds synthesized by members of the Phaeophyta have been demonstrated to have properties that may have potential applications as anti-fouling agents (see Armstrong et al. 2000). Wang et al. (2009) studied the bacteriostatic and bactericidal effects of phlorotannins, as isolated from *Ascophyllum nodosum* on *Escherichia coli* strains and found a strong inhibition at 50 or 100 µg/ml; their anti-bacterial efficacy was greater than tannins purified from the extracts of Quebracho (*Schinopsis lorentzii*) and sumach (*Rhus semialata*) plants, known for their anti-bacterial activity (Costabile et al. 2011). Various aqueous and organic extracts of *A. nodosum* inhibited growth of *Pseudomonas syringae* pv. *tabaci* (*Pst*) DC3000 (Subramanian et al. 2011). However, no anti-microbial activity was observed against this pathogen when aqueous extracts of brown seaweeds *Cystoseira myriophylloides*, *Laminaria digitata*, and *Fucus spiralis* were tested *in vitro*, although the disease was suppressed *in planta* (Esserti et al. 2017). Seaweed extract treatment could also be useful to reduce the virulence factors of the bacterium, which in itself may potentiate disease suppression by the seaweed extracts (Prithiviraj et al. 2005).

Organic extracts (e.g., 80% ethanol, methanol, and acetone) of several green algae: that is, *Ulva fasciata*, *Ulva* (*Enteromorpha*) *intestinalis*, and *Chaetomorpha* sp. inhibited the growth of several Gram negative bacteria, that is, *Pseudomonas aeruginosa* and *Klebsiella pneumoniae* and Gram positive bacteria such as *Staphylococcus aureus* (Seenivasan et al. 2010). Additionally, extracts of the red algae *Hypnea valentiae* and *Rosenvingea intricata* and the green *U. intestinalis* were applied against the bacterium *Xanthomonas oryzae*, a causal organism of leaf blight disease on rice (Manimala and Rangasamy 1993). The *U. intestinalis* extract strongly inhibited the bacterium, followed by *R. intricata* and *H. valentiae*. Among the six different extracts of *U. intestinalis*, the diethyl ether extract demonstrated the greatest inhibition, followed by those prepared with ethanol and methanol. Interestingly, the extracts obtained with water, acetone, and chloroform were not effective. In another *in vitro* experiment, Arunkumar and Rengasamy (2000) rated the anti-bacterial potential of several seaweeds extracts against *Xanthomonas oryzae* pv. *oryzae*. Three seaweeds, viz. *Gracilaria edulis*, *Sargassum wightii*, and *U. flexuosa* showed the highest anti-bacterial activity from the eleven species tested. In a recent investigation, various extraction procedures were tested to determine the anti-microbial activity of extracts from the red alga *Kappaphycus alvarezii* against *Xanthomonas oryzae* pv. *oryzae* (Venkatesh et al. 2011). The results revealed a considerable inhibition of the pathogen growth *in vitro*.

Paulert et al. (2007) investigated cell-wall polysaccharides and crude extracts from the green alga *Ulva fasciata* against *Xanthomonas campestris, Erwinia carotovora*, and the fungus *Colletotrichum lindemuthianum*. Both methanol soluble and -insoluble (i.e., the remaining fraction) extracts were active against *X. campestris* and *E. carotovora*. Using an agar diffusion assay and broth dilution method, the minimal inhibitory concentration (MIC) was identified as being greatest against *E. carotovora* (e.g., the MIC was 1 mg/ml). Interestingly, no *in vitro* activity of the ulvans (a water soluble polysaccharide in green seaweeds), was observed against the test organisms. Tobacco plants injected with laminarin, purified from the brown alga *Laminaria digitata*, when inoculated with *Erwinia carotovora* subsp. *Carotovora*, five days after treatment, strongly reduced the infection, as compared to the control (Klarzynski et al. 2000). Yi et al. (2001) used extracts from 23 species of seaweeds belonging to the divisions Chlorophyta, Phaeophyta, and Rhodophyta and demonstrated that ethanolic extracts of selected rhodophytes (i.e., *Laurencia okamurai, Dasya scoparia, Grateloupia filicina*, and *Plocamium telfairiae*) showed a wide spectrum of anti-bacterial activity; *Pseudomonas solancearum* was more sensitive to the algal extracts, as compared to the other microbes tested.

Fungi: One of the earliest reports of the effects of seaweed extracts against fungal pathogens was on the anti-fungal substances extracted from the brown alga, *Dictyopteris zonarioides* (Fenical et al. 1973). Barreto et al. (1997) investigated the *in vitro* activity of several seaweeds against two phytopathogenic fungi, that is, *Verticillium* sp. and *Rhizoctonia solani*. Extracts of six seaweeds, including the greens *Caulerpa filiformis* and *Ulva rigida*, the brown *Zonaria tornefortii* and the reds *Hypnea spicifera, Gelidium abottiorium* and *Osmundaria serrata*, inhibited fungal growth by more than 50%. In contrast, the extracts of the red alga *Spyridia cupressans* and *Beckerella pinnatifida* demonstrated minimal anti-fungal activity, under the conditions tested. Anti-fungal activity of extracts of the brown seaweed *Cystoseira tamariscifolia* was reported against the plant pathogens *Botrytis cinerea, Fusarium oxysporum*, and *Verticillium albo-atrum* (Abourriche et al. 1999) and also against the food spoiling pathogen *Aspergillus* spp. (Bhosale et al. 1999). Martinez-Lozano et al. (2000) examined the anti-fungal properties of an extract of *Sargassum filipendula* on *Aspergillus niger, A. flavus, A. parasiticus, Penicillium* sp., *Fusarium oxysporum, Candida albicans*, and *C. rugosa*. An extract produced from the phaeophyte *S. filipendula* inhibited the growth of all fungi tested, even at different concentrations. Aqueous and ethanolic extracts from *Gracilaria chilensis* reduced the growth of *Phytophthora cinnamonni* under *in vitro* conditions (Jiménez et al. 2011).

Modulations of anti-microbial activity by the various seaweed extracts were dependent on the type of seaweed, the extraction procedure, the solvents used for preparation of bioactive fractions and also the target species tested. Chloroform and methanolic fractions of an ethanol extract of the phaeophyte *Spatoglossum asperum* demonstrated antifungal activity against the plant pathogen, *Macrophomina phaseolina*, whereas an n-hexane fraction suppressed *Rhizoctonia solani* and *Fusarium solani* (Ara et al. 2005). Using different solvents, Yi et al. (2001) identified differential activity from of the extracts of 23 species of marine algae belonging to the divisions Chlorophyta, Phaeophyta, and Rhodophyta against selected fungi. In this case, the ethanol extract was highly effective against the fungus tested, for example, *Penicilium citrinum*, a common filamentous fungus present in soil, cereals, spices, and other environments, was reported to be the most sensitive to the extracts. Manilal et al. (2009) compared the crude extracts of fresh and dried seaweeds prepared using different polar and non-polar solvents against pathogenic fungi and found that methanol was the best solvent, in those experiments and conditions, for extraction of anti-microbial metabolites from the dried seaweed samples. However, it is not surprising that contrasting results were presented by Khanzada et al. (2007) who investigated the anti-fungal activity of various fractions of *Solieria robusta* (Rhodophyta) against five fruit-spoiling fungi; that is, *Aspergillus flavus, A. niger, A. ochraceus K, Penicillium funiculosum*, and *Phytophthora infestans*, all of which had been isolated from fruits. In this case, the aqueous fraction was demonstrated to have the largest inhibition ratio, this was followed by methanol, ethyl acetate, chloroform, and ethanolic solvents. Nevertheless, bioactivity of the extracts also depended on the target species; some microbes are more resistant, even though seaweed extracts have been shown to be effective against resistant strains of pathogens (Shanmughapriya et al. 2008).

Jayaraj et al. (2011) provided economically important evidence on the positive applications of an *Ascophyllum* extract against fungal diseases of greenhouse cucumbers. Abkhoo and Sabbagh (2016) published evidence for the effects of a similar product (an *Ascophyllum* extract) against damping-off, also in cucumbers. Most recently, Ramikissoon et al. (2017) produced extracts of three Caribbean seaweeds (i.e., green—*Ulva lactuca*, brown—*Sargassum filipendula*, and red—*Gracilaria serrulata*) to suppress pathogenic infections (i.e., *Alternaria solani* and *Xanthomonas campestris* pv *vesicatoria*) in tomatoes. A discussion on salicylic and jasmonic acid pathways is included in their modes of action discussion.

Paulert et al. (2009) tested a crude seaweed extract and the polysaccharide ulvan (at rates of 0.1; 1; and 10 mg ml⁻¹), against conidial germination and mycelial growth of *Colletotrichum lindemuthianum*. The authors reported inhibition of mycelial growth *in vitro* with a soluble methanolic extract. However, mycelial growth and conidial germination of the fungus was increased *in vitro* with ulvan, whereas greenhouse plants sprayed with 10 mg ml⁻¹ ulvan reduced the anthracnose severity by 38%. In addition, Cluzet et al. (2004) reported that polysaccharides from *Ulva armoricana* had no effect, *in vitro*, on *Colletotrichum trifolii* development, although it protected alfalfa (*Medicago truncatula*) against the fungus. Borsato et al. (2010) showed that the polysaccharide ulvan did not inhibit germination of the fungus on leaf discs, nor alter the activity of peroxidases, but still protected the plants, suggesting induced plant resistance. The extracts of *S. zonale, L. dendroidea, P. canaliculata, S. muticum, A. nodosum,* and *F. spiralis* significantly inhibited *Colletotrichum lagenarium,* whereas the same extracts did not inhibit *Aspergillus flavus* growth *in vitro*. Similar results were reported by Sangha et al. (2010) who also demonstrated that various red algal polysaccharides had no direct effects on the fungus *Sclerotinia sclerotiorum, in vitro*, but it was effective through induced plant resistance.

It is interesting to note that many of the seaweed extracts tested have also shown variations in bioactivity, simply due to seasonal effects (Arunkumar and Rangasamy 2000). Moreau et al. (1988) identified anti-fungal substances from some members of the Dictyotales (brown algae) collected around the French Mediterranean coast and reported that seasonal variations were found in the bioactivity of the extracts. Manilal et al. (2009) investigated *in vitro* anti-microbial activity against four species of plant pathogenic fungi using extracts of 15 seaweeds, belonging to 13 families and six orders of the Rhodophyta, sampled for a year between April 2007 and March 2008, along the southwest coast of India. The crude extracts of both fresh and dried samples, prepared from different polar and non-polar solvents, were used. Four species of red algae (e.g., *Asparagopsis taxiformis, Laurencia ceylanica, L. brandenii,* and *Hypnea valentiae*) were found to be highly active, although a seasonal variation was observed for antimicrobial activity which was most prominent in the green coenocyte *A. taxiformis* between December and January. Jiménez et al. (2011) demonstrated potential anti-phytopathogenic activities of aqueous and ethanolic extracts obtained from nine Chilean marine macroalgae, collected at different seasons, in *in vitro* conditions, and showed differential suppression of microbes with the extracts.

Virii: Various seaweed extracts have been shown to have broad spectrum viral suppression (Shi et al. 2017). In particular, algal oligosaccharides have been shown to stimulate plant defence responses against tobacco mosaic virus (TMV) in whole plants, or suspension cell cultures (Klarzynski et al. 2003; Laporte et al. 2007; Fu et al. 2011). The oligosaccharides, especially laminarin or chemically sulphated laminarin (PS3), can prime or activate defence mechanisms, increase the expression of genes, promote oxidative bursts and other defense mechanisms in cell suspensions, indicating that plant defense activity may be induced against viral pathogens (also discussed in a later section). For example, tobacco cell suspensions, treated with laminarin or PS3, increased molecular and biochemical defense response that was directly related to the induction of resistance in treated plants, when challenged with pathogens (Menard et al. 2005). Reunov et al. (2011) also reported that fucoidan (extracted from *Fucus evanescens*) treatment, to detached *Datura stramonium* leaves, resulted in reduced accumulation of potato virus X (PVX) in the mesophyll cells, as compared to the control.

Indirect bioactivity of seaweed extracts against microbes

In addition to *in vitro* activities against plant pathogens, seaweed extracts have also been shown to enhance plant resistance to various pests and diseases (Allen et al. 2001; Sangha et al. 2010; Jiménez et al. 2011). Such effects could be effected indirectly through altering plant-pathogen interactions, activating plant defence pathways, as well as promoting plant health by modulation of the rhizosphere microbial community, or by improving the soil environment by direct antagonism to the pathogen (see Mercier et al. 2001; Sultana et al. 2005; Arunkumar et al. 2010). Furthermore, the rich composition of various crude algal extracts may themselves promote the growth of favourable microbes, which could further out-compete or antagonize pathogenic organisms (Dixon and Walsh 2002; Sultana et al. 2005). It is most often hypothesized that plant responses are elicited via seaweed components such as: oligo- or polysaccharides, peptides, proteins, lipids, cell wall debris, carrageenans, laminarin, β 1–3 glucans or sulphated fucans, or combinations thereof (Kobayashi et al. 1993; Lizzy et al. 1998; Mercier et al. 2001; Subramanian et al. 2011). A selective literature review on seaweed extract-induced plant resistance is discussed in the following section:

Bacteria: The use of various seaweed extracts has been shown to suppress infections in plants caused by certain bacteria. Polysaccharides, such as green algal ulvans, when sprayed on tomato leaves (0.1–10 mg/ml), protected the plants against the bacterium *Pseudomonas syringae* (Jaulneau et al. 2010). The same treatment was also shown to impart protection against a mite (*Tetranichus urticae*) in strawberry and a parasitic nematode (*Meloidogyne incognita*) also in tomato; the forgoing observations indicated a broad spectrum activity of the seaweed extracts. Subramanian et al. (2011) also reported that infection with *Pseudomonas syringae* pv. *tomato* DC3000 on Arabidopsis plants was reduced with root treatment of commercial *A. nodosum* extract. Another study reported the suppression of bacterial spot of tomatoes caused by *Xanthomonas campestris*, using two different formulations of *A. nodosum* seaweed extract. Further field studies with these two formulations in Virginia, USA, in 2007, demonstrated suppression of bacterial spot in tomatoes, under low disease pressure conditions, although the results did not show a significant difference (R. Ross unpublished data). Caccamese et al. (1980) studied extracts from more than twenty different algae on the growth of the plant pathogenic bacterium *Xanthomonas malvacearum* and reported a considerable suppression of that plant pathogen.

A commercial, aqueous seaweed extract from *Sargassum wightii*, was evaluated against bacterial blight caused by *X. campestris* pv. *malvacearum* in cotton (Raghavendra et al. 2007). Cotton seeds were soaked with commercial extract before being sown and subsequently, the seedlings were sprayed at 10 day intervals, starting 10 days after sowing. These applications resulted in a reduction in the incidence of bacterial blight on plants (ranging from 66–74%), that was observed between 40–80 days after sowing. Seeds of soybean: cv. SibNIIK-315, treated with aqueous solutions of laminarin (from *Laminaria cichorioides*; most recent name - *Saccharina cichorioides*), fucoidan (from *Fucus evanescens*), and polymannuronic acid to soybeans resulted in a reduced incidence of both bacterial and fungal diseases on the seedlings, buds, and flowers in the field (Zaostrovnykh et al. 2009). The incidence of seedling bacteriosis (caused by *Pseudomonas savastanoi* pv. *glycinea*) on seedlings treated with oligosaccharides from *Laminaria* was the lowest (i.e., 10.2%) as compared to 25.9% in the control. Kumar et al. (2008) screened the anti-microbial activity of 12 different seaweeds, including the greens: *Chaetomorpha antennina*, *Halimeda tuna*, *Ulva lactuca*, the reds: *Laurencia obtusa*, *Gracilaria corticata*, *G. verrucosa*, *Grateloupia lithophila* and the browns: *Padina boergesenii*, *Sargassum wightii*, and *Turbinaria conoides*, against the phytopathogenic bacterium *Pseudomonas syringae* (i.e., causative agent for leaf spot), in the medicinal plant *Gymnema sylvestre*. The results showed that methanolic extracts, followed by ethyl acetate fractions, of *Sargassun wightii* were more effective than the other organic solvent extracts when tested in this study.

Fungi: The suppression of plant diseases, caused by fungal pathogens, by various seaweed extracts has been observed on several ornamental, agricultural and horticultural plants, both in the field and also greenhouse studies. Foliar applications of commercial *A. nodosum* extract on Capsicum (peppers) decreased infection caused by *Phytophthora capsici* (Lizzi et al. 1998). The same study also reported that the incidence of downy mildew (*Plasmopara viticola*) on grapes was decreased with seaweed extract application. In greenhouse experiments, a biostimulant formulation, containing seaweed extract (of *A. nodosum*), showed strong suppressive activity against tomato late blight caused by *P. infestans*, although, interestingly it was found that the *A. nodosum* extract alone was not effective (Portillo et al. 2007), indicating the roles of synergism and integrated disease management. Interestingly, this biostimulant was also fungistatic *in vitro* and suppressed mycelial growth. A commercial product (i.e., BioFeed Basis) containing a complex mixture of plant proteins and extracts of seaweeds, especially *A. nodosum* and *Fucus* spp., when used to fertilize tomatoes, was effective in reducing late blight

(*P. infestans*) severity, as compared to plants given chemical fertilizer alone (Sharma et al. 2012). Al-Mughrabi (2007) reported that seaweed extract, in combination with compost tea, reduced the incidence and severity of late blight by 36% in comparison to the control. Extracts of *A. nodosum* when used alone, or in combination, with other amendments showed some limited potential to reduce the total incidence of grey mold (*Botrytis cinerea*), but not leather rot (*Phytophthora cactorum*), nor anthracnose caused by *Colletotrichum acutatum* (Washington et al. 1999). Jeandet et al. (1996) reported that a commercial product, containing a mixture of a seaweed extract plus AlCl2 hexahydrate, increased the efficacy of iprodione for the control of grey mold on grapes. Abreu et al. (2008) studied extracts from 17 marine seaweeds against bean anthracnose and observed that an extract of *Bryothamnion seaforthii* reduced the disease severity in beans by up to 35%. In this study, a residual effect of *Ulva fasciata* extract was also reported, as the application of seaweed extracts was able to reduce the anthracnose severity by 22%, 12 d post-inoculation. Interestingly, extracts of the water plant (duckweed) *Lemna* sp. and *U. fasciata* caused the reduction in disease severity at 7 d, post-inoculation by 55 and 44%, respectively.

Goicoechea et al. (2004) used a combination of an *A. nodosum* extract, along with other organic amendments of natural origin, to investigate their ability to suppress diseases in greenhouse peppers; seedlings, cv. "Piquillo", were grown for three months in an organic medium ('COA H'), that contained salicylic acid (SA), soluble ammonium salts, and *A. nodosum* extract, and were then inoculated with *Verticillium dahliae* Kleb. The pre-treated seedlings showed delayed appearance of disease symptoms and the disease index, on day 33, was almost 40% lower than that measured in the control plants. Jayaraj et al. (2008) used a commercial *A. nodosum* extract to suppress *Alternaria radicana* and *Botrytis cinerea* on carrots in greenhouse trials. Pre-treatment of the plants with a commercial *A. nodosum* extract suppressed foliar symptoms caused by *A. radicana* on carrots at 10 and 25 d post-inoculation. The authors also observed a similar trend with *B. cinerea* infection after using the *A. nodosum* extract. In another trial, the same *A. nodosum* extract treatment was used on cucumbers to compare soil drench against foliar applications. It was reported that foliar, or drench and a combination of foliar and drench applications (at 0.5% or 1% concentrations) of the *A. nodosum* extract, suppressed infection with three different, potential cucumber pathogens (e.g., *B. cinerea*, *F. oxysporum*, and *Didymella applanata*). In growth room trials, treatments with seaweed extract (*A. nodosum*) significantly reduced *Verticillium* wilt incidence and severity (i.e., 70–80% reduction) in potatoes. The same treatment also reduced the severity of *Verticillium* wilt under field conditions (Uppal et al. 2008). Meszka and Bielenin (2009) tested a commercial product ("Remedier WP" containing mycelium and spores of two antagonistic fungi, e.g., *Trichoderma harzianum* and *T. viride*), and plant extracts (from pine and spruce needles, horsetail, herbs and a seaweed, probably *A. nodosum*) under field conditions, to test the protection of strawberry cultivars against *V. dahliae*. Amongst the tested extracts, seaweed and herbs were the most effective at reducing the wilt disease (i.e., 43–76% reduction, depending on the severity of the disease).

Application of seaweed extract reduced the incidence of gray mold (*B. cinerea*) on strawberries, powdery mildew (*Erysiphe polygoni*) on turnips, and damping-off of tomato seedlings (Kulik 1995). Spray treatment of ethanolic extracts of the phaeophyte, *Lessonia trabeculata*, on tomato leaves, reduced the severity of the disease caused by *B. cinerea* (Jiminez et al. 2011). This effect was evident in a dose- and seasonal dependent manner. Foliar applications of *K. alvarezii* reduced bacterial wilt and leaf curl disease in tomato (Zodape et al. 2011). Various seaweed extracts have also been reported to be effective in suppressing plant pathogens in some graminaceous crops, for example, the incidence of disease in rice caused by *Alternaria padwickii* and *Bipolaris oryzae* was significantly reduced after seed treatment with a commercial seaweed product (Dravya), prepared from *Sargassum wightii* (Sathyanarayana et al. 2006). Reduction in diseases of greenhouse grown plants treated with seaweed based products has also been observed. Flora and Rani (2012) reported the use of aqueous concentrates of *Padina pavonia*, *Acanthophora spicifera* and *Ulva lactuca* as foliar sprays to suppress rice blast caused by *Pyricularia oryzae* and observed a reduction in disease severity with all of the seaweed extracts tested, although a higher activity was evident with the *A. spicifera* extract. A commercial *A. nodosum* extract, showed anti-fungal activity on the bentgrass (*Agrostis stolonifera*) against dollar spot disease, caused by the fungus, *Sclerotinia homoeocarpa* (Zhang et al. 2003). An extract from the same brown seaweed was also shown to suppress infection of the model plant Arabidopsis by *S. sclerotiorum* (Subramanian et al. 2011) which implicated the potential application of seaweed extract to manage other pathogens in other commercial crops, particularly the mustard family.

A few studies have shown that direct application of selected seaweed extracts to the soil has resulted in indirect protection of plants against pathogens, possibly through an improved soil environment. Soil applications of liquid seaweed extract to cabbage (*Brassica oleracea* var. *capitata*) stimulated microbes that were antagonistic to *Pythium ultimum*, resulting in the reduced incidence of damping-off disease in seedlings (Dixon and Walsh 2002). Soil amendments using dry powder from brown, green, and red seaweeds was shown to have broad spectrum activity to control root infecting fungal pathogens of different crops (see Sultana et al. 2005, 2008, 2009, 2011a for details). Extracts of the seaweeds *Stokeyia indica* (*Cystoseira indica*), *Padina pavonia* (brown), and *Solieria robusta* (red), at concentrations of 1% w/w, effectively reduced infection by *Macrophomina phaseolina*, *Rhizoctonia solani* and *Fusarium solani* on okra roots. In contrast, an extract of *Codium iyengarii* (Chlorophyta) was effective at the lower dose of 0.5% w/w against *F. solani* (Sultana et al. 2005). The application of *Solieria indica* and *S. robusta,* used as soil amendments, was able to suppress root-infecting fungi, that is, *M. phaseolina, R. solani*, and *F. solani* in chili roots, thus suggesting a broad applicability of these treatments (Sultana et al. 2008). Similar results were observed in tomato using the powdered brown seaweeds *Spatoglossum asperum* and *Sargassum swartzii* as soil amendments on these root-rotting fungi (Sultana et al. 2009). Sultana et al. (2011a) compared the efficacy of the extract of the red seaweed, *S. robusta* in combination with chemical fertilizers and pesticides on soybeans. Soil amendments using *S. robusta* was more effective at suppressing the root-rot fungus *F. solani,* than the commercial fungicide Topsin-M. Although the seaweed was less effective against *M. phaseolina* and *R. solani* than Topsin-M, there seemed to be promising aspects of using seaweeds in control management practices against soil pathogens and also a role in enhancement of the activity of beneficial organisms (i.e., a "prebiotic" effect on beneficial soil microorgansims). Rekanović et al. (2010) evaluated the activity of a commercial seaweed concentrate from the kelp *Ecklonia maxima* against wilt in peppers caused by *Verticillium dahliae* and compared extract treatment with two conventional fungicides, that is, thiophanate-methyl and carbendazim. Although pre-treatment of the pepper plants with carbendazim was the most efficient fungicide (69.64%), the seaweed concentrates also proved to be effective when applied at a 1.0% concentration (41.96%). In contrast, the activity of thiophanate-methyl against *Verticillium*-induced wilt was 60.71%. The results showed that although the seaweed extract was less effective than thiophanate-methyl and carbendazim, it showed activity against the pathogen, as compared to the control. These observations are important when non-synthetic solutions to plant disease are sought.

Marine algae in general, are a rich source of bioactive compounds, such as cell wall polysaccharides, ulvans, laminarins, etc., which are present in specific groups (e.g., green, red, and brown) of seaweeds. Polysaccharides from seaweeds are extensively used to suppress plant pathogens. Ulvan, carrageenan, laminorin, and fucans are some of the notable polysaccharide compounds with reported anti-microbial activities, be it as direct suppressants or by induced plant responses. *Ascochyta phaseolorum* and *Peronosopora manshurica* infections were reduced with pre-application of soybean seeds with extracts of *Laminaria* (or *Saccharina*), or associated fucoidans (from *Laminaria cichorioides*) and polymannuronic acid and fucoidan from *Fucus evanescens* (Zaostrovnykh et al. 2009). Pre-treatment of plants with an oligosaccharide extract from the *L. cichorioides,* or the fucoidan fraction derived from *L. cichorioides* and polymannuronic acid and fucoidan from *F. evanescens,* showed a reduced (5%) infection by *A. phaseolorum* and *P. manshurica,* as compared to 15% in the controls. Trouvelot et al. (2008) showed that sulphated laminarin (PS3) could elicit defense responses in grapevine (*Vitis vinifera*) against downy mildew (*Plasmopara viticola*) through induction of H_2O_2 production at the infection sites, up-regulation of defence-related genes, callus and phenol depositions, and the hyper-sensitive response. PS3 was also shown to induce defence responses in tobacco and *Arabidopsis* against tobacco mosaic virus. Aziz et al. (2003) studied the ability of laminarin to elicit defense responses in grapevines against *B. cinerea* and *P. viticola*; pre-treatment reduced infections by approximately 55 and 75%, respectively.

Extensive research has been published on the use of ulvan, a polysaccharide extracted from green macroalgae, for the induction of plant defences to pathogens. Paulert et al. (2010) reported on the priming activity of ulvan, against the fungal pathogen *Blumeria graminis*. Ulvan pre-treatment of whole plants significantly reduced the severity of *B. graminis* infection, by 45% in wheat and 80% in barley.

Montealegre et al. (2010) used ulvan to control the grey rot of apple fruits and found that post-harvest applications reduced *B. cinerea* infections on apple by 56%. Yet another study, showed the potential of ulvan to protect apples from pathogenic infections (Araújo et al. 2008). Two consecutive treatments of extracted, sulphated polysaccharide from *U. armoricana* protected alfalfa (*Medicago truncatula*) against anthracnose (*Colletotrichum trifolii*); the mechanisms were related to induced expression of several pathogenesis-related genes such as PR-10 (Cluzet et al. 2004). Similarly, Paulert et al. (2007), using 2 mg ml^{-1} of ulvan, demonstrated induced plant resistance to these pathogens. Paulert et al. (2011) evaluated the potential of crude extracts and fractionated ulvan from *U. fasciata* to control bean (*Phaseolus vulgaris*) anthracnose caused by *C. lindemuthianum,* under greenhouse conditions. The soluble, methanolic extract did not reduce disease severity, however ulvan, sprayed at 10 mg ml^{-1}, reduced disease severity by 38%. These results indicated that ulvan itself can be a strong elicitor of plant resistance to *Colletotrichum* sp. attacks. Borsato et al. (2010) showed that an ulvan spray treatment (10 mg mL^{-1}) to bean plants, 3–6 days before infection, induced resistance to rust caused by *Uromyces appendiculatus.* The diameter of the pustules on the bean cultivars was reduced, and the activity of glucanases within the treated plants was increased in moderately susceptible cultivars two days, post-inoculation. However, treatments were not effective against fungal germination on leaf discs and the activity of plant peroxidases was unaffected.

Jaulneau et al. (2011) studied the effect of an aqueous extract of *Ulva armoricana*, rich in ulvan, against three powdery mildew pathogens, for example, *Erysiphe polygoni, E. necator*, and *Sphareotheca fuliginea* on the common bean, grapevine, and cucumber; they reported a noticeable difference in efficacy based on the concentrations used. Weekly spraying at 3 g/l dry matter provided 50% protection, whereas the severity of the symptoms was reduced by up to 90% at double the concentration (i.e., 6 g/l dry matter). A reporter gene, tagged to a defence-gene promoter, in a transgenic tobacco line, was elicited by treatment with extracts of *U. armoricana* that probably contributed to induced plant defence mechanisms against the powdery mildew pathogen. *U. armoricana* is a reproducible source of active compounds which can be used to efficiently protect crop plants against powdery mildew diseases. Similar results were observed with ethanolic fractions of *Ulva* spp. which activated plant defence enzymes in *Arabidopsis* (Jaulneau et al. 2010). Paulert et al. (2010) showed that ulvan could prime the chitin- and chitosan-elicited oxidative burst in wheat and rice cells. The pre-treatment of wheat cells with the ulvan increased the chitin-elicited oxidative burst by about five- to six-fold, as compared to chitosan (two-fold). Similarly, in rice cells, the elicitation of H_2O_2 production by chitin or chitosan was increased by 150 and 80 times, respectively after ulvan pre-treatment. Furthermore, ulvan-treated plants showed reduced symptoms of *Blumeria graminis* infection, by 45% in wheat and by 80% in barley.

Freitas et al. (2011) evaluated the effectiveness of ulvan, to induce resistance of *Phaseolus vulgaris* to *Colletotrichum lindemuthianum*, in combination with kaolinite, amorphous silica, or attapulgite clay. The formulations were applied twice to the bean plants (cv. Uirapuru), one at six days and again three days before the foliage was inoculated with *C. lindemuthianum*. The application of ulvan (control) or its formulations with amorphous silica or kaolinite reduced anthracnose severity by 45%. The efficiency of ulvan in controlling anthracnose was maintained throughout 12 months of storage when it was formulated with amorphous silica or kaolinite, but not with attapulgite. Araújo et al. (2008) revealed that resistance to 'Gala' leaf spot in apples was induced with ulvan.

Virii and viroids: Caccamese et al. (1981) reported on the effectiveness of lipid extracts of more than twenty algae against tobacco mosaic virus (TMV). Menard et al. (2005) studied the effect of PS3 on tobacco, challenged with TMV. Pre-treatment of plants with PS3 decreased both the lesion number and the lesion size, eight days after application, whereas the laminarin reduced only the lesion number. Interestingly this study did not show induction of systemic acquired resistance to TMV, although previous studies reported that PS3 could activate the salicylic acid (SA) signaling pathway in infiltrated tobacco and *Arabidopsis thaliana* leaves (Menard et al. 2004). Laminarin used at 200–500 ug/mL was able to inhibit TMV infection on tobacco by 66.48% and 66.66%, respectively (Fu et al. 2011). Lapshina et al. (2007) showed that fucoidan from the brown alga *Fucus evanescens* inhibited infection of tobacco plants with TMV by inhibiting the formation of intra-cellular, tubular inclusions, presumably formed from the granular ones on the last stages of the infection process. Reunov et al. (2011) reported that pre-treatment of *Datura stramonium* plants with fucoidan, extracted from *Fucus evanescens,* suppressed potato virus X (PVX) in the leaves. Fucans purified from the fucoid alga *Pelvetia canaliculata* activated both local and systemic resistance in tobacco against TMV resulting in suppression of multiplication of the virus (Klarzynski et al. 2003). Similar results were reported with the use of oligosaccharides obtained from Chilean marine macroalgae (Laporte et al. 2007). The application of oligosaccharides, for example, guluronic acid (Poly-Gu) extracted from the blades of *Lessonia trabeculata*, mannuronic acid (Poly-Ma) from *L. vadosa* and sulphated galactan (Poly-Ga) from the blades of *Schizymenia binderi* to tobacco plants, stimulated plant defences against tobacco mosaic virus, seven and 15 days after spray treatment. Vera et al. (2011a) also studied the anti-viral effect of the Poly-Ga in tobacco leaves infected with TMV. The number of necrotic lesions was lower in the treatment than the control, with increasing number of applications and concentrations of Poly-Ga. The levels of TMV-capsid protein (CP) transcripts decreased in the distal leaves, indicating that Poly-Ga induced systemic protection against TMV. It was shown that increased activity of the defence enzymes correlated with decreased numbers of necrotic lesions and TMV-CP transcript levels. Pre-treatment of tomato plants with λ-carrageenan significantly reduced the symptoms of the Tomato Chlorotic Dwarf Viroid (TCDVd) eight weeks after inoculation (Sangha et al. 2015). Taken together, the above studies suggested that certain seaweed extracts have considerable potential to protect plants against virii and viroids.

Mechanisms of seaweed extract-induced plant defences to pathogens

The response of plants to pathogens involves various pathways that can be regulated by molecules such as salicylic acid (SA), jasmonic acid (JA), and ethylene (ETH) (Grant and Lamb 2006). Plant disease responses to various seaweed extracts can be due to the elicitor-like abilities of peptides, proteins, lipids, cell wall debris, oligo- or polysaccharides, carrageenans, ulvans, laminarin, β 1–3 glucans or sulphated fucans, or some combination of these, within the extract (Kobayashi et al. 1993; Mercier et al. 2001; Subramanian et al. 2011). Different mechanisms have been reported behind disease resistance associated with elicitation with various seaweed extracts. Using the model plant Arabidopsis, it has been shown that the jasmonic acid pathway plays a critical role in plant defence responses elicited by a commercial extract of *Ascophyllum nodosum* (Subramanian et al. 2011). Inoculation of *Pst* DC3000 on *A. nodosum* extract-treated plants (by root irrigation), showed a JA-dependent, induced systemic resistance. This was evident when the genes associated in JA-signaling, such as allene oxide synthase (AOS) and 'plant defensin 1.2' (PDF1.2) were strongly expressed in the *A. nodosum*-treated plants. Pathogens which are inhibited by the JA-pathway are most often necrotrophs. Many devastating diseases in agricultural crops are caused by necrotrophic fungal pathogens such as *Alternaria*, *Fusarium*, *Botrytis*, *Verticillium*, *Phytophthora* and *Pythium* and bacterial pathogens such as *Xanthomonas*, *Pseudomonas*, and *Erwinia*. Thus, seaweed extracts such as *A. nodosum* which can elicit JA-response in plants may offer protection to many of these phytopathogens.

The oligo- and polysaccharides of seaweed extracts are reported to affect a broad spectrum and multiple plant defence responses to pathogens (Klarzynsci et al. 2003; Cluzet et al. 2004; Stadnik and Freitas 2014; de Freitas and Stadnik 2015; Van Oosten et al. 2017). Phytohormones have been detected in some seaweed extracts which might modulate the bio-stimulatory and stress tolerance plant responses (Górka and Wieczorek 2017). The seaweeds also activate a variety of phytohormone responses in the treated plants to the different stressors. Ulvans from green algae have been shown to activate plant defences through the JA-signaling pathway, but not via a SA-dependent pathway, as shown in *Nicotianae tabacum* (Solanaceae), *Arabidopsis thaliana* (Brassicaceae), and *Medicago truncatula* (Cluzet et al. 2004; Jaulneau et al. 2010). The response of ulvan-induced gene expression in *Medicago truncatula* was similar to that observed with methyl jasmonate (Me-JA)-treatment; there was increased proteinase inhibitory activity, which is a marker for the Me-JA response. Although, exogenous application of λ-carrageenan at low concentrations, increased salicylic acid (SA) levels in the plant, the JA and ethylene levels were also increased (Mercier et al. 2001). In addition, ulvan-induced expression of JA-dependent genes, such as PDF1.2 ('defensin') in *Arabidopsis thaliana* and the lipoxygenase ('NtLOX1') promoter in *Nicotiana tabacum*, fucan, and carrageenans, also induced lipoxygenase gene expression in tobacco (Mercier et al. 2001; Klarzynski et al. 2003), suggesting that activation of the JA-pathway could be a general feature of sulphated oligosaccharides. Nevertheless, the role of other defence mechanisms is also entirely possible.

Treatment with laminarin and PS3 synergistically promoted an oxidative burst in tobacco cell suspensions and induced the SA-signaling pathway in infiltrated tobacco and *Arabidopsis thaliana* leaf tissues (Menard et al. 2005). This treatment was suggested to impart resistance in tobacco against TMV. In transgenic PR1–β-glucuronidase (GUS) tobacco plants, PS3 increased the GUS activity in treated tissues, as compared to the untreated (control) leaves. Interestingly, PS3 did not induce systemic acquired resistance (SAR) to TMV, as reported in other studies. There was, however, increased expression of genes encoding O-methyltransferases of the phenylpropanoid pathway in the tobacco plants. However, the SA-dependent, acidic PR1 gene, or the ethylene-dependent, basic PR5 gene, were not affected in tobacco plants. On the contrary, oligofucans prepared by enzymatic hydrolysis of fucan from the brown alga *Pelvetia canaliculata*, induced both local and systemic defence responses in tobacco cell suspensions and marked alkalinization of the extracellular medium, with associated release of hydrogen peroxide (Klarzynsci et al. 2003). Oligo-fucans are also reported to induce phenylalanine ammonia-lyase and lipoxygenase activity. Additionally, local accumulation of salicylic acid (SA) and the phytoalexin scopoletin, were observed, along with expression of several pathogenesis-related (PR) proteins. It was suggested that SA is required for the establishment of oligo-fucan-induced resistance in plants such as tobacco against TMV. Vera et al. (2011a) also showed that applications of sulphated galactan to tobacco and subsequent infection with TMV, resulted in induction of the defence enzymes phenylalanine ammonia lyase and lipoxygenase, with a decrease in the TMV-CP transcript level. Increased activity of defence-related enzymes with seaweed extract treatment is another mechanism that operates to enhance plant resistance to pathogens. Disease resistance reported after the use of a commercial aqueous extract from *Sargassum wightii,* against *Xanthomonas campestris* pv. *malvacearum* in cotton, was associated with higher levels of total phenols and peroxidase activity in the plant (Raghavendra et al. 2007). Goicoechea et al. (2004) further demonstrated that pepper grown in an organic medium (COA H, containing SA, soluble ammonium salts, and *A. nodosum* extract) resulted in an early accumulation of phenolics which were suggested to have contributed to disease resistance and/or tolerance, exhibited by the seedlings against *Verticillium dahliae*. However, it was not clear which component of the growing medium contributed the most towards disease suppression. Antioxidant properties of polyphenols contained in some seaweed extracts may have acted against certain target pathogens (Zhang et al. 2006). Phytoalexins produced in grapevines with seaweed extract treatment were associated with reduced severity of grey mold (Jeandet et al. 1996). Laminarin from *A. nodosum* was also shown to elicit a plant defence response by increasing anti-microbial phytoalexins (Patier et al. 1993).

Another study revealed that grapevine cells treated with laminarin resulted in an increased potential calcium influx, alkalinization of the extracellular medium, production of an oxidative burst, activation of mitogen-activated protein kinases (MAPKs), expression of defence-related genes, increased levels of enzymes such as chitinase and B-1,3-glucanase and phytoalexins (i.e., resveratrol and E-viniferin). This elicited defence response was attributed with protection of treated grapevines against *B. cinerea* and *Plasmopara viticola* (Aziz et al. 2003). An oxidative burst was also attributed to be associated in the protection of ulvan-primed wheat and rice cells that also correlated with a decrease of disease symptoms in the infected plants (Paulert et al. 2010).

Treatment of carrots and cucumber, with a commercial *A. nodosum* extract, to suppress different pathogens, indicated that enhanced activity of various defence-related enzymes and genes played various roles at the molecular and biochemical level (Jayaraj et al. 2008). The plant defence mechanisms induced by other macroalgal extracts, including polysaccharides, could also follow a similar response through induction of defence genes or enzymes against pathogens. The *A. nodosum* extract and kappa, oligo-carrageenan induced an increase in activity of the defence enzyme β-1,3-glucanase which is known to have anti-fungal properties (Vera et al. 2011b). The polysaccharide fucoidan extracted from the blades of *Lessonia vadosa* (Phaeophyta) showed significant activation of various defence enzymes such as phenylalanine-ammonia lyase, lipooxygenase, and glutathione-S-transferase in tobacco (Chandia and Matsuhiro 2008). Ethanol-soluble extracts from the red alga *Gracilaria chilensis* and *L. trabeculata,* contained active compound(s) having polar characteristics, which either acted directly on the mycelial growth of *P. cinnamonni,* or activated plant defence mechanisms to protect the plant tissues against

B. cinerea infection (Jiménez et al. 2011). Increased activity of superoxide dismutase (SOD), was observed in bentgrass (*Agrostis stolonifera*) given a combined application of *A. nodosum* extract and humic acid that resulted in reduced dollar spot (*S. homoeocarpa*) disease (Zhang et al. 2003). Peroxidase activity was enhanced in rice seeds treated with a commercial seaweed extract of *Sargassum wightii* that resulted in enhanced resistance to pathogens, *Alternaria padwickii* and *Bipolaris oryzae* (Sathyanarayana et al. 2006). The activity of defence-related enzymes, viz, peroxidase, phenylalanine lyase and polyphenol oxidase was also increased in potato, with a treatment of the commercial extract of *S. wightii* (Raghavendra et al. 2008). Using the Arabidopsis model, the role of reactive oxygen species derived from the respiratory burst oxidase homologue D (RBOHD) NADPH oxidase was implicated in resistance against *A. brassicicola* (de Freitas and Stadnik 2015). Higher peroxidase activity and phenylalanine ammonia lyase activity was also attributed to the resistance of rice blast (*Pyricularia oryzae*) when given foliar applications of aqueous concentrates of *Padina pavonia*, *Acanthophora spicifera* and *Ulva lactuca* (Flora and Rani 2012). Foliar and root applications of *A. nodosum* extract were reported to suppress *P. capsici* infections of peppers as a result of the increased activity of soluble peroxidases, phytoalexin and capsidiol, indicating an elicited defence response (Lizzy 1998).

It has been postulated that various seaweeds and their extracts can stimulate beneficial microbial antagonists, as revealed from the efficacy of various extracts against different plant root pathogens. It has also been suggested that components of the seaweed extracts may serve as food sources and promote the proliferation of favourable microbes which out-compete or antagonize pathogenic organisms (Dixon and Walsh 2002), in a manner similar to a gut health prebiotic in animals and humans. Direct antagonism of the pathogen (Sultana et al. 2005; Arunkumar et al. 2010) and components of some extracts, such as betaines, may also suppress plant disease (Blunden et al. 2010).

Seaweed extracts as biostimulants and plant health agents in integrated disease management

Biological control and induced resistance are part of an integrated pest management approach, which has emerged as a promising system for disease control which exerts low environmental impacts and reduces the need for synthetic chemicals. Certain seaweed extracts contain bioactive, natural products which are able to elicit molecular and biochemical defence responses in plants to protect against pathogens (Subramanian et al. 2011). As a result, bioactive seaweed products can be integrated with disease management strategies. This is further simplified with the availability of commercial products that have an elevated "eco-friendly" potential for successful application in the integrated management of crop diseases.

There is an increasing trend towards the use of milled seaweeds and their extracts as soil amendments to control soil-borne plant diseases. This may be due to improvements in the activity of beneficial microorganisms in the soil which are antagonistic to plant pathogens (Sultana et al. 2005; Sultana et al. 2008). For example, the use of *Solieria robusta* as a soil amendment to suppress the root-rotting fungi, *Fusarium solani*, *Macrophomina phaseolina*, and *Rhizoctonia solani* which infect soybeans (Sultana et al. 2011a). Integration of extracts of green (*Halimeda tuna*), brown (*Spatoglossum variabile*), and red (*Melanothamnus afaqhusainii*) seaweeds with fungicides reduced the infection of *M. phaseolina*, *R. solani*, and *F. solani* on sunflower (Sultana et al. 2011b). However, *S. variabile* alone or *H. tuna*, in combination with Topsin-M or carbofuran, completely inhibited *M. phaseolina* infection on sunflower roots. Kuwada and co-workers demonstrated that seaweed extracts helped to improve soil mycorrhizal activity which further contributed to improved plant growth (Kuwada et al. 1999, 2006). The application of methanolic extracts of the red seaweeds, *Gracilaria verrucosa*, *Gelidium amansii*, and *Eucheuma cottonii*, and a green microalga, *Chlorella pyrenoidosa* either *in vitro*, or to the soil stimulated the growth of the arbuscular mycorrhizal fungi, *Gigaspora margarita* and *Glomus caledonium* which form a symbiotic relationship with the roots of their host. Root colonization of papaya and passionfruit with these mycorrhizal fungi was markedly stimulated and the plant growth was improved.

The efficacy and compatibility of selected seaweed extracts have been tested in combination with other disease control methods. Tuber soaking and three foliar sprays of a commercial product of a *Sargassum wightii,* product (at 0.4%), in combination with mancozeb (0.3%), at 15 day intervals, reduced disease incidence by up to 80%, as compared to the control (Raghavendra et al. 2008). An extract obtained from the red alga *S. robusta* amended at 0.5 or 1% w/w, individually or in combination, with the plant growth-promoting bacterium *Pseudomonas aeruginosa* showed significant control of plant pathogenic fungi such as *M. phaseolina, R. solani,* and *F. solani* in pepper roots (Sultana et al. 2005). In fact, a symbiotic combination of *P. aeruginosa* and an extract of *S. robusta* showed better control of *M. phaseolina* infection than when they were used individually. A similar observation was made with extracts of the brown seaweed *Padina pavonia,* when used in combination with *P. aeruginosa* and against *R. solani.* Combined use of compatible strains of *P. aeruginosa* with whole seaweeds, and or their extracts, holds tremendous promise for symbiotic associations, of which we will see and hear more in future. A commercial seaweed product, with oligosaccharides extracted from *Ascophyllum nodosum* (Basak 2008), was used in combination with field (*Bacillus amyloliquefaciens* KPS46 and *Paenibacillus pabuli* SW01/4) and commercial biological control agents (*Trichoderma harzianum,* and *B. subtilis*) on disease epidemics of soybean (Thowthampitak and Prathuangwong 2007). The combined treatments significantly reduced diseases such as *Sclerotium* root-rot (*Sclerotinia sclerotiorum*), damping-off (*Rhizoctonia solani*), anthracnose (*Colletotrichum* spp.), bacterial pustule (*Xanthomonas axonopodis* pv. *glycines*), soybean mosaic virus SMV and soybean crinkle leaf virus (SCLV) severity by 34–90%. This was suggested as good management strategy to be adopted in sustainable agriculture systems.

Advances in nanotechnology have been emerging as potential eco-friendly methods of suppressing pathogens that could incorporate seaweed components with anti-pathogenic properties. This has been demonstrated with examples such as bio-nanoparticles prepared with a crude ethyl acetate extract of *Ulva fasciata* that showed strong inhibition of *Xanthomonas campestris* pv. *malvacearum* with a minimum inhibitory concentration of 40.00 ± 5.77 μg/mL (Rajesh et al. 2012). Such advances will help towards integrated management of plant diseases and minimal application of synthetic pesticides.

Conclusions

There is compelling evidence that disease resistance in plants, imparted by various seaweed extracts is associated with the priming and activation of plant defence pathways that are often associated with resistance to plant pathogens. Since many diseases of agricultural crops are caused by such pathogens as *Alternaria, Fusarium, Phytophthora, Pythium, Botrytis, Verticillium,* extracts of seaweeds, or their fractionated components, may offer a valuable tool to improve the health and productivity of commercial agriculture. There is a tremendous amount of work which remains to be completed on synergies of these components and also their interactions with beneficial soil-living and endophytic plant bacteria. Moreover, many marine algae, out of necessity, produce a large number of anti-bacterial and anti-fungal compounds, which have no apparent ecotoxicity and many of these seaweeds, though not all, can be sustainably harvested or grown in quantity using well-established techniques of mass culture. The acceptance of seaweed components as disease management tools might be accelerated using new technologies with an enhanced ability to control plant diseases. There is a need to carefully evaluate further algal products for their potential and potent roles in future biocontrol of plant pathogens.

Acknowledgements

We thank Dr R. Loureiro for constructive comments on the manuscript.

References

Abouraïcha, E.F., E.Z. Alaoui-Talibi, A. Daoulas-Ouafi, R.E. Boutachfaiti, E. Petit, A. Douira, B. Courtois, J.Courtois and C.E. Modafar. 2017. Glucuronan and oligoglucuronans isolated from green algae activate natural defence responses in apple fruit and reduce postharvest blue and gray mold decay. J. Appl. Phycol. 29: 471–480.

Abkhoo, J. and S.K. Sabbagh. 2016. Control of *Phytophthora melonis* damping-off, induction of defense responses, and gene expression of cucumber treated with commercial extract from *Ascophyllum nodosum.* J. Appl. Phycol. 28: 1333–1342.

Abourriche, A., M. Charrouf, M. Berrada, A. Bennamara, N. Chaib and C. Francisco. 1999. Antimicrobial activities and cytotoxicity of the brown alga *Cystoseira tamariscifolia*. Fitoterapia. 70: 611–614.

Abreu, G.F.D., V. Talamini and M.J. Stadnik. 2008. Bioprospecção de macroalgas marinhas e plantas aquáticas para o controle da antracnose do feijoeiro. Summa Phytopathol. 34: 78–82.

Allen, V.G., K.R. Pond, K.E. Saker, J.P. Fontenot, C.P. Bagley, R.L. Ivy, R.R. Evans, R.E. Schmidt, J.H. Fike, X. Zhang, J.Y. Ayad, C.P. Brown, M.F. Miller, J.L. Montgomery, J. Mahan, D.B. Wester and C. Melton. 2001. Tasco: influence of a brown seaweed on antioxidants in forages and livestock—a review. J. Anim. Sci. 79(E Suppl): E21–E31.

Al-Mughrabi, K.I. 2007. Suppression of *Phytophthora infestans* in potatoes by foliar application of food nutrients and compost tea. Aust. J. Basic Appl. Sci. 1: 785–792.

Ara, J., V. Sultana, R. Qasim, S. Ehteshamul-Haque and V.U. Ahmad. 2005. Biological activity of *Spatoglossum asperum*: a brown alga. Phytother. Res. 19: 618–623.

Araújo, L., M.J. Stadnik, L.C. Borsato and R. Valdebenito-Sanhueza. 2008. Fosfito de potássio e ulvana no controle da mancha foliar da gala em macieira. Trop. Plant Pathol. 33: 148–152.

Armstrong, E., K.G. Boyd and J.G. Burgess. 2000. Prevention of marine biofouling using natural compounds from marine organisms. Biotechnol. Annu. Rev. 6: 221–241.

Arunkumar, K. and R. Rengasamy. 2000. Antibacterial activities of seaweed extracts/fractions obtained through a TLC profile against the phytopathogenic bacterium *Xanthomonas oryzae* pv. *oryzae*. Bot. Mar. 43: 417–421.

Arunkumar, K., N. Selvapalam and R. Rengasamy. 2005. The antibacterial compound sulphoglycerolipid 1-0 palmitoyl-3-0(6'-sulpho-α-quinovopyranosyl)-glycerol from *Sargassum wightii* Greville (Phaeophyceae). Bot. Mar. 40: 441–445.

Arunkumar, K., S.R. Sivakumar and R. Rengasamy. 2010. Review on bioactive potential in seaweeds (Marine Macroalgae): A special emphasis on bioactivity of seaweeds against plant pathogens. Asian J. Plant Sci. 9: 227–240.

Aziz, A., B. Poinssot, X. Daire, M. Adrian, A. Bézier, B. Lambert, J.M. Joubert and A. Pugin. 2003. Laminarin elicits defense responses in grapevine and induces protection against *Botrytis cinerea* and *Plasmopara viticola*. Mol. Plant Microbe Interact. 16: 1118–1128.

Bahar, B., J.V. O'Doherty, T.J. Smyth and T. Sweeney. 2016. A comparison of the effects of an Ascophyllum nodosum ethanol extract and its molecular weight fractions on the inflammatory immune gene expression *in vitro* and *ex vivo*. Innov. Food Sci. Emerg. Technol. https://doi.org/10.1016/j.ifset.2016.07.027.

Bandara, B.M.R., A.A.I. Gunatilaka, N.S. Kumar, W.R. Wimalasiri, N.K.B. Adikaram and S. Balasubramaniam. 1988. Antimicrobial activity of some marine algae of Sri Lanka. J. Natl. Sci. Council Sri Lanka. 16: 209–221.

Barreto, M., C.J. Straker and A.T. Critchley. 1997. Short note on the effects of ethanolic extracts of selected south African seaweeds on the growth of commercially important plant pathogens, *Rhizoctonia solani* Kuhn and *Verticellium* sp. S. Afr. J. Bot. 63: 521–523.

Basak, A. 2008. Biostimulators definitions, classification and legislation. *In*: Biostimulators in Modern Agriculture. General Aspects, (ED: Helena Gawronska). Warsaw, pp.7–17.

Bassolé, I.H.N. and H.R. Juliani. 2012. Essential oils in combination and their antimicrobial properties. Molecules 17: 3989–4006.

Bennanmara, A., A. Abourriche, M. Berrada, M. Charrout, N. Chaib, M. Boundouma and F.X. Garjeall. 1999. MethoxyBifurcarenone: An antifungal and antibacterial meroditerpeoid from the Brown alga *Cystoseria tamariscifolia*. Phytochemistry 52: 37–40.

Battacharyya, D., M.Z. Babgohari, P. Rathor and B. Prithiviraj. 2015. Seaweed extracts as biostimulants in horticulture. Sci. Hort. 196: 39–48.

Bhosale, S.H., T.G. Jagtap and C.G. Naik. 1999. Antifungal activity of some marine organisms from India, against food spoilage *Aspergillus* strains. Mycopathologia 147: 133–138.

Bianco, É.M., V.L. Teixeira and R.C. Pereira. 2010. Chemical defenses of the tropical marine seaweed *Canistrocarpus cervicornis* against herbivory by sea urchins. Braz. J. Oceanogr. 58: 213–218.

Blunden, G., P.F. Morse, I. Mathe, J. Hohmann, A.T. Critchleye and S. Morrell. 2010. Betaine yields from marine algal species utilized in the preparation of seaweed extracts used in agriculture. Natural Product Communication 5: 581–585.

Blunt, J.W., B.R. Copp, W-P Hu, M.H.G. Munro, P.T. Northcote and M.R. Prinsep. 2007. Marine natural products. Nat. Prod. Rep. 24: 31–86.

Borsato, L.C., R.M. Di piero and M.J. Stadnik. 2010. Mechanisms of defense elicited by ulvan against *Uromyces appendiculatus* in three bean cultivars. Trop. Plant Pathol. 35: 318–322.

Burketova, L., L. Trda, P.G. Ott and O. Valentova. 2015. Bio-based resistance inducers for sustainable plant protection against pathogens. Biotech. Adv. 33: 994–1004.

Caccamese, S., R. Azzolina, G. Furnari, M. Cormaxi and S. Grasso. 1980. Antimicrobial and antiviral activities of extracts from Mediterranean algae. Bot. Mar. 23: 285–288.

Caccamese, S., R. Azzolina, G. Furnari, M. Cormaxi and S. Grasso. 1981. Antimicrobial and antiviral activities of some marine algae from eastern Sicily. Bot. Mar. 24: 365–367.

Cardozo, K.H.M., G. Thais, P.B. Marcelo, R.F. Vanessa, P.T. Angela, P.L. Norberto, C. Sara, A.T. Moacir, O.S. Anderson, C. Pio and P. Ernani. 2007. Metabolites from algae with economical impact. Comp. Biochem. Physiol. C Toxicol. Pharmacol. 146: 60–78.

Chandia, N.P. and B. Matsuhiro. 2008. Characterization of a fucoidan from *Lessonia vadosa* (Phaeophyta) and its anticoagulant and elicitor properties. Int. J. Biol. Macromol. 42: 235–240.

Charoensiddhi, S., M.A. Conlon, M.S. Vuaran, C.M.M. Franco and W. Zhang. 2017. Polysaccharide and phlorotannin-enriched extracts of the brown seaweed *Ecklonia radiata* influence human gut microbiota and fermentation *in vitro*. J. Appl. Phycol. https://doi.org/10.1007/s10811-017-1146-y.

Cluzet, S., C. Torregrosa, C. Jacquet, C. Lafitte, J. Fournier, L. Mercier, S. Salamagne, X. Briand, M.T. Esquerré-Tugayé and B. Dumas. 2004. Gene expression profiling and protection of *Medicago truncatula* against a fungal infection in response to an elicitor from green algae *Ulva* sp. Plant Cell Environ. 27: 917–928.

Contreras-Porcia, L., D. Thomas, V. Flores and J.A. Correa. 2011. Tolerance to oxidative stress induced by desiccation in *Porphyra columbina* (Bangiales, Rhodophyta). J. Exp. Bot. 62: 1815–1829.

Costabile, A., S. Sanghi, S. Martin-Pelaez, I. Mueller-Harvey, G.R. Gibson, R.A. Rastall and A. Klinder. 2011. Inhibition of *Salmonella typhimurium* by tannins *in vitro*. Journal of Food Agriculture and Environment 9: 119–124.

Courtois, J. 2009. Oligosaccharides from land plants and algae: production and applications in therapeutics and biotechnology. Curr. Opin. Microbiol. 12: 261–273.

Cox, S., N. Abu-Ghannam and S. Gupta. 2010. An assessment of the antioxidant and antimicrobial activity of six species of edible irish seaweeds. Int. Food Res. J. 17: 205–220.

Craigie, J.S. 2010. Seaweed extract stimuli in plant science and agriculture. J. Appl. Phycol. doi:10.1007/s10811-010–9560-4.

Dawczynski, C., R. Schubert and G. Jahreis. 2007. Amino acids, fatty acids, and dietary fibre in edible seaweed products. Food Chemistry 103: 891–899.

de Freitas, M.B. and M.J. Stadnik. 2015. Ulvan-induced resistance in *Arabidopsis thaliana* against *Alternaria brassicicola* requires reactive oxygen species derived from NADPH oxidase. Physiological and Molecular Plant Pathology 90: 49–56.

de Freitas, M.B., L.G. Ferreira, C. Hawerroth, M.E.R. Duarte, M.D. Noseda and M.J. Stadnik. 2015. Ulvans induce resistance against plant pathogenic fungi independently of their sulfation degree. Carbohydr. Polym. 133: 384–390.

Desbois, A.P. and V.J. Smith. 2010. Antibacterial free fatty acids: activities, mechanisms of action and biotechnological potential. Appl. Microbiol. Biotechnol. 85: 1629–1642.

Dixon, G.R. and U.F. Walsh. 2002. Suppressing *Pythium ultimum* induced damping-off in cabbage seedlings by biostimulation with proprietary liquid seaweed extracts. Acta Hortic. (ISHS) 635: 103–106.

Eom, S.H, Y.M. Kim and S.K. Kim. 2012. Antimicrobial effect of phlorotannins from marine brown algae. Food Chem. Toxicol. 50: 3251–3255.

Esserti, S., S. Amal, L.A. Rifai, T. Koussa, K. Makroum, M. Belfaiza, E.M. Kabil, L. Faize, L. Burgos, N. Alburquerque and M. Faize. 2017. Protective effect of three brown seaweed extracts against fungal and bacterial diseases of tomato. J. Appl. Phycol. 29: 1081–1093.

Esserti, S., A. Smaili, K. Makroum, M. Belfaiza, L.A. Rifai, T. Koussa, I. Kasmi and F. Mohamed. 2017. Priming of *Nicotiana benthamiana* antioxidant defences using brown seaweed extracts. J. Phytopath. 166: 86–94.

Fenical, W., J.J. Sim, D. Squatrito, R.M. Wing and P. Radlick. 1973. Zonarol and Isozonarol fungitoxic hydroquinones from the brown seaweed *Dictyopteris zonarioides*. J. Org. Chem. 38: 2383–2385.

Fenical, W. and V.J. Paul. 1984. Antimicrobial and cytotoxic terpenoids from tropical green algae of the family Udoetaceae. Hydrobiologia 116: 135–140.

Flora, G. and S.M.V. Rani. 2012. An approach towards control of blast by foliar application of seaweed concentrate. Sci. Res. Rep. 2: 213–217.

Freitas, M.B.D., C.C. Medugno, R.F. Schons and M.J. Stadnik. 2011. Effectiveness of formulas based on ulvan in inducing resistance in *Phaseolus vulgaris* against *Colletotrichum lindemuthianum*. Trop. Plant Pathol. 36: 45–49.

Fu, Y., H. Yin, W. Wang, M. Wang, H. Zhang, X. Zhao and Y. Du. 2011. β-1,3-Glucan with different degree of polymerization induced different defense responses in tobacco. Carbohydr. Polym. 86: 774–782.

Ghazala, B., B. Naila, M. Shameel, S. Shahzad and S.M. Leghari. 2004. Phycochemistry and bioactivity of two stonewort algae (Charophyta) of Sindh. Pak. J. Bot. 36: 733–743.

Goicoechea, N., J. Aguirreolea and J.M. García-Mina. 2004. Alleviation of verticillium wilt in pepper (*Capsicum annuum* L.) by using the organic amendment COA H of natural origin. Sci. Hort. 101: 23–37.

Gonzalez, del Val A., G. Platas, A. Basilio, A. Cabello, J. Garrochategui, I. Suay, F.Vicente, E. Portillo, M. Jimenez del Rio, G. Reina and F. Pelaez. 2001. Screening of antimicrobial activities in red, green and brown macroalgae from Gran Canaria (Canary Islands, Spain). Int. Microbio. 4: 35–40.

Górka, B. and P.P. Wieczorek. 2017. Simultaneous determination of nine phytohormones in seaweed and algae extracts by HPLC-PDA. J. Chromatogr. B Analyt. Technol. Biomed. Life Sci. https://doi.org/10.1016/j.jchromb.2017.04.048.

Grant, M. and C. Lamb. 2006. Systemic immunity. Current Opinion in Plant Biology 9: 414–420.

Guiry, M.D. 2012. How many species of algae are there? J. Phycol. 48: 1057–1063.

Guiry, M.D. and G.M. Guiry. 2018. Algae Base. World-Wide Electronic Publication, National University of Ireland, Galway. http://www.algaebase.org; searched on 16 January 2018.

Gupta, S. and N. Abu-Ghannam. 2011a. Bioactive potential and possible health effects of edible brown seaweeds. Trends Food Sci. Technol. 22: 315–326.

Gupta, S. and N. Abu-Ghannam. 2011b. Recent developments in the application of seaweeds or seaweed extracts as a means for enhancing the safety and quality attributes of foods. Innov. Food Sci. Emerg. Technol. https://doi.org/10.1016/j.ifset.2011.07.004.

Hernández-Herrera, R.M., G. Virgen-Calleros, M. Ruiz-López, J. Zañudo-Hernández, J.P. Délano-Frier and C. Sánchez-Hernández. 2014. Extracts from green and brown seaweeds protect tomato (*Solanum lycopersicum*) against the necrotrophic fungus *Alternaria solani*. J. Appl. Phycol. 26: 1607–1614.

Jaulneau, V., C. Lafitte, M.F. Corio-Costet, M.J. Stadnik, S. Salamagne, X. Briand, M.T. Esquerré-Tugayé and B. Dumas. 2011. An *Ulva armoricana* extract protects plants against three powdery mildew pathogens. Eur. J. Plant Pathol. 131: 393–401.

Jaulneau, V., C. Lafitte, C. Jacquet, S. Fournier, S. Salamagne, X. Briand, M.T. Esquerré-Tugayé and B. Dumas. 2010. Ulvan, a sulphated polysaccharide from green algae, activates plant immunity through the jasmonic acid signaling pathway. J. Biomed. Biotechnol. 525291.

Jayaraj, J., A. Wan, M. Rahman and Z.K. Punja. 2008. Seaweed extract reduces foliar fungal diseases on carrot. Crop Prot. 27: 1360–1366.

Jayaraj, J., J. Norrie and Z.K. Punja. 2011. Commercial extract from the brown seaweed *Ascophyllum nodosum* reduces fungal diseases in greenhouse cucumber. J. Appl. Phycol. 23: 353–361.

Jeandet, P., M. Adrian, J.M. Joubert, F. Hubert and R. Bessis. 1996. Stimulating the natural defences of grape. A complement to phytosanitary control of *Botrytis*. Phytoma 488: 21–5.

Jiao, G., G. Yu, J. Zhang and S.H. Ewart. 2011. Chemical structures and bioactivities of sulphated polysaccharides from marine algae. Marine Drugs 9: 196–223.

Jiménez, E., F. Dorta, C. Medina, A. Ramírez, I. Ramírez and H. Peña-Cortés. 2011. Anti-phytopathogenic activities of macro-algae extracts. Mar. Drugs 9: 739–756.

Kamenarska, Z., M.J. Gasic, M. Zlatovic, A. Rasovic, D. Sladic, Z. Kljajic and S. Popov. 2002. Chemical composition of the brown alga *Padina pavonia* (L.) Gaill. from the Adriatic Sea. Bot. MAr. 45: 339–345.

Katayama, T. 1962. Volatile constituents. pp. 467–473. *In*: R.A. Lewin (ed.). Physiology and Biochemistry of Algae. Academic Press, New York.

Katayama, T. 1964. Biochemical significance of the existence of acrylic acid in seaweeds. Nippon Sorui Gakkai (Japan) 12: 14–19.

Khan, W., U.P. Rayirath, S. Subramanian, M.N. Jithesh, P. Rayorath, D.M. Hodges, A.T. Critchley, J.S. Craigie, J. Norrie and P. Prithiviraj. 2009. Seaweed extracts as biostimulants of plant growth and development. J. Plant Growth Regul. 28: 386–399.

Khanzada, A.K., W. Shaikh, T.G. Kazi, S. Kabir and S. Soofia. 2007. Antifungal activity, elemental analysis and determination of total protein of seaweed, *Solieria robusta* (Greville) Kylin from the coast of Karachi. Pak. J. Bot. 39: 931–937.

Khotimchenko, S.V., V.E. Vaskovsky and T.V. Titlyanova. 2002. Fatty acids of marine algae from the Pacific coast of North California. Bot. Mar. 45(1): 17–22.

Klarzynski, O., B. Plesse, J.M. Joubert, J.C. Yvin, M. Kopp, B. Kloareg and B. Fritig. 2000. Linear beta-1,3 glucans are elicitors of defense responses in tobacco. Plant Physiol. 124: 1027–1038.

Klarzynski, O., V. Descamps, B. Plesse, J.C. Yvin, B. Kloareg and B. Fritig. 2003. Sulfated fucan oligosaccharides elicit defense responses in tobacco and local and systemic resistance against tobacco mosaic virus. Mol. Plant Microbe Interact. 16: 115–122.

Kobayashi, A., A. Tai, H. Kanzaki and K. Kawazu. 1993. Elicitor-active oligosaccharides from algal laminaran stimulate the production of antifungal compounds in alfalfa. Z. Naturforsch. 48c: 575–579.

Kubanek, J., P.R. Jensen, P.A. Keifer, M.S. Cameron, D.O. Collins and W. Fenical. 2003. Seaweed resistance to microbial attack: A targeted chemical defense against marine fungi. Proc. Natl. Acad. Sci. 100: 6916–6921.

Kulik, M.M. 1995. The potential for using cyanobacteria (blue–green alge) and algae in the biological control of plant pathogenic bacteria and fungi. Eur. J. Plant Pathol. 101: 585–599.

Kumar, C.D.S., V.L. Sarada and R. Rengasamy. 2008. Seaweed extracts control the leaf spot disease of the medicinal plant *Gymnema sylvestre*. Indian J. Sci. Technol. 39: 93–94.

Kuwada, K., T. Ishii, I. Matsushita, I. Matsumoto and K. Kadoya. 1999. Effect of seaweed extracts on hyphal growth of vesicular–arbuscular mycorrhizal fungi and their infectivity on trifoliate orange roots. J. Jpn. Soc. Hortic. Sci. 68: 321–326.

Kuwada, K., L.S. Wamocho, M. Utamura, I. Matsushita and T. Ishii. 2006. Effect of red and green algal extracts on hyphal growth of arbuscular fungi, and on mycorrhizal development and growth of papaya and passion fruit. Agron. J. 98: 1340–1344.

Laporte, D., J. Vera, N.P. Chandía, E.A. Zúñiga, B. Matsuhiro and A. Moenne. 2007. Structurally unrelated algal oligosaccharides differentially stimulate growth and defense against tobacco mosaic virus in tobacco plants. J. Appl. Phycol. 19: 79–88.

Lapshina, L.A., A.V. Reunov, V.P. Nagorskaya, T.N. Zvyagintseva and N.M. Shevchenko. 2007. Effect of fucoidan from brown alga *Fucus evanescens* on a formation of TMV-specific inclusions in the cells of tobacco leaves. Russ. J. Plant Physiol. 54: 111–114.

Liu, H. and L. Gu. 2012. Phlorotannins from brown algae (*Fucus vesiculosus*) inhibited the formation of advanced glycation endproducts by scavenging reactive carbonyls. J. Agric. Food Chem. 60: 1326–1334.

Lizzi, Y., C. Coulomb, C. Polian, P.J. Coulomb and P.O. Coulomb. 1998. Seaweed and mildew: what does the future hold? Laboratory tests have produced encouraging results [L'algue face au mildiou: quel avenir? Des resultats de laboratoire tres encourageants]. Phytoma 508: 29–30.

Lordan, S., R.P. Ross and C. Stanton. 2011. Marine bioactives as functional food ingredients: potential to reduce the incidence of chronic diseases. Mar. Drugs 9: 1056–1100.

Manilal, A., S. Sujith, J. Selvin, C. Shakir and G.S. Kiran. 2009. Antibacterial activity of *Falkenbergia hillebrandii* (Born) from the Indian coast against human pathogens. Phyton (Buenos Aires) 78: 161–166.

Manimala, K. and R. Rengasamy. 1993. Effect of bioactive compound of seaweed on the phytopathogen *Xanthomonas oryzae*. Phykos 32: 77–83.

Marfaing, H. 2017. Qualités nutritionnelles des algues, leur présent et futur sur la scène alimentaire. Cahiers de Nutrition et de Dietetique. https://doi.org/10.1016/j.cnd.2017.05.003.

Martinez-Lozano, S.L., S. Garcia and N. Heredia. 2000. Antifungal activity of extract of *Sargassum filipendula*. Phyton (Buenos Aires) 66: 179–182.

McLachlan, J. 1985. Macroalgae (seaweeds): industrial resources and their utilization. Plant and Soil 89: 137–157.

Menard, R., P. De Ruffray, B. Fritig, J.C. Yvin and S. Kauffmann. 2005. Defense and resistance-inducing activities in tobacco of the sulfated beta-1,3 glucan PS3 and its synergistic activities with the unsulfated molecule. Plant Cell Physiol. 46: 1964–1972.

Ménard, R., S. Alban, P. de Ruffray, F. Jamois, G. Franz, B. Fritig, J.C. Yvin and S. Kauffmann. 2004. β-1,3 glucan sulfate, induces salicylic acid signaling pathway in tobacco and Arabidopsis. Plant Cell. 16: 3020–3032.

Mercier, L., C. Lafitte, G. Borderies, X. Briand, M.T. Esquerré-Tugayé and J. Fournier. 2001. The algal polysaccharide carrageenans can act as an elicitor of plant defence. New Phytol. 149: 43–51.

Meszka, B. and A. Bielenin. 2009. Bioproducts in control of strawberry verticillium wilt. Phytopathologia 52: 21–27.

Montealegre, J.R., C. Lopez, M.J. Stadnik, J.L. Henríquez, R. Herrera, R.Polanco, R.M.D. Piero and L.M. Pérez. 2010. Control of grey rot of apple fruits by biologically active natural products. Trop. Plant Pathol. 35: 271–276.

Moreau, J., D. Pesando, P. Bernard, B. Caram and J.C. Pionnat. 1988. Seasonal variations in the production of antifungal substances by some dictyotales (brown algae) from the French Mediterranean coast. Hydrobiologia 162: 157–162.

Nabil, S. and J. Cosson. 1996. Seasonal variations in sterol composition of *Delesseria sanguinea* (Ceramiales, Rhodophyta). Hydrobiologia 326: 511–514.

Navarro, G., J.J. Fernández and M. Norte. 2004. Novel meroditerpenes from the brown alga *Cystoseira* sp. J. Nat. Prod. 67: 495–499.

Patier, P., J.C. Yvin, B. Kloareg, Y. Liénart and C. Rochas. 1993. Seaweed liquid fertilizer from *Ascophyllum nosodosum* contains elicitors of plant glycanases. J. Appl. Phycol. 5: 343–349.

Paulert, R., A. Smânia Jr., M. Stadnik and M. Pizzolatti. 2007. Antimicrobial properties of extracts from the green seaweed *Ulva fasciata* Delile against pathogenic bacteria and fungi. Algol. Stud. 123: 123–130.

Paulert, R., V. Talamini, J.E.F. Cassolato, M.E.R. Duarte, M.D. Noseda, J.A. Smania and M.J. Stadnik. 2009. Effects of sulfated polysaccharide and alcoholic extracts from green seaweed *Ulva fasciata* on anthracnose severity and growth of common bean (*Phaseolus vulgaris* L.). J. Plant Dis. Protect. 6: 263–270.

Paulert, R., D. Ebbinghaus, C. Urlass and B.M. Moerschbacher. 2010. Priming of the oxidative burst in rice and wheat cell cultures by ulvan, a polysaccharide from green macroalgae, and enhanced resistance against powdery mildew in wheat and barley plants. Plant Pathol. 59: 634–642.

Peres, J.C.F., L.R.D. Carvalho, E. Gonçalez, L.O.S. Berian and J.D. Felicio. 2012. Evaluation of antifungal activity of seaweed extracts. Ciênc. Agrotec. 36: 294–299.

Portillo, I., M. Collina and A. Brunelli. 2007. Activity of biostimulants towards *Phytophthora infestans* on tomato. Tenth Workshop of an European Network for development of an Integrated Control Strategy of potato late blight. Bologna (Italy).

Pratt, R., R.H. Mautner, G.M. Gardner, Y. Sha and J. Dufrenoy. 1951. Reports on antibiotic activity of sea weeds extracts. J. Am. Pharm. Assoc. 40: 575–579.

Prithiviraj, B., H.P. Bais, T. Weir, B. Suresh, H.P. Schweizer and J.M. Vivanco. 2005. Down regulation of virulence factors of pseudomonas aeruginosa by salicylic acid attenuates its virulence on *Arabidopsis thaliana* and *Caenorhabditis elegans*. Infect. Immun. 73. 3319–3328.

Ramikissoon, A., A. Ramsubhag and J. Jayaraman. 2017. Phytoclicitor activity of three Caribbean seaweed species on suppression of pathogenic infections in tomato plants. J. Appl. Phycol. 29: 353–361.

Raghavendra, V.B., S. Lokesh and H.S. Prakash. 2007. Dravya, a product of seaweed extract (*Sargassum wightii*), induces resistance in cotton against *Xanthomonas campestris* pv. *malvacearum*. Phytoparasitica 35: 442–449.

Raghavendra, V.B., S.J. Siddagangaiah, S. Lokesh and S.R. Niranjana. 2008. Phyton-T: an extract of seaweed (*Sargassum wightii*) induces defense enzymes against late blight and enhances quality of potato. Mycol. Plant Pathol. 38: 27–32.

Rajesh, S., D.P. Raja, J.M. Rathi and K. Sahayaraj. 2012. Biosynthesis of silver nanoparticles using *Ulva fasciata* (Delile) ethyl acetate extract and its activity against *Xanthomonas campestris* pv. *malvacearum*. J. Biopestic. 5(Supplementary): 119–128.

Rasher, D.B., E.P. Stout, S. Engel, J. Kubanek and M.E. Hay. 2011. Macroalgal terpenes function as allelopathic agents against reef corals. Proc. Natl. Acad. Sci. 108: 17726–17731.

Reichelt, J.L. and M.A. Borowitzka. 1984. Antimicrobial activity from marine algae: Results of a large-scale screening programme. Hydrobiologia 116/117: 158–168.

Rekanović, E., I. Potočnik, S. Milijašević-Marčić, M. Stepanović, B.Todorović and M. Mihajlović. 2010. Efficacy of seaweed concentrate from *Ecklonia maxima* (Osbeck) and conventional fungicides in the control of *Verticillium* wilt of pepper. Pestic. Phytomed. 25: 319–324.

Reunov, A., L. Lapshina, V. Nagorskaya, T. Zvyagintseva and N. Shevchenko. 2011. Effect of fucoidan from the brown alga *Fucus evanescens* on the development of infection induced by potato virus X in *Datura stramonium* L. leaves. J. Plant Dis. Protect. 2: 49–54.

Ross, H. 1957. Untersuchungen uber das Vorkommen antimikrobieller substanzen in Meersalgen, Kiel. Meeresf 13: 41–58.

Sánchez-Machado, D.I., J. López-Hernández, P. Paseiro-Losada and J. López-Cervantes. 2004. An HPLC method for the quantification of sterols in edible seaweeds. Biomed. Chromatogr. 18: 183–190.

Sangha, J.S., S. Ravichandran, K. Prithiviraj, A.T. Critchley and B. Prithiviraj. 2010. Sulfated macroalgal polysaccharides λ-carrageenan and ι-carrageenan differentially alter *Arabidopsis thaliana* resistance to *Sclerotinia sclerotiorum*. Physiol. Mol. Plant Pathol. 75: 38–45.

Sangha, J.S., S. Kandasamy, W. Khan, N.S. Bahia, R.P. Singh, A.T. Critchley and B. Prithiviraj. 2015. λ-Carrageenan suppresses Tomato Chlorotic Dwarf Viroid (TCDVd) replication and symptom expression in tomatoes. Marine Drugs 13: 2875–2889.

Sathyanarayana, S.G., S. Lokesh, T.V. Kumar and H.S. Shetty. 2006. Dravya: a putative organic treatment against *Alternaria padwickii* infection in paddy. Integr. Biosci. 10: 21–25.

Schmid, M, F. Guihéneuf and D.B. Stengel. 2017. Ecological and commercial implications of temporal and spatial variability in the composition of pigments and fatty acids in five Irish macroalgae. Mar. Biol. https://doi.org/10.1007/s00227-017-3188-8.

Seenivasan, R., H. Indu, G. Archana and S. Geetha. 2010. The antibacterial activity of some marine algae from south east coast of India. J. Pharmacol. Res. 8: 1907–1912.

Seth, A. and M. Shanmugam. 2016. Seaweeds as agricultural crops in India: New vistas. *In*: Innovative Saline Agriculture. https://doi.org/10.1007/978-81-322-2770-0_20.

Shanmughapriya, S., A. Manilal, S. Sujith, J. Selvin, G. Segal-Kiran and K. Natarajaseenivasan. 2008. Antimicrobial activity of seaweeds extracts against multiresistant pathogens. Ann. Microbiol. 58: 535–541.

Shi, Q., A. Wang, Z. Lu, C. Qin, J. Hu and J. Yin. 2017. Overview on the antiviral activities and mechanisms of marine polysaccharides from seaweeds. Carbohydr. Res. 453-454: 1–9.

Sharma, K., C. Bruns, A.F. Butz and M.R. Finckh. 2012. Effects of fertilizers and plant stratheners on the susceptibility of tomatoes to single and mixed isolates of Phytophthora infestans. Eur. J. Plant Pathol. 133: 739–751.

Sieburth, J.M. and J.T. Conover. 1965. Sargassum tannin an antibiotic which retards fouling. Nature 208: 52–53.

Srikong, W., N. Bovornreungroj, P. Mittraparparthorn and P. Bovornreungroj. 2017. Antibacterial and antioxidant activities of differential solvent extractions from the green seaweed *Ulva intestinalis*. ScienceAsia. https://doi.org/10.2306/scienceasia1513-1874.2017.43.088.

Stadnik, M.J. and M.B. de Freitas. 2014. Algal polysaccharides as source of plant resistance inducers. Trop. Plant Pathol. 39: 111–118.

Subramanian, S., J.S. Sangha, B.A. Gray, R.P. Singh, D. Hiltz, A.T. Critchley and B. Prithiviraj. 2011. Extracts of the marine brown macroalga, *Ascophyllum nodosum*, induce jasmonic acid dependent systemic resistance in *Arabidopsis thaliana* against *Pseudomonas syringae* pv. *tomato* DC3000 and *Sclerotinia sclerotiorum*. Eur. J. Plant Pathol. 131: 237–248.

Sultana, V., E.S. Haque, J. Ara and M. Athar. 2005. Comparative efficacy of brown, green and red seaweeds in the control of root infecting fungi and okra. Int. J. Environ. Sci. Technol. 2: 129–132.

Sultana, V., J. Ara and S.E. Haque. 2008. Suppression of root rotting fungi and root knot nematode of chili by seaweed and *Pseudomonas aeruginosa*. J. Phytopathol. 156: 390–395.

Sultana, V., S.E. Haque, J. Ara and M. Athar. 2009. Effect of brown seaweeds and pesticides on root rotting fungi and root-knot nematode infecting tomato roots. J. Appl. Bot. Food Qual. 83: 50–53.

Sultana, V., S.E. Haque, G.N. Baloch and J. Ara. 2011a. Comparative efficacy of a red alga *solieria robusta*, chemical fertilizers and pesticides in managing the root diseases and growth of soybean. Pak. J. Bot. 43: 1–6.

Sultana, V., G.N. Baloch, J. Ara, S.E. Haque, M.R. Tariq and M. Athar. 2011b. Seaweeds as an alternative to chemical pesticides for the management of root diseases of sunflower and tomato. J. Appl. Bot. Food Qual. 84: 162–168.

Thowthampitak, J. and S. Prathuangwong. 2007. Disease suppression and crop improvement of green soybean with potential biocontrol agent and fertilizer. Proceedings of the 45th Kasetsart University Annual Conference, Bangkok, Thailand, 30 January–2 February 2007. Subject: Plants, pp. 638–646.

Trouvelot, S., A.L. Varnier, M. Allègre, L. Mercier, F. Baillieul, C. Arnould, V.G. Pearson, O. Klarzynski, J.M. Joubert, A. Pugin and X. Daire. 2008. A β-1,3 glucan sulfate induces resistance in grapevine against *Plasmopara viticola* through priming of defense responses, including HR-like cell death. Mol. Plant Microbe Interact. 21: 232–243.

Uppal, A.K., A.E. Hadrami, L.R. Adam, M. Tenuta and F. Daayf. 2008. Biological control of potato Verticillium wilt under controlled and field conditions using selected bacterial antagonists and plant extracts. Biol. Control 44: 90–100.

Van Oosten, M.J, O. Pepe, S. De Pascale, S. Silletti and A. Maggio. 2017. The role of biostimulants and bioeffectors as alleviators of abiotic stress in crop plants. Chem. Biol. Technol. Agric. https://doi.org/10.1186/s40538-017-0089-5.

Venkatesh, R., S. Shanthi, K. Rajapandian, S. Elamathi, S. Thenmozhi and N. Radha. 2011. Preliminary study on antixanthomonas activity, phytochemical analysis, and characterization of antimicrobial compounds from *Kappaphycus alvarezii*. Asian J. Pharm. Clin. Res. 4: 46–51.

Vera, J., J. Castro, A. González, H. Barrientos, B. Matsuhiro, P. Arce, G. Zuñiga and A. Moenne. 2011a. Long-term protection against tobacco mosaic virus induced by the marine alga oligo-sulphated-galactan Poly-Ga in tobacco plants. Mol. Plant Pathol. DOI: 10.1111/j.1364-3703.2010.00685.x.

Vera, J., J. Castro, A. Gonzalez and A. Moenne. 2011b. Seaweed polysaccharides and derived oligosaccharides stimulate defense responses and protection against pathogens in plants. Mar. Drugs 9: 2514–2525.

Wang, Y., Z. Xu, S.J. Bach and T.A. McAllister. 2009. Sensitivity of *Escherichia coli* to seaweed (*Ascophyllum nodosum*) phlorotannins and terrestrial tannins. Asian-Australas. J. Anim. Sci. 22: 238–245.

Wang, K., M. Senthil-Kumar, C.M. Ryu, L. Kang and K.S. Mysore. 2012. Phytosterols play a key role in plant innate immunity against bacterial pathogens by regulating nutrient efflux into the apoplast. Plant Physiol. 158: 1789–1802.

Washington, W.S., S. Engleitner, G. Boontjes and N. Shanmuganathan. 1999. Effect of fungicides, seaweed extracts, tea tree oil, and fungal agents on fruit rot and yield in strawberry. Aust. J. Exp. Agric. 39: 487–94.

Weinberger, F. 2007. Pathogen-induced defense and innate immunity in macroalgae. Biological Bulletin 213: 290–302.

Welch, A.M. 1962. Preliminary survey of fungistatic properties of marine algae. J. Bacteriol. 83: 97–99.

Yakhin, O.I, A.A. Lubyanov, I.A. Yakhin and P.H. Brown. 2017. Biostimulants in plant science: a global perspective. Front. Plant Sci. https://doi.org/10.3389/fpls.2016.02049.

Yi, Z., C. Yin-shan and L. Hai-sheng. 2001. Screening for antibacterial and antifungal activities in some marine algae from the Fujian coast of China with three different solvents. Chin. J. Oceanol. Limn. 19: 327–331.

Zaostrovnykh, V.I., T.F. Trofimova, N.M. Shevchenko, E.L. Chaĭkina, T.N. Zvyagintseva and M.M. Anisimov. 2009. Effect of carbohydrate-containing biopolymers from seaweeds on the resistance of soyabean to diseases. Zas. Karantin. Rast. No. 12 pp. 24.

Zhang, X., E.H. Ervin and R.E. Schmidt. 2003. Physiological effects of liquid applications of a seaweed extract and a humic acid on creeping bentgrass. J. Am. Soc. Hortic. Sci. 128: 492–496.

Zhang, Q., J. Zhang, A. Silva, D.A. Dennis and C.J. Barrow. 2006. A simple 96 – well microplate method for estimation of total polyphenol content in seaweeds. J. Appl. Phycol. 18: 445–450.

Zodape, S.T., A. Gupta, S.C. Bhandari, U.S. Rawat, D.R. Chaudhary, K. Eswaran and J. Chikara. 2011. Foliar application of seaweed sap as biostimulant for enhancement of yield and quality of tomato (*Lycopersicon esculentum* Mill.). J. Sci. Ind. Res. 70: 215–219.

11

The Cosmeceutical Properties of Compounds Derived from Marine Algae

Snezana Agatonovic-Kustrin[1,*] and *David W. Morton*[2]

Introduction

Even though the ocean covers two thirds of the Earth's surface, the marine environment largely remains an unexplored source of bioactive compounds, many of which have been shown to possess new and unique chemical structures. Marine organisms produce and secrete a wide range of biologically active compounds that have a broad range of biological activity. Over the last few decades, marine algae have been extensively investigated for their active metabolites and as a source of potential cosmeceutical compounds. They can also be harvested economically as they are mostly found in shallow ocean water.

Marine algae are simple, chlorophyll-containing organisms with extremely diverse morphological and reproductive features, that can produce a range of compounds with unique physiological and biochemical properties (Li and Chen 2001). They thrive in extreme habitats resulting from changes in sea level due to tidal effects along the coast that results in relatively large changes in light intensity, drying out at low tide, etc. Like all photosynthetic plants, algae are exposed to light and high oxygen concentrations, which encourages formation of free radicals and other potent oxidizing agents that have the potential to damage cell structure and function (Sukenik et al. 1993). The absence of structural damage suggests that algae are able to produce compounds that protect them against harmful free radical oxidation (Matsukawa et al. 1997; Jiménez-Escrig et al. 2001; Lim et al. 2002). Seawater has a unique nutrient composition that is able to support and help maintain cellular activity. The similarity between seawater and human plasma is so close that white blood cells can live and function in seawater for a long time. The human body needs daily replenishment of minerals such as calcium, magnesium zinc, and phosphorous. As seawater contains the body's ideal balance of minerals, it is the perfect medium for restoring such cellular levels. Many of the minerals and trace elements present in seawater are important catalysts, able to activate cellular enzymes

---
[1] School of Pharmacy, Monash University Malaysia, Jalan Lagoon Selatan, Bandar Sunway, 47500, Selangor Darul Ehsan, Malaysia.

[2] School of Pharmacy and Applied Science, La Trobe Institute of Molecular Sciences, La Trobe University, Edwards Rd, Bendigo 3550, Australia.

* Corresponding author: snezana.agatonovic@monash.edu

and increase the rate of virtually all the chemical reactions within cells. Maintaining ideal mineral levels will improve overall cell functions and prevent cellular imbalances. Not only does seawater provide a balanced supplement of minerals for human cells, it also supplies the ideal nutrient-rich environment for marine algae growth. As marine plants grow, they absorb and concentrate nutrients that are also beneficial to the human body. Algae are able to produce a variety of compounds in response to their environment, with some acting as defence mechanisms against predators.

Distinguishing between the different algal groups can be difficult as they often lack gross morphological features. There are more than 17000 marine algae species recorded to date, which may generally be divided into two main categories, macroalgae and microalgae. Macroalgae are a diverse family of marine plants that are for simplicity, classified according to the color of the pigments they use for photosynthesis: brown (*Phaeophyta*), red (*Rhodophyta*), and green (*Chlorophyta*) algae. Microalgae, such as blue-green algae (Rasmussen and Morrissey 2007), have traditionally been considered as algae. However, they are now commonly identified as cyanobacteria. From an ecological perspective they are algae, but from the cell-structural perspective, they are Gram-negative bacteria (Berner 1993).

Algal extracts have been used in cosmetic formulations for many decades as an excipient (i.e., viscosity-controlling ingredient), or as a therapeutic agent (i.e., as a moisturizer, emollient, or skin conditioning agent) due to their intrinsic stability and physical and bioactive properties (Kim et al. 2005). They require minimal light, water, and nutrients for growth, and do not require the use of arable land for cultivation. The range of chemicals, agrichemicals, and pharmaceuticals that can be isolated in commercially significant quantities from macroalgae is remarkable. Given that there is a marked similarity between the structure of human skin and the cellular structure of algae, it is not surprising that many compounds derived from marine algae are also beneficial in improving human skin function (Athukorala et al. 2006). Presently, compounds from red and brown algae are the ones that are most commonly used in cosmetic formulations and toiletries.

Environmental factors such as UV radiation, wind, and smoke, combined with chronological ageing and degeneration of the skin barrier, contribute to fine lines, wrinkles, pigmentation, sunspots, and increased skin coarseness. Some of the bioactive ingredients of microalgae, such as alguronic acid, a heterogeneous mix of exopolysaccharides, have been claimed to exhibit anti-ageing benefits that hinder and combat these signs of aging (Coragliotti et al. 2012). Microalgae exopolysaccharides are high molecular weight carbohydrate polymers that make up a substantial component of the extracellular polymers surrounding most microbial cells in the marine environment.

Many components from marine algae have been found to have both pharmaceutical and cosmetic like benefits, giving rise to the term "cosmeceuticals". The term "cosmeceutical" describes skin care products that fall in between the categories of cosmetics and drugs. At a fundamental level, cosmetics are products that affect the appearance of the skin, while drugs affect the structure and function of the skin. Hence, products considered as cosmeceuticals do physiologically affect the structure and function of the skin (drug-like effects), but are marketed using skin appearance based claims. Cosmeceutical formulations contain active ingredients such as vitamins, phytochemicals, enzymes, antioxidants, and essential oils, which are incorporated into creams, lotions, and ointments. Since a diverse range of these types of compounds have been found in marine algae, there is an increasing interest in marine algae as sources of effective active ingredients for cosmeceutical skin care products (Kim et al. 2008).

There are a large number of active components in macroalgae, including bioactive carbohydrates such as fucoidan and laminarin, pigments such as polyphloroglucinols and fucoxanthin (Yan et al. 1999; Shiratori et al. 2005), and minerals, including iodine. Many of these compounds have antioxidant properties and contribute to the antioxidant nature of aqueous and non-aqueous extracts. Glutathione, an antioxidant sometimes used as an orally delivered skin whitening agent, is found in all macroalgae. Kakinuma et al. reported the glutathione content of 37 species of macroalgae. Most had glutathione concentrations ranging from 0.1–200 mg/100 g (dry weight). However, two brown algae, *Sargassum thunbergii* and *Ishige okamurai*, had exceptionally high of glutathione concentrations (1432 and 3082 mg/100 g (dry weight) respectively) (Kakinuma et al. 2001). Additionally, omega-3 fatty acids such as stearidonic acid and hexadecatetraenoic acid are found in such edible marine algae as *Undaria pinnatifida* and *Ulva*, contributing up to 40% of the plants' total fatty acid content (Ishihara et al. 2000). Fucoidan fractions, alginates, and phloroglucinols isolated from marine algae have been found to have enzyme inhibitory properties against hyaluronidase, heparanases, phospholipase A2, tyrosine kinase, and collagenase expression (Wessels et al. 1999; Shibata et al. 2003; Joe et al. 2006). Sulfated polysaccharides found in macroalgae also show significant anti-viral activity against coated viruses, such as herpes and HIV (Schaeffer and Krylov 2000; Thompson and Dragar 2004).

Terpenoids

Terpenoids (isoprenoids) are the largest and most widespread class of secondary metabolites found in abundance in higher plants, including marine algae. They are also found in insects and microorganisms. Although, terpenoids and isoprenoids are sometimes referred to as terpenes, chemically terpenes are hydrocarbons (composed only of carbon and hydrogen), while terpenoids and isoprenoids are oxygen-containing analogs of terpenes. Recent research into marine natural products has shown marine algae are a rich source of terpenoids with unique and unusual structures. Many substituents rarely found in terrestrial terpenoids occur in marine terpenoids (for example, bromo- and chloro-substituents are found in algal terpenoids).

Novel marine terpenoids show great promise as a source for new antioxidant agents in cosmetic preparations (Kang et al. 2004; Paduch et al. 2007), due to their good skin penetration enhancing abilities, low systemic toxicity, and low skin irritation. Fucosterol (Fig. 1) is a steroidal terpenoid extracted from brown marine algae (*Ecklonia stolonifera, Pelvetia siliquosa, Sargassum carpophyllum*) (Tang et al. 2002; Lee et al. 2003; Jung et al. 2006). Fucosterol is usually the major component in the non-polar fraction of an algae extract (Jung et al. 2006). This compound shows strong antioxidant activity by increasing the concentration of antioxidant enzymes superoxide dismutase (SOD), catalase, and glutathione peroxidase (GSH-px)—enzymes involved in the fine control of cellular H_2O_2 concentration. Fucosterol can help in cellular defense mechanisms by preventing cell membrane oxidation as it has an important role in scavenging hydrogen peroxide and restoring SOD activity. Similarly, an increase in GSH-px activity indicates that fucosterol also helps in the restoration of vital endogenous antioxidants such as glutathione (Lee et al. 2003; Pillai et al. 2005).

Extrinsic factors, such as environmental pollution and UVB radiation (Pontius and Smith 2011) initiate a full cascade of biochemical reactions in human skin causing depletion of enzymatic and non-enzymatic antioxidants. Hence, the skin of aged and photo-aged individuals has a reduced capacity to fight reactive oxygen species (ROS) and free radicals that lead to ageing, and in turn results in further production of highly reactive *free radicals.* Chronic free radical assault leads to the appearance of uneven, blotchy pigmentation, and disrupts the structural matrix of the skin, giving rise to wrinkles and sagging skin (Fisher et al. 2009). ROS have a major role in the photo-aging of human skin *in vivo* (Lavker 1979; Rhie et al. 2001; Pillai et al. 2005), by causing oxidative damage to DNA, proteins, membrane lipids, and carbohydrates, which accumulate in the dermal and epidermal compartments (Tapiero et al. 2004). The inflammatory process in the skin, resulting from accumulation of ROS, might lead to the progression of photodermatoses, erythema development, and skin cancer emergence (Stahl and Sies 2005). ROS have been shown to induce matrix metalloproteinases (MMPs) expression in various cells. Degradation of

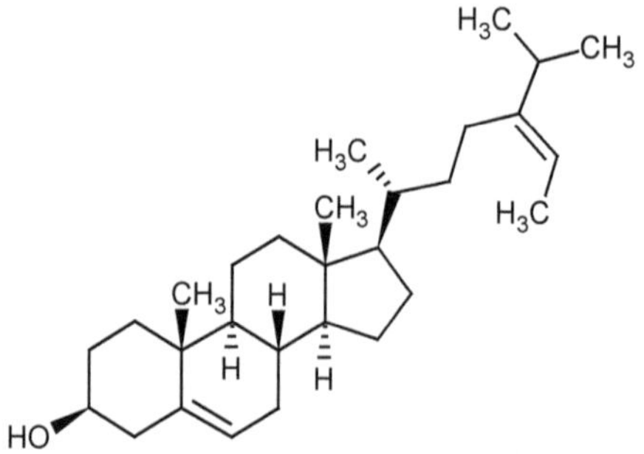

Fig. 1. The structure of fucosterol.

fibrillar collagen that occurs in photo damaged skin is a consequence of upregulation of MMPs (Klein and Bischoff 2011). MMPs are major enzymes involved in the remodeling of the extracellular matrix, by proteolytic degradation of collagen and elastic fibres, and loss of the skin's ability to resist stretching. In normal skin, MMPs are expressed in very low levels and are kept inactive (Pillai et al. 2005).

In recent years, experimental research has led to the discovery of a new type of carotenoid from seaweed and similar plants, called fucoxanthin (Fig. 2). It is a reddish-brown pigment present only in brown algae, and is a type of carotenoid, similar to vitamin A and β-carotene. Fucoxanthin has the ability to protect against oxidative stress and UVB induced cell injury in human fibroblast cells. It has been reported that fucoxanthin isolated from *Laminaria japonica* suppresses tyrosinase activity in UVB-irradiated guinea pigs and melanin synthesis in UVB-irradiated mice. Oral doses of fucoxanthin significantly suppressed skin mRNA expression related to melanogenesis, suggesting that fucoxanthin negatively regulates the melanogenesis factor at the transcriptional level (Shimoda et al. 2010).

Fucoxanthin has remarkable biological properties based on its unique molecular structure when compared to other carotenoids, with an unusual allenic bond and epoxy group in its molecule (Fig. 2) (Nomura et al. 1997; Yan et al. 1999). It was the first allenic carotenoid found in brown seaweeds (Dembitsky and Maoka 2007), with the allenic group thought to be responsible for its higher antioxidant properties (Sachindra et al. 2007).

Astaxanthin is another example of a carotenoid that is found in marine algae (and also other marine organisms), with superior antioxidant properties to carotenoids such as vitamin A and β-carotene. The presence of the ketone and hydroxy groups on the rings attached at the ends of the molecule (Fig. 3) are thought to be responsible for its higher antioxidant activity (Shibata et al. 2001; Riccioni et al. 2011). Recently, Tominaga et al. (2012) conducted an 8 wk open label non-controlled study involving both topical and oral administration of astaxanthin on a group of 30 women. Improvements in skin wrinkle, age spot size, skin texture, moisture content of the corneocyte layer/condition were observed. They also conducted a randomized double-blind placebo study on 36 male subjects for a period of 6 wk, with similar improvements in skin quality also observed for men (Tominaga et al. 2012).

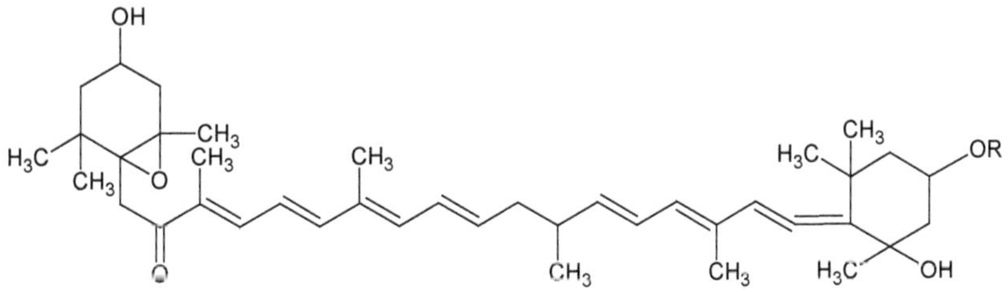

Fig. 2. The structure of fucoxanthin.

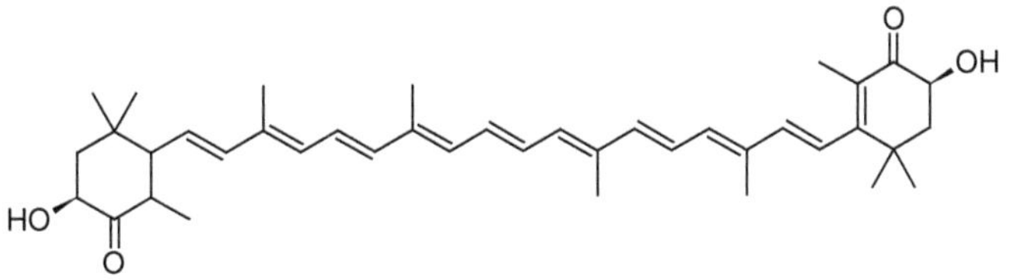

Fig. 3. The structure of astaxanthin.

Phenolic compounds

There is increasing interest in the use of marine plants as a source of antioxidants. It has been found that marine macroalgae are a rich source of phenolic compounds (Radhir et al. 2004; Smit 2004; Kim et al. 2006; Shibata et al. 2008; Heo et al. 2010), and possess a wide range of physiological properties, such as anti-allergenic, anti-artherogenic, anti-inflammatory, anti-microbial, anti-thrombotic, cardioprotective, vasodilatory, and antioxidant effects (Athukorala et al. 2006; Balasundram et al. 2006; Kim et al. 2006; Shibata et al. 2008; El Gamal 2009).

Most naturally occurring phenolic compounds are associated with mono- or polysaccharides (glycosides), linked to one or more of the phenolic groups. They may also occur as functional derivatives such as esters and methyl esters. Phlorotannins (Fig. 4a), a group of phenolic compounds restricted to polymers of phloroglucinol (1,3,5-trihydroxybenzene) (Fig. 4b), have been identified in several brown algal families such as *Alariaceae, Fucaceae*, and *Sargassaceae* (Pavia and Brock 2000; Jormalainen and Honkanen 2004). Some species of brown algae contain up to 20% of the phlorotannins in its dried form (Koivikko et al. 2005; Koivikko et al. 2007). Phlorotannins purified from several different brown algae have up to eight interconnected rings. Antioxidant activity is strongly related to the number of phenolic rings present, which act as electron traps to scavenge peroxy, superoxide-anions, and hydroxyl radicals (Athukorala et al. 2006; Balasundram et al. 2006).

Methanol extracts of marine brown algae are rich in phlorotannins, exhibiting inhibitory effects on melanogenesis, and providing protection against photo-oxidative stress induced by UVB radiation (Heo et al. 2010). Melanin, the major pigment responsible for the color of human skin, may be over stimulated by chronic sun exposure or other hyperpigmentation diseases. Tyrosinase is a copper-containing monooxygenase that catalyzes melanin synthesis in melanocytes (Kim et al. 2006) by hydroxylation of tyrosine into dihydroxyphenylalanine (DOPA) and other intermediates (Heo et al. 2010). Phlorotannins possess tyrosinase inhibitory activity due to their ability to complex the copper present in this enzyme. Inhibition of tyrosinase activity can prevent melanogenesis. Overexposure to UV radiation causes an acute sunburn reaction, clinically manifested as erythema. Chronic sun exposure is associated with abnormal cutaneous reactions; that is, epidermal hyperplasia, accelerated breakdown of collagen, and inflammation. Photo-oxidative stress has been directly linked to the onset of skin photodamage (Pillai et al. 2005). Hence, regular skin treatment with products containing antioxidants is thought be a useful strategy for preventing/reducing UV induced damage.

UVB radiation (280–320 nm) is highly oxidative and directly induces photodamage in skin cells by overproduction of ROS that interact with cellular DNA, proteins and lipids to alter their cellular functions (Heo et al. 2010). Brown algae are more tolerant to UVB radiation due to the presence of phlorotannins. The inhibitory effect of phlorotannins on melanin synthesis (melanogenesis) and their protective effect against photo-oxidative stress induced by UVB radiation were studied using three phlorotannins (phloroglucinol, eckol, and dieckol) isolated from the brown algae, *Ecklonia cava*. The UVB radiation-induced DNA damage in cultured human fibroblast cells was reported as 51.5%. However, the DNA damage was reduced by the addition of phlorotannins to cells exposed to the UVB radiation (Balasundram et al. 2006). Dieckol was found to be most potent phlorotannin in protecting the cells from UVB radiation by reducing the DNA damage induced by UVB radiation (Ko et al. 2011). Dieckol (the hexamer of phloroglucinol) had greater activity than eckol (trimer) and phloroglucinol (monomer). Dieckol (Fig. 3b) was found to inhibit cellular pigmentation more effectively than kojic acid, a commercial tyrosinase inhibitor, but was much less effective than retinol. Recent studies have demonstrated that the protective effect against oxidative stress induced by ROS and UV radiation is connected to the number and position of hydrogen-donating hydroxyl groups on the aromatic ring of the phenolic molecules. Therefore, as dieckol has more hydroxyl groups than other phlorotannins, it would be the phlorotannin of choice for use as a natural antioxidant in

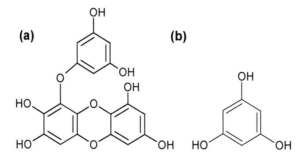

Fig. 4. The structure of: (a) phlorotannin; and (b) phloroglucinol.

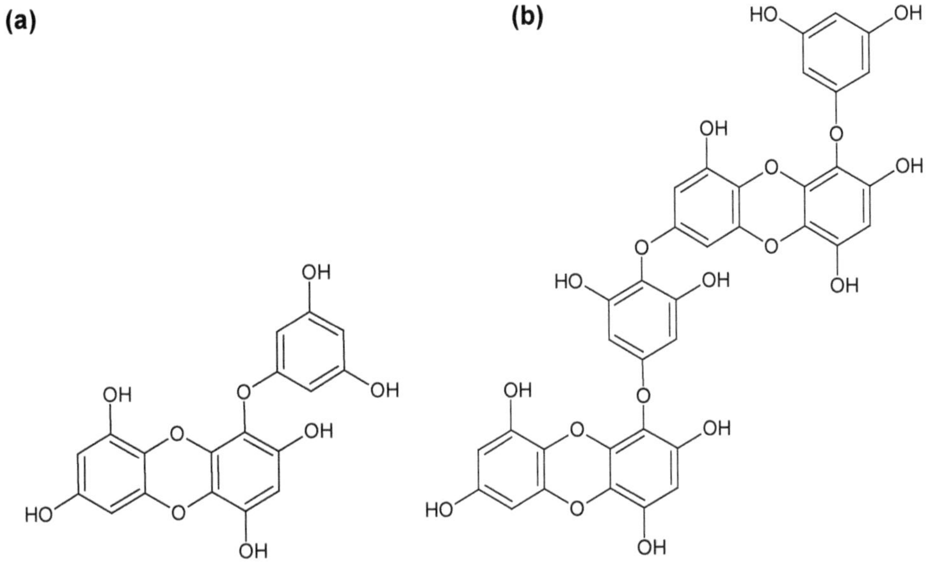

(a)

(b)

Fig. 5. The structure of: (a) eckol; and (b) dieckol.

cosmeceuticals (Athukorala et al. 2006; El Gamal 2009; Li et al. 2009). Due to its strong inhibitory activity upon tyrosinase and melanin synthesis, it might also be useful as a natural whitening agent (Heo et al. 2009).

Fucoidans

Fucoidans (or fucans) is the generic name for a class of gelatinous bioactive polysaccharides that were first isolated by Kylin in 1913 from brown algae (Kylin 1913). Fucoidans are sticky and viscous substances that are mostly found in the fibrillar cell walls and intercellular spaces of various species of brown algae (*Phaeophytae*), such as *Laminaria japonica, Fucus vesiculosus, Undaria pinnatifida, Cladosiphon okamuranus,* and *Hizikia fusiforme*. They reinforce cell walls and protect them from dehydration when the seaweed is exposed at low-tide. Experimental studies have suggested that fucoidans have antioxidant (Li et al. 2008), skin protective, and anti-aging properties, in addition to their anti-bacterial (Zapopozhets et al. 1995), anti-viral (Hayashi et al. 2008), anti-inflammatory, anti-coagulant (Cumashi et al. 2007), anti-tumor properties (Kim et al. 2005), and immune modulatory effects (Zapopozhets et al. 1995). Fucoidans show great potential in cosmeceutical formulations as they: (1) Inhibit matrix metalloproteases (collagenase) and the serine protease elastase; (2) enable contraction of fibroblast-populated collagen gels, increasing integrin expression on fibroblasts; (3) increase skin thickness and elasticity; (4) modulate growth factor activity; (5) have anti-inflammatory properties; and (6) inhibit tyrosinase (Thomas and Kim 2013).

Fucoidans are a complex and extremely diverse group of compounds. They are L-fucose rich, long chained highly branched polysaccharides with species specific sugar compositions (Zyvagintseva et al. 2003), and generally contain sulfated and acetylated groups. The side branches contain fucose or other glycosyl units that are bound together by differing glycosidic bonds (Percival and Ross 1950; Holtkamp et al. 2009). They contain high amounts of fucose and varying amounts of galactose, xylose, and glucuronic acid. The location of minor monosaccharide constituents (galactose, mannose, xylose, glucose) is still unknown. Fucoidans differ in their structure between algal species with structural variations even found within the same species. Due to the large variation of fucoidan structure within algae and the effect of the differing extraction procedures used, many distinct fucoidan forms have been isolated (Li et al. 2008).

Using GC/MS methylation data, Patankar et al. (1993) suggested that the core region of fucoidan was primarily a polymer of (1→3) linked fucose with sulfate groups substituted at the C-4 position on some of the fucose residues. Fucose was also attached to this polymer to form branched points, one for every 2–3 fucose residues within the chain (Fig. 6) (Li et al. 2008).

Each brown algae contains its own specific fucoidan with specific sugar composition, molecular weight, and level of sulfation. Fucoidans are often classified into two major groups based on differences in structural characteristics that originate from different algal species. The first group of fucoidans are isolated from *Laminaria cichorioides* and *L. japonica*. Fucoidans from *Laminaria cichorioides* (Anastyuk et al. 2010) consist of α (1–3) linked fucopyranose residues, while those from *L. japonica* are primarily α (1–3) linked fucopyranose residues (75%) with a few α (1–4) fucopyranose linkages (25%) (Wang et al. 2010). Sulfate groups (SO_3^{2-}) occupy mainly C-2 and sometimes C-4, although 3,4-diglycosylated and some terminal fucose residues may be non-sulfated. Acetate (O-acetyl group) occupies C-4 of 3-linked α-L-fucopyranose and C-3 of 4-linked α-L-fucopyranose in a ratio of around 7:3. The second group are fucoidans isolated from *Ascophyllum nodosum* and *Fucus* species (*Fucus serratus* (Bilan et al. 2006)), *F. evanescens* (Anastyuk et al. 2009) and *F. distichus* (Bilan et al. 2004) with a backbone consisting of alternating (1–3) and (1–4) linked α-L-fucopyranose residues (Bilan et al. 2004; Bilan and Usov 2008). A fucoidan isolated from *Turbinaria conoides* was shown to be highly complex, with terminals and/or branched in the (1→3) linked main chain (Chattopadhyay et al. 2010). Each brown macroalgae contains its own specific fucoidan (Patel et al. 2002). The molecules tend to vary in their natural sugar composition, molecular weight, and level of sulfation.

Since fucoidans are present in many edible seaweeds as a dietary fiber and have been found to be nontoxic in cell culture, they are considered to be of very low toxicity and hence, safe for use as ingredients in cosmeceutical products (Holtkamp et al. 2009). Safety trials in cancer patients have demonstrated that ingestion of up to 6 g of *Undaria* per day, containing 10% w/v fucoidan had no observable side effects and hence, was considered safe for use as a therapeutic agent (Fujimura et al. 2002). Fucoidans are easily incorporated (dispersed) into cosmeceutical formulations. Pure fucoidan extracts are generally off-white or brown water soluble powders with no strong odor or taste. They form relatively non-viscous solutions in water and unlike alginates or agar, do not add significant body to formulations. Suitable formulation concentrations providing biological activity are found to be around 0.1% w/v–1% w/v, although this depends on other factors such as molecular weight (Fujimura et al. 2000).

There are three types of commercially available fucoidans, U, F, and G fucoidan (Takara-Bio Inc. Japan) that are marketed as a cosmeceutical bioactive. Classification is based on the monosugar composition. U-fucoidan has greater amount (20%) of glucuronic acid (containing glucuronic acid besides sulfated fucose), F-fucoidan is mostly composed of sulfated L-fucose, while G-fucoidan is a sulfated fucogalactan (containing galactose) (Mizunum et al. 2010; Kim et al. 2013).

Traditionally, the extraction of fucoidan has relied on ethanol precipitation or high temperatures, resulting in extracts with unpredictable molecular weights and solvent residues. A newer and more efficient method involves a solvent free cold water process that yields extracts with defined molecular weight ranges and high levels of purity (Fujimura et al. 2002). Currently, skin care products using fucoidan are generally composed of a partially hydrolyzed fucoidan dispersed in a suitable base. Partially hydrolyzed fucoidan

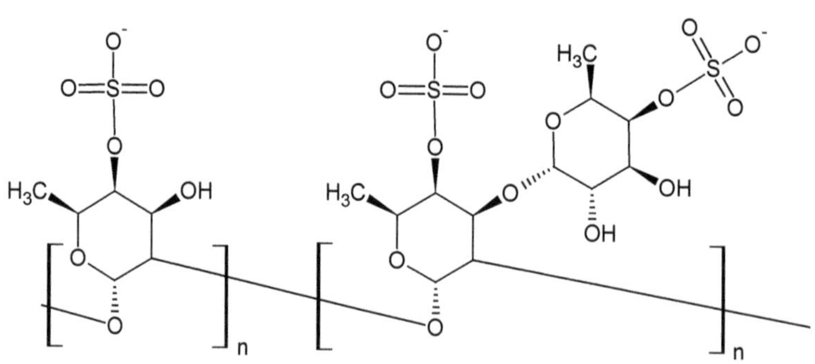

Fig. 6. Patankar model for the average structure of fucoidans.

is usually extracted from heated algae. Extraction methods may be controlled to produce high molecular weight materials that can further be fractionated to lower molecular weight derivatives. Levels of sulfation and acetylation can be controlled to enable formation of new substances with increased efficacy. Wang et al. (2008) reported that there is a positive correlation between the ratio of sulfate content to fucose content and the level of antioxidant activity in a fucoidan extract.

Of particular interest are the purified fucoidan extracts of *Fucus vesiculosus*. When formulated into creams and lotions, they are used as anti-ageing and anti-wrinkle products, inhibiting matrix enzymes, such as hyaluronidase, heparanase, phospholipase A2, tyrosine kinase, and collagenase and also act as anti-inflammatory agents. These products were also found to increase the number of dermal fibroblasts and deposition of collagen, collagen tightness, and facial elasticity. As such, it demonstrates soothing, smoothing, emollient, and skin conditioning properties (Fujimura et al. 2000). The fucoidan extract of *Fucus vesiculosus* ($M \leq 30,000$ g mol^{-1}), promotes contraction of fibroblast-populated collagen gels (a simplified *in vitro* model of the dermis), thus contributing to its epithelizing properties (Fujimura et al. 2000). The promotion of collagen gel contraction is caused by an increased expression of cell surface integrin α_2 and β_1 subunit molecules on the surface of the fibroblasts (Fujimura et al. 2002). Integrins mediate interactions between fibroblasts and extracellular matrix proteins (including collagen fibers) in the dermis. This suggests the possibility that fucoidan extracts could improve the thickness and the mechanical properties of human skin *in vivo* by enhancing the integrin expression of skin fibroblasts. A significant decrease in skin thickness, together with a significant improvement in elasticity was recorded, when compared to skin treated with a placebo gel (Fujimura et al. 2002). Given that the thickness normally increases and the elasticity usually decreases with age in cheek skin, these results demonstrate the anti-ageing activity of the fucoidan extracted from *Fucus vesiculosus*. Moreover, it was shown that fucoidan extracts aid in the promotion of collagen gel contraction. Such contraction is usually caused by an increased expression of cell surface integrins, which mediate interactions between fibroblast and extracellular matrix proteins in the dermis. Integrins are a large family of transmembrane glycoproteins that attach cells to the extracellular matrix or to ligands on other cells. Hence, *Fucus vesiculosus* extracts can be beneficial in a wide variety of cosmetics due to their effects on skin tightening, anti-sagging, and wrinkle smoothing properties. Fucoidans are also of interest due to their inhibitory effects on aging and photo-damaged skin when applied topically. Studies have demonstrated that fucoidan enhances dermal fibroblast proliferation and deposition of collagen (Senni et al. 2006). Matrix metalloproteases (MMPs), which modulate connective tissue breakdown, and hyaluronidase, are inhibited by low molecular weight fucoidan from *Undaria* and a smaller molecular weight fucoidan from *Sargassum*.

Skin wrinkling is normally attributed to the ROS released upon oxidative stress. ROS activated protein kinases phosphorylates transcription factor activator protein 1, which, in turn, increases expression of MMP that contributes to the degradation of skin collagen, ultimately leading to skin ageing (Fisher et al. 1996). Low molecular weight fucoidans enhance dermal fibroblast proliferation and deposition of collagen and other matrix factors. They inhibit activity of MMPs due to increased association with their inhibitors, resulting in the protection of human skin elastic fiber network (Senni et al. 2006). MMPs are proteolytic enzymes that can be classified into subclasses of collagenases, gelatinases, stromelysins, matrilysins, and membrane-type MMPs (MT-MMPs). MMPs are expressed in response to stimuli such as ultraviolet radiation, cytokines, and growth factors. While their increased activation accelerates skin ageing, they are essential in the prevention of wound scars. They are involved in degradation of the extracellular matrix which is responsible for tissue repair (healing of wounds). MMPs are essential for remodelling the extracellular matrix. Alterations in collagen and elastin present in the extracellular matrix are responsible for the clinical manifestations of skin ageing (i.e., wrinkles, sagging and laxity) (Philips et al. 2007). The atrophy of collagen and elastin fibers are the result of unregulated expression of their degradative enzymes, collagenases (MMP-1), gelatinases (MMP-2 and -9), and elastases. Acute exposure of human skin to UVB radiation can induce expression of the fibroblast collagenase-1 (MMP-1), thereby enhancing the rate of loss of collagen from the skin (Brenneisen et al. 2002; Moon et al. 2008). The collagenases (MMP-1, -8, -13, and -18) cleave structural (interstitial) collagens, with MMP-1 being the predominant one. Fucoidan extracts have been shown to inhibit expression of MMP-1, and thus can be used to restore elasticity of human skin (Moon et al. 2008). Floridoside and d-isofloridoside, isolated from the edible red alga *Laurencia undulata*, show significant antioxidant capacity and are potential inhibitors of MMP-2 and MMP-9 (gelatinase A and gelatinase B). Gelatinases, especially MMP-2 and MMP-9, degrade basement membrane collagens and denatured structural collagens (Philips et al. 2011).

The gelatinases, which include MMP-2 and MMP-9, promote UV-induced skin damage. It is reported that sun-damaged skin shows significantly elevated levels of active gelatinases (MMP-2 and -9) when compared to intrinsically aged skin (Chung et al. 2001). Ryu et al. (2009) conducted *in vitro* studies on methanol extracts from marine alga *Corallina pilulifera* and showed they were able to prevent UV-induced oxidative stress, and also expression of MMP-2 and MMP-9 in human dermal fibroblast (HDF) cells. This suggests that many phenolic compounds from marine algae may be useful as actives in sunscreen and anti-photoageing skin formulations (Ryu et al. 2009).

Hyperpigmentation disorders that are associated with abnormal accumulation of melanin pigments can be improved by treatment with de-pigmenting agents. Melanin reducing compounds that inhibit tyrosinase, an oxidase responsible for the first step in melanin biosynthesis, are the most promising for preventing and treating pigmentation disorders, as well as acting as effective skin-whitening agents. Fucoidan fractions have been found to inhibit tyrosinase (Kang et al. 2006; Wang et al. 2012; Song et al. 2015). Recent work has shown that the melanin synthesis and tyrosinase activity in B16F10 melanoma cells were decreased in a dose-dependent manner, when treated by fucoidans, as compared to the control (Jung et al. 2009).

Low molecular weight fucoidan fractions, obtained using acidic hydrolysis, are also considered to be plant homologues of mammalian heparin due to their heparin-like properties. Although low molecular weight fucoidans and mammalian heparin belong to different polysaccharide families with different sugar back bones, both molecules are highly sulfated and they both exhibit fibrinolytic and anticoagulant activities (Berteau and Mulloy 2003) and are able to potentiate activity of the transforming growth factor-β1 (TFG-β1) protein (Soeda et al. 1992; McCaffrey et al. 1994; Halloran 1996; O'Leary et al. 2004). The two families of polysaccharide, however, are different. For example, both fucoidan and heparin exhibit anti-proliferative effects on vascular smooth muscle, but the activity follows subtly different mechanisms. Heparin is usually formulated into topical lotions for the treatment of bruises where it acts as a fibrinolytic agent. Its activity in skin is in part modulated by growth factor potentiation. Heparin is an essential cofactor for growth factor activities, including that of fibroblast growth factor. TFG-β proteins are important regulators of wound healing and exert diverse and potent effects on proliferation, differentiation, and extracellular matrix synthesis. Fucoidans have been shown to modulate the effects of a variety of growth factors through mechanisms of action similar to heparin. Irregular wound healing will cause either scarring or chronic wounds. Therefore, there is a considerable interest in agents which can modulate certain aspects of the wound healing process. It has been found that preparations of fucoidan, as well as heparin inhibit fibroblast proliferation at concentrations from 0.01 to 100 mg/mL through interaction with transforming growth factor TGF β1. These results show the potential of fucoidans as important wound healing agents. When compared to heparin, fucoidans may be a better choice in cosmeceuticals, such as after sun products, allergic skin condition soothing products, or specialty postsurgical products due to their greater stability (Teixeira and Hellewell 1997). Fucoidans are also effective inhibitors of chemokine receptor type 4 (CXCR4), critical regulators of cell migration in the context of immune surveillance, inflammation, and development; and inhibit accumulation of eosinophils in models of allergic skin inflammation. Topical heparin is useful as a treatment in burns, where modulation of the CXCR4 interaction may result in superior healing.

Due to their anti-inflammatory properties, fucoidans have potential for use in both topical and oral anti-inflammatory products as they are a potent inhibitor of enzyme phospholipases A2 (Marcussi et al. 2007). Phospholipases A2 PLA2s catalyze the hydrolysis of fatty esters in the 2-position of 3-phospholipid to release fatty acid and lysophospholipid. The fatty acid so formed may act as either second messenger or a precursor of eicosanoids, which are lipid mediators of inflammatory reactions, such as prostaglandins (PGs) and leukotrienes (LTs).

Alginates

Alginate is a polysaccharide that was first extracted from brown algae by Stanford in 1881 (Stanford 1883). He named the substance algin after the Latin word for alga, meaning seaweed. Alginic acid

(Fig. 7) and alginate salts are structural components of the cell wall of brown algae (*Phaeophyceae*), mainly *Laminaria* species (*Laminaria hyperborean, Laminaria digitata, Laminaria japonica*), and also from *Macrocystis pyrifera, Ascophyllum nodosum, Ecklonia maxima, Lessonia nigrescens, Durvillea antarctica*, and *Sargassum* spp. These polysaccharides provide the algae with mechanical strength and flexibility, enabling them to adjust to the range of water movements in which they grow. Alginates also allow them to swell in water, which make them able to resist hydration when exposed to air (Rinaudo 2008). All brown seaweeds contain alginate, but there is a large variation in the quantity and quality of the alginate present. Carefully selected species that are dried and pulverized can contain up to 20–40% of alginic acid.

Alginates are composed of alginic acid and its salts (i.e., sodium, potassium, magnesium, and calcium). Sodium alginates are water soluble, while heavy metal alginates are not. Sodium alginate is widely used for its thickening, gel-forming, and stabilizing properties to form mucilages and gels of controllable consistency. Alginic acid forms a high-viscosity "acid gel" at low pH, while alginate is also easily gelled in the presence of a divalent cation such as calcium. Alginic acid is a high molecular weight linear long-chained polymeric salt of β-D-mannuronic acid (M block) and its C5 epimer, α-L-guluronic acid (G block). Mannuronic acid and guluronic acid units are arranged as homopolymeric G blocks, M blocks, alternating GM or random heteropolymeric G/M stretches. Although these units only differ at C5, they possess very different conformations; D-mannuronic acid being $^4C_1$ with diequatorial links between them and L-guluronic acid being $^1C_4$ with diaxial links between them. The proportion as well as the distribution of the two monomers determine to a large extent the physicochemical properties of alginate. The M/G composition varies from one species of brown alga to another.

Alginates rich in mannuronic acid that are found in the brown algae *Durvillea* and *Ascophyllum,* form soft, flexible gels, with added elasticity and low porosity, whereas alginates which are rich in guluronic acid (found in *Laminaria hyperborea*) form firmer rigid gels with high porosity (Kim et al. 2008). Alginates have been widely used in cosmetics as a foundation for face masks, applications for the body, and as a broad spectrum body wash ingredient. They are a valuable component in cosmetics due to their role in repairing skin structure and function (Podkorytova et al. 2007), their outstanding capacity to preserve water, and their desirable gelling, viscosity enhancing, and stabilizing characteristics (Prasad et al. 2007). Alginate solubility and water-holding capacity depends on pH (precipitating below about pH 3.5), molecular weight (lower molecular weight calcium alginate chains with less than 500 residues showing increasing water binding with increasing size), ionic strength (low ionic strength increasing the extended nature of the chains), and the nature of the ions present. At low pH, alginates are extremely efficient hydrocolloids that are used to solidify and stabilize emulsions (Kim et al. 2008). Due to their linear molecular arrangement and high molecular weight, alginates form strong films and good fibres in the solid state (Rinaudo 2008). A gel network is formed by the selective cross linking of two G-blocks of adjacent polymer chains with multivalent cations (e.g., Ca^{2+} or Ba^{2+}) through interaction of the carboxylic groups in the sugars (Augst et al. 2006). Alginates also form acidic gels stabilized by hydrogen bonds at low pH. Although alginic acid and its calcium salt (calcium alginate) are water-insoluble, they can swell and absorb more than several hundred times their weight in water. Alginic acid is used as a thickening agent and to form a moisture-retaining surface film. It can also bind heavy metal ions that are involved in oxidative processes and formation of radicals. A slight tightening effect is also experienced during the superficial filming process. Hence, it fulfills several cosmetic functions at the same time. Alginic acid also

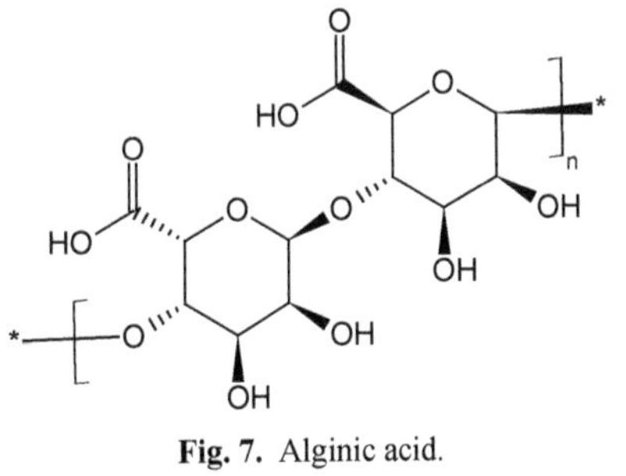

Fig. 7. Alginic acid.

stabilizes the oil phase in emulsifier-free cosmetics by increasing its viscosity. It is important to note that alginates are not absorbed into the skin. Propylene glycol alginate, an ester formed from propylene glycol and alginic acid, has similar properties to alginic acid.

The main use of the alginates are as thermally stable cold setting gelling agents and are prepared by the addition of calcium ions, with gelling occurring at much lower concentrations than when compared to gelatin. Such gels can be heat treated without melting, although they may eventually undergo degradation. The choice of gelling ions has a significant effect on the final gel properties. Low molecular weight alginate with a low concentration of gelling ions (like Ca^{2+}) generally exhibits the highest inhomogeneity (Skjåk-Bræk et al. 1989). Alginate with high G content produces strong brittle gels with good heat stability, but prone to water separation on freeze-thaw. Alginate with high M content produces weaker more-elastic gels with good freeze-thaw behavior. High MGMG content and a high concentration of gelling ions (Ca^{2+} ions) are found to reduce shear (Donati et al. 2005).

The uses of alginates are based on their ability to: (a) increase the viscosity of solutions when dissolved in water; (b) their ability to form gels, when a calcium salt is added to a solution of sodium alginate in water; and (c) their ability to form films of sodium or calcium alginate and fibres of calcium alginates. In contrast to the agar gels, where the water must initially be heated to about 80°C in order to dissolve the agar and the gel forms when cooled below about 40°C, no heat is required to form an alginate gel. Also, these gels do not melt when heated.

Alginic acid is also structurally related to hyaluronic acid or hyaluronan, and shows molecular weight dependent inhibition of hyaluronidase (Asada et al. 1997). Hyaluronidases are a group of enzymes that degrade hyaluronic acid polymers (Fraser et al. 1997). Hyaluronic acid is a major constituent of the extracellular matrix (ECM) of most tissues, including skin, synovial fluid, and vitreous humor (Fraser et al. 1997). Therefore, it is highly compatible and suitable for use in many medical applications. The name "hyaluronic acid" comes from the Greek "hyalos" (vitreous) and uronic acid (Meyer and Palmer 1934). It is a high molecular weight linear polysaccharide polymer of the family of glycosaminoglycans, one of the largest matrix molecules, and is a polymer with a length of 2–25 µm and a molecular weight 106–107 Da. The apparent size of hyaluronic acid is even greater, because it can incorporate a large volume of water. It is composed of repeating disaccharide units of β-1,3 linked N-acetyl-D-glucosamine and D-glucuronic acid, which are connected by β-1,4 glycosidic bonds (Lapcik et al. 1998).

Hyaluronic acid is frequently referred to as "hyaluronan", since it exists *in vivo* as a polyanion, because the carboxylic groups of the glucuronic acid moieties are deprotonated at physiological pH (pKa 3–4). Hyaluronic acid can form highly viscous solutions and can influence the properties of the extracellular matrix (Kreil 1995).

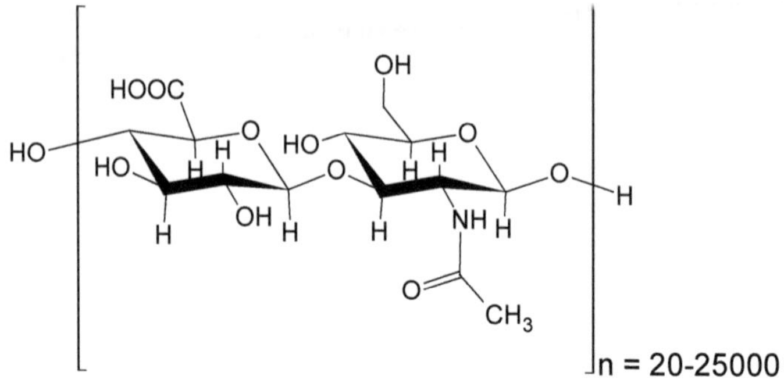

Fig. 8. Chemical structure of hyaluronic acid.

Polyunsaturated fatty acids (PUFA)

Polyunsaturated fatty acids (PUFAs) are fatty acids with two or more methylene-interrupted double bonds in the hydrocarbon chain, with a methyl group at one end and a carboxyl group at the other (Funk 2001). The reactive carboxyl group readily combines with alcohol groups to form triglycerides and phospholipids. PUFAs are classified primarily by the number and position of their double bonds, into two distinct groups; ω-3 and ω-6 PUFA (or omega-3 and omega-6 PUFA). In particular, there is increasing interest in a ω-3 PUFA, named eicosapentaenoic acid (EPA, C20:5 $\Delta^{5,8,11,14,17}$, 20:5 ω-3). EPA is a fatty acid with 20 carbon atoms and five double bonds with the last double bond located at the third carbon from the methyl end of the hydrocarbon chain (ω-3) (Nettleton 1995).

ω-3 PUFAs are synthesized from the essential fatty acid α-linolenic acid (ALA, 18:3, ω-3), whilst ω-6 PUFAs are synthesized from the essential fatty acid precursor linoleic acid (LA, 18:2, ω-6) (Das 2006). Plants can insert double bonds between carbons 12 and 13 via conversion of oleic acid to LA by Δ^{12}-desaturase. Further desaturation can occur via the insertion by Δ^{15}-desaturase of a double bond between carbons 15 and 16 to form ALA. LA and ALA are essential fatty acids, since they cannot be synthesized *de novo* in mammals, due to a genetic lack of Δ^{12} and Δ^{15}-desaturase. Therefore, LA and ALA must be obtained from the diet. Marine microalgae and macroalgae (seaweeds) and certain microorganisms are capable of *de novo* PUFA synthesis, and thus represent a good source of these fatty acids.

Pereira et al. recently investigated the PUFA content of 17 species of green, brown and red macroalgae. They found the major PUFAs mainly consisted of C18 and C20 molecules (arachidonic, eicosapentaenoic, and linoleic acids) with higher concentrations of PUFAs observed in brown and red macroalgae (Pereira et al. 2012). Marine microalgae are an attractive source of PUFA as many of them have a high PUFA content (Wood 1988), especially in the lipophilic extracts (Wood 1988; Li et al. 2002). PUFAs comprise a large proportion of the total lipids in marine algae. For example, the PUFA concentration in the total lipids in microalgae *Tetraselmis suecica*, *Porphyridium cruentum*, and *Isochrysis galbana* is 20.9%, 17.1%, and 17.0%, respectively (Servel et al. 1994). However, the actual fatty acid and PUFA concentration can vary widely. The microalgae *Tetraselmis suecica*, *Porphyridium cruentum* and *Isochrysis galbana* have PUFA concentrations of 0.5%, 0.3%, and 4.3% (dry weight), respectively. Examples of other microalgae species with high PUFA content are *Isochrysis galbana*, *Pavlova* sp., and *Nannochloropsis oceanica* which have concentrations of 3.99%, 3.98%, and 3.78% (dry weight), respectively (Patil et al. 2007).

PUFAs have essential structural and functional roles in tissues. They form the core structure of the phospholipid bilayer in all cell membranes, regulating membrane fluidity and rigidity, electron and oxygen transport, as well as thermal adaptation and cell signaling.

PUFAs are of interest in cosmetics as components of sun lotions and as regenerating and anti-wrinkle agents in cosmetic products. They function by restoring the skin permeability barrier, preventing scaly dermatitis and skin dehydration associated with a lack of unsaturated fatty acids in the skin (Servel et al. 1994). Some of the PUFAs, such as linoleic acid and arachidonic acid, are necessary for growth and protection of the skin (Mansour et al. 1999). ω-3 and ω-6 PUFAs are known to facilitate cell regeneration and improvement in skin health. Most unsaturated fatty acids have good antioxidant activities (Henry et al. 2002). Hence, given that algae is an abundant source of PUFAs, their use in cosmeceuticals as natural antioxidative compounds has attracted much attention. However, the topical use of PUFAs in cosmetics and topical skin formulations is still limited due to the formation of malodorous secondary oxidation products.

Mycosporine-like amino acids

Continuous depletion of the stratospheric ozone layer has resulted in an increase in the intensity of UV radiation; that is, UVB irradiation (280–315 nm) and to some extent UVA irradiation (315–400 nm). Exposure from UVB and UVA radiation is considered unfavorable for living matter, and different protecting strategies have accordingly been developed to cope with their impact (Sachindra et al. 2007). As a result, a number of photosynthetic organisms have developed mechanisms to reduce the toxicity and damaging effects of UV radiation (Singh et al. 2008). These mechanisms include repair of UV induced damage of DNA, accumulation of carotenoids, detoxifying enzymes, radical scavengers and antioxidants, and synthesis of UV absorbing compounds, such as mycosporine-like amino acids (MAAs). MAAs form the most common class of UVA absorbing compounds, and occur widely in various marine organisms (Yang et al. 2012). The synthesis of scytonemin, a predominant UVA photoprotective pigment, is mostly reported in cyanobacteria. Carotenoids have both light-harvesting and photoprotective potential, either by direct quenching of the singlet oxygen and other ROS, or as a UVB absorbing compound.

Mycosporine like amino acids (MAAs) (e.g., mycosporine-glycine) are a group of over 20 UV absorbing compounds that are found in a diverse range of marine organisms where they act as sunscreens to reduce UV induced damage. Their main role is in screening against energetic UVA radiation. Besides their role as a sunscreen in aquatic organisms, it has been suggested that some MAAs can act as antioxidants (Dunlap and Yamamoto 1995). Their antioxidant nature increases their therapeutic effectiveness. MAAs also play a role in protecting against sunlight damage by acting as antioxidant molecules scavenging toxic oxygen radicals (Dunlap and Yamamoto 1995) and providing protection against photo-oxidative stress induced by ROS. They also act to protect cells against salt stress, against desiccation or thermal stress, and as an intracellular nitrogen reserve (Oren and Gunde-Cimerman 2007).

MAAs are colorless, water soluble compounds with low molecular weights (< 400 g mol^{-1}). They have a strong UV absorption maxima between 310–362 nm, a very high molar absorptivity ($\varepsilon = 28,000$–$50,000$ M^{-1} cm^{-1}) (Conde et al. 2000), and are photostable in both distilled and sea water in the presence of photosensitizers. This photoprotective effect is the primary function of MAAs (Carignan et al. 2009). They are composed of a cyclohexenone (3-aminocyclohexen-1-one) or a cyclohexenimine (1,3-diaminocyclohexene) chromophore that is conjugated with the nitrogen substituent of an amino acid or amino alcohol (Bandaranayake 1998; Cardozo et al. 2008). Some MAAs also contain sulfate esters or glycosidic linkages through the imino substituent (Bohm et al. 1995) to oligosaccharides consisting of galactose, glucose, xylose, glucuronic acid, and glucosamine (Fig. 9).

Incorporation of various amino acids or imino alcohol groups results in a diversity of about 20 MAAs. The red alga *Porphyra umbilicalis,* produces the MAAs, Porphyra-334 and Shinorine, with molar absorption coefficients at 334 nm of 42,300 and 44,700 M^{-1} cm^{-1}, respectively. Their filter capacity is therefore similar to that of synthetic UVA sunscreens. It has been shown that a cream with 0.005% MAAs can neutralize UVA effects as efficiently as a cream with 1% synthetic UVA filters and 4% UVB filters (Schmid et al. 2004). Studies on their photodegradation and photophysical characteristics have shown that MAAs are stable and effective as sunscreen compounds. However, MAAs have low photodynamic reactivity when compared with several commercially available sunscreen agents. Diverse synthetic analogues of MAAs have been developed for commercial purposes (Dunlap et al. 1998). Analogues of mycosporine-glycine (3-alkylamino-2-methoxycyclohex-2-enones) were too hydrolytically reactive and oxidatively unstable for practical applications. Tetrahydropyridine derivatives (1-alkyl-3-alkanoyl-1,4,5,6-tetrahydropyridines) developed from the natural MAAs chromophore model were sufficiently stable for commercial application as suncare products. After evaluation, a liposomal formulation Helioguard® 365 with mycosporine-like amino acids isolated from the red alga, *Porphyra umbilicalis*, was developed (Cardozo et al. 2007). Helioguard® 365 is the first natural UVA sunscreen formulation to come onto the market.

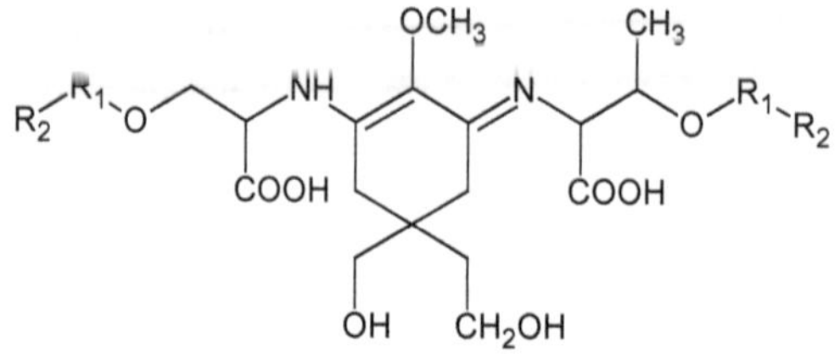

Fig. 9. A structure of the chromophore E335 conjugated with the amino acids serine and threonine, and the two saccharides R$_1$ (galactose, xylose, or glucuronic acid) and R$_2$ (galactose, glucose, or glucosamine).

Conclusion

Natural products play an invaluable role in the drug discovery process. Traditionally, terrestrial plants were used as the main source for the discovery of new compounds for skin care products. However, there is an increased interest in the marine environment as a new source of natural products. Investigation of new compounds from marine algae has proven to be promising for pharmaceutical study, due to their ability to produce unique metabolites unlike those found in terrestrial species, with high complexity and unlimited diversity of pharmacological and/or biological properties. This is due in part to the differences in the physicochemical nature of the sea environment where high pressures, low temperatures, lack of light, and high ionic concentrations are present, and may lead to the biosynthesis of highly functionalized and unusual molecules in marine organisms. Marine algae are important sources for a number of the high value chemicals, with unique beneficial skin properties that are used in cosmetic products. Chemicals such as β-carotene, astaxanthin, unsaturated fatty acids derived from algae, fucoidans, and mycosporine-like amino acids, have been found to provide antioxidant activity, UV protection, and an inhibitory effect on melanogenesis as demonstrated by *in vitro* studies. Many compounds found in algae have not been fully investigated in relation to their cosmeceutical benefits, but current evidence shows that they will have an increasingly important role as actives in cosmeceutical products. With the increasing demand for products aimed at slowing down and treating skin ageing, marine algae will become an increasingly important source of actives for these types of products.

References

Anastyuk, S.D., N.M. Shevchenko, E.L. Nazarenko, P.S. Dmitrenok and T.N. Zvyagintseva. 2009. Structural analysis of a fucoidan from the brown alga *Fucus evanescens* by MALDI-TOF and tandem ESI mass spectrometry. Carbohydr. Res. 344: 779–787.

Anastyuk, S.D., N.M. Shevchenko, E.L. Nazarenko, T.I. Imbs, V.I. Gorbach, P.S. Dmitrenok and T.N. Zvyagintseva. 2010. Structural analysis of a highly sulfated fucan from the brown alga *Laminaria cichorioides* by tandem MALDI and ESI mass spectrometry. Carbohydr. Res. 345: 2206–2212.

Asada, M., M. Sugie, M. Inoue, K. Nakagomi, S. Hongo, K. Murata, S. Irie, T. Takeuchi, N. Tomizuka and S. Oka. 1997. Inhibitory effect of alginic acids on hyaluronidase and on histamine release from mast cells. Biosci. Biotechnol. Biochem. 61: 1030–1032.

Athukorala, Y., K.N. Kim and Y.J. Jeon. 2006. Antiproliferative and antioxidant properties of an enzymatic hydrolysate from brown alga, *Echlonia cava*. Food Chem. Toxicol. 44: 1065–1074.

Augst, A.D., H.J. Kong and D.J. Mooney. 2006. Alginate hydrogels as biomaterials. Macromol. Biosci. 6: 623–633.

Balasundram, N., K. Sundram and S. Samman. 2006. Phenolic compounds in plants and agri-industrial by-products: Antioxidant activity, occurrence, and potential uses. Food Chem. 99: 191–203.

Bandaranayake, W.M. 1998. Mycosporines: are they nature's sunscreens? Nat. Prod. Rep. 15: 159–172.

Berner, T. 1993. Ultrastructure of Microalgae. CRC Press, Boca Raton.

Berteau, O. and B. Mulloy. 2003. Sulfated fucans, fresh perspectives: structures, functions, and biological properties of sulfated fucans and an overview of enzymes active toward this class of polysaccharide. Glycobiology 13: 29R–40R.

Bilan, M.I., A.A. Grachev, N.E. Ustuzhanina, A.S. Shashkov, N.E. Nifantiev and A.I. Usov. 2004. A highly regular fraction of a fucoidan from the brown seaweed *Fucus distichus* L. Carbohydr. Res. 339: 511–517.

Bilan, M.I., A.A. Grachev, A.S. Shashkov, N.E. Nifantiev and A.I. Usov. 2006. Structure of a fucoidan from the brown seaweed *Fucus serratus* L. Carbohydr. Res. 341: 238–245.

Bilan, M.I. and A.I. Usov. 2008. Structural analysis of fucoidans. Nat. Prod. Commun. 3: 1639–1648.

Bohm, G.A., W. Pfleiderer, P. Boger and S. Scherer. 1995. Structure of a novel oligosaccharide-mycosporine-amino acid ultraviolet A/B sunscreen pigment from the terrestrial cyanobacterium Nostoc commune. J. Biol. Chem. 270: 8536–8539.

Brenneisen, P., H. Sies and K. Scharffetter-Kochanek. 2002. Ultraviolet-B irradiation and matrix metalloproteinases: from induction via signaling to initial events. Ann. N. Y. Acad. Sci. 973: 31–43.

Cardozo, K.H., T. Guaratini, M.P. Barros, V.R. Falcao, A.P. Tonon, N.P. Lopes, S. Campos, M.A. Torres, A.O. Souza, P. Colepicolo and E. Pinto. 2007. Metabolites from algae with economical impact. Comp. Biochem. Physiol. C Toxicol. Pharmacol. 146: 60–78.

Cardozo, K.H.M., R. Vessecchi, V.M. Carvalho, E. Pinto, P.J. Gates, P. Colepicolo, S.E. Galembeck and N.P. Lopes. 2008. A theoretical and mass spectrometry study of the fragmentation of mycosporine-like amino acids. Int. J. Mass Spectrom. 273: 11–19.

Carignan, M.O., K.H. Cardozo, D. Oliveira-Silva, P. Colepicolo and J.I. Carreto. 2009. Palythine-threonine, a major novel mycosporine-like amino acid (MAA) isolated from the hermatypic coral *Pocillopora capitata*. J. Photochem. Photobiol. B. 94: 191–200.

Chattopadhyay, N., T. Ghosh, S. Sinha, K. Chattopadhyay, P. Karmakar and B. Ray. 2010. Polysaccharides from *Turbinaria conoides*: Structural features and antioxidant capacity. Food. Chem. 118: 823–829.

Chung, J.H., J.Y. Seo, H.R. Choi, M.K. Lee, C.S. Youn, G. Rhie, K.H. Cho, K.H. Kim, K.C. Park and H.C. Eun. 2001. Modulation of skin collagen metabolism in aged and photoaged human skin *in vivo*. J. Invest. Dermatol. 117: 1218–1224.

Conde, F.R., M.S. Churio and C.M. Previtali. 2000. The photoprotector mechanism of mycosporine-like amino acids. Excited-state properties and photostability of porphyra-334 in aqueous solution. J. Photochem. Photobiol. B. 56: 139–144.

Coragliotti, A., S. Franklin, A.G. Day and S.M. Decker. 2012. Microalgal polysaccharide compositions. US20120202768A1.

Cumashi, A., N.A. Ushakova, M.E. Preobrazhenskaya, A. D'Incecco, A. Piccoli, L. Totani, N. Tinari, G.E. Morozevich, A.E. Berman, M.I. Bilan, A.I. Usov, N.E. Ustyuzhanina, A.A. Grachev, C.J. Sanderson, M. Kelly, G.A. Rabinovich, S. Iacobelli and N.E. Nifantiev. 2007. A comparative study of the anti-inflammatory, anticoagulant, antiangiogenic, and antiadhesive activities of nine different fucoidans from brown seaweeds. Glycobiology 17: 541–552.

Das, U.N. 2006. Essential fatty acids: biochemistry, physiology and pathology. Biotechnol. J. 1: 420–439.

Dembitsky, V.M. and T. Maoka. 2007. Allenic and cumulenic lipids. Prog. Lipid Res. 46: 328–375.

Donati, I., S. Holtan, Y.A. Morch, M. Borgogna, M. Dentini and G. Skjak-Braek. 2005. New hypothesis on the role of alternating sequences in calcium-alginate gels. Biomacromolecules 6: 1031–1040.

Dunlap, W.C. and Y. Yamamoto. 1995. Small-molecule antioxidants in marine organisms: Antioxidant activity of mycosporine-glycine. Comp. Biochem. Physiol. B. 112: 105–114.

Dunlap, W.C., B.E. Chalker, W.M. Bandaranayake and J.J. Wu Won. 1998. Nature's sunscreen from the great barrier reef, Australia. Int. J. Cosmet. Sci. 20: 41–51.

El Gamal, A.A. 2009. Biological importance of marine algae. Saudi Pharm. J. 18: 1–25.

Fisher, G.J., S.C. Datta, H.S. Talwar, Z.Q. Wang, J. Varani, S. Kang and J.J. Voorhees. 1996. Molecular basis of sun-induced premature skin ageing and retinoid antagonism. Nature 379: 335–339.

Fisher, G.J., T. Quan, T. Purohit, Y. Shao, M.K. Cho, T. He, J. Varani, S. Kang and J.J. Voorhees. 2009. Collagen fragmentation promotes oxidative stress and elevates matrix metalloproteinase–1 in fibroblasts in aged human skin. Am. J. Pathol. 174: 101–114.

Fraser, J.R., T.C. Laurent and U.B. Laurent. 1997. Hyaluronan: its nature, distribution, functions and turnover. J. Intern. Med. 242: 27–33.

Fujimura, T., K. Tsukahara, S. Moriwaki, T. Kitahara and Y. Takema. 2000. Effects of natural product extracts on contraction and mechanical properties of fibroblast populated collagen gel. Biol. Pharm. Bull. 23: 291–297.

Fujimura, T., K. Tsukahara, S. Moriwaki, T. Kitahara, T. Sano and Y. Tatema. 2002. Treatment of human skin with an extract of *Fucus vesiculosus* changes thickness and mechanical properties. J. Cosmet. Sci. 53: 1–9.

Funk, C.D. 2001. Prostaglandins and leukotrienes: Advances in eicosanoid biology. Science 294: 1871–1875.

Halloran, P.F. 1996. Molecular mechanisms of new immunosuppressants. Clin. Transplant. 10: 118–123.

Hayashi, K., T. Nakano, M. Hashimoto, K. Kanekiyo and T. Hayashi. 2008. Defensive effects of a fucoidan from brown alga *Undaria pinnatifida* against herpes simplex virus infection. Int. Immunopharmacol. 8: 109–116.

Henry, G., R. Momin, M. Nair and D. Dewitt. 2002. Antioxidant and cyclooxygenase activities of fatty acids found in food. J. Agric. Food Chem. 50: 2231–2234.

Heo, S.J., S.C. Ko, S.H. Cha, D.H. Kang, H.S. Park, Y.U. Choi, D. Kim, W.K. Jung and Y.J. Jeon. 2009. Effect of phlorotannins isolated from *Ecklonia cava* on melanogenesis and their protective effect against photo-oxidative stress induced by UV-B radiation. Toxicol. *In Vitro*. 23: 1123–1130.

Heo, S.J., S.C. Ko, S.M. Kang, S.H. Cha, S.H. Lee, D.H. Kang, W.K. Jung, A. Affan, C. Oh and Y.J. Jeon. 2010. Inhibitory effect of diphlorethohydroxycarmalol on melanogenesis and its protective effect against UV-B radiation-induced cell damage. Food Chem. Toxicol. 48: 1355–1361.

Holtkamp, A.D., S. Kelly, R. Ulber and S. Lang. 2009. Fucoidans and fucoidanases-focus on techniques for molecular structure elucidation and modification of marine polysaccharides. Appl. Microbiol. Biotechnol. 82: 1–11.

Ishihara, K., M. Murata, M. Kaneniwa, H. Saito, W. Komatsu and K. Shinohara. 2000. Purification of stearidonic acid (18:4(n-3)) and hexadecatetraenoic acid (16:4(n-3)) from algal fatty acid with lipase and medium pressure liquid chromatography. Biosci. Biotechnol. Biochem. 64: 2454–2457.

Jiménez-Escrig, A., I. Jiménez-Jiménez, R. Pulido and F. Saura-Calixto. 2001. Antioxidant activity of fresh and processed edible seaweeds. J. Sci. Food Agr. 81: 530–534.

Joe, M.J., S.N. Kim, H.Y. Choi, W.S. Shin, G.M. Park, D.W. Kang and Y.K. Kim. 2006. The inhibitory effects of eckol and dieckol from Ecklonia stolonifera on the expression of matrix metalloproteinase-1 in human dermal fibroblasts. Biol. Pharm. Bull. 29: 1735–1739.

Jormalainen, V. and T. Honkanen. 2004. Variation in natural selection for growth and phlorotannins in the brown alga *Fucus vesiculosus*. J. Evolution. Biol. 17: 807–820.

Jung, H.A., S.K. Hyun, H.R. Kim and J.S. Choi. 2006. Angiotensin-converting enzyme I inhibitory activity of phlorotannins from Ecklonia stolonifera. Fisheries Sci. 72: 1292–1299.

Jung, S.-H., M.-J. Ku, H.-J. Moon, B.-C. u, M.-J. Jeon and Y.-H. Lee. 2009. Effects of fucoidan on melanin synthesis and tyrosinase activity. J. Life Sci. 19: 75–80.

Kakinuma, M., C.S. Park and H. Amano. 2001. Distribution of free L-cysteine and glutathione in seaweeds. Fisheries Sci. 67: 194–196.

Kang, H., H. Chung, J. Kim, B. Son, H. Jung and J. Choi. 2004. Inhibitory phlorotannins from the edible brown alga *Ecklonia stolonifera* on total reactive oxygen species (ROS) generation. Arch. Pharm. Res. 27: 194–198.

Kang, X.J., F.X. Wang, C.M. Sheng and Y. Zhu. 2006. Undaria pinnatifida stem fucoidan biological activity of the composition and bioactivity. Chin. J. Pharmacol. Toxicol. 41: 1748–1750.

Kim, H.H., C.M. Shin, C.H. Park, K.H. Kim, K.H. Cho, H.C. Eun and J.H. Chung. 2005. Eicosapentaenoic acid inhibits UV induced MMP-1 expression in human dermal fibroblasts. J. Lipid Res. 46: 1712–1720.

Kim, M.M., Q. Van Ta, E. Mendis, N. Rajapakse, W.K. Jung, H.G. Byun, Y.J. Jeon and S.K. Kim. 2006. Phlorotannins in *Ecklonia cava* extract inhibit matrix metalloproteinase activity. Life Sci. 79: 1436–1443.

Kim, S.-K., D. Ravichandran, S.B. Khan and Y.T. Kim. 2008. Prospective of the cosmeceuticals derived from marine organisms. Biotechnol. Bioprocess Eng. 13: 511–523.

Kim, S.-K., T.-S. Vo and D.-H. Ngo. 2013. Fucoidan: a potent ingredient of marine nutraceuticals. *In*: S.-K. Kim (ed.). Marine Nutraceuticals: Prospects and Perspectives. CRC Press, Hoboken.

Klein, T. and R. Bischoff. 2011. Physiology and pathophysiology of matrix metalloproteases. Amino Acids 41: 271–290.

Ko, S.C., S.H. Cha, S.J. Heo, S.H. Lee, S.M. Kang and Y.J. Jeon. 2011. Protective effect of *Ecklonia cava* on UVB-induced oxidative stress: *in vitro* and *in vivo* zebrafish model. J. Appl. Phycol. 23: 697–708.

Koivikko, R., J. Loponen, T. Honkanen and V. Jormalainen. 2005. Contents of soluble, cell-wall-bound and exuded phlorotannins in the brown alga *Fucus vesiculosus,* with implications on their ecological functions. J. Chem. Ecol. 31: 195–212.

Koivikko, R., J. Loponen, K. Pihlaja and V. Jormalainen. 2007. High-performance liquid chromatographic analysis of phlorotannins from the brown alga *Fucus vesiculosus*. Phytochem. Anal. 18: 326–332.

Kreil, G. 1995. Hyaluronidases—a group of neglected enzymes. Protein Sci. 4: 1666–1669.

Kylin, H. 1913. Biochemistry of sea algae. Z. Physiol. Chem. 83: 171–197.

Lapcik, L., Jr., L. Lapcik, S. De Smedt, J. Demeester and P. Chabrecek. 1998. Hyaluronan: preparation, structure, properties, and applications. Chem. Rev. 98: 2663–2684.

Lavker, R. 1979. Structural alterations in exposed and unexposed aged skin. J. Invest. Dermatol. 73: 59–66.

Lee, S., Y. Lee, S. Jung, S. Kang and K. Shin. 2003. Anti-oxidant activities of fucosterol from the marine algae Pelvetia siliquosa. Arch. Pharm. Res. 26: 719–722.

Li, B., F. Lu, X. Wei and R. Zhao. 2008. Fucoidan: structure and bioactivity. Molecules 13: 1671–1695.

Li, H.B. and F. Chen. 2001. Preparative isolation and purification of astaxanthin from the green microalga *Chlorococcum* sp. by high-speed counter-current chromatography. pp. 127–134. F. Chen and Y. Jiang (eds.). *In*: Algae and Their Biotechnological Potential: Proceedings of the 4th Asia-Pacific Conference on Algal Biotechnology, 3–6 July 2000 in Hong Kong. Springer Netherlands, Dordrecht.

Li, X.C., X. Fan, L.J. Han, X.J. Yan and Q.X. Lou. 2002. Fatty acids of common marine macrophytes from the Yellow and Bohai Seas. Oceanol. Limnol. Sin. 33: 215–223.

Li, Y., Z.J. Qian, B. Ryu, S.H. Lee, M.M. Kim and S.K. Kim. 2009. Chemical components and its antioxidant properties *in vitro*: An edible marine brown alga, *Ecklonia cava*. Bioorg. Med. Chem. 17: 1963–1973.

Lim, S.N., P.C.K. Cheung, V.E.C. Ooi and P.O. Ang. 2002. Evaluation of antioxidative activity of extracts from a brown seaweed, *Sargassum siliquastrum*. J. Agric. Food Chem. 50: 3862–3866.

Mansour, M.P., J.K. Volkman, D.G. Holdsworth, A.E. Jackson and S.I. Blackburn. 1999. Very-long-chain (C28) highly unsaturated fatty acids in marine dinoflagellates. Phytochemistry 50: 541–548.

Marcussi, S., C.D. Sant'Ana, C.Z. Oliveira, A.Q. Rueda, D.L. Menaldo, R.O. Beleboni, R.G. Stabeli, J.R. Giglio, M.R. Fontes and A.M. Soares. 2007. Snake venom phospholipase A2 inhibitors: medicinal chemistry and therapeutic potential. Curr. Top. Med. Chem. 7: 743–756.

Matsukawa, R., Z. Dubinsky, E. Kishimoto, K. Masaki, Y. Masuda, T. Takeuchi, A.M. Chihar, Y. Yamamoto, E. Niki and I. Karube. 1997. A comparison of screening methods for antioxidant activity in seaweeds. J. Appl. Phycol. 9: 29–35.

McCaffrey, T.A., D.J. Falcone, D. Vicente, B. Du, S. Consigli and W. Borth. 1994. Protection of transforming growth factor-β1 activity by heparin and fucoidan. J. Cell Physiol. 159: 51–59.

Meyer, K. and J.W. Palmer. 1934. The polysaccharide of the vitreous humor. J. Biol. Chem. 107: 629–634.

Mizutani, S., S. Deguchi, E. Kobayashi, E. Nishiyama, H. Sagawa and I. Kato. 2010. Fucoidan-containing cosmetics. US 7678368 B2.

Moon, H.J., S.R. Lee, S.N. Shim, S.H. Jeong, V.A. Stonik, V.A. Rasskazov, T. Zvyagintseva and Y.H. Lee. 2008. Fucoidan inhibits UVB-induced MMP-1 expression in human skin fibroblasts. Biol. Pharm. Bull. 31: 284–289.

Nettleton, J.A. 1995. Omega-3 Fatty Acids and Health. Chapman & Hall, New York.

Nomura, T., M. Kikuchi, A. Kubodera and Y. Kawakami. 1997. Proton-donative antioxidant activity of fucoxanthin with 1,1-diphenyl-2-picrylhydrazyl (DPPH). Biochem. Mol. Biol. Int. 42: 361–370.

O'Leary, R., M. Rerek and E.J. Wood. 2004. Fucoidan modulates the effect of transforming growth factor (TGF)-β₁ on fibroblast proliferation and wound repopulation in *in vitro* models of dermal wound repair. Biol. Pharm. Bull. 27: 266–270.

Oren, A. and N. Gunde-Cimerman. 2007. Mycosporines and mycosporine-like amino acids: UV protectants or multipurpose secondary metabolites? FEMS Microbiol. Lett. 269: 1–10.

Paduch, R., M. Kandefer-Szersze, M. Trytek and J. Fiedurek. 2007. Terpenes: substances useful in human healthcare. Arch. Immunol. Ther. Exp. (Warsz.). 55: 315–327.

Patankar, M.S., S. Oehninger and T. Barnett. 1993. A revised structure for fucoidan may explain some of its biological activities. J. Biol. Chem. 268: 21770–21776.

Patel, M.K., B. Mulloy, K.L. Gallagher, L. O'Brien and A.D. Hughes. 2002. The antimitogenic action of the sulphated polysaccharide fucoidan differs from heparin in human vascular smooth muscle cells. Thromb. Haemost. 87: 149–154.

Patil, V., T. Källqvist, E. Olsen, G. Vogt and H.R. Gislerød. 2007. Fatty acid composition of 12 microalgae for possible use in aquaculture feed. Aquacult. Int. 15: 1–9.

Pavia, H. and E. Brock. 2000. Extrinsic factors influencing phlorotannin production in the brown alga *Ascophyllum nodosum*. Mar. Ecol. Prog. Ser. 193: 285–294.

Percival, E.G.V. and A.G. Ross. 1950. Fucoidin. Part I. The isolation and purification of fucoidin from brown seaweeds. J. Chem. Soc. 717–720.

Pereira, H., L. Barreira, F. Figueiredo, L. Custódio, C. Vizetto-Duarte, C. Polo, E. Rešek, A. Engelen and J. Varela. 2012. Polyunsaturated fatty acids of marine macroalgae: Potential for nutritional and pharmaceutical applications. Mar. Drugs. 10: 1920–1935.

Philips, N., T. Keller, C. Hendrix, S. Hamilton, R. Arena, M. Tuason and S. Gonzalez. 2007. Regulation of the extracellular matrix remodeling by lutein in dermal fibroblasts, melanoma cells, and ultraviolet radiation exposed fibroblasts. Arch. Dermatol. Res. 299: 373–379.

Philips, N., S. Auler, R. Hugo and S. Gonzalez. Beneficial Regulation of Matrix Metalloproteinases for Skin Health Enzyme Res. [Online], 2011. http://dx.doi.org/10.4061/2011/427285.

Pillai, S., C. Oresajo and J. Hayward. 2005. Ultraviolet radiation and skin aging: roles of reactive oxygen species, inflammation and protease activation, and strategies for prevention of inflammation-induced matrix degradation—a review. Int. J. Cosmetic Sci. 27: 17–34.

Podkorytova, A.V., L.H. Vafina, E.A. Kovaleva and V.I. Mikhailov. 2007. Production of algal gels from the brown alga, Laminaria japonica Aresch., and their biotechnological applications. J. Appl. Phycol. 19: 827–830.

Pontius, A.T. and P.W. Smith. 2011. An antiaging and regenerative medicine approach to optimal skin health. Facial Plast. Surg. 27: 29–34.

Prasad, K., A.K. Siddhanta, M. Ganesan, B.K. Ramavat, B. Jha and P.K. Ghosh. 2007. Agars of *Gelidiella acerosa* of west and southeast coasts of India. Bioresour. Technol. 1907–1915.

Radhir, R., Y.T. Lin and K. Shetty. 2004. Phenolics, their antioxidant and antimicrobial activity in dark germinated fenugreek sprouts in respose to peptide and phytochemical elicitors. Asia Pac. J. Clin. Nutr. 13: 295–307.

Rasmussen, R.S. and M.T. Morrissey. 2007. Marine biotechnology for production of food ingredients. Adv. Food Nutr. Res. 52: 237–292.

Rhie, G., M. Shin, J. Seo, W. Choi, K. Cho, K. Kim, K. Park, H. Eun and J. Chung. 2001. Aging-and photoaging-dependent changes of enzymic and nonenzymic antioxidants in the epidermis and dermis of human skin *in vivo*. J. Invest. Dermatol. 117: 1212–1217.

Riccioni, G., N. D'Orazio, S. Franceschelli and L. Speranza. 2011. Marine carotenoids and cardiovascular risk markers. Mar. Drugs 9: 1166–1175.

Rinaudo, M. 2008. Main properties and current applications of some polysaccharides as biomaterials. Polym. Int. 57: 397–430.

Ryu, B., Z.-J. Qian, M.-M. Kim, K.W. Nam and S.-K. Kim. 2009. Anti-photoaging activity and inhibition of matrix metalloproteinase (MMP) by marine red alga, *Corallina pilulifera* methanol extract. Radiat. Phys. Chem. 78: 98–105.

Sachindra, N.M., E. Sato, H. Maeda, M. Hosokawa, Y. Niwano, M. Kohno and K. Miyashita. 2007. Radical scavenging and singlet oxygen quenching activity of marine carotenoid fucoxanthin and its metabolites. J. Agric. Food Chem. 55: 8516–8522.

Schaeffer, D.J. and V.S. Krylov. 2000. Anti-HIV activity of extracts and compounds from algae and cyanobacteria. Ecotox. Environ. Safety 45: 208–227.

Schmid, D., C. Schürch and F. Zülli. 2004. UV-A sunscreen from red algae for protection against premature skin aging. Cosmet. Toilet. Manuf. Worldw. 139–143.

Senni, K., F. Gueniche, A. Foucault-Bertaud, S. Igondjo-Tchen and F. Fioretti. 2006. Fucoidan a sulfated polysaccharide from brown algae is a potent modulator of connective tissue proteolysis. Arch. Biochem. Biophys. 445: 56–64.

Servel, M.O., C. Claire, A. Derrien, L. Coiffard and Y. De Roeck-Holtzhauer. 1994. Fatty acid composition of some marine microalgae. Phytochemistry 36: 691–693.

Shibata, A., Y. Kiba, N. Akati, K. Fukuzawa and H. Terada. 2001. Molecular characteristics of astaxanthin and β-carotene in the phospholipid monolayer and their distributions in the phospholipid bilayer. Chem. Phys. Lipids 113: 11–22.

Shibata, T., K. Nagayama, R. Tanaka, K. Yamaguchi and T. Nakamura. 2003. Inhibitory effects of brown algal phlorotannins on secretory phospholipase A₂s, lipoxygenases and cyclooxygenases. J. Appl. Phycol. 15: 61–66.

Shibata, T., K. Ishihara, S. Kawaguchi, H. Yoshikawa and Y. Hama. 2000. Antioxidant activities of phlorotannins isolated from Japanese Laminariaceae. J. Appl. Phycol. 20: 705–711.

Shimoda, H., J. Tanaka, S.J. Shan and T. Maoka. 2010. Anti-pigmentary activity of fucoxanthin and its influence on skin mRNA expression of melanogenic molecules. J. Pharm. Pharmacol. 62: 1137–1145.

Shiratori, K., K. Ohgami, I. Ilieva, X.H. Jin, Y. Koyama, K. Miyashita, K. Yoshida, S. Kase and S. Ohno. 2005. Effects of fucoxanthin on lipopolysaccharide-induced inflammation *in vitro* and *in vivo*. Exp. Eye Res. 81: 422–428.

Singh, S.P., S. Kumari, R.P. Rastogi, K.L. Singh and R.P. Sinha. 2008. Mycosporine-like amino acids (MAAs): chemical structure, biosynthesis and significance as UV-absorbing/screening compounds. Indian J. Exp. Biol. 46: 7–17.

Skjåk-Bræk, G., H. Grasdalen and O. Smidsrød. 1989. Inhomogeneous polysaccharide ionic gels. Carbohydr. Polym. 10: 31–54.

Smit, A.J. 2004. Medicinal and pharmaceutical uses of seaweed natural products. A review. J. Appl. Phycol. 16: 245–262.

Soeda, S., S. Sakaguchi, H. Shimeno and A. Nagamatsu. 1992. Fibrinolytic and anticoagulant activities of highly sulfated fucoidan. Biochem. Pharmacol. 43: 1853–1858.

Song, Y.S., M.C. Balcos, H.-Y. Yun, K.J. Baek, N.S. Kwon, M.-K. Kim and D.-S. Kim. 2015. ERK activation by fucoidan leads to inhibition of melanogenesis in Mel-Ab cells. Korean J. Physiol. Pharmacol. 19: 29–34.

Stahl, W. and H. Sies. 2005. Bioactivity and protective effects of natural carotenoids. Biochim. Biophys. Acta. 1740: 101–107.

Stanford, E.C.C. 1883. On algin; a new substance obtained from some of the commoner species of marine algae. Chem. News 47: 254–257.

Sukenik, A., O. Zmora and Y. Carmeli. 1993. Biochemical quality of marine unicellular algae with special emphasis on lipid composition: II. *Nannochloropsis* sp. Aquaculture 117: 313–326.

Tang, H.-F., Y.-H. Yi, X.-S. Yao, Q.-Z. Xu, S.-Y. Zhang and H.-W. Lin. 2002. Bioactive steroids from the brown alga *Sargassum carpophyllum*. J. Asian Nat. Prod. Res. 4: 95–101.

Tapiero, H., D.M. Townsend and K.D. Tew. 2004. The role of carotenoids in the prevention of human pathologies. Biomed. Pharmacother. 58: 100–110.

Teixeira, M.M. and P.G. Hellewell. 1997. The effect of the selectin binding polysaccharide fucoidin on eosinophil recruitment *in vivo*. Br. J. Pharmacol. 120: 1059–1066.

Thomas, N.V. and S.-K. Kim. 2013. Beneficial effects of marine algal compounds in cosmeceuticals. Mar. Drugs 11: 146–164.

Thompson, K.D. and C. Dragar. 2004. Antiviral activity of Undaria pinnatifida against herpes simplex virus. Phytother. Res. 18: 551–555.

Tominaga, K., N. Hongo, M. Karato and E. Yamashitae. 2012. Cosmetic benefits of astaxanthin on humans subjects. Acta Biochim. Pol. 59: 43–47.

Wang, J., Q. Zhang, Z. Zhang and Z. Li. 2008. Antioxidant activity of sulfated polysaccharide fractions extracted from Laminaria japonica. Int. J. Biol. Macromolec. 42: 127–132.

Wang, J., Q. Zhang, Z. Zhang, H. Zhang and X. Niu. 2010. Structural studies on a novel fucogalactan sulfate extracted from the brown seaweed *Laminaria japonica*. Int. J. Biol. Macromol. 47: 126–131.

Wang, Z.J., Y.X. Si, S. Oh, J.M. Yang, S.J. Yin, Y.D. Park, J. Lee and G.Y. Qian. 2012. The effect of fucoidan on tyrosinase: computational molecular dynamics integrating inhibition kinetics. J. Biomol. Struct. Dyn. 30: 460–473.

Wessels, M., G.M. Konig and A.D. Wright. 1999. A new tyrosine kinase inhibitor from the marine brown alga Stypopodium zonale. J. Nat. Prod. 62: 927–930.

Wood, B.J.B. 1988. Lipids of algae and protozoa. pp. 807–867. *In*: C. Ratledge and S.G. Wilkinson (eds.). Microbial Lipids. Academic Press, London.

Yan, X., Y. Chuda, M. Suzuki and T. Nagata. 1999. Fucoxanthin as the major antioxidant in Hijikia fusiformis, a common edible seaweed. Biosci. Biotechnol. Biochem. 63: 605–607.

Yang, B., X. Lin, X.-F. Zhou, X.-W. Yang and Y. Lin. 2012. Chemical and biological aspects of marine cosmeceuticals. *In*: S.-K. Kim (ed.). Marine Cosmeceuticals: Trends and Prospects. CRC Press, Boca Raton.

Zapopozhets, T.S., N.N. Besednova and N. Loenko Iu. 1995. Antibacterial and immunomodulating activity of fucoidan. Antibiot. Khimioter. 40: 9–13.

Zyvagintseva, T.N., N.M. Shevchenko, A.O. Chizhov, T.N. Krupnova, E.V. Sundukova and V.V. Isakov. 2003. Water-soluble polysaccharides of some far-eastern brown seaweeds. Distribution, structure, and their dependence on the developmental conditions. J. Exp. Mar. Biol. Ecol. 294: 1–13.

12

Dinoflagellates and Toxin Production

Joana Assunção[1] and F. Xavier Malcata[1,2]

Introduction

Marine life possesses huge structural and chemical diversity, which may support development of promising new drugs with higher efficacy than its terrestrial counterparts (Malve 2016; Joseph 2016). Therefore, the interest for novel bioactive compounds from marine sources has boomed in recent years, namely with regard to treatment of a few human diseases. More than 25,000 structurally diverse bioactive products have indeed been isolated from marine species since 1965 (Blunt et al. 2016). Dinoflagellates (a type of microalgae) are an important group which has already contributed to this number, and is likely to contribute even further in terms of pharmacological roles (Gallardo-Rodríguez et al. 2012a). This complex taxon is estimated to include over 2300 living species (Gomez 2012)—of which more than 50 have been found to produce marine toxins (Gallardo-Rodríguez et al. 2012a; Daranas et al. 2001). Besides their intrinsic ecological role in aquatic environments, dinoflagellates have adapted to a broad variety of environments thus reflecting their extraordinary flexibility; fossil records thereof date back several hundred million years (Taylor 1987; Wisecaver and Hackett 2011; Gomez 2012). Occasional and sudden proliferation of dinoflagellates in marine environments has led to a phenomenon called Harmful Algal Blooms (HABs); they are likely to produce toxins that can negatively distress marine life, especially via poisoning fish and shellfish (Hallegraeff 2003). Moreover, HABs and associated toxin production can impact human health if direct consumption of contaminated shellfish (or fish) takes place afterwards; and disturb economic activities, with unfavorable implications upon fisheries or tourism (Anderson 1995; Smayda 1997).

Despite associated risks, the aforementioned toxins possess remarkable biotechnological features—thus justifying an effort to still seek more toxin biocompounds, and eventually produce them at large scale. Unfortunately, only a meager quantity of those molecules has reached the market—and still with several limitations. The restricted availability thereof from natural sources (Haefner 2003), and the limited and unfeasible routes for chemical synthesis, or the scarce advances in genetic engineering of dinoflagellates

[1] LEPABE – Laboratory of Process Engineering, Environment, Biotechnology and Energy, College of Engineering, University of Porto, Rua Dr. Roberto Frias, s/n, P-4200-465 Porto, Portugal.
[2] Department of Chemical Engineering, University of Porto, Rua Dr. Roberto Frias, s/n, P-4200-465 Porto, Portugal.
* Corresponding author: fmalcata@fe.up.pt

(Gallardo-Rodríguez et al. 2012a) have greatly hampered development of toxin-derived leads for pharmacores. There are at present many difficulties to grow dinoflagellates in laboratory cultures (using conventional reactors), and several attempts have failed to obtain the intended compounds (Wynn et al. 2010; Gallardo-Rodríguez et al. 2012a). Nevertheless, production of such potent bioactive compounds in relatively high quantities (and in a safe mode) is an important issue before pharmacological studies and pre-clinical trials can be developed (Glaser and Mayer 2009; Zittelli et al. 2013). The production of toxin by dinoflagellate cultures in photobioreactors is still the preferred approach—despite the underlying constraints regarding fastidious growth and extreme sensitiveness to shear stress (Gallardo-Rodríguez et al. 2009). Most cultivation systems grow cells in suspension (i.e., vessels or classical tubular reactors), with conventional operation conditions (i.e., continuous light supply, or CO_2 supply turbulent bubbling); this can lead to cell damage due to high shear stress, thus raising a difficulty for dinoflagellate cultivation at a large scale. Stirring is necessary to avoid gradients of CO_2 and O_2, while enhancing mass transfer to cells for performance of photosynthesis. A uniform rate of supply of light should also be assured—thus preventing photoinhibition or photoxidation. It is believed that such susceptibility is caused by a complex cell organization, and the nature of stimuli that can affect (in a transient or permanent way) the ability of biomass to divide and grow.

Several improvements have been done in this area of dinoflagellate cultivation, yet the biotoxin titers remain very low (in the order of picogram). This chapter includes a brief revision on the diversity and main features of dinoflagellates, ecology, and HABs; how and why toxins are synthesized, and their putative pharmacological applications; the major difficulties encountered to produce toxins via synthetic routes; the ongoing efforts pertaining to genetic and metabolic engineering; and discuss the main cause affecting dinoflagellate growth, that is, shear stress, arising from turbulence and agitation throughout cultivation in conventional photobioreactors. The main advances anticipated as necessary for dinoflagellate culture in bioreactor will be discussed last, with the goal of producing such biotoxins in a more efficient manner.

Diversity and features

Dinoflagellates (Phylum: *Dinoflagellata*; Division: Phyrrophyta) are a large group of eukaryotic, biflagellate organisms exhibiting great diversity in morphology, cellular organization, and behavior (Taylor 1987; Gomez 2012). Ubiquitous in all types of ecosystems (i.e., marine, freshwater, benthic, brackish, ice sea), dinoflagellate populations distribute according to temperature, salinity, and depth (Taylor et al. 2008). Ecologically speaking, dinoflagellates can occur both in the water column, as a component of the plankton (ca. 90% responsible for primary productivity), and at the bottom of water bodies, as part of benthos (Gomez 2012). The unique diversity of this taxon reflects most trophic types: from photosynthetic and pure autotrophically-growing species (Galius and Elbraehter 1987; Schnepf and Elbrächter 1992), to mixotrophic species that acquire nutrients from both photosynthesis and dissolved organic matter or organic particulates (phagotrophy); and from pure heterotrophy to parasitic or symbiotic ways of living (Bralewska and Witek 1995; Gomez 2012). Roughly half of the species possess photosynthetic pigments, and thus play a major role as primary producers in freshwater and marine habitats (Hickman et al. 2008). Their relatively slow proliferation among unicellular algae may justify why these organisms are particularly well-suited for symbiotic relationships (Hackett et al. 2004). They can indeed be found associated with several marine organisms, for example, sea anemones, protozoa, certain invertebrates, and stony corals (Coffroth and Santos 2005). Interestingly, only corals with symbiotic dinoflagellates-in with *Symbiodinium* genus accounting for the most representative group, can form coral reefs, thus conveying bright colors (Fensome et al. 1995; Wong and Kwok 2005).

Belonging to the ancient eukaryotic lineage *Alveolata*, dinoflagelaltes are composed by unicellular microalgae species with complex and unique morphological features (Leander and Keeling 2004). The *Alveolata* is in fact one of the most biologically vast supergroups of single-celled eukaryotic microorganisms, consisting of ciliates, dinoflagellates, and apicomplexans (Kellmann et al. 2010). Typically exhibiting between 10 and 100 μm in length (or, in more extreme situations, 2 μm–2 mm), dinoflagellates differ from other two groups for possessing two dissimilar flagella (at some stage of their life) (Taylor 1987; Lin 2011). The major difference on flagella lies on their relative insertion into the cell: one flagellum is transversal and lies in a surface groove (cingulum), and the other is longitudinal and emerges from a ventral furrow (sulcus) (Taylor 1980). The exception to this rule is the family of protocentroid dinoflagellates (*Prorocentraceae*), which have their flagellum inserted in a specific region called periflagellar area (Faust 1990, 1991).

In a combined action, the flagellum provides cells with the necessary propulsion and mobility, which may be a competitive advantage in obtaining nutrients and harvesting light at different levels in the water column (Taylor 1987; Lin 2011). Their mechanism of motility allows them to survive in other parts of the water column, as they are able to move in response to stressful conditions (i.e., turbulence or shear forces, quick changes in temperature, nutrient limitation, or intense light intensity). It should be emphasized that this behavior is regulated not only by the aforementioned factors, but also by the cell age (Steidinger and Baden 1984). The predation and hostile environmental conditions are also handled via production of dormant or resilient cysts. Most of these microalgae chiefly reproduce by binary fission (asexual reproduction), but vegetative motile cells can recombine sexually and produce hipnozygotes or resting cysts that sink to the benthic layers and remain latent therein. When the surroundings become more auspicious, resting cysts are able to germinate and thrive, thus restarting their life cycle and recolonizing the water column (Lewis et al. 1999; Bravo and Figueroa 2014).

Compared to normal eukaryotic cytology, dinoflagellate cells exhibit numerous outstanding features concerning organization of plastids (Zhang et al. 1999; Speckhard 2010), mitochondria (Waller and Jackson 2009) and genomes (Rizzo 2003; Wisecaver and Hackett 2011). One of the most striking difference is their unusual nucleus (called dinokaryon), lacking nucleosomes and histones, and a high chromosomal DNA content, present in a liquid crystalline form (Rill et al. 1989). Dinoflagellates present the largest genomes of any known organism; their DNA content is estimated to range within 3–250 pg per cell, whereas most eukaryotes contain an average about 0.54 pg DNA per cell (Spector 1984; Rizzo 2003). In addition, they possess extranuclear spindles where circular and permanently condensed chromosomes are attached, even during mitosis (Fensome et al. 1995; Wong and Kwok 2005). A large number of genes is encoded in tandem gene arrays; lack of common eukaryotic transcription sites (e.g., TATA box) and other types of post-transcriptional regulation mechanisms (Beam et al. 1984; Hackett et al. 2004), materialize their unconventional generic organization and regulation of gene expression.

Another distinguishing characteristic is their cell wall cover. The nine major orders (Peridiniales, Gonyaulacales, Gymnodiniales, Prorocentrales, Suessiales, Dinophysiales, Phytodiniales, Blastodiniales, and Noctilucales), recognized within the dinoflagellate group, can be distinguished in terms of major morphological features and life cycles by referring to their membrane (Not et al. 2012). They may be naked, or covered with cellulose (or other polysaccharide) plates or valves, or create a sort of armor named theca (Fensome et al. 1999; Hackett et al. 2004). Theca has a variety of shapes, arrangements, and sizes, depending on species and cell life stage—thus conferring a more rigid and inflexible wall. The naked or unarmored forms have an outer plasmalemma, surrounded by a single layer of flattened vesicles (amphiesma)—apparently easy to distort (Wong and Kwok 2005). Although all dinoflagellates share certain physiological and structural characteristics, they display a surprising miscellany of forms in terms of external morphology. Some cells are small and smoothly spherical, whereas others have elaborated structures such as horn and spikes (Fensome et al. 1999).

Photosynthetic species may possess plastids (i.e., chloroplasts), probably incorporated by secondary or tertiary endosymbiosis (Cavalier-Smith et al. 1999; Yoon et al. 2002). Acquisition, loss, and replacement of such organelles are quite common, and may depend on their life cycle. They are also able to harbor those foreign plastids by a long period of time, but not in a fully integrated way (Steidinger and Baden 1984).

Another trait characterizing dinoflagellates relates to their slow proliferation among unicellular algae—generally believed to be an attribute related to their low chlorophyll to carbon ratio (Tang 1996). Theoretically, dinoflagellates express a type II ribulose-1,5-bisphosphatecarboxylase-oxygenase (*Rubisco*) enzyme, probably originating from anaerobic proteobacteria, yet this enzyme is known to poorly discriminate between CO_2 and O_2 (Wong and Kwok 2005). This support the idea that *Rubisco* may contribute in general to the very low growth rates at stake, while other carbon concentration mechanisms are implicated in CO_2 biofixation of dinoflagellates (Rost et al. 2006).

Complex circadian systems control the behavior of dinoflagellates *in vivo*—as daylight and changes in nutrient levels condition vertical migration (Doblin et al. 2006). Cell growth has to be coordinated with cell division, not only to produce a homeostasis of cell size but also to allow cells respond properly to nutrient availability. The cell size may affect buoyancy and sinking rates, and hence the period in the photic zone (Wong and Kwok 2005). Phototactic vertical migrations of dinoflagellates, in response to changing light intensity, are quite common and referred to as positive photoaxis (Erga et al. 2015). Depending on the species, one may find aggregates of dinoflagellates in the upper layer of the ocean during daylight (where they photosynthesize), which at night tent to move or sink as a response to gravity (or positive geotaxis)—to deeper layers, where nutrient concentrations are higher (García-Camacho et al. 2007; Erga et al. 2015). It is presumed that their alternative modes of coupling cell cycle progression with cell growth may play a role upon such slow growth (Wong and Kwok 2005).

HABs and toxin-related issues

Dinoflagellates are one of the major responsible for the so-called harmful algal blooms (HABs)—which have implications upon aquatic faunal mortalities, marine food web chains, and ultimately, human health. These high densities of dinoflagellates apparently kill fish and/or shellfish—either indirectly, because of cell accumulation in animal gills and depletion of oxygen, or directly by toxin production (Hallegraeff 1993; Smayda 1997). Such toxins can be readily transferred into higher trophic levels through the food chain (Wang 2008); and episodic mortalities of other animals, such as birds and other marine mammals, are not uncommon (Scholin et al. 2000; Flewelling et al. 2005; Landsberg et al. 2009). How and why these natural phenomena occur is not fully understood, but hydrographic and weather conditions have been implicated (Wells et al. 2015). The proliferation of toxic species in non-endemic areas can be a consequence of dragging of water currents, or even interaction with bivalves that can harbor viable cells and cysts—thus aiding in their marine dispersion (Matsuoka et al. 2003). Furthermore, the high densities of dinoflagellates in water environments near coastal areas also seems to be influenced by anthropogenic action (i.e., discharges of industrial, domestic, and agriculture wastes), as crucial nutrient ratios (i.e., nitrogen:phosphorous) are essential for their metabolism (Hodgkiss and Ho 1997; Gobler and Sañudo-Wilhelmy 2001; Hallegraeff 2003).

Depending on the period and extent of the occurrence, HABs economic impact may be more or less severe. Those events are continuously monitored by several countries, which usually prohibit shellfish harvesting and restrict human consumption thereof. Several local activities may be constrained: from seafood distributors to restaurants and supermarkets, or even fish farms that can be forced to discontinue production. In addition, HABs may cause discoloration (usually red, and less frequently yellow or brown) in seaside water, and affect other tourism attractions in coastal areas. Poor advertisement of the affected area may case deep negative impacts on tourism and associated activities (Steidinger and Baden 1984).

The consumption of contaminated shellfish (i.e., soft shell clams, mussels, oysters, scallops, hard clams), crustaceans (i.e., shrimps, crabs, lobsters) or fish can be potentially lethal to humans. There are six major toxic poisoning syndromes reported in the literature in regard to humans: paralytic shellfish poisoning (PSP) (i.e., saxitoxin), neurologic shellfish poisoning (NSP) (i.e., brevetoxin), diarrheic shellfish poisoning (DSP) (i.e., okadaic acid, pectenotoxin), ciguatera fish poisoning (CFP) (i.e., ciguatoxin, maitotoxin), azaspiracid shellfish poisoning (azaspiracid), and amnesic shellfish poisoning (ASP) (i.e., domoic acid)—the latter not directly due to dinoflagellates, but to diatoms (Hackett et al. 2004; Wong and Kwok 2005; Wang 2008; Lin 2011). In general, such intoxications can lead to neurological and gastrointestinal illness. Some of them are so potent that affect normal nerve function, can act as modulators of higher neurological processes (i.e., saxitoxin, domoic acid), and can ultimately lead to death (Wang 2008). The symptoms associated to such type of intoxications are diverse and nonspecific (i.e., headache, dizziness, nausea, vomiting, abdominal cramps)—which many times leads to misleading symptoms and thus inaccurate medical diagnosis.

Due to the impacts on both public health and economy, extensive studies have been devoted to HABs and environmental outcomes of dinoflagellate toxins in the latests decades (Hallegraeff 1993). The production of toxins by different dinoflagellate is well documented, but the mechanisms regulating the amount of such toxins—associated with a given phytoplankton bloom, are comparably less well understood (Hackett et al. 2004). The ecological reason why dinoflagellates produce potent biotoxins is not clear. Maybe such secondary metabolites act as allellochemical agents against other species competing for a specific niche, as defense against predation, as enzyme regulation, or as sexual response induction (feromones) (Cembella 2003; Sheng et al. 2010).

Toxin biosynthesis is apparently related to cell cycle and light-dependent events, but can also be affected by other factors (i.e., nutrient availability, temperature, salinity). In the benthic dinoflagellate *Prorocentrum lima*, dinophysistoxin-4 toxin synthesis is initiated in the G1 phase of the cell cycle but persists into the S phase. On the other hand, the biosynthesis of spirolide toxin produced by *Alexandrium ostenfeldii* is governed by such light-dependent mechanisms—since toxin concentration per cell quota increases in the beginning of the dark period (probably corresponding to the G1 or S phase cell cycle) (John et al. 2001). The biosynthesis of saxitoxin by *Alexandrium fundyense* occurs for 8–10 h during the G1 phase, and its levels remain almost undetectable during the other phases of the cell cycle (Taroncher-Oldenburg et al. 1997).

Under nutrient-balanced conditions, toxin production is normally low—while increased production is associated with recovery from various types of nutrient stress. Therefore, biosynthesis of both alkaloids (e.g., saxitoxin) and polyketides (e.g., brevetoxins) appear to be related to nitrogen and carbon metabolism deficiency. The synthesis of PSP toxins in *Alexandrium fundyense* was observed to increase when the medium is enriched in nitrogen and reduced in phosphorous. The opposite, that is, very low content of PSP toxins, was observed when the medium was rich in phosphate and poor in nitrogen (John and Flynn 2000). Several studies support the idea that nitrogen-limiting conditions increase PSP toxin levels (Wang et al. 2002; Frangópulos et al. 2004; Han et al. 2016). This seems to be related to the increased availability of intracellular arginine (an important precursor for biosynthesis of PSP toxins), due to a reduced demand of competing phosphorous-dependent pathways involved in cell division. High N:P ratios apparently influence higher toxin production of ciguatoxin in *Gamberdiscus* (Chinain et al. 2010) and brevetoxin by *Karenia brevis*—which is also influenced by limited-phosphorous concentration (Hardison et al. 2013). However, the influence of enhanced phosphate supply upon toxin content is still a matter of debate. In axenic *Alexandrium tamarense*-cultured in low-level phosphate conditions and supplemented with nitrate and phosphate, toxin production was obtained when compared to the outcome under high nutrient levels (Hu et al. 2006).

Induction or cellular accumulation of algal toxins during unbalanced growth/cell stress has become a paradigm in attempts to understand modulation of phytoplankton toxicity by environmental factors. Nonetheless, different patterns may extensively vary between species (Kibler et al. 2012)—and even among strains with similar geographic distribution (Etheridge and Roesler 2005). The high degree of variability in toxicity observed for a number of HAB species reflects the complex relationship between environmental and genetic factors (Adolf et al. 2009; Hardison et al. 2013). Changes in total cellular toxicity of a dinoflagellate cell, in response to environmental factors, depends upon both the absolute toxin concentration (or the rate of production of toxin) and the relative toxin composition (Etheridge and Roesler 2005). Some cells may exhibit higher toxicity (even when they have lower toxin concentration) than others, depending on the established toxin profile. *Alexandrium fundyense* was found to have higher toxicity (at 5°C) owing to its toxin profile containing a higher fraction of saxitoxins and gonyatoxins, but a lower total cell concentration at 20°C (Etheridge and Roesler 2005).

Cellular toxicity among different isolates in laboratory cultures may be substantially lower when compared to fresh bloom samples. *Karlodinium veneficum* cultivated in laboratory exhibited lower content of karlotoxin (10 times less) than their counterparts from bloom samples; possibly culturing conditions favor low toxin genotypes, or conditions *in situ* lead to enhanced toxicity (Adolf et al. 2009).

In addition, increased cellular toxicity in dinoflagellates may relate to the presence of bacteria in the media, or via establishment of an endosymbiotic relationship (Steidinger and Baden 1984). Bacteria have been associated with HABs events in the case of both planktonic and benthic dinoflagellates (Schweikert and Meyer 2001); hence, truly axenic cultures of dinoflagellates are extremely difficult to obtain. The bottom line is that bacteria may contribute to variation of toxin production, either inside the cell or secreted to the culture medium. Some studies referred to *de novo* synthesis of some toxins, as shown in axenic cultures of *Gambierdiscus toxicus* (Bomber et al. 1989) and *Prorocentrum convaum* (Carlson et al. 1984). It is known that dinoflagellates (at least some species) have the ability to engulf other cells by a cascade of endosymbiotic events, and associate bacteria to their nucleus or cytoplasm. In addition, presence of bacteria in the medium may affect toxicity of cultures. For instance, *Gyrodinium instriatum* was found to be toxic when co-cultured with *Pseudomonas* sp., previously collected from a culture of *Alexandrium tamarensis* (Silva 1990). It was suggested that a close relationship between dinoflagellates and bacteria may induce toxicogenesis; this may also explain why non-toxic dinoflagellates were developed toxicity under certain conditions (Steidinger and Baden 1984). However, it should be stressed that not all toxic species possess intracellular bacteria, yet further studies are still warranted; for instance, several studies on *Karenia brevis* were unable to show any bacterial profiles (Steidinger and Baden 1984).

Dinoflagellate toxin applications

Although dinoflagellates are associated to HABs, their ability to synthesize some of the largest and most complex polyketides (and other secondary metabolites) discovered to date (Rein and Borrone 1999; Kellmann et al. 2010) make them a valuable natural source of marine biotoxins. With a wide spectrum of chemical structures, and an even wider number of biological activities to different levels of potency, they have found (or will likely find) potential applications in human and veterinary medicine (Gallardo-Rodríguez et al. 2012a). Examples include saxitoxin and tetrodotoxin—produced by dinoflagellates causing PSP, useful as topical anesthetics (Kohane et al. 2000; Duncan et al. 2001) and pain management (Nieto et al. 2012). In fact a powerful drug from the potent neurotoxin tetrodotoxin is currently being developed in Canada by Wex Pharmaceuticals, and undergoing Phase III clinical trials with great success as pain controller in cancer patients (Berde et al. 2011). However, tetrodotoxin used to produce this drug has been sourced from pufferfish—since the production by dinoflagellates has not yet attained sufficiently high levels (Gallardo-Rodríguez et al. 2012a). Okadaic acid, known to be associated with DSP species, is a neurotoxin used as model to study the therapeutical effects in some neurodegenerative disorders (e.g., Alzheimer's, or memory-impairment) (Kamat et al. 2013); yessotoxin, isolated from DSP-causing agents, as appears to be a potential bioactive agent against melanoma cancer (*in vivo* and *in vitro*), to possess anti-fungal activity and possibly having a role in immune regulation (asthma treatment) (Tobío et al. 2016); gonyautoxin, produced by dinoflagellate *Amphydinium* spp., is now safely used as part of a therapeutical approach against acute or chronic anal fissures (Garrido et al. 2005); neosaxitoxin as a topical formulation to prolonged anesthesia from PSP dinoflagellates (Rodriguez-Navarro et al. 2007); amphidinolides and colopsinols, derived from *Amphidium* genus, are reported as potential antitumoral agents against lymphoma and epidermic carcinoma (Kobayashi and Tsuda 2004); gymnocin-A, a complex polyether toxin isolated from the red tide dinoflagellate *Gymnodinium mikimotoi*, has revealed to be cytotoxic against P388 mouse leukemia cells (Tsukano and Sasaki 2006); goniodomin-A, reported in dinoflagellates causing PSP, is considered as effective antifungal and antiangiogenic agent (Abe et al. 2002); pectenotoxins, derived from dinoflagellates causing DSP, can act as chemotherapeutic agents against p53-defficient tumors (Chae et al. 2005); ostreocins and palytoxins, isolated from the *Ostreopsis* genus, include several pharmacological actions, such as modulation of a few neurotransmitters (e.g., norepinephrine and/or acetylcholine) and activation of pro-inflammatory signaling cascades (Pelin et al. 2016); symbiospirols (symbioimine and neosymbioimine), obtained from cultivated symbiotic marine dinoflagellate *Symbiodinium* sp., have been claimed to be antiosteoporosis agents (Kita et al. 2005); karlotoxins, produced by species *Karlodinium veneficum*, has apparently a hypocholestermic role and an anti-tumoral role (Waters et al. 2011); gambieric acid and gambierol, both unequivocally isolated from species *Gambierdiscus toxicus*, revealed potential as anti-filamentous fungal agent and modulator of Alzheimer's disease, respectively (Nagai et al. 1993); and brevetoxins and their derivatives, mainly reported in *Karenia brevis* as major contributor of NSP, are regulators of immune system and pulmonary diseases (i.e., cystic fibrosis) (Abraham et al. 2005).

Toxin production approaches

Dinoflagellate-generated toxins unfold a wide applicability in the health sector; however, most such molecules have not progressed beyond discovery stages because of limited availability from natural sources—and almost insurmountable difficulty to obtain via chemical synthesis, owing to their structural complexity. Laboratory cultures of dinoflagellates have also proven hard to establish, or failed to yield the intended compounds (Gallardo-Rodríguez et al. 2012a).

Lack of sufficient quantities has systematically hampered further biochemical investigation and clinical testing, thus compromising eventual development into commercial products. Very few commercial biotoxins are available for purchase, and the existing ones reach outrageous prices—ranging from 1,000 to 500,000 € per milligram. This is the case of commercial okadaic acid produced from *Prorocentrum* spp., a cytotoxic inhibitor of protein phosphatase-2A, able to alter the phosphorylation state of cellular proteins, thus leading to collapse of normal regulatory pathways—for example, induced downregulation of T-cell receptor expression compromising T-cell activation (Valdiglesias et al. 2013). Another example is azaspiracid, a polyether characterized by a cyclic-amine or aza group, produced by *Azadinium* spp. (Tillmann et al. 2009); azaspiracid functions as an activator of c-Jun-N-terminal kinase and caspases (implicated in stress signaling pathways) (Cao et al. 2010), but may also interfere with gene expression and inhibit cell cholesterol levels—especially in T-lymphocytes (Twiner et al. 2008). Azaspiracid-3—an analogue of this bioactive compound, can reach prices of 500 to 600 € just for 1 µg of product. In any case, Care should be taken when these type of commercial substances are purchased—as their producton may be easily discontinued, and the purity/quantity allegedly claimed by companies may be questioned (Quilliam 2003).

Different strategies have been applied to enhance production of target dinoflagellate derived-toxins. Nevertheless, some of those have shown to be complex, raising several challenges ahead.

Chemical synthesis and genetic engineering

The potential of dinoflagellate biotoxins and derivatives has been severely constrained by inability to attain acceptable productivities, sufficient to respond to the increase in demand for investigational purposes (i.e., for assessment of pharmacological potential and activity), as well as for preclinical studies and clinical trials. Strategies including chemical synthesis of most dinoflagellate toxins are theoretically possible, but still excessively expensive (except okadaic acid)—while most of them comprise several laborious and intricate steps (more than 100 steps may actually be required); more practical synthetic routes remain a challenge, unlikely to succeed in the short run. However, it is important to highlight that this approach has allowed a few advances concerning elucidation of structure and mode of action of some complex metabolites derived from dinoflagellates [e.g., brevetoxin A (Gawley et al. 1995), gambierol (Fuwa et al. 2002), gymnocin A (Tsukano and Sasaki 2006), and azaspiracid-1 (Nicolaou et al. 2006)]. On the other hand, genetic improvements and metabolic engineering are very hard, since the genome of dinoflagellates is complex (Jaeckisch et al. 2011)—and available genetic tools have chiefly been developed for nondinoflagellate microalgae and target biofuel production (Radakovits et al. 2010).

Metabolic engineering is not easy in dinoflagellate cells, as they present astonishingly large and complex genomes—with a great many introns, bearing redundant repetitive noncoding sequences (McEwan et al. 2008) and a high proportion of unusual bases with a fifth base replacing uracyl in their DNA (Wisecaver and Hackett 2011). DNA content makes it difficult to perform simple genomic hybridization, like Southern blots, and it is impractical to construct genomic libraries or to consider sequencing their genome in the first place. As dinoflagellates are lacking a general transformation system and experience difficulties to grow on solid media, cloning will prove a very complicated procedure. Genetic transformation was only reported in two different dinoflagellates (*Symbiodinium* sp. and *Amphidinium* sp.) (ten Lohuis and Miller 1998), and no further attempts have meanwhile been reported. Furthermore, dinoflagellate genes lack recognizable promoter features and common eukaryotic transcription factor binding sites (Jaeckisch et al. 2011), and their toxin production capacity apparently results from multiple independent evolutionary origins (Lin 2011)—what may turn out hampering the identification of toxin-related genes.

The identification of genes and enzymes involved in the biosynthesis of toxins by dinoflagellates has been limited so far, in spite of considerable efforts (Kellmann et al. 2010). Gene-function mapping would be essential for engineering improved production of dinoflagellate derived-bioactives (Gallardo-Rodríguez et al. 2012a). To date, all data regarding gene regulation mechanisms in dinoflagellates has emerged inconsistently, from studies of specific genes that are of interest for a particular function (Hackett et al. 2004). The molecular genetics underlying biosynthesis of dinoflagellate able to produce toxins is scarcely understood. However, it is known that the genes responsible for toxin synthesis are in their chromosomes, with some exceptions. For instance, polyketide synthase (PKS) genes—a sequence closely related to the toxigenic cyanobacteria PKS genes, has been found in *Karenia brevis* chloroplast (López-Legentil et al. 2010). This clearly suggests multiple independent origins regarding toxin competence of some dinoflagellates (Lin 2011).

The majority of dinoflagellates apparently produce their toxins via the polyketide pathway, which involves a polyketide synthase that may be combined with some functional segments—such as non-ribosomal peptide synthase (Kellmann et al. 2010). The enzymes carrying out polyketide synthesis, PKSs, have been classified into three types, depending on their domain organization. According to chemical structure, it is suggested that type I PKS are the enzymes involved in its production (Kathleen and Snyder 2006). In addition, the PSP toxins are believed to be synthesize via a pathway involving arginine, S-adenosylmethionine (SAM), and acetate. Interestingly, homologs of SAM synthetase gene have been identified in both toxic and non-toxic dinoflagellates (Harlow et al. 2007).

Despite many difficulties, new high-throughput omic tools are moving towards exploring toxin genes and proteins related to dinoflagellate-mediated toxin production, and may provide some insights about their biosyntheses (Wang et al. 2016).

Bioreactor culture and shear-stress

Despite their remarkable ecological importance and productivity in nature, dinoflagellates are of limited biotechnological importance due mainly to the difficulty in establishing such groups in conventional reactors (i.e., closed fermentors) or even in adequate synthetic media (Not et al. 2012). Nevertheless, general attempts have focused on photoautotrophic dinoflagellate cultures (e.g., Gallardo-Rodríguez et al. 2010; Fuentes-Grünewald et al. 2016; López-rosales et al. 2016). As dinoflagellates exhibit an intricate metabolism, and exhibit substantially lower growth rates than common microalgae (Wong and Kwok 2005), toxin concentrations have rarely gone over the microgram/liter threshold (Gallardo-Rodríguez et al. 2012a).

The sensitivity of dinoflagellates to shear stress and turbulence when in suspension apperar to be a critical issue (García-Camacho et al. 2007). Dinoflagellates obey a strict circadian cycle, where cells divide at the end of the dark period and grow during the light phase (corresponding to the G1 phase of the cell cycle), precisely when production of many toxins occurs (Pan et al. 1999). This can also be problematic when they are cultivated in conventional photobioreactor configurations and operating conditions. As an example, several simplistic modes of light supply (i.e., continuous illumination) combined with the typical uniform levels of nutrients may break down natural rhythms, and metabolic behavior may accordingly depart from the original one. Any type of agitation, shaking, aeration and stirring imposed in laboratory cultures by typical photobioreactor configurations are largely reported to affect dinoflagellate cells (Sullivan et al. 2003; Berdalet et al. 2007). Even in natural settings, turbulent conditions may discourage bloom formation—because of physical dispersion and lower time-integrated light exposure of individual cells (Juhl et al. 2001).

High levels of turbulence translate to high hydrodynamic shear forces, especially very small eddies (Chisti 2000; Berdalet et al. 2007) that lead to inhibition of cell growth or cell damage—and, consequently, low biotoxin-associated productivities (Gallardo-Rodríguez et al. 2009).

High liquid motions (turbulent regimes) inside photobioreactors are needed to potentiate gas exchanges into media to sufficiently high transfer rate so as to promote photosynthetic growth (Carvalho et al. 2006). Due to low solubility of CO_2 in water, a high mass-liquid transfer is needed for suitable CO_2 supply and to stripe dissolved oxygen—which may cause also harmful effects on microalga cells (Gouveia 2011; Kumar et al. 2011).

As dinoflagellates are so sensitive, agitation and mixing become major issues—because a compromise is to be reached between slow growth rates by inhibition due to shear forces and increased speed of growth by improvement of transport features in the bioreactor. The situation becomes more complex when the (sensitive) cells to be cultivated are photoautotrophic and require illumination for their growth. In this scenario, mixing and thus fluid dynamics acquire extraordinary relevance, as they are responsible for transport of cells from less illuminated areas in the center of the photobioreactor to areas near the surface that are well lit. Hence, the massive and productive cultivation of delicate photoautotrophic cells may present a triple restriction associated with stirring: cell damage by shear forces, nutrient restriction due to poor mixing, and poor internal illumination (also due to insufficient mixing).

The sensitivity threshold of microorganisms grown under such aggressive conditions may vary according to species, or even strain (Gallardo-Rodríguez et al. 2012a). Such phenomena trigger a prompt range of physiological effects on dinoflagellate cells: from bioluminescence (Chen et al. 2003) to cell membrane fluidization, production of peroxisomes and reactive oxygen species to interference in calcium mobilization, morphology alteration (e.g., increasing in cell size), alteration of cell cycle and inhibition of cell proliferation/growth to metabolite synthesis (Jaouen et al. 1999; Gallardo Rodríguez et al. 2009).

The actual mechanisms by which cells are damaged, or turn susceptible to shear as a result of hydrodynamic forces are not exactly understood (Hu et al. 2011). The molecular basis remains unclear—as it is hypothesized that the cortical cytoskeleton (a particular apparatus present on dinoflagellate cell) may be implicated in the transmission of mechanical stimuli relevant to arrest cell cycle and/or growth inhibition (Wong and Kwok 2005; Gallardo-Rodríguez et al. 2015). Essentially due to the large nucleus that occupies half of the volume of cell, there is an increasing probability of transduction of external mechanical forces to the DNA in the cells. In addition, chromosomes are attached to the nuclear envelope—and this can potentiate susceptibility to mechanical stimuli, via surface-adhesion receptors associated to cytoskeleton. For that reason, external mechanical alterations may entail changes in DNA conformation—and ultimately gene expression alterations (Alam et al. 2014).

The intensity of cell responses to shear stress, and ultimately cell damage, depends on magnitude, duration, frequency of exposure to shear field, and how the experimental set up is established to generate the motion regimes (Berdalet et al. 2007). Surprisingly, some responses can be attenuated, or even reversed in specific conditions without causing cell damage or death (Yeung and Wong 2003). An interesting study with *Protoceratium reticulatum*, subjected to "lethal" agitation, reported increase in cell membrane fluidity within a few minutes—yet the effect was totally reversed within a few hours by stopping agitation (Gallardo-Rodríguez et al. 2012b). Similar effects of restoration of normal cell characteristics were also reported in *Alexandrium minutum*, *Prorocentrum triestinum*, and *Akashiwo sanguinea* (Berdalet et al. 2007). At lower levels of turbulence, the cell cycle can be affected—but instead of arrested, it can merely progress in a slower way (and without causing any cell damage) (Yeung and Wong 2003). The magnitude of shear stress also appears to influence production of toxins in at least some dinoflagellates (Juhl et al. 2001; Gallardo-Rodríguez et al. 2011). Juhl et al. have demonstrated that quantified shear toxin concentration per cell can increase up to three fold that of control cultures in *Alexandrium fundyense*, a PSP producer (Juhl et al. 2001). In addition, cell shear tolerance is also apparently influenced by light/dark cycle. It is plausible that dinoflagellates do not withstand shear forces during the dark cycle because of the synchronizing cell division during that time (García-Camacho et al. 2007).

No clear link has been established between flow-sensitivity, and size, shape and taxonomy of dinoflagellate species (Sullivan et al. 2003; Berdalet et al. 2007). Dinoflagellates generally have large cells—up to 2 mm in diameter. In photosynthetic species associated to HABs, the size ranges from ca. 10 up to 60 μm. Some species are thecate, meaning that they have a cell wall of cellulosic plates. "Armored" species had been suggested as less susceptible to turbulent conditions than "naked" ones. Theca was speculated to conferr a sort of protection against shear (Smayda 2010). However, this hypothesis is hardly supported because both thecate and athecate species are susceptible to turbulent conditions (e.g., White 1976; Berdalet 1992; Juhl et al. 2001; Sullivan et al. 2003). Even among thecate species, difference concerning resistance to shear stress exists. For instance, *Protoceratium reticulatum* is a thecate organism highly sensitive to shear, whereas *Crypthecodinium cohnii*—bearing a similar structure, is more robust. The latter is able to support values of energy dissipation rates (EDRs) of 5.8×10^5 cm^2 s^{-3} (Hu et al. 2007), without visible damage—while the former cannot exceed the marginal value of 0.8 cm^2 s^{-3} (García-Camacho et al. 2007). Therefore, shear sensitivity of such type of microalgae is apparently species-dependent (Berdalet et al. 2007; Gallardo-Rodríguez et al. 2015). Several studies, encompassing various orders of dinoflagellates, have shown that EDRs values in the range [0.011, 10] cm^{2}s^{-3} generally inhibit dinoflagellate growth (Gallardo-Rodríguez et al. 2012a). Only EDR under 1 cm^2 s^{-3} (~ 0.1 W m^{-3}) seems not to trigger detrimental effects upon dinoflagellate cells (Berdalet et al. 2007).

The levels of turbulence stood by dinoflagellates are generally one to two orders of magnitude smaller than those that produce damage in most suspended plant and animal cells; the latter is among the most sensitive to shear in bioreactors (Gallardo-Rodríguez et al. 2012a). Previous strategies proven successful in commercial large-scale cultivation of these types of cells (Varley and Birch 1999; Juhl et al. 2001) cannot be extrapolated to dinoflagellates—due to complex circadian cycles and unique metabolism.

Biotechnological cultivation of dinoflagellates—some advances

For many decades, studies related to dinoflagellates remained focused on systematic of bloom dynamics and HABs. Conversely, scare attempts were devoted to development of controlled cultures of dinoflagellates in bioreactors (Gallardo-Rodríguez et al. 2012a). Scientists have recently invested a lot of effort into this latter issue—motivated by the outstanding potential of dinoflagellate-derived compounds (e.g., Gallardo-Rodríguez et al. 2010; García-Camacho et al. 2011; Beuzenberg et al. 2012; López-Rosales et al. 2015; Fuentes-Grünewald et al. 2016; López-Rosales et al. 2016).

In this regard, controlled cultivation is a must, and safety issues must be taken into account for culture scale up. Cultivation the dinoflagellates in open systems (i.e., raceways, natural lagoons)—despite being easily operated and having low input requirements, does not seem a feasible option for safety reasons and/or environmental contamination. Only dinoflagellates producing interesting non-toxic or relatively benign compounds might be candidates for such reactor configurations (Gallardo-Rodríguez et al. 2012a). For instance, *Karenia brevis*—a producer of brevetoxins is known to lyse when cultures are aerated. Releasing such type of toxins in the environment can be extremely dangerous for humans or other organisms, as it works like an aerosol (Abraham et al. 2005; Flewelling et al. 2005; Fleming et al. 2007). Moreover, production of toxins as final product—with potential application as medicines, would be improbable under this system; contamination of metabolites would raise several risks for drug development and/or human treatment. On the other hand, the direct effect of weather would probably constrain dinoflagellate growth, as many species are deeply sensitive to physico-chemical alterations and possess a complex metabolism (Carvalho et al. 2006).

Taking such drawbacks into consideration, the logical alternative is employing enclosed systems, as is the case of photobioreactors (PBRs). PBRs tend to be more complex and expensive than open systems, but permit better control of culture environment (Chisti 2007). Many studies on dinoflagellate cultivation have resorted to just flask or bottle cultures at bench scale; a few studies have, however, surfaced dealing with larger scale (Zittelli et al. 2013). Depending on the bioreactor technologies employed, volumes ranging from 4 L to 700 L and distinct cultivation strategies have been attempted, with the goal of achieving high biomass productivities, as well as higher biotoxin concentrations. The culture systems described for those purposes range from carboys (a type of container made of glass or plastic), chemostats or stirred-tanks, to classical airlift, bubble column, tubular reactor or flat-plate PBR—chiefly the designs used for conventional microalga mass culture. For instance, Pan et al. (1999) produced a toxic *Prorocentrum lima* in 36 L-glass carboys with 18 L-working volume, at about 90 μmol photons m^{-2} s^{-1}, 14 h/10 h light/dark cycle and gentle aeration, and obtained an increase from ca. 2,000 to 11,000 cells mL^{-1}, and from ca. 20 to 220 nMol DSP toxins (okadaic acid and dinophysis toxins) for 40 d (Pan et al. 1999). Also, Loader and coworkers (2007) have tested, in polycarbonate carboys of 14 L (in a total of 226 L), growth of *Protoceratium reticulatum* and production of yessotoxin and furanoyessotoxin with the light/dark cycles applied in batch culture for 43 d. Between 200 to 15,000 cells mL^{-1} were achieved, as well as 250 to 2,500 µg L^{-1} of yessotoxins (Loader et al. 2007). Parker et al. (2002) successfully cultured dinoflagellate *Alexandrium minutum*, implementing a classical PBR design—the flat alveolar panel (Tredici et al. 1991), but resorting to smaller volumes (4 L), with a light path length of 12 mm, at ca. 100 µmol photons m^{-2} s^{-1}, a light cycle of 12 h/12 h and low aeration (0.2 vvm), and attained a productivity of 14,000 cells mL^{-1} d^{-1} and 20 µgL^{-1} d^{-1} of gonyautoxin (Parker et al. 2002). An improved classical stirred PBR of 15 L was tested by Camacho et al. (2011) for cultivation of *Protoceratium reticulatum*. The bioreactor was made of 5 mm-thick borosilicate glass, with internal diameter of 19.3 cm, bearing a spin filter with diameter of 7 cm, and height of 22 cm. A perforated pipe air sparger was placed within the spin filter, in order to prevent direct gas sparging of algal cells. The culture mixing/stirring was performed by resorting to three-bladed marine propeller, located 5 cm above the bottom of the vessel. PBR illumination required up to five fluorescent lamps, mounted around its periphery. Temperature control was affected by resorting to circulating, thermostatted water through the glass jacket surrounding the PBR. The pH was controlled by automatic injection of CO_2 within the spin filter. Experiments included perfusion culture (i.e., the cells remained in the bioreactor), fed-batch, and semicontinuous modes, and average cell and yessotoxin productivities of 5,228 cells mL^{-1} d^{-1} and 9.16 µgL^{-1} d^{-1}, respectively, were reported (García-Camacho et al. 2011).

Jauffrais et al. (2012) have successfully conducted an experiment with toxic dinoflagellate *Azadinium spinosum*, using two column-stirred-PBRs in series (about 100 L each), operated continuouswise with light intensity of 200 µmol photons m^{-2}·s^{-1}, under light cycles of 16 h/8 h, low stirring (40 rpm) and air bubbling—and obtained 190,000 and 210,000 cell mL^{-1} (steady state), with a maximum azaspiracid production of 475 ± 17 µg·d^{-1} (Jauffrais et al. 2012).

Medhioub et al. (2011) have conducted another study with *Alexandrium ostenfeldii* in flat-bottom flasks (8 L), in batch mode and continuous column-stirred-PBR (100 L), to produce spirolide toxins. The batch cultures ran with a photon irradiance rate of 155 µmol m^{-2} s^{-1}, under a 12 h/12 h light/dark regime with gentle aeration for 18 d. In continuous-flow culture (stirred-PBR), experiments were carried out under an irradiance rate of 188 µmol m^{-2} s^{-1}, 16h/8h light and dark cycles, automated pH and temperature control, with gentle aeration and stirring provided by four impeller turbines (50 rpm) for ~ 130 d. The latter conditions allowed values of cell concentration up to 70,000 cells mL^{-1}, for more than 60 d, and the maximum value attained in the batch mode was 16,788 cells mL^{-1} and spirolide toxin content of 34 to 50 µg L^{-1} (Medhioub et al. 2011).

Wang et al. (2015) also developed a large-scale production method for benthic dinoflagellate *Prorocentrum lima*, using a vertical flat PBR with dimensions 60 cm x 60 cm x 28 cm and volume of 100 L. Light was provided by 4 W fluorescent lamps (white light), placed at the bottom of PBR, with intensity thereof controlled by changing number of lamps. The PBR was equipped with a gas filled tube in the four corners, and the center of the reactor received the aeration system, with an aeration rate of 0.08 vvm that also provided appropriate mixing. The experiment was performed under batch conditions for 35 d, and a maximum cell count of 24,600 cells mL^{-1} was attained, ca. 3.2 g of dry algal powder and 115.2 mg of PSP toxins (okadaic acid plus dinophysis toxins) (Wang et al. 2015).

Rahman et al. (2016) have likewise employed a vertical column PBR for three benthic dinoflagellates, *Amphidinium carterae*, *Prorocentrum rathymum*, and *Symbiodinium* sp., with a total capacity of ~ 700 L (12 columns with 60 L each, connected to each other). The light intensities at stake varied between 40–50 µmol m^{-2}s^{-1}, the light/night cycle was 12 h/12 h and moderate aeration and mixing facilitated by circulation of the culture between columns. Biomass yields attained 0.226 g L^{-1}, 0.183 g L^{-1}, 0.173 g L^{-1} for *Amphidinium carterae*, *Prorocentrum rathymum*, and *Symbiodinium* sp., respectively (Rahman et al. 2016).

Fuentes et al. (2016) have studied only the production of biomass of *Amphidinium carterae* and several metabolites (i.e., pigments, lipids) thereby, using airlift column PBR indoors and outdoors—and proved that this dinoflagellate can be exploited in a larger scale. Different culture volumes were tested: a PBR with 540 L was tested indoors, with light path of 10 cm, comprising three columns connected to a central tube that allow the mixing of the medium; then cultures were transferred to a PBR with a working volume of 320 L, a shorter light path of 5 cm but the same configuration; an airlift column PBR was finally tested outdoors with a small working volume (48 L) and a light path of 5 cm. All PBRs were continuously aerated with a mixture of 2.5% (v/v) CO_2 enriched-air, at a flow rate set at 6.25 L·min^{-1}. The indoor systems were operated under semi-continuous mode, with a photoperiod of 18 h/6 h and an average light irradiance of 158 µmol m^{-2}s^{-1} for 230 d, while the outdoors system was operated under ~ 460 µmol m^{-2}s^{-1}. Average biomass productivities were 0.052 mg L d^{-1} and 0.036 g L d^{-1} indoors and outdoors, respectively (Fuentes-Grünewald et al. 2016).

In a previous study, Fuentes et al. (2013) have tested a large-scale method, with two different toxic dinoflagellates, *Alexandium minutum* and *Karlodinium veneficum*; nine bubble column PBRs, with a total of 350 L, were also used indoors and outdoors. The working volume was established as 315 L (~ 35 L each column), and agitation was provided by continuously injecting pre-filtered air at a flow rate of 0.1 vvm, concomitant with a well-mixed supply of nutrients (thus avoiding O_2 accumulation). Light intensity indoors was 110 µE m^{-2} s^{-1}, with a light and dark regime of 12 h/12 h, while outdoors it varied between 200 to 4,000 µE m^{-2} s^{-1}. Indoor experiments were run in batch mode, with biomass productivities not surpassing 0.16 g L^{-1} d^{-1} for both microalgae; the outdoor cultures were run in semi-continuous mode, and attained biomass productivities of 0.35 g L^{-1} d^{-1} and 0.22 g L^{-1} d^{-1} for *Alexandium minutum* and *Karlodinium veneficum*, respectively (Fuentes-Grünewald et al. 2013).

A LED-illuminated bubble column PBR was also designed and built by López-Rosales et al. (2016), with a working volume of 80 L. The culture was mixed by sparging the vessel with pre-filtered air, at a superficial aeration velocity below 2.74 x 10^3 m s^{-1} (to prevent cell damage) and operated in a sequential batch. Cell concentrations achieved were ~ 120,0000 cells mL^{-1} at the final stage (López-Rosales et al. 2016).

Another approach was attempted by Benstein et al. (2014), namely the use of a biofilm photobioreactor—a type of bioreactor that relies on immobilization of microalgae in a biofilm on sheet-like surfaces. *Symbiodinium voratum* was chosen, for implementation in an optimized Twin-Layer PBR consisting of a thin membrane of polycarbonate with a total area 414 cm^2 placed in transparent poly(methyl methacrylate) tubes, with inner diameter of 11.4 cm. Several cultivation discs (where microalgae were inoculated) were produced by punching out discs of 47 mm in diameter, from the outer paper layer of a thin layer sheet. The cultures were aerated by membrane pumping, at 75 L min^{-1}, of filter-sterilized ambient air. The reactor was placed in a climatic chamber and a rooftop greenhouse, under different light conditions—the former with light Biolux lamps, intensity of 26 µmol m^{-2}s^{-1} and photoperiod of 14 h/10 h; and the latter with natural light and external sodium discharge lamp providing ~ 73 µmol m^{-2}s^{-1} photon irradiance. Biomass concentrations (in dry matter) have reached 1 g m^{-2} d^{-1} in the former conditions, but more than doubled in the latter—thus achieving ca. 2.6 g m^{-2} d^{-1}. An optimized version of this PBR was tested by employing polycarbonate membranes instead of paper as substrate for immobilization but using the same conditions, in the greenhouse chamber; 4.3 g m^{-2} d^{-1} was attained, and higher light intensity (~ 417 µmol m^{-2} s^{-1}) led to 11 g m^{-2} d^{-1} (Benstein et al. 2014).

Several constrains have to be taken into account concerning scalability of cultivation. Large volumes are commonly associated to decreased productivities when compared to bench-scale experiments. For that reason, optimization to attain high toxin titers is not easy, and further investigation is needed on the topic of operation and PBR design—without forgetting safety issues. Furthermore, the understanding of how a species can turn a higher-producer of this type of value-added products is crucial; special features required, ecology, behavior, metabolic pathways and genetic factors would all reinforce knowledge on how to rationally improve toxin production, so as to eventually help expand this niche market.

Concluding remarks

Dinoflagellates are a versatile group of microalgae possessing outstanding features, morphology, and ecology—exceptionally well adapted to a wide range of environments. They have an impact on the ecosystem of water columns, and are majorly responsible for occurrence of HABs concomitant with toxin production. Such events are quite unfavorable for the marine life, and well-established human activities in coastal areas can be severely affected—with detrimental economic implications. Several aspects of HABs and toxin production remain to be elucidated: how and why such events take place, and why and how such toxins are produced. The effective influence of abiotic factors (i.e., nutrition, light, circadian cycle) and the role of bacteria may be critical issues to fully understand the inter- and intraspecific variability in toxin composition and content.

The ability of dinoflagellates to produce several toxins with a wide spectrum of bioactivities has provided the impetus for potential drug development therefrom. Most said toxins may have noteworthy application in treatment human or animal treatment diseases, for example, cancer, respiratory, neurological, or immunologic degenerative pathologies; and they can also play the role of antibiotic, analgesic, anticholesterol, and/or citotoxic agents. However, several difficulties remain in attempts to implement and commercialize the aforementioned substances, namely sufficient amounts for clinical trials. Only small quantities of such compounds are indeed available, with extra limitations arising from their outrageous prices and frequent discontinued distribution by the supplier. The limited availability of natural sources, along with main complications for *de novo* synthesis, genetic and metabolic engineering have been a restraining dinoflagellate biotechnology development. In addition, the difficulties in growing them in the laboratory, have been humpering bulk uses of their unique metabolites for pharmacological pursposes. Although chemical synthesis may be possible, this is not probable to succeed in short run. In praticte, it will entail laborious and hard-demanding steps with low cost-effectiveness. Despite several efforts, the peculiar DNA of dinoflagellates, and their complex and specific metabolism have greatly constrained applicability of genetic and metabolic tools already available for other organisms. Lack of a general transformation system, and scarce information about gene-mapping have severely hampered the improvement of the synthetic pathways regarding toxin production. Menwhile, high throughput technologies are evolving rather fast, and may constitute a useful tool in a near future. As for now, the only realistic approach to improve toxin production is resorting to autotrophic growth of dinoflagellates in photobioreactors. However, shear stress, cell damage and growth inhibition remain an issue during such cultivations, due to turbulence arising from stirring or bubbling. Dinoflagellate cells are extremely sensitive to external mechanical forces with obvious consequences upon morphology, growth, and toxin production. These phenomena are not well understood at molecular level—yet it has been claimed that their cytoskeleton is involved in stimuli transduction, which may lead to alterations in DNA conformation. The range of outcomes (i.e., growth inhibition, cell damage) is apparently species- and strain-dependent. Obviously, the duration and magnitude of forces also have an effect on shear consequences. Even species with theca—once thought to possess additional protection against shear, are themselves sensitive to turbulent liquid motion.

Despite the above difficulties, dinoflagellate cultivation has undergone considerable advances, even at large scale. It should be noted that such cultures must be confined to closed reactors, since open systems may raise an environmental contamination issue. Additionally, toxins—for pharmaceutical application—have to be obtained with good and consistent quality, and this may be uncompatible with growth under somewhat unpredictable conditions. Advances to date encompass development of several types of photobioreactors (carboys, chemostats, airlift, bubble column, flat-plate, biofilm PBRs), from bench to large scale (700 L); however, further improvements are still a must to maximize biomass production. In this regard, it is crucial to improve design and operation of photobioreactors along with better knowledge of mechanism of synthesis *in vivo* of a given toxin—in terms of morphology and/or metabolic requirements. It is expected that such efforts will eventually overcome the aforementioned limitations upon dinoflagellate cultivation, by achieving higher volumetric titers and toxin productivities suitable for performance of clinical experimentation and thus, pharmacological development afterwards.

Acknowledgments

This research was partially supported by project DINOSSAUR—PTDC/BBB-EBB/1374/2014 - POCI-01-0145-FEDER-016640, funded by FEDER funds through COMPETE2020—Programa Operacional Competitividade e Internacionalização (POCI), and by national funds through FCT—Fundação para a Ciência e a Tecnologia, I.P., coordinated by author F.X.M.; and also partially supported by project POCI-01-0145-FEDER-006939 (Laboratory for Process Engineering, Environment, Biotechnology and Energy—UID/EQU/00511/2013), funded by the European Regional Development Fund (ERDF), through COMPETE2020—Programa Operacional Competitividade e Internacionalização (POCI) and by national funds through FCT and project NORTE-01-0145-FEDER-000005—LEPABE-2-ECO-INNOVATION, supported by North Portugal Regional Operational Programme (NORTE 2020), under Portugal 2020 Partnership Agreement, through European Regional Development Fund (ERDF). Part of this work was also supported by the COST Action ES1408 European network for algal-bioproducts (EUALGAE).

References

Abe, M., D. Inoue, K. Matsunaga, Y. Ohizumi, H. Ueda, T. Asano, M. Murakami and Y. Sato. 2002. Goniodomin A, an antifungal polyether macrolide, exhibits antiangiogenic activities via inhibition of actin reorganization in endothelial cells. J. Cell. Physiol. 190: 109–116.

Abraham, W.M., A.J. Bourdelais, J.R. Sabater, A. Ahmed, T.A. Lee, I. Serebriakov and D.G. Baden. 2005. Airway responses to aerosolized brevetoxins in an animal model of asthma. Am. J. Respir. Crit. Care Med. 171: 26–34.

Adolf, J.E., T.R. Bachvaroff and A.R. Place. 2009. Environmental modulation of karlotoxin levels in strains of the cosmopolitan dinoflagellate, *Karlodinium veneficum* (dinophyceae). J. Phycol. 45: 176–192.

Alam, S., D.B. Lovett, R.B. Dickinson, K.J. Roux and T.P. Lele. 2014. Nuclear forces and cell mechanosensing. Prog. Mol. Biol. Transl. Sci. 126: 205–215.

Anderson, D.M. 1995. Toxic red tides and harmful algal blooms: a practical challenge in coastal oceanography. Rev. Geophys. 33: 1189–1200.

Beam, C.A. and M. Himes. 1984. Dinoflagellate genetics. pp. 5685–5697. *In*: D.L. Spector (ed.). Dinoflagellates. Academic Press, Orlando.

Benstein, R.M., Z. Çebi, B. Podola and M. Melkonian. 2014. Immobilized growth of the peridinin-producing marine dinoflagellate symbiodinium in a simple biofilm photobioreactor. Mar. Biotechnol. 16: 621–628.

Berdalet, E. 1992. Effects of turbulence on marine dinoflagellate Gymnodinium nelsonii. J. Phycol. 28: 267–272.

Berdalet, E., F. Peters, V.L. Koumandou, C. Roldán, Ò. Guadayol and M. Estrada. 2007. Species-specific physiological response of dinoflagellates to quantified small-scale turbulence. J. Phycol. 43: 965–977.

Berde, C.B., U. Athiraman, B. Yahalom, D. Zurakowski, G. Corfas and C. Bognet. 2011. Tetrodotoxin-bupivacaine-epinephrine combinations for prolonged local anesthesia. Mar. Drugs 9: 2717–2728.

Beuzenberg, V., D. Mountfort, P. Holland, F. Shi and L. Mackenzie. 2012. Optimization of growth and production of toxins by three dinoflagellates in photobioreactor cultures. J. Appl. Phycol. 24: 1023–1033.

Blunt, J.W., B.R. Copp, R.A. Keyzers, M.H.G. Munro and M.R. Prinsep. 2016. Marine natural products. Nat. Prod. Rep. 33: 382–431.

Bomber, J.W., M.G. Rubio and D.R. Norris. 1989. Epiphytism of dinoflagellates associated with the disease ciguatera: substrate specificity and nutrition. Phycologia 28: 360–368.

Bošnjaković, M. 2013. Biodiesel from algae. J. Mechan. Eng. and Autom. 3: 179–188.

Bralewska, J.M. and Z. Witek. 1995. Heterotrophic dinoflagellates in the ecosystem of the Gulf of Gdansk. Mar. Ecol. Prog. Ser. 117: 241–248.

Bravo, I. and R. Figueroa. 2014. Towards an ecological understanding of dinoflagellate cyst functions. Microorganisms. 2:11–32

Cao, Z., K.T. LePage, M.O. Frederick, K.C. Nicolaou and T.F. Murray. 2010. Involvement of caspase activation in azaspiracid-induced neurotoxicity in neocortical neurons. Toxicol. Sci. 114: 323–334.

Carlson, R.D., G. Morey-Ganes, D.R. Tindall and R.W. Dickey. 1984. Ecology of toxic dinoflagellates from the caribbean sea: Effects of macroalgal extracts on growth in culture. pp. 271–287. *In*: E.P. Ragelis (ed.). Seafood Toxins. American Chemical Society, Washigton DC.

Carvalho, A.P., L.A. Meireles and F.X. Malcata. 2006. Microalgal reactors: a review of enclosed system designs and performances. Biotechnol. Prog. 22: 1490–506.

Cavalier-Smith, T., Z. Zhang and B.R. Green. 1999. Single gene circles in dinoflagellate chloroplast genomes. Nature 400: 155–159.

Cembella, A.D. 2003. Chemical ecology of eukaryotic microalgae in marine ecosystems. Phycologia 42: 420–447.

Chae, H.D., T.S. Choi, B.M. Kim, J.H. Jung, Y.J. Bang and D.Y. Shin. 2005. Oocyte-based screening of cytokinesis inhibitors and identification of pectenotoxin-2 that induces Bim/Bax-mediated apoptosis in p53-deficient tumors. Oncogene 24: 4813–4819.

Chen, A.K., M.I. Latz and J.A. Fringes. 2003. The use of dinoflagellate bioluminescence to characterize cell stimulation in bioreactors. Biotechnology. Boeing. 83: 93–103.

China in, M., H.T. Darius, A. Dung, P. Crochet, Z. Wang, D. Pontoon, D. Laurent and S. Papilla. 2010. Growth and toxin production in the ciguatera-causing dinoflagellate Gambier discus polynesiensis (Dinophyceae) in culture. Toxicon. 56: 739–750.

Chisti, Y. 2000. Animal-cell damage in sparged bioreactors. Trends Biotechnology 18: 420–432.

Coffroth, M. and S. Santos. 2005. Genetic diversity of symbiotic dinoflagellates in the genus *Symbiodinium*. Protist. 156: 19–34.

Daranas, A.H., M. Norte and J.J. Fernández. 2001. Toxic marine microalgae. Toxicon. 39: 1101–1132.

Doblin, M.A., P.A. Thompson, A.T. Revill, E.C. V. Butler, S.I. Blackburn and G.M. Hallegraeff. 2006. Vertical migration of the toxic dinoflagellate *Gymnodinium catenatum* under different concentrations of nutrients and humic substances in culture. Harmful Algae 5: 665–677.

Duncan, K.G., J.L. Duncan and D.M. Schwartz. 2001. Saxitoxin: an anesthetic of the deepithelialized rabbit cornea. Cornea 20: 639–642.

Erga, S.R., C.D. Olseng and L.H. Aarø. 2015. Growth and diel vertical migration patterns of the toxic dinoflagellate *Protoceratium reticulatum* in a water column with salinity stratification: The role of bioconvection and light. Mar. Ecol. Prog. Ser. 539:47–64.

Etheridge, S.M. and C.S. Roesler. 2005. Effects of temperature, irradiance, and salinity on photosynthesis, growth rates, total toxicity, and toxin composition for *Alexandrium fundyense* isolates from the Gulf of maine and Bay of fundy. Deep. Res. Part II Top. Stud. Oceanogr. 52: 2491–2500.

Faust, M.A. 1990. Morphologic details of six benthic species of *Prorocentrum* (Pyrrophyta) from a mangrove island, twin cays, Belize, including two new species. J. Phycol. 26: 548–558.

Faust, M.A. 1991. Morphology of Ciguatera-causing *Prorocentrum lima* (pyrrophyta) from widely differing sites. J. Phycol. 27: 642–648.

Fensome, R.A., R.A. MacRae and G.L. Williams. 1995. Dinoflagellate evolution and diversity through time. Mar Micropaleontol. 4: 1–12.

Fensome, R.A., J.F. Saldarriaga and T. Fensome. 1999. Dinoflagellate phylogeny revisited: reconciling morphological and molecular based phylogenies. Grana. 38: 66–80.

Fleming, L.E., B. Kirkpatrick, L.C. Backer, J.A. Bean, A. Wanner, A. Reich, J. Zaias, Y.S. Cheng, R. Pierce, J. Naar, W.M. Abraham and D.G. Baden. 2007. Aerosolized red-tide toxins (brevetoxins) and asthma. Chest. 131: 187–94.

Flewelling, L.J., J.P. Naar, J.P. Abbott, D.G. Baden, N.B. Barros, G.D. Bossart, M.Y.D. Bottein, D.G. Hammond, E.M. Haubold, C.A. Heil, M.S. Henry, H.M. Jacocks, T.A. Leighfield, R.H. Pierce, T.D. Pitchford, S.A. Rommel, P.S. Scott, K.A. Steidinger, E.W. Truby, F.M.V. Dolah and J.H. Landsberg. 2005. Brevetoxicosis: red tides and marine mammal mortalities. Nature 435: 755–756.

Frangópulos, M., C. Guisande, E. deBlas and I. Maneiro. 2004. Toxin production and competitive abilities under phosphorus limitation of *Alexandrium* species. Harmful Algae 3: 131–139.

Fuentes-Grünewald, C., C. Bayliss, F. Fonlut and E. Chapuli. 2016. Long-term dinoflagellate culture performance in a commercial photobioreactor: *Amphidinium carterae* case. Bioresour. Technol. 218: 533–540.

Fuentes-Grünewald, C., E. Garcés, E. Alacid, S. Rossi and J. Camp. 2013. Biomass and lipid production of dinoflagellates and raphidophytes in indoor and outdoor photobioreactors. Mar. Biotechnol. 15: 37–47.

Fuwa, H., N. Kainuma, K. Tachibana and M. Sasaki. 2002. Total Synthesis of (−)-Gambierol. J. Am. Chem. Soc. 124: 14983–14992.

Gaines, G. and M. Elbrachter. 1987. Heterotrophic nutrition. pp. 224–247. *In*: F.J.R. Taylor (ed.). The Biology of Dinoflagellates, Blackwell Scientific Publications, Oxford.

García-Camacho, F., J.J. Gallardo-Rodríguez, A. Sánchez-Mirón, M.C. Cerón-García, E.H. Belarbi, Y. Chisti and E. Grima-Molina. 2007a. Biotechnological significance of toxic marine dinoflagellates. Biotechnol. Adv. 25: 176–194.

García-Camacho, F., J.J. Gallardo-Rodríguez, A. Sánchez-Mirón, M.C. García-Cerón, E.H. Belarbi and E. Molina-Grima. 2007b. Determination of shear stress thresholds in toxic dinoflagellates cultured in shaken flasks Implications in bioprocess engineering. Process Biochem. 42: 1506–1515.

García-Camacho, F., J.J. Gallardo-Rodríguez, A. Sánchez-Mirón, E.H. Belarbi, Y. Chisti and E. Molina-Grima. 2011. Photobioreactor scale-up for a shear-sensitive dinoflagellate microalga. Process Biochem. 46: 936–944.

Gallardo-Rodríguez, J.J., A. Sánchez-Mirón, F. García-Camacho, M.C. Cerón-García, E.H. Belarbi, Y. Chisti and E. Molina-Grima. 2009. Causes of shear sensitivity of the toxic dinoflagellate *Protoceratium reticulatum*. Biotechnol. Prog. 25: 792–800.

Gallardo-Rodríguez, J.J., A. Sánchez-Mirón, F. García-Camacho, M. Cerón-García, E.H. Belarbi and E. Molina-Grima. 2010. Culture of dinoflagellates in a fed-batch and continuous stirred-tank photobioreactors: Growth, oxidative stress and toxin production. Process Biochem. 45: 660–666.

Gallardo-Rodríguez, J.J., A. Sánchez Mirón, F. García Camacho, M.C. Cerón García, E.H. Belarbi, Y. Chisti and E. Molina Grima. 2011. Carboxymethyl cellulose and Pluronic F68 protect the dinoflagellate *Protoceratium reticulatum* against shear-associated damage. Bioprocess Biosyst. Eng. 34: 3–12.

Gallardo-Rodríguez, J.J., F. García-Camacho, A. Sánchez-Mirón, L. López-Rosales, Y. Chisti and E. Molina-Grima. 2012a. Shear-induced changes in membrane fluidity during culture of a fragile dinoflagellate microalga. Biotechnol. Prog. 28: 467–473.

Gallardo-Rodríguez, J.J., A. Sánchez-Mirón, F. García-Camacho, L. López-rosales, Y. Chisti and E. Molina-Grima. 2012b. Bioactives from microalgal dinoflagellates. Biotechnol. Adv. 30: 1673–1684.

Gallardo-Rodríguez, J.J., L. López-Rosales, A. Sánchez-Mirón, F. García-Camacho and E. Molina-Grima. 2015. New insights into shear-sensitivity in dinoflagellate microalgae. Bioresour. Technol. 200: 699–705.

Garrido, R., N. Lagos, K. Lattes, M. Abedrapo, G. Bocic, A. Cuneo, H. Chiong, J. Christian, R. Azolas, A. Henriquez and C. Garcia. 2005. Gonyautoxin: New treatment for healing acute and chronic anal fissures. Dis. Colon Rectum. 48: 335–340.

Gawley, R.E., K.S. Rein, G. Jeglitsch, D.J. Adams, E.A. Theodorakis, J. Tiebes, K.C. Nicolaou and D.G. Baden. 1995. The relationship of brevetoxin 'length' and A-ring functionality to binding and activity in neuronal sodium channels. Chem. Biol. 2: 533–541.

Glaser, K.B. and A.S. Mayer. 2009. A renaissance in marine pharmacology: From preclinical curiosity to clinical reality. Biochem. Pharmacol. 78: 440–448.

Gobler, C.J. and S.A. Sañudo-Wilhelmy. 2001. Effects of organic carbon, organic nitrogen, inorganic nutrients, and iron additions on the growth of phytoplankton and bacteria during a brown tide bloom. Mar. Ecol. Prog. Ser. 209: 19–34.

Gomez, F. 2012. A quantitative review of the lifestyle, habitat and trophic diversity of dinoflagellates (Dinoflagellata, Alveolata). Syst. Biodivers 10: 267–275.

Gouveia, L. 2011. Microalgae and biofuels production. pp. 1–69. *In*: Microalgae as a Feedstock for Biofuels. SpringerBriefs in Microbiology. Springer Berlin.

Hackett, J.D., D.M. Anderson, D.L. Erdne and D. Bhattacharya. 2004. Dinoflagellates: a remarkable evolutionay experiment. Am. J. Bot. 91: 1523–1534.

Haefner, B. 2003. Drugs from the deep: marine natural products as drug candidates. Drug Discov. Today. 8: 536–544.

Hallegraeff, G. M. 1993. A review of harmful algal blooms and their apparent global increase. Phycologia 32: 79–99.

Hallegraeff, G. M.2003. Harmful algal blooms: a global overview. pp. 25–45. *In*: G.M. Hallegraeff, D.M. Anderson and A.D. Cembella (eds.). Manual of Harmful Marine Microalgae. UNESCO Publishing, Landais.

Han, M., H. Lee, D.M. Anderson and B. Kim. 2016. Paralytic shellfish toxin production by the dinoflagellate *Alexandrium pacificum* (Chinhae Bay, Korea) in axenic, nutrient-limited chemostat cultures and nutrient-enriched batch cultures. Mar. Pollut. Bull. 104: 34–43.

Hardison, D.R., W.G. Sunda, D. Shea and R.W. Litaker. 2013. Increased toxicity of *Karenia brevis* during phosphate limited growth: ecological and evolutionary implications. PLoS One 8: 1–15.

Harlow, L.D., A. Koutoulis and G.M. Hallegraeff. 2007. S-adenosylmethionine synthetase genes from eleven marine dinoflagellates. Phycologia 46: 46–53.

Hickman, C.P., L.S. Roberts, A.L. Larson and H.l'Anson. 2008. Protozoan Groups. *In*: Hickman, S., Larry Roberts, L. Allan and H.I. Larson (eds.). Integrated Principles of Zoology, McGraw-Hill, New York.

Hodgkiss, I.J. and K.C. Ho. 1997. Are changes in N:P ratios in coastal waters the key to increased red tide blooms? Hydrobiologia 352: 141–147.

Hu, H., W. Chen, Y. Shi and C. Wei. 2006. Nitrate and phosphate supplementation to increase toxin production by the marine dinoflagellate Alexandrium tamarense. Mar. Pollut. Boll. 52: 756–760.

Hu, W., C. Berdugo and J.J. Chalmers. 2011. The potential of hydrodynamic damage to animal cells of industrial relevance: current understanding. Cytotechnology 63: 445–60.

Hu, W., R. Gladue, J. Hansen, C. Wojnar and J.J. Chalmers. 2007. The sensitivity of the dinoflagellate Crypthecodinium cohnii to transient hydrodynamic forces and cell-bubble interactions. Biotechnol. Prog. 23: 1355–1362.

Jaeckisch, N., I. Yang, S. Wohlrab, G. Glöckner, J. Kroymann, H. Vogel, A. Cembella and U. John. 2011. Comparative genomic and transcriptomic characterization of the toxigenic marine dinoflagellate *Alexandrium ostenfeldii*. PLoS One 6: 1–15.

Jaouen, P., L. Vandanjon and F. Quéméneur. 1999. The shear stress of microalgal cell suspensions (Tetraselmis suecica) in tangential flow filtration systems: the role of pumps. Bioresour. Technol. 68: 149–154.

Jauffrais, T., J. Kilcoyne, V. Séchet, C. Herrenknecht, P. Truquet, F. Hervé, J. B. Bérard, Cíara Nulty, S. Taylor, U. Tillmann, C. O. Miles and P. Hess. 2012. Production and isolation of azaspiracid-1 and -2 from *Azadinium spinosum* culture in pilot scale photobioreactors. Mar. Drugs 10: 1360–1382.

John, E.H. and K.J. Flynn. 2000. Growth dynamics and toxicity of *Alexandrium fundyense* (Dinophyceae): the effect of changing N:P supply ratios on internal toxin and nutrient levels. Eur. J. Phycol. 35: 11–23.

John, U., M.A. Quilliam, L. Medlin and A.D. Cembella. 2001. Spirolide production and photoperiod-dependent growth of the marine dinoflagellate *Alexandrium ostenfeldii*. pp. 299–302. *In*: G.M. Hallegraeff, S.I. Blackburn, C.J. Bolch and R.J. Lewis (eds.). Harmful Algal Bloom 2000. International Oceanographic Commission (UNESCO), Paris.

Joseph, A. 2016. Oceans: Abode of nutraceuticals, pharmaceuticals, and biotoxins. pp. 493–554. *In*: A. Joseph (ed.). Investigating Seafloors and Oceans. Candice Janco, Goa.

Juhl, A.R., V.L. Trainer and M.I. Latz. 2001. Effect of fluid shear and irradiance on population growth and cellular toxin content of the dinoflagellate *Alexandrium fundyense*. Limnol. Oceanogr. 46: 758–764.

Kamat, P.K., S. Rai and C. Nath. 2013. Okadaic acid induced neurotoxicity: An emerging tool to study Alzheimer's disease pathology. Neurotoxicology 37: 163–172.

Kellmann, R., A. Stüken, R.J.S. Orr, H.M. Svendsen and K.S. Jakobsen. 2010. Biosynthesis and molecular genetics of polyketides in marine dinoflagellates. Mar. Drugs 8: 1011–1048.

Kibler, S.R., R.W. Litaker, W.C. Holland, M.W. Vandersea and P.A. Tester. 2012. Growth of eight *Gambierdiscus* (Dinophyceae) species: Effects of temperature, salinity and irradiance. Harmful Algae 19: 1–14.

Kita, M., N. Ohishi, K. Washida, M. Kondo, T. Koyama, K. Yamada and D. Uemura. 2005. Symbioimine and neosymbioimine, amphoteric iminium metabolites from the symbiotic marine dinoflagellate *Symbiodinium* sp. Bioorg. Med. Chem. 13: 5253–5258.

Kobayashi, J. and M. Tsuda. 2004. Amphidinolides, bioactive macrolides from symbiotic marine dinoflagellates. Nat. Prod. Rep. 21: 77–93.

Kohane, D.S., N.T. Lu, A.C. Gökgöl-Kline, M. Shubina, Y. Kuang, S. Hall, G.R. Strichartz and C.B. Berde. 2000. The local anesthetic properties and toxicity of saxitonin homologues for rat sciatic nerve block *in vivo*. Reg. Anesth. Pain Med. 25: 52–59.

Kumar, K., C. Nag, B. Nayak, P. Lindblad and D. Das. 2011. Development of suitable photobioreactors for CO_2 sequestration addressing global warming using green algae and cyanobacteria. Bioresour. Technol. 102: 4945–4953.

Landsberg, J.H., L.J. Flewelling and J. Naar. 2009. *Karenia brevis* red tides, brevetoxins in the food web, and impacts on natural resources: Decadal advancements. Harmful Algae 8: 598–607.

Leander, B.S. and P.J. Keeling. 2004. Early evolutionary history of dinoflagellates and apicomplexans (Alveolata) as inferred from hsp90 and actin phylogenies. J. Phycol. 40: 341–350.

Lewis, J., A.S. Harris, K.J. Jones and R.L. Edmonds. 1999. Long-term survival of marine planktonic diatoms and dinoflagellates in stored sediment samples. J. Plankton Res. 21: 343–354.

Lin, S. 2011. Genomic understanding of dinoflagellates. Res. Microbiol. 162: 551–569.

Loader, J.I., A.D. Hawkes, V. Beuzenberg, D.J. Jensen, J.M. Cooney, A.L. Wilkins, J.M. Fitzgerald, L.R. Briggs and C.O. Miles. 2007. Convenient large-scale purification of yessotoxin from *Protoceratium reticulatum* culture and isolation of a novel furanoyessotoxin. J. Agric. Food Chem. 55: 11093–11100.

ten Lohuis, M.R. and D.J. Miller. 1998. Genetic transformation of dinoflagellates (*Amphidinium* and *Symbiodinium*): expression of GUS in microalgae using heterologous promoter constructs. Plant J. 13: 427–435.

López-Legentil, S., B. Song, M. DeTure and D.G. Baden. 2010. Characterization and localization of a hybrid non-ribosomal peptide synthetase and Polyketide synthase gene from the toxic dinoflagellate *Karenia brevis*. Mar. Biotechnol. 12: 32–41.

López-rosales, L., A. Sánchez-Mirón, A. Contreras-Gomes, F. García-camacho and E. Molina-grima. 2015. An optimisation approach for culturing shear-sensitive dinoflagellate microalgae in bench-scale bubble column photobioreactors. Bioresour. Technol. 197: 375–382.

López-rosales, L., F. García-Camacho, A. Sánchez-Mirón, E.M. Beato and Y. Chisti. 2016. Pilot-scale bubble column photobioreactor culture of a marine dinoflagellate microalga illuminated with light emission diodes. Biosource Technol. 216: 845–855.

Malve, H. 2016. Exploring the ocean for new drug developments: Marine pharmacology. J. Pharm. Bioallied Sci. 8: 83–91.

Matsuoka, K., L.B. Joyce, Y. Kotani and Y. Matsuyama. 2003. Modern dinoflagellate cysts in hypertrophic coastal waters of Tokyo Bay, Japan. J. Plankton Res. 25: 1461–1470.

McEwan, M., R. Humayun, C.H. Slamovotis and P.J. Keeling. 2008. Nuclear genome sequence survey of the dinoflagellate *Heterocapsa triquetra*. J. Eukaryot. Microbiol. 55: 530–535.

Medhioub, W., V. Séchet, P. Truquet, M. Bardouil, Z. Amzil, P. Lassus and P. Soudant. 2011. *Alexandrium ostenfeldii* growth and spirolide production in batch culture and photobioreactor. Harmful Algae 10: 794–803.

Nagai, H., Y. Mirakami, K. Yazawa, T. Gonoi and T. Yasumoto. 1993. Biological activities of novel polyether antifungals, gamberic acids A and B from a marine dinogflagellate *Gambierdiscus toxicus*. J. Antibiot. (Tokyo). 46: 520–522.

Nicolaou, K.C., Theocharis V. Koftis, S. Vyskocil, G. Petrovic, W. Tang, M.O. Frederick, D.Y.K. Chen, Y. Li, T. Ling and Y.M.A. Yamada. 2006. Total synthesis and structural elucidation of azaspiracid-1. Final assignment and total synthesis of the correct structure of azaspiracid-1. J. Am. Chem. Soc. 128: 2244–2257.

Nieto, F.R., E.J. Cobos, M.Á. Tejada, C. Sánchez-Fernández, R. González-Cano and C.M. Cendán. 2012. Tetrodotoxin (TTX) as a therapeutic agent for pain. Mar. Drugs 10: 281–305.

Not, F., R. Siano, W. Kooistra, N. Simon, D. Vaulot and I. Probert. 2012. Diversity and ecology of eukaryotic marine phytoplankton. pp. 1–53 *In*: G. Piganeau (ed.). Advances in Botanical Research: Genomic Insights into the Biology of Algae. Academic Press, Oxford.

Pan, Y., A.D. Cembella and M.A. Quilliam. 1999. Cell cycle and toxin production in the benthic dinoflagellate Prorocentrum lima. Mar. Biol. 134: 541–549.

Parker, N.S., A.P. Negri, D.M.F. Frampton, L. Rodolfi, M.R. Tredici and S.I. Blackburn. 2002. Growth of the toxic dinoflagellate *Alexandrium minutum* (Dinophyceae) using high biomass culture systems. J. Appl. Phycol. 14: 313–324.

Pelin, M., C. Florio, C. Ponti, M. Lucafò, D. Gibellini, A. Tubaro and S. Sosa. 2016. Pro-inflammatory effects of palytoxin: an *in vitro* study on human keratinocytes and inflammatory cells. Toxicol. Res. 5: 1172–1181.

Quilliam, M.A. 2003. Chemical methods for lipophilic shellfish toxins. pp. 211–246. *In*: G. Hallegraeff, D. Anderson and A. Cembella (eds.). Manual on Harmful Marine Microalgae. Intergovernmental Oceanographic Comission (UNESCO), Paris.

Radakovits, R., R.E. Jinkerson, A. Darzins and M.C. Posewitz. 2010. Genetic engineering of algae for enhanced biofuel production. Eukaryot. Cell. 9: 486–501.

Rahman Sha, M.M., K.W. Samarakoon, S.J. An, Y.J. Jeon and J.B. Lee. 2016. Growth characteristics of three benthic dinoflagellates in mass culture and their antioxidant properties. J. Fish. Aquat. Sci. 11: 268–277.

Rein, K.S. and J. Borrone. 1999. Polyketides from dinoflagellates: origins, pharmacology and biosynthesis. Comp. Biochem. Physiol. B. Biochem. Mol. Biol. 124: 117–31.

Rein, K.S. and R.V. Snyder. 2006. The biosynthesis of polyketide metabolites by dinoflagellates. Adv. Appl. Microbiol. 59: 93–125.

Rill, R.L., F. Livolant, H.C. Aldrich and M.W. Davidson. 1989. Electron microscopy of liquid crystalline DNA: direct evidence for cholesteric-like organization of DNA in dinoflagellate chromosomes. Chromosoma 98: 280–286.

Rizzo, P.J. 2003. Those amazing dinoflagellate chromosomes. Cell Res. 13: 215–217.

Rodriguez-Navarro, A.J., N. Lagos, M. Lagos, I. Braghetto, A. Csendes, J. Hamilton, C. Figueroa, D. Truan, C. Garcia, A. Rojas, V. Iglesias, L. Brunet and F. Alvarez. 2007. Neosaxitoxin as a local anesthetic. Anesthesiology 106: 339–345.

Rost, B.B., K.U. Richter, U.L. Riebesell and P.E. Hansen. 2006. Inorganic carbon acquisition in red tide dinoflagellates. Plant, Cell Environ. 29: 810–822.

Schnepf, E. and M. Elbrächter. 1992. Nutritional strategies in dinoflagellates: a review with emphasis on cell biological aspects. Eur. J. Protistol. 28: 3–24.

Scholin, C.A., F. Gulland, G.J. Doucette, S. Benson, M. Busman, F.P. Chavez, J. Cordaro, R. DeLong, A. De Vogelaere, J. Harvey, M. Haulena, K. Lefebvre, T. Lipscomb, S. Loscutoff, L.J. Lowenstine, R. Marin, P.E. Miller, W.A. McLellan, P.D.R. Moeller, C.L. Powell, T. Rowles, P. Silvagni, M. Silver, T. Spraker, V. Trainer and F.M.V. Dolah. Mortality of sea lions along the central California coast linked to a toxic diatom bloom. Nature 403: 80–84.

Schweikert, M. and B. Meyer. 2001. Characterization of intracellular bacteria in the freshwater dinoflagellate *Peridinium cinctum*. Protoplasma 217: 177–184.

Sheng, J., E. Malkiel, J. Katz, J.E. Adolf and A.R. Place. 2010. A dinoflagellate exploits toxins to immobilize prey prior to ingestion. Proc. Natl. Acad. Sci. 107: 2082–2087.

Silva, E.S. 1990. Intracellular bacteria: the origin of dinoflagellate toxicity. J. Environ. Pathol. Toxicol. Oncol. 10: 124–128.

Smayda, T.J. 1997. What is a bloom? A commentary. Limnol. Ocean. 42: 1132–1136.

Smayda, T.J. 2010. Adaptations and selection of harmful and other dinoflagellate species in upwelling systems. 2. Motility and migratory behaviour. Prog. Oceanogr. 85: 71–91.

Speckhard, A. 2010. Evolution of the Dinoflagellates: from the origin of the group to thir genes. ICHA14 Conference Proceedings. Crete. 12–20.

Spector, D.L. 1984. Dinoflagellates: an introduction. pp. 1–15. *In*: D.L. Spector (ed.). Dinoflagellates. Academic Press, Orlando.

Steidinger, K.A. and D.G. Baden. 1984. Toxic Marine Dinoflagellates. pp. 201–248. *In*: D.L. Spector (ed.). Dinoflagellates. Academic Press, Oxford.

Sullivan, J.M., E. Swift, P.L. Donaghay and J.E. Rines. 2003. Small-scale turbulence affects the division rate and morphology of two red-tide dinoflagellates. Harmful Algae 2: 183–199.

Tang, E.P.Y. 1996. Why do dinoflagellates have lower growth rates? J. Phycol. 32: 80–84.

Taroncher-Oldenburg, G., D.M. Kulis and D.M. Anderson. 1997. Toxin variability during the cell cycle of the dinoflagellate *Alexandrium fundyense*. Limnol. Oceanogr. 42: 1178–1188.

Taylor, F.J.R. 1980. On dinoflagellate evolution. Biosystems 13: 65–108.

Taylor, F.J.R. 1987. The Biology of Dinoflagellates. Blackwell Scientific Publications, Oxford.

Taylor, F.J.R., M. Hoppenrath and J.F. Saldarriaga. 2008. Dinoflagellate diversity and distribution. Biodivers. Conserv. 17: 407–418.

Tillmann, U., M. Elbrächter, B. Krock, U. John and A.D. Cembella. 2009. *Azadinium spinosum* gen. et sp. nov. (Dinophyceae) identified as a primary producer of azaspiracid toxins. Eur. J. Phycol. 44: 63–79.

Tobío, A., A. Alfonso, I. Madera-Salcedo, L.M. Botana and U. Blank. 2016. Yessotoxin, a marine toxin, exhibits anti-allergic and anti-tumoural activities inhibiting melanoma tumour growth in a preclinical model. PLoS One 11: 1–14.

Tredici, M.R., P. Carlozzi, G. Chini Zittelli and R. Materassi. 1991. A vertical alveolar panel (VAP) for outdoor mass cultivation of microalgae and cyanobacteria. Bioresour. Technol. 38: 153–159.

Tsukano, C. and M. Sasaki. 2006. Structure-activity relationship studies of gymnocin-A. Tetrahedron Lett. 47: 6803–6807.

Twiner, M.J., J.C. Ryan, J.S. Morey, K.J. Smith, S.M. Hammad, F.M. Van Dolah, P. Hess, T. McMahon, M. Satake, T. Yasumoto and G.J. Doucette. 2008. Transcriptional profiling and inhibition of cholesterol biosynthesis in human T lymphocyte cells by the marine toxin azaspiracid. Genomics 91: 289–300.

Valdiglesias, V., M.V. Prego-Faraldo, E. Pasaro, J. Mendez and B. Laffon. 2013. Okadaic acid: more than a diarrheic toxin. Mar. Drugs 11: 4328–4349.

Varley, J. and J. Birch. 1999. Reactor design for large scale suspension animal cell culture. Cytotechnology 29: 177–205.

Waller, R.F. and C.J. Jackson. 2009. Dinoflagellate mitochondrial genomes: Stretching the rules of molecular biology. BioEssays 31: 237–245.

Wang, D.Z. 2008. Neurotoxins from marine dinoflagellates: a brief review. Mar. Drugs 6: 349–371.

Wang, D.Z., A.Y.T. Ho and D.P.H. Hsieh. 2002. Production of C2 toxin by *Alexandrium tamarense* CI01 using different culture methods. J. Appl. Phycol. 14(6): 461–468.

Wang, D.Z., S.F. Zhang, Y. Zhang and L. Lin. 2016. Paralytic shellfish toxin biosynthesis in cyanobacteria and dinoflagellates: A molecular overview. J. Proteomics 135: 132–140.

Wang, S., J. Chen, Z. Li, Y. Wang, B. Fu, X. Han and L. Zheng. 2015. Cultivation of the benthic microalga *Prorocentrum lima* for the production of diarrhetic shellfish poisoning toxins in a vertical flat photobioreactor. Bioresour. Technol. 179: 243–248.

Waters, A.L., R.T. Hill, A.R. Place and M.T. Hamann. 2011. The expanding role of marine microbes in pharmaceutical. Curr. Opin. Biotechnol. 21: 780–786.

Wells, M.L., V.L. Trainer, T.J. Smayda, B.S. Karlson, C.G. Trick, R.M. Kudela, A. Ishikawa, S. Bernard, A. Wulff, D.M. Anderson and W.P. Cochlan. 2015. Harmful algal blooms and climate change: Learning from the past and present to forecast the future. Harmful Algae 49: 68–93.

White, A.W. 1976. Growth inhibition caused by turbulence in the toxic marine dinoflagellate *Gonyaulax excavata*. J. Fish. Res. Board Canada 33: 2598–2602.

Wisecaver, J.H. and J.D. Hackett. 2011. Dinoflagellate genome evolution. Annu. Rev. Microbiol. 65: 369–387.

Wong, J.T. and A.C. Kwok. 2005. Proliferation of dinoflagellates: blooming or bleaching. BioEssays. 27: 730–740.

Wynn, J., P. Behrens, A. Sundararajan, J. Hansen and K. Apt. 2010. Production of single cell oils by dinoflagellates. pp. 115–129. *In*: Z. Cohen and R. Colin (eds.). Single Cell Oils (2nd Ed.). AOCS Press, Illinois.

Yeung, P.K. and J.T. Wong. 2003. Inhibition of cell proliferation by mechanical agitation involves transient cell cycle arrest at G1 phase in dinoflagellates. Protoplasma 220: 173–178.

Yoon, H.S., J.D. Hackett and D. Bhattacharya. 2002. A single origin of the pyridine- and fucoxanthin-containing plastids in dinoflagellates through tertiary endosymbiotic. Proc. Natl. Acad. Sci. 99: 11724–11729.

Zittelli, G.C., L. Rodolfi, N. Bassi, N. Biondi and M.R. Tredici. 2013. Photobioreactors for microalgal biofuel production. *In*: M. Borowitzka and N. Moheimani (eds.). Algae for Biofuels and Energy. Developments in Applied Phycology, vol. 5. Springer, Dordrecht.

Integrated Multitrophic Aquaculture

An Overview

Isabel C. Azevedo,[1,*] *Tânia R. Pereira,*[1] *Sara Barrento*[2] *and*
Isabel Sousa Pinto[1,3]

Introduction

Aquaculture is generally considered as the cultivation of aquatic organisms in freshwater or saltwater, from microscopic algae to large fish, mainly for food purposes. It can be viewed from a food system approach, consisting of a set of activities ranging from production to consumption, which usually includes: growing, harvesting, processing, packaging, transporting, marketing, consumption, and disposal of food and food-related items. With global food security becoming an increasing concern, aquaculture is and will probably remain the most rapidly increasing food production system worldwide.

Integrated Multi-Trophic Aquaculture (IMTA) has the potential to increase the sustainability of aquaculture across the globe. The basic concept of IMTA is the farming of several species at different trophic levels, that is, species that occupy different positions in the food chain. This allows one species' uneaten feed and wastes, nutrients, and by-products to be recaptured and converted into fertilizer, feed, and energy for the other crops (Chopin 2012). As an example, we can combine the cultivation of fed species (e.g., finfish, shrimp, etc.) with organic extractive species (e.g., oysters, mussels, and other invertebrates and inorganic extractive species (e.g., seaweeds or other photosynthetic organisms)). The different types of aquaculture are integrated, that is, operating in proximity to each other, but not necessarily right at the same location (Chopin 2006).

This chapter will address IMTA as an aquaculture production system showing the concept shift that has recently been occurring toward viewing waste nutrients as a resource that can be recycled through

---

[1] Coastal Biodiversity Lab, Interdisciplinary Centre of Marine and Environmental Research (CIIMAR), University of Porto, Novo Edifício do Terminal de Cruzeiros do porto de Leixões. Avenida General Norton de Matos, S/N. 4450-208 Matosinhos, Portugal.
  Email: ispinto@fc.up.pt; pereirataniar@gmail.com
[2] Department of Biosciences, Swansea University, Wallace Building, Singleton Campus, UK.
  Email: s.i.barrento@swansea.ac.uk
[3] Department of Biology, Faculty of Sciences, University of Porto, Rua do Campo Alegre s/n, 4150-181 Porto, Portugal.
* Corresponding author: iazevedo@ciimar.up.pt

macroalgae in a range of aquaculture systems. Several examples will be given of the incorporation of macroalgae to increase the environmental and economic sustainability of land-based aquaculture systems and also aquaculture at sea.

IMTA an historical perspective

IMTA can be traced back to the origins of aquaculture. In China, the integration of fish with aquatic plants and vegetable production and the development of cage culture has been described as early as 2000 BC by You Hou Bin. Moving fast forward in time and to the Fertile Crescent region, there is evidence of tilapia grown in integrated agriculture-aquaculture drainable ponds on bas-reliefs in Egyptian tombs from about 1550–1070 B.C. In Europe, during the French Renaissance, an IMTA system was built at the Château de Fontainebleau, the Etang aux Carpes (Carp Pond). Moving again to China, Xu Guangqi wrote the Nong Zheng Quan Shu, an outstanding agricultural treatise (The Complete Book on Agriculture) published posthumously in 1639. The treatise covered many topics, including irrigation and the rotation of fish and aquatic plant production, the integration of fish with livestock and the effects of manure on pond production, and the integration of mulberry trees, rice paddies, and fish ponds (Chopin 2013). So, when rice/fish culture started to be popular in Europe in the 19th–early 20th centuries, it had already been practiced in China for millennia (Fernando 2002).

In the West, this type of integration never fully developed into commercial scale and aquaculture only started its expansion in the 20th century. From then to now more species were brought into culture and the industry continued to expand both in area and in quantity of production. Only in the late 20th century aquaculture started to work with economies of scale. A new trend to select species that are most profitable to culture was adopted by operators in the industry: for example, *Penaeid* shrimps and high value finfishes (seabass/groupers), and seaweeds and related species started to become interesting aquaculture products (Rabanal 1988). In many places aquaculture developed into an industrial aquaculture food system, based on a monoculture system with high tech methods, and intensification of production towards more efficiency, higher production with lower costs.

However, with the rapid expansion of intensive monoculture systems several environmental and socio-economic problems also started to arise. A major issue, pointed out by several authors, is that Western-oriented aquaculture has been managed as an isolated part of its supporting environment (Folke and Kautsky 1992). In the 1970s, the eutrophication problem, with oxygen depletion, biodiversity modifications and pollution of the surrounding waters, was one of the first concerns leading to research on IMTA. According to Chopin (2013), John Ryther and co-authors reignited the interest in IMTA and can be considered the grandfathers of modern IMTA for their pivotal work entitled "integrated waste-recycling marine polyculture systems" (Ryther et al. 1975). The main aim was to recreate a cultivating system based on a food chain, capable of providing an effluent virtually free of inorganic nitrogen, thus avoiding the risk of eutrophication in the receiving waters (Ryther et al. 1975).

This work was followed by three productive decades addressing this issue under many different designations including polyculture, integrated mariculture, integrated aquaculture, ecologically engineered aquaculture, and ecological aquaculture (Chopin 2013). Then in 2004, Thierry Chopin and Jack Taylor, understanding the need to harmonize all these names, combined integrated aquaculture and multi-trophic aquaculture into the term integrated multi-trophic aquaculture—IMTA (Chopin 2013). Even though the designation is fairly recent, the concept is ancient. Nevertheless, the designation brought with it the research debate, which has evolved and is developing to new commercial and legal perspectives.

The role of seaweeds in IMTA—research approach

In 1975, Ryther and co-authors described the development and testing of a combined tertiary sewage treatment/marine aquaculture system. In this work, red seaweeds were included as a final inorganic nutrient bio filtration unit in a complex system. This used domestic wastewater effluent from secondary sewage treatment mixed with seawater as a source of nutrients for growing unicellular marine algae. These microalgae served as feed for oysters, clams, and other bivalve mollusks. Polychaete worms, amphipods, and other small invertebrates fed on the solid wastes from the shellfish and were fed upon by flounder and lobsters. Commercial red seaweeds (*Chondrus, Gracilaria, Agardhiella, Hypnea*) then removed the dissolved wastes excreted by the shellfish and other animals and any nutrients not initially removed by the unicellular algae (Ryther et al. 1975). This resulted in a low nutrient discharge that was not suitable for the proliferation of algae and, thus, did not contribute to the eutrophication of adjacent wet areas. Since then many studies have been done on the utilization of seaweeds as a biofilter in IMTA systems.

Between 1975 and 1993, the cultivation of seaweeds in integrated systems was mainly investigated for treating effluents from enclosed land-based mariculture systems. This line of research was initiated in the mid-1970s (Haines 1975; Ryther et al. 1975; Langton et al. 1977). But it was only in the 1990s, that renewed and increased research into the development of seaweed-based integrated techniques took place, mainly as a consequence of the rapid expansion of intensive offshore mariculture systems (i.e., fish farming and shrimp cultivation) and the concern for negative effects on the environment from such practices (Vandermeulen and Gordin 1990; Cohen and Neori 1991; Neori et al. 1991; Neori et al. 1996).

Kautsky and Folke (1991) introduced the concept of "integrated open sea aquaculture", in which coastal waters, made eutrophic by fish net pens, agricultural runoff, and sewage discharge, are used to supply cultured seaweed with dissolved nutrients and shellfish with plankton. One year later, the same authors suggested that a coastal culturing system of seaweeds, mussels, and salmon could be developed as an ecologically and economic viable option to integrate culturing activities in coastal areas. In their own words, "a successful aquaculture system does not have wastes, only by-products, to be used as positive contributors to the surrounding ecosystems and the economy" (Folke and Kautsky 1992). In 1993, Hirata and co-authors, tested the cultivation of *Ulva* sp. in crab cages placed in a yellowtail farm at sea at a depth of about 50 cm. The authors concluded that the sterile *Ulva* sp. is a suitable aquatic plant for introducing a polyculture of seaweed and fish in the sea (Hirata et al. 1993).

However, during this period and until 2004, most studies were still mainly focused on demonstrating that wastewater from intensive and semi-intensive mariculture in land based systems is a suitable nutrient source for the intensive production of seaweed, thereby reducing the discharge of dissolved nutrients to the environment. The seaweed species tested were mainly *Ulva* sp. and *Gracilaria* sp. (Table 1).

Western countries are committed to do research and advance technology for development of an industrial IMTA at sea and offshore. Whereas is China, Indonesia, Ecuador, India, the Philippines, Taiwan, Thailand, Japan, and more recently in Vietnam, earthen marine ponds, integrated with natural or agriculture plants (such as mangroves and rice), are used on a wide scale for extensive shrimp farming (in Binh et al. 1997; Alongi et al. 2000).

Table 1. Main species tested for use in bioremediation of aquaculture wastewater.

Species	System type	Co-species	Reference
Ulva lactuca	Inland	N/A	Vandermeulen and Gordin 1990
	Inland	*Sparus aurata*	Neori et al. 1996
	Inland	*Sparus aurata* and *Haliotis discus hannai*	Neori et al. 2000
Ulva rigida	Inland	N/A	del Rio et al. 1996
Gracilaria chilensis	Inland	*Oncorhynchus kisutch*	Buschman et al. 1994
	Coast	N/A	Buschman et al. 1996
	Coast	*Oncorhynchus mykiss* and *O. kisutch*	Troell et al. 1997
Gracilaria conferta	Inland	*Sparus aurata* and *Haliotis discus hannai*	Neori et al. 2000
Porphyra yezoensis	Coast	*Salmo salar*	Chopin et al. 1999
	Coast	scallop	
Nereocystis leutkeana	Inland	N/A	Ahn et al. 1998
Saccharina latissima	Inland	N/A	Ahn et al. 1998
Saccharina latissima	Coast	Salmon	Subandar et al. 1993

Table 2. Biofiltration efficiency of seaweed co-cultivated with other trophic species in land based IMTA systems.

Co-cultured species / Fed species	Volume & water renovation	Recirculation (%)	Density (fw)	Duration	Production g_{dw} m^2 d^{-1}	N removal (%)	P removal (%)	References
Ulva lactuca *Gracilaria conferta* *Haliotis discus hannai* *Sparus aurata*	1200 L	0%	2.5g L^{-1} (*U.l.*) 8–21 g L^{-1} (*G.c.*)	1 yr Oct-Oct	233 (fw *U.l.*) 38 (fw *G.c.*)	67% TAN (*U.l.*) 16% TAN (*G.c.*)	-	Neori et al. 2000
Ulva reticulata *Gracilaria crassa finfish*	tidal outflow channels	0%	3 kg.m^{-2}	6 wk May-Oct	72 ± 53 *U.r.* 105 ± 60 *G.c.*	0.43 g m^{-2} d^{-1} (*U.r.*) 0.15 g m^{-2} d^{-1} (*G.c.*)	-	Msuya and Neori 2002
Ulva rotunda *S. aurata*	1900 L 0.6 vol/h			2 mon May & Sep	48	60% TAN	-	Mata and Santos 2003
Ulva lactuca *H. discus hannai* *Paracentrotus lividus* *S. aurata*	12,000 L 0.3 vol/h	50%	1 kg.m^{-2}	47 d (Nov-Dec) 24 d (Feb-Mar)	94 (fw) 117 (fw)	70% TAN	20	Schuenhoff et al. 2003
Ulva rigida *S. aurata*	110 L 2 vol/h	0%	4 g.L^{-1}	1 wk (Dec) 1 wk (May)	44 73	2.7 g m^{-2} d^{-1} 5.1 g m^{-2} d^{-1}	-	Mata et al. 2010
Ulva pertusa *Saccharina japonica* *Gracilariopsis chorda* *Sebastes schlegeli*	600 L	100%	10 g.L^{-1}	7 d	-	22% TNO; 84% NH$_4^+$ (*U.p.*) 65% TNO; 46% NH$_4^+$ (*S.j.*) 35% TNO; 59% NH$_4^+$ (*G.c.*)	30.6% (*U.p.*) 20.2% (*S.j.*) 38.1% (*G.c.*)	Kang et al. 2011
Asparagopsis armata *S. aurata*	110 L			9 mon Out-Jul	63.7	85% TAN		Schuenhoff et al. 2006
Asparagopsis armata *S. aurata*	110 L 2 vol/h	0%	5 g L^{-1}	1 wk (Dec) 1 wk (May)	71 125	2.8 g m^{-2} d^{-1} TAN 6.5 g m^{-2} d^{-1} TAN	-	Mata et al. 2010
Chondrus crispus *Scophthalmus maximus* *Dicentrarchus labrax*	1500 L 0.1 vol/h			1 mon: May Jul	8.4 36.6	14% TAN (May) 41% TAN (Jul)		Matos et al. 2006
Gracilaria vermiculophylla *Oncorhynchus kisutch* *Orhynchus mykiss*	-		2.5 kg/m^3 (Aut/Win) 4.0 kg/m^3 (Spr/Sum)	13 mon	134	64% TAN	32	Buschmann et al. 1996

Species	System	Flow	Density	Duration				Reference
Graciaria parvispora L. conceptionis C. gigas L. albus Control	200 L 50 L/h	0%	7 kg/m²	10 mon Aut-Sum	51.3 ± 25.1 23.9 ± 17.1 16.2 ± 2.4 18.6 ± 2.4	60% all NS 80% Aut, Win, Spr	60% Spr 60% all 20% Win	(Chow et al. 2001)
Gracilaria vermiculophylla Scophthalmus maximus Dicentrarchus labrax	1500 L 0.1 vol/h			1 mon Jul Oct	31.2 7.3	33% TAN 75% TAN	–	Matos et al. 2006
Gracilaria lemaneiformis S. fuscescens	2150 L		350 550 700 850	31 d Autumn		NH_4–N: 60–80% 3.28 mol g⁻¹ $_{dw}$ h⁻¹A	Max 80%	Zhou et al. 2006
G. lemaneiformis Chlamys farreri	3000 L	Static	69 139 264 348 g m⁻³	3 wk Autumn	0.73 2.52 1.25 0.55%B	21.4 63.0 70.7 73.9%D	40.0 65.5 70.4 58.4%D	Mao et al. 2009
Gracilaria vermiculophylla & ME	1200 L 0.2 vol/h			9 mon (Nov-Aug)	23.3	83% TAN 17% NO_3^-	70 PO_4^{3-}	Abreu et al. 2011
Gracilariopsis chorda Ulva pertusa Saccharina japonica Sebastes schlegeli			See in the green algae section, under *Ulva pertusa*					Kang et al. 2011
Mastocarpus stellatus	1200 L	0%	3 Kg m⁻³	4 weeks (May) 4 weeks (June)	29.6 (May) 38.9 (June)	43.19 ± 1.61% NH_3 (May) 49.00 ± 1.59% NH_3 (June)	–	Domingues et al. 2015
Palmaria palmata Scophthalmus maximus Dicentrarchus labrax	300 L 0.5 vol/h			1 mon May		40.0	41% TAN	Matos et al. 2006
Saccharina latissima salmon						26–40% DIN		Subandar et al. 1993
Saccharina japonica Gracilariopsis chorda Ulva pertusa Sebastes schlegeli			See in the green algae section, under *Ulva pertusa*					Kang et al. 2011

NR, nutrient removal measurement type; NW, measurement of nutrients from the water; NT, measurement of nutrients from the algae tissues; Ft, Flow through; Fw, fresh weight; A, Mean N uptake rates based on N content in the thalli; B Specific Growth Rate; C, N uptake rates based on N content in the thalli for each density; D, reduction efficiency; TAN, Total ammonia nitrogen; TNO, Total nitric oxide: NO_3^- & NO_2^-; DIN, Dissolved Inorganic Nitrogen.

Table 3. Biofiltration efficiency of seaweed co-cultivated with other trophic species open sea IMTA systems.

Co-cultured species	Location	Distance from fed production	Average current (cm s⁻¹)	Duration	Production g_{dw} m⁻²d⁻¹	N removal	P removal	References
Ulva rigida *Sparus aurata*	Israel	0.5 22.5 45 m	20.2 cm s⁻¹	12 d (Aug)	16.8% d⁻¹ (22.5 m)			Korzen et al 2015
Ulva ohnoi *Pagrus major* *Seriolo quinqueradiata*	Gokasho Bay, Central Japan	-	10 cm s⁻¹	2 wk	4.2–23.6%	4.2–13.9 mg δ15N DW⁻¹ d⁻¹	-	Yokoyama and Ishihi 2010
Gracilaria chilensis Salmon (species unknown)	Los Lagos, Southern Chile	100 m 800 m	7.6 (flood) 2.4 (ebb) cm s⁻¹	Summer (Jan-Feb) Autumn (Mar-Apr)	Max: over 1680 g m⁻¹ (800 m from fed production)	9.3 g N m⁻¹ mon⁻¹A	-	Abreu et al. 2009
Gracilaria chouae *Sparus macrocephalus*	Xiangshan Bay, East China	-	50 to 60 cm s⁻¹	May to June (47 d)	7.43 ± 0.37% d⁻¹ to 8.47 ± 0.41% d⁻¹	2.10% (dry weight)	0.31% (dry weight)	Wu et al. 2015
Gracilaria bursa-pastoris *Sparus aurata*	Israel	0.5 22.5 45 m	13.6 cm s⁻¹	14 d (Aug)				Korzen et al 2015
Kappaphycus alvarezii *Litopenaeus vennamei*	Ubatuba, S. Paulo, Brazil	-	-	33 ± 6 d	17.47 ± 5.71 Kg m⁻² yr⁻¹	-	-	Lombardi et al. 2006
Gracilaria lemaneiformis *Epinepheleus akkara*	Shenao Bay Quingdao, Japan	-	-	26 d	7.07–11.71% d⁻¹			Yang et al. 2006
E. awoara *Crassostrea gigas* *Perna viridis*	Jiaozhou bay Quingdao, Japan	-	-	28 d	11.2–13.9% d⁻¹			
Palmaria palmata *Saccharina latissima* Salmon	Badcall bay and Calbha. NW Scotland	N/A		See in the brown algae section, below				Sanderson et al. 2012
Saccharina latissima *Mytilus galloprovincialis*	Ria de Ares y Betanzos, NW Spain	N/A	N/A	27 Nov–5 Feb 5 Feb–10 Mar 10 Mar–17 Apr	11.37 (Feb) 4.52 (Mar) 2.66% (Apr)	8.3–8-5 7.8 7.8% (δ15N)	-	Freitas et al. 2006
Saccharina latissima *Palmaria palmata* Salmon	Badcall bay and Calbha. NW Scotland	-	-	4–6 months	27% d⁻¹ (*S. latissima*) 63% d⁻¹ (*P. palata*)	-	-	Sanderson et al. 2012

A, Mean N uptake rates based on N content in the thalli.

From research to the commercial scale

Since the very beginning of integrated aquaculture development, researchers have been concerned with its commercial application. In 1972, Ryther and co-authors pointed out that there is the need to combine the "small-scale and usually inadequately-supported basic research of the laboratory scientist with the bold, often extravagant, and usually spectacularly unsuccessful empirical approach of industry if intensive aquaculture is to succeed" (Ryther et al. 1972).

In the following decades, in Europe, consortia integrating research centers and companies from a number of different countries developed and implemented projects to boost the knowledge transfer between academia and industry and drive the IMTA concept to commercial scale (Table 4). Some of these projects were more focused on seaweed cultivation, such as SEAPURA (SEAPURA 2004), whereas others aimed at species diversification and novel integrated farming systems to increase sustainability, such as the SEAFARE project (2010–2013, www.seafareproject.eu). **IDREEM**, a collaborative research project, was launched in 2012 with 15 partners, to move IMTA beyond the current state of the art and to demonstrate its viability for the European aquaculture sector. Seven pilot scale IMTA operations were developed across Europe, which provided sufficient evidence that, even though the conditions are not yet fully in place in Europe for the wide scale adoption of IMTA, there is a growing commercial interest and a consumer interest. These pilot installations were established in Scotland (Ardtoe and Loch Fyne), Norway (Oldervika), Ireland (Bantry Bay), Italy (Lavagna), Cyprus (Vasiliko/Zygi), and Israel (Ashdod), associating fish farms (turbot, cod, Atlantic salmon, sea bream, and sea bass), oyster producers (in Scotland) and research partners. IMTA systems were tested using a diverse set of economically interesting organisms including seaweeds, sea urchins, sea cucumber, scallops, mussels, abalones, and sponges (Hughes et al. 2016).

Table 4. Examples of past and present European projects concerned with development of Integrated Multi-Trophic Aquaculture.

Acronym	Title	Start date	End date	Coordination	Participants	Funding	Budget
IDREEM	Increasing Industrial Resource Efficiency In European Aquaculture	2012	2016	SAMS, UK	15 partners, from seven countries: Cyprus, Ireland, Israel, Italy, Norway, The Netherlands, and UK	FP7 Program	€ 5.7 M
IMTA-EFFECT	Integrated Multi Trophic Aquaculture for Efficiency and Environmental ConservaTion	2016	2019	INRA, France	9 partners, from six countries: Portugal, Greece, France, Romania, Madagascar, and Indonesia	COFASP	€ 750 K
INTEGRATE	Integrate Aquaculture: an eco-innovative solution to foster sustainability in the Atlantic Area	2017	2020	CTAQUA, Spain	19 partners, from five countries: Spain, France, Ireland, Portugal, and the United Kingdom	INTERREG Atlantic Area	€ 2 M
SEAFARE	Sustainable and Environmentally friendly Aquaculture For the Atlantic Region of Europe	2010	2013	Bangor university, UK	5 partners and 9 associated partners, from 5 countries: United Kingdom, France, Ireland, Portugal, and Spain	European Union Atlantic Area Transnational Programme	€ 3.18 M
SEAPURA	Species diversification and improvement of aquatic production in seaweeds purifying effluents from integrated fish farms and from other waste sources	2001	2004	ALFRED-WEGENER INSTITUT, Germany	8 partners, from 5 countries: Germany, France, UK, Spain, and Portugal	FP 5	€ 1.45 M

These funding efforts are continuing in Europe and other projects focusing on IMTA development are ongoing, such as the **IMTA EFFECT** project (2016–2019), and **INTEGRATE** (Integrate Aquaculture: an eco-innovative solution to foster sustainability in the Atlantic Area), launched in 2017. The **IMTA EFFECT** project, funded by COFASP is assessing the efficiency of different IMTA systems, analyzing nutrient and energy flows, and developing tools based on modelling to predict systems functioning. Multitrophic marine systems and freshwater polyculture systems will be considered. This project also addresses the perception of IMTA by stakeholders, through economic and social evaluation of ecological services (http://www.agence-nationale-recherche.fr/Project-ANR-15-COFA-0001). The **INTEGRATE** project gathered 19 partners from five countries to develop tools to improve competitiveness, as well as the quality and public perception of IMTA aquaculture products. It includes the creation of a platform for sectorial collaboration in the Atlantic area, for developing methods and technology, assessing environmental impact, identifying bottlenecks, and designing a strategy for industrial upscaling of IMTA in this region (http://www.atlanticarea.eu/news/40).

Other western countries such as Canada have been actively working to develop IMTA in order to increase aquaculture profitability, environmental sustainability, and societal acceptability through economic diversification by co-cultivating several value-added marine crops. Over the last decade, the Canadian IMTA Network (CIMTAN) at the University of New Brunswick has been working with an industrial partner, Cooke Aquaculture Inc., in Atlantic Canada to develop seaweed cultivation in IMTA systems for their valorization using the integrated sequential biorefinery (ISBR) approach. This research/industrial partnership is developing markets for the use of seaweed (kelp) in areas such as human consumption, cosmetics, or fish feed, and also implementing eco-labeling and organic certification (Chopin 2015).

Some of the conclusions from research projects are apparently rather straightforward, such as the importance of choosing species that are endemic and present a good growth performance. In the case of seaweed, species selection probably needs to be based on site specific testing, although some general criteria may be used for pre-selection such as high nutrient uptake rate and efficiency, high growth rates, the ability of sustaining stress due to cultivation conditions, and ease of cultivation (Neori et al. 2004; Kang et al. 2013).

The reasons why IMTA is not still widely applied in western countries are manifold and related to biological, ecological, technical, and socio-economic issues. Biological constraints are related to stabilization of cultivation methods for seaweed, an important component in IMTA, namely development of cultivars adapted to local conditions (Kim et al. 2017) and also a better understanding of cultivation methods for benthic organisms (Hughes et al. 2016). One ecological issue is, for example, kelp cultivation as the inorganic extractor since a seasonal mismatch exists between highest growth periods in fish and seaweed. Fish grow faster in summer when the temperatures are higher whereas kelp present higher growth rates in early spring when they need more nutrients (Broch et al. 2013). The solution may be to select different seaweed species that grow well in the summer. Technical issues are related to structure stability in more exposed off-shore sites, logistics for deployment and harvesting of seaweed and other organisms, and drying of large amounts of seaweed biomass. Finally, another important issue that needs to be handled is policy and regulations, by providing information based on scientific data to regulatory bodies aiming to implement the IMTA licensing procedures, which account for IMTA specificities such as species interactions and ecosystem-based management approaches to aquaculture (Chopin 2017).

If in western countries the concept of IMTA was developed as a strategy to overcome the negative environmental impacts of intensive fish farming and at the same time diversify production, in the East, as in China, integrated aquaculture has been practiced at the scale of whole coastal bays, with dozens of species farmed in close proximity to each other. Sanggou Bay, located on the eastern tip of Shandong Peninsula, China, is one example of this whole bay aquaculture. In this bay, more than 100 km^2 of a total of 163 km^2, are used for aquaculture, producing above 240,000 ton of seafood per year, including kelp, scallops, oysters, abalone, and sea cucumbers (Fang et al. 2016). Although this practice started as unintentional IMTA systems, integration of aquaculture using different species at the water body scale are now viewed as a useful approach for IMTA in western regions, facilitating the management of technical issues, such as spatial planning and species integration (Hughes et al. 2016). Nevertheless, management of large-scale IMTA areas still needs improvement, requiring a deeper understanding of the interactions between the different components and between species and the environment, in order to ensure sustainable development (Shi et al. 2013; Fang et al. 2016).

Although commercial IMTA is not yet widely applied in western countries, some companies are already using this concept to produce seaweed, since the nutrients produced by fed aquaculture are a valuable resource, increasing yields and productivity. Examples are AlgaPlus, in Portugal, a land-based seaweed aquaculture company that is established in association with a semi-intensive fish aquaculture, receiving water from the fish earthen tanks (www.algaplus.pt) and Hjarno Havbrug (www.havbrug.dk), in Denmark, which produces sugar kelp in open water with organic certification, and also mussels, to reduce the nitrogen in the water originated by fish aquaculture.

In fact, a driver for IMTA development in western countries is the growing interest in seaweed aquaculture, both for their contribution to nutrient reduction and for the multiple high value applications of their biomass, from high quality healthy food to source of bioactive compounds for the cosmetic and pharmaceutical industries (Hafting et al. 2015). Nevertheless, for the seaweed component of IMTA to fulfill its role in nutrient mitigation, and to supply the market for the above mentioned applications, biomass production must be up-scaled, and the logistics for processing and marketing must be developed for this increased biomass. Thus, besides the need to resolve the regulatory constraints, the economic stability is still to be demonstrated, especially over the long term (Chopin 2017), in order to attract investment for IMTA development and move IMTA from a research concept to a mainstream commercial activity.

Challenges and opportunities

Policy and regulation are, as stated above, the main challenges that IMTA has to meet to reach a commercial level. In order to inform licensing and regulatory authorities, it is fundamental to establish clear definitions for IMTA systems and products. Once this is established, added value will be placed on IMTA products, based on their quality as well as their sustainable way of production. Consumers are increasingly willing to support environmentally sustainable production of food and other goods, if properly demonstrated, including IMTA originated food products (Barrington et al. 2010), allowing the development of markets for differentiated premium IMTA products. Thus, it is important to declare this differenciation to the consumers, informing them about the environmental sustainability of IMTA products through organic certification or environmental labelling.

In order to meet the demand of those markets, and support the environmental sustainability claim, upscaling the production of extractive species (e.g., mussels or oysters and seaweed) is needed. In the case of mussels or oysters, their value is already recognized and the market and supply chain are already in place. In the case of seaweed, the interest in seaweed based products has pushed the market forward with the demand and prices rising in the last years, currently for food, food ingredients, cosmetics, fertilizers, and hydrocolloids industries (Nayar and Bott 2014). Besides these established uses, promising new markets for seaweed based products include the bioactive extracts for incorporation in food products (nutraceuticals), premium feed products, soil conditioners and biofertilizers, and even biofuels. Seaweed processing for these applications is being considered under the cascade biorefinery concept, which allows for a more efficient use of seaweed biomass by sequentially extracting different compounds such as protein, carbohydrates, and minerals (Francavilla et al. 2015; Bikker et al. 2017).

The establishment of larger markets for IMTA products, including seaweed, implies increasing biomass production, which raises another issue, marine space availability for farm expansion. Coastal areas are used by very different and sometimes conflicting activities, such as fisheries, navigation, and recreation, demanding development of spatial planning strategies that accommodate these different interests. Factors to be considered include physical characteristics such as temperature and hydrodynamics, water transparency, and distance to harbor infrastructures. An important issue to account for is the carrying capacity of the site, which is an opportunity for IMTA. For the IMTA concept to work, balanced production must be achieved, through careful management, which is a recognized challenge for IMTA implementation. The need for a balanced production, in terms of cultivated organisms, underlying the

IMTA concept is an argument in favor of its implementation which must be also transferred to the public, since the potential to exceed the carrying capacity of the water is one of the issues of concern considering aquaculture intensification (Froehlich et al. 2017). Another concern is the excessive occupation of coastal areas, which may also lead to a rejection by the local populations due to negative aesthetic impacts on the landscapes. These issues are being addressed in the context of spatial planning, and possible solutions may rely on moving aquaculture offshore or on the implementation of multi-use platforms, associating different activities such as wind energy production and aquaculture (Troell et al. 2009; Jansen et al. 2016; Holm et al. 2017).

The argument for producing seaweed in IMTA systems, under the production point of view, is the high nutrient availability in areas associated (in land-based systems) or near fish or shellfish aquaculture (in open water) that increases productivity and yield. From an environmental point of view, other arguments may be made on the benefits of seaweed production, related to the ecosystem services they provide, besides nutrient bioremediation. Natural seaweed assemblages, including kelp forests, provide a number of ecosystem services, that may be classified in provisioning, regulating, and cultural services (MEA 2005). Provisioning services are biomass production for different uses, such as food, feed, extraction of phycocolloids, and bioactive compounds among other uses. Regulating services include their role in carbon and nutrient cycling, and cultural services include aesthetic and recreational values. Other ecosystem services, with indirect benefits to humans are the habitat/supporting services, which include habitat and food for associated species (Table 5). Most of these services may be valued economically, directly, as the market value of biomass for those different applications, market value of associated species with economic importance, and indirectly, such as value of scientific information, biodiversity repository, climate buffer, or cultural heritage (Vásquez et al. 2014).

Table 5. Examples of seaweed ecosystem services, categorized according to MEA (2005) and TEEB (2010).

Provisioning	Regulating	Cultural	Habitat/Supporting
Nutrition Animal feed Human food Fertilizers	**Water Quality** Nutrient bioremediation Nitrogen cycle Oxygen cycle Carbon cycle	**Aesthetic** Seaweeds beds Kelp forests	**Nutrition** Dissolved nutrients Particulate nutrients Symbiosis
Health Bio actives Nutraceuticals Pharmaceuticals Well-being		**Recreational** Diving Fishing	**Habitats** Provide niches Provide substrate Stabilize seashore
Chemicals Biopolymers Aminoacids Minerals			
Energy Biogas Biofuel Alcohols Firewood			
Other Building material Bio plastics Paper substitutes			

Acknowledgements

This work was partially supported by the Structured R&D&I Project INNOVMAR – "Innovation and Sustainability in the Management and Exploitation of Marine Resources" (ref. NORTE-01-0145-FEDER-000035) within the research line "INSEAFOOD – Innovation and valorization of seafood products: meeting local challenges and opportunities", founded by the Portuguese Northern Regional Operational Programme (NORTE 2020) through the European Regional Development Fund (ERDF); and also by the project GENIALG - GENetic diversity exploitation for Innovative macro-ALGal biorefinery, funded by H2020 (EC Grant agreement no: 727892).

References

Abreu, M.H., D.A. Varela, L. Henríquez, A. Villarroel, C. Yarish, I. Sousa-Pinto and A.H. Buschmann. 2009. Traditional vs. integrated multi-trophic aquaculture of *Gracilaria chilensis* CJ Bird, J. McLachlan and EC Oliveira: productivity and physiological performance. Aquaculture 293(3): 211–220.

Abreu, M.H., R. Pereira, C. Yarish, A.H. Buschmann and I. Sousa-Pinto. 2011. IMTA with *Gracilaria vermiculophylla*: Productivity and nutrient removal performance of the seaweed in a land-based pilot scale system. Aquaculture 312(1-4): 77–87.

Ahn, O., R.J. Petrell and P.J. Harrison. 1998. Ammonium and nitrate uptake by *Laminaria saccharina* and *Nereocystis luetkeana* originating from a salmon sea cage farm. J. Appl. Phycol. 10(4): 333–340.

Alongi, D.M., D.J. Johnston and T.T. Xuan. 2000. Carbon and nitrogen budgets in shrimp ponds of extensive mixed shrimp–mangrove forestry farms in the Mekong Delta, Vietnam Aquac. Res. 31: 387–399.

Barrington, K. et al. 2010. Social aspects of the sustainability of integrated multi-trophic aquaculture. Aquacult. Int. 18(2): 201–211.

Bikker, P., M.M. van Krimpen, P. van Wikselaar, B. Houweling-Tan, N. Scaccia, J.W. van Hal, W.J. Huijgen, J.W. Cone and A.M. López-Contreras. 2016. Biorefinery of the green seaweed *Ulva lactuca* to produce animal feed, chemicals and biofuels. J. Appl. Phycol. 28(6): 3511–3525.

Binh, C.T., M.J. Phillips and H. Demaine. 1997. Integrated shrimp–mangrove farming systems in the Mekong Delta of Vietnam Aquac. Res. 28: 599–610.

Broch, O.J., I.H. Ellingsen, S. Forbord, X. Wang, Z. Volent, M.O. Alver, A. Handå, K. Andresen, D. Slagstad, K.I. Reitan and Y. Olsen. 2013. Modelling the cultivation and bioremediation potential of the kelp *Saccharina latissima* in close proximity to an exposed salmon farm in Norway. Aquac Environ Interact. 4(2): 187–206.

Buschmann, A.H., O.A. Mora, P. Gómez, M. Böttger, S. Buitano, C. Retamales, P.A. Vergara and A. Gutierrez. 1994. *Gracilaria chilensis* outdoor tank cultivation in Chile: Use of land-based salmon culture effluents. Aquac. Eng. 13: 283–300.

Buschmann, A.H., M. Troell, N. Kautsky and L. Kautsky. 1996. Integrated tank cultivation of salmonids and *Gracilaria chilensis* (Gracilariales, Rhodophyta). Hydrobiologia 326-327(1): 75–82.

Buschmann, A.H., D.A. López and A. Medina. 1996. A review of the environmental effects and alternative production strategies of marine aquaculture in Chile. Aquac. Eng. 15: 397–421.

Chopin, T. 2006. Integrated multi-trophic aquaculture. Northern Aquaculture 12(4): 4.

Chopin, T. 2013. Integrated Multi-Trophic Aquaculture Ancient, Adaptable Concept Focuses On Ecological Integration. Global Aquaculture Advocate March/April, 16–19.

Chopin, T. 2015. Marine aquaculture in Canada: Well-established monocultures of finfish and shellfish and an emerging Integrated Multi-Trophic Aquaculture (IMTA) approach including seaweeds, other invertebrates, and microbial communities. Fisheries 40(1): 28–31.

Chopin, T. 2017. Challenges of moving integrated multi-trophic aquaculture along the R&D and commercialization continuum in the western world. J. Ocean Techno. 112: 34–47.

Chopin, T., C. Yarish, R. Wilkes, E. Belyea, S. Lu and A. Mathieson. 1999. Developing *Porphyra*/salmon integrated aquaculture for bioremediation and diversification of the aquaculture industry. J. Appl. Phycol. 11(5): 463.

Chow, F., J. Macchiavello, S. Cruz, E. Fonck and J. Olivares. 2001. Utilization of *Gracilaria chilensis* (Rhodophyta: Gracilariaceae) as a biofilter in the depuration of effluents from tank cultures of fish, oysters, and sea urchins. J. World Aquacult. Soc. 32(2): 215–220.

Cohen, I. and A. Neori. 1991. *Ulva lactuca* biofilters for marine fishpond effluents. I. Ammonia uptake kinetics and nitrogen content. Bot. Mar. 34: 475–482.

del Río, M.J., Z. Ramazanov and G. García-Reina. 1996. *Ulva rigida* (Ulvales, Chlorophyta) tank culture as biofilters for dissolved inorganic nitrogen from fishpond effluents. In Fifteenth International Seaweed Symposium, 61–66. Springer Netherlands.

Domingues, B., M.H. Abreu and I. Sousa-Pinto. 2015. On the bioremediation efficiency of *Mastocarpus stellatus* (Stackhouse) Guiry, in an integrated multi-trophic aquaculture system. J. Appl. Phycol. 27(3): 1289–1295.

Fang, J., J. Zhang, T. Xiao, D. Huang and S. Liu. 2016. Integrated multi-trophic aquaculture (IMTA) in Sanggou Bay, China. Aquacult Environ Interact 8: 201–205. https://doi.org/10.3354/aei00179.

Fernando, C. 2002. Bitter harvest-rice fields and fish culture. World Aquac. Mag. 33: 23–24.

Folke, C. and N. Kautsky. 1992. Aquaculture with its environment: Prospects for sustainability. Ocean Coast Manage. 17(1): 5–24.

Francavilla, M., P. Manara, P. Kamaterou, M. Monteleone and A. Zabaniotou. 2015. Cascade approach of red macroalgae *Gracilaria gracilis* sustainable valorization by extraction of phycobiliproteins and pyrolysis of residue. Bioresour. Technol. 184: 305–313.

Freitas, J.R., J.M.S. Morrondo and J.C. Ugarte. 2016. *Saccharina latissima* (Laminariales, Ochrophyta) farming in an industrial IMTA system in Galicia (Spain). J. Appl. Phycol. 28(1): 377–385.

Froehlich, H.E., R.R. Gentry, M.B. Rust, D. Grimm and B.S. Halpern. 2017. Public perceptions of aquaculture: evaluating spatiotemporal patterns of sentiment around the world. PloS One 12(1): p.e0169281.

Haines, K.C. 1975. Growth of the carrageenan-producing tropical red seaweed *Hypnea musciformis* in surface water, 870 m deep water effluent from a clam mariculture system and in deep water enriched with artificial fertilizers or domestic sewage. pp. 207–20. *In*: G. Persson and E. Jaspers (eds.). Proc. Xth Europ. Symp. Mar. Biol., Vol. 1. University Press, Wetteren.

Hafting, J.T., J.S. Craigie, D.B. Stengel, R.R. Loureiro, A.H. Buschmann, C. Yarish, M.D. Edwards and A.T. Critchley. 2015. Prospects and challenges for industrial production of seaweed bioactives. J. Phycol. 51(5): 821–837.

Hirata, H., E. Kohirata, F. Guo and B. Xu. 1993. Culture of the Sterile *Ulva* sp. (Chlorophyceae) in a Mariculture Farm. Aquaculture Science 41(4): 541–545. (https://www.jstage.jst.go.jp/article/aquaculturesci1953/41/4/41_4_541/_article).

Holm, P., B.H. Buck and R. Langan. 2017. Introduction: New approaches to sustainable offshore food production and the development of offshore platforms. pp. 1–20. *In*: Aquaculture Perspective of Multi-Use Sites in the Open Ocean. Springer, Cham.

Hughes, A.D., R.A. Corner, M. Cocchi, K.A. Alexander, S. Freeman, D. Angel, M. Chiantore, D. Gunning, J. Maguire, A.M. Beltran, J. Guinée, J. Ferreira, R. Ferreira, C. Rebours and D. Kletou. 2016. Beyond fish monoculture, Developing Integrated Multi-trophic Aquaculture in Europe. Final report. ETA-Florence Renewable Energies srl. Italy.http://www.idreem.eu/cms/wp-content/uploads/2016/09/IDREEM_FINALREPORT_2109.pdf.

Jansen, H.M., S. Van Den Burg, B. Bolman, R.G. Jak, P. Kamermans, M. Poelman and M. Stuiver. 2016. The feasibility of offshore aquaculture and its potential for multi-use in the North Sea. Aquac. Int. 24(3): 735–756.

Jiménez del Río, M., Z. Ramazanov and G. García-Reina. 1996. *Ulva rigida* (Ulvales, Chlorophyta) tank culture as biofilters for dissolved inorganic nitrogen from fishpond effluents. *In*: S.C. Lindstrom and D.J. Chapman (eds.). Fifteenth International Seaweed Symposium. Developments in Hydrobiology, vol 116. Springer, Dordrecht.

Kang, Y.H., S.R. Park and I.K. Chung. 2011. Biofiltration efficiency and biochemical composition of three seaweed species cultivated in a fish-seaweed integrated culture. Algae 26(1): 97.

Kautsky, N. and C. Folke. 1991. Integrating open system aquaculture: ecological engineering for increased production and environmental improvement through nutrient recycling. pp. 320–334. *In*: C. Etnier and B. Guterstam (eds.). Ecological Engineering for Wastewater Treatment. Gothenburg, Bokskogen.

Kim, J.K., C. Yarish, E.K. Hwang, M. Park and Y. Kim. 2017. Seaweed aquaculture: cultivation technologies, challenges and its ecosystem services. Algae 32(1): 1–13.

Korzen, L., A. Abelson and A. Israel. 2016. Growth, protein and carbohydrate contents in *Ulva rigida* and *Gracilaria bursa-pastoris* integrated with an offshore fish farm. J. Appl. Phycol. 28(3): 1835–1845.

Langton, R.W., K.C. Haines and R.E. Lyon. 1977. Ammonia-nitrogen production by the bivalve mollusc Tapes japonica and its recovery by the red seaweed *Hypnea musciformis* in a tropical mariculture system. Helgol. Wiss. Meeres. 30: 217–229.

Lombardi, J.V., H.L. de Almeida Marques, R.T.L. Pereira, O.J.S. Barreto and E.J. de Paula. 2006. Cage polyculture of the Pacific white shrimp *Litopenaeus vannamei* and the Philippines seaweed *Kappaphycus alvarezii*. Aquaculture 258(1): 412–415.

Mao, Y., H. Yang, Y. Zhou, N. Ye and J. Fang. 2009. Potential of the seaweed *Gracilaria lemaneiformis* for integrated multi-trophic aquaculture with scallop *Chlamys farreri* in North China. J. Appl. Phycol. 21(6): 649.

Mata, L. and R. Santos. 2003. Cultivation of *Ulva rotundata* (Ulvales, Chlorophyta) in raceways using semi-intensive fishpond effluents: yield and biofiltration. pp. 237–242. *In*: Proceedings of the 17th International Seaweed Symposium. Cape Town, South Africa, Oxford University Press.

Mata, L., A. Schuenhoff and R. Santos. 2010. A direct comparison of the performance of the seaweed biofilters, *Asparagopsis armata* and *Ulva rigida*. J. Appl. Phycol. 22(5): 639–644.

Matos, J., S. Costa, A. Rodrigues, R. Pereira and I.S. Pinto. 2006. Experimental integrated aquaculture of fish and red seaweeds in northern Portugal. Aquaculture 252(1): 31–42.

MEA. 2005. Ecosystems and Human Well-being: A Framework for Assessment. Millennium Ecosystem Assessment. Washington, D.C., USA, Island Press.

Msuya, F.E. and A. Neori. 2002. *Ulva reticulata* and *Gracilaria crassa*: macroalgae that can biofilter effluent from tidal fishponds in Tanzania. Western Indian Ocean Journal of Marine Science 1(2): 117–126. (https://www.jstage.jst.go.jp/article/aquaculturesci1953/41/4/41_4_541/_article).

Nayar, S. and K. Bott. 2014. Current status of global cultivated seaweed production and markets. World aquaculture. WAS Magazine 15: 32–37

Neori, A., T. Chopin, M. Troell, A.H. Buschmann, G.P. Kraemer, C. Halling, M. Shpigel and C. Yarish. 2004. Integrated aquaculture: rationale, evolution and state of the art emphasizing seaweed biofiltration in modern mariculture. Aquaculture 231(1-4): 361–391.

Neori, A., M.D. Krom, S.P. Ellner, C.E. Boyd, D. Popper, R. Rabinovitch, P.J. Davison, O. Dvir, D. Zuber, M. Ucko and D. Angel. 1996. Seaweed biofilters as regulators of water quality in integrated fish-seaweed culture units. Aquaculture 141(3): 183–199.

Neori, A., M. Shpigel and D. Ben-Ezra. 2000. A sustainable integrated system for culture of fish, seaweed and abalone. Aquaculture 186(3-4): 279–291.

Rabanal, H.R. 1988. History of Aquaculture. Manila.

Ryther, J.H., W.M. Dunstan, K.R. Tenore and J.E. Huguenin. 1972. Controlled eutrophication—increasing food production from the sea by recycling human wastes. Bioscience 22(3): 144–152.

Ryther, J.H., J.C. Goldman, C.E. Gifford, J.E. Huguenin, A.S. Wing, J.P. Clarner, L.D. Williams and B.E. Lapointe. 1975. Physical models of integrated waste recycling-marine polyculture systems. Aquaculture 5(2): 163–177.

Sanderson, J.C., M.J. Dring, K. Davidson and M.S. Kelly. 2012. Culture, yield and bioremediation potential of *Palmaria palmata* (Linnaeus) Weber & Mohr and *Saccharina latissima* (Linnaeus) CE Lane, C. Mayes, Druehl & GW Saunders adjacent to fish farm cages in northwest Scotland. Aquaculture 354: 128–135.

Schuenhoff, A., L. Mata and R. Santos. 2006. The tetrasporophyte of *Asparagopsis armata* as a novel seaweed biofilter. Aquaculture 252(1): 3–11.

Schuenhoff, A., M. Shpigel, I. Lupatsch, A. Ashkenazi, F.E. Msuya and A. Neori. 2003. A semi-recirculating, integrated system for the culture of fish and seaweed. Aquaculture 221(1): 167–181.

SEAPURA. 2004. Species diversification and improvement of aquatic production in seaweeds purifying effluents from integrated fish farms. SEAPURA final report. http://www.cbm.ulpgc.es/seapura/index.html.

Shi, H., W. Zheng, X. Zhang, M. Zhu and D. Ding. 2013. Ecological–economic assessment of monoculture and integrated multi-trophic aquaculture in Sanggou Bay of China. Aquaculture 410: 172–178.

Subandar, A., R.J. Petrell and P.J. Harrison. 1993. Laminaria culture for reduction of dissolved inorganic nitrogen in salmon farm effluent. J. Appl. Phycol. 5(4): 455–463.

TEEB. 2010. The Economics of Ecosystems and Biodiversity Ecological and Economic Foundations. Edited by Pushpam Kumar. Earthscan, London and Washington.

Troell, M., C. Halling, A. Nilsson, A.H. Buschmann, N. Kautsky and L. Kautsky. 1997. Integrated marine cultivation of *Gracilaria chilensis* (Gracilariales, Rhodophyta) and salmon cages for reduced environmental impact and increased economic output. Aquaculture 156(1-2): 45–61.

Troell, M., A. Joyce, T. Chopin, A. Neori, A.H. Buschmann and J.G. Fang. 2009. Ecological engineering in aquaculture—potential for integrated multi-trophic aquaculture (IMTA) in marine offshore systems. Aquaculture 297(1-4): 1–9.

Vandermeulen, H. and H. Gordin. 1990. Ammonium uptake using *Ulva* (Chlorophyta) in intensive fishpond systems: mass culture and treatment of effluent. J. Appl. Phycol. 2: 363–374.

Vásquez, J.A., S. Zuniga, F. Tala, N. Piaget, D.C. Rodríguez and J.M.A. Vega. 2014. Economic valuation of kelp forests in northern Chile: values of goods and services of the ecosystem. J. Appl. Phycol. 26: 1081–1088.

Wu, H., Y. Huo, F. Han, Y. Liu and P. He. 2015. Bioremediation using *Gracilaria chouae* co-cultured with *Sparus macrocephalus* to manage the nitrogen and phosphorous balance in an IMTA system in Xiangshan Bay, China. Mar. Pollut. Bull. 91(1): 272–279.

Yang, Y.F., X.G. Fei, J.M. Song, H.Y. Hu, G.C. Wang and I.K. Chung. 2006. Growth of *Gracilaria lemaneiformis* under different cultivation conditions and its effects on nutrient removal in Chinese coastal waters. Aquaculture 26 1(1): 248–255.

Yokoyama, H. and Y. Ishihi. 2010. Bioindicator and biofilter function of *Ulva* spp. (Chlorophyta) for dissolved inorganic nitrogen discharged from a coastal fish farm—potential role in integrated multi-trophic aquaculture. Aquaculture 310(1): 74–83.

Zhou, Y., H. Yang, H. Hu, Y. Liu, Y. Mao, H. Zhou, X. Xu and F. Zhang. 2006. Bioremediation potential of the macroalga *Gracilaria lemaneiformis* (Rhodophyta) integrated into fed fish culture in coastal waters of north China. Aquaculture 252(2-4): 264–276.

14

Alternative Green Biofuel from Microalgae

A Promising Renewable Resource

Katkam N. Gangadhar,[1,2,3] *João Varela*[1] and *F. Xavier Malcata*[2,3,*]

Introduction

Energy is essential and an important factor of production in the global economy, and approximately 90% of the commercially produced energy is from fossil fuels such as crude oil (36%), coal (27.4%), and gas (23%) (Demirbas 2010; Chen et al. 2011), of which 58% is consumed by the transport sector (Daroch et al. 2013). The world energy consumption, including oil, natural gas, nuclear, and coal has decreased around 1.1% (BP 2010). At the same time, the production of natural gas and crude oil has lessened about 2.1% and 7.3%, respectively (Emma and Rosalam 2012). This suggests that the world energy supply met by fossil fuels is not sustainable and will diminish in the middle of the century (Chisti 2007). Fossil fuels are non-renewable in nature and were formed millions of years ago from the remains of ancient algae (e.g., diatoms), plants, and animals, a process that usually requires long periods of time. In addition, fossil fuel is directly related to air pollution, water degradation, and climate change due to greenhouse gas (GHG) emissions, which can significantly contribute to global warming (Rojas-Downing et al. 2017). The constant rising of global demand of transportation fuels and electricity generation, in which these two sectors have played an important role in improving human living standards. Most of the energy sources, such as solar, wind, hydroelectric, geothermal, and nuclear power which are mainly used to generate electricity and heat. Electricity is considered to be a secondary energy resource that is used to convey and distribute energy over large distances. Moreover, these renewable resources such as solar, wind, and hydrothermal have several drawbacks. For instance, solar energy can only be produced in places with ample sunlight; solar energy can only be generated at daytime; wind technologies generate noise pollution and require suitable locations where wind is frequent. Due to the reduction of oil resources and

---

[1] CCMAR – Centre of Marine Sciences, University of Algarve, Campus de Gambelas, P-8005-139 Faro, Portugal.
[2] Department of Chemical Engineering, University of Porto, Rua Dr. Roberto Frias, s/n, 4200-465 Porto, Portugal.
[3] LEPABE – Laboratory of Process Engineering, Environment, Biotechnology and Energy, Rua Dr. Roberto Frias, s/n, P-4200-465 Porto, Portugal.
* Corresponding author: fmalcata@fe.up.pt

consequent increasing price of oil (Yanan et al. 2010) has motivated researchers to find alternate sources of energy from renewable feedstocks (Jasvinder and Gu 2010). Therefore, biofuels from renewable feedstocks can thus be an alternative to reduce our dependency on fossil fuel and help safeguard the environment and economic sustainability (Mussgnug et al. 2010). For example, energy produced from combustible renewables and waste biomass has the highest potential than that of other renewable sources, and accounted for 10.0% of the total primary energy supply, compared to hydro energy (2.2%) and other energy sources including geothermal, solar, wind, and heat (0.7%) (IEA 2010). Hence, renewable combustible energy sources such as liquid biofuels will likely play a crucial role as an alternative to fossil fuels in the near future, contributing to the effort of diversifying global energy sources rather than electricity or hydrogen (Zhao 2017). In this regard, it is crucial to explore renewable and cost-effective sources of energy in the near future. Phototrophic biomass is one of the main renewable energy resources available. In contrast to other renewables, biomass represents the only source of liquid, solid, and gaseous fuels that can be treated in a number of different ways to provide such biofuels. Biofuels are also a suitable option due to their biodegradability, renewability, and the ability of producing exhaust gases of acceptable quality (Bhatti et al. 2008). Most transportation (bio) fuels are liquids, as vehicles usually require high energy densities fuels (Demirbas 2008) that are easy to transport and handle.

Environmental advantages of biofuels

Biodiesel properties

Biofuels such as biodiesel, bioethanol, biomethane, and biohydrogen, but also propanol and butanol, among others are being obtained from renewable resources like sugar cane, corn, vegetable oils, lignocellulosic waste, and microalgal biomass. Biodiesel is typically defined as a mixture of fatty acid alkyl esters. "Bio" stands for being produced from a biological source in contrast to the traditional petroleum-based diesel fuel. "Diesel" refers to liquid fuel having a cetane number suitable for ignition and its use in diesel engines. Technically, biodiesel must follow international specifications for its use in diesel engines as diesel fuel. Biodiesel refers to the pure fuel designated as B100 before blending with diesel fuel. Biodiesel blends are represented as "BXX" in which "XX" represents the percentage of biodiesel contained in the blend. For example, B20 is 20% biodiesel and 80% petroleum diesel (Knothe et al. 1996).

In view of environmental considerations, biodiesel is referred to as 'carbon neutral' because all the carbon dioxide (CO_2) released during consumption has been sequestered from the atmosphere for the growth of terrestrial plants, microalgae, and others (Barnwal and Sharma 2005). Biodiesel forms thus a closed carbon cycle, as CO_2 released in the atmosphere upon biodiesel combustion is recycled by growing photosynthetic organisms, which are later processed into biofuel (Kumar et al. 2010).

The main advantage of biodiesel is that it is a renewable, non-toxic, and biodegradable liquid biofuel (Gerpen 2005). This biofuel also combusts in a clean way, possessing a heating value of 39–41 MJ/Kg, which is comparable with that of petrodiesel (43 MJ/Kg). Other parameters like the cetane number, flash point, and kinematic viscosity are mostly similar to those of petrodiesel (Knothe and Steidley 2005; Demirbas 2009) and appear to reduce emissions of air pollutants and carcinogens (Shakeel et al. 2009). One of the most important advantage of biodiesel compared to many other alternative transportation fuels is that it can be used in existing diesel engines with little or no modification (Lam and Lee 2012). Using biodiesel instead of petrodiesel will significantly reduce acid rain formation and will most probably mitigate global warming, by lowering net carbon monoxide, sulphur (SOx) and nitrogen (NOx) oxides, and hydrocarbon emissions, displaying also a very low risk of explosion from vapours (Antolin et al. 2002; Abreu et al. 2005; Barnwal and Sharma 2005). The lack of toxic and carcinogenic aromatic compounds in biodiesel can also lead to a reduced impact on human health and the environment by the resulting combustion gases. When the fuel is switched from low-sulphur petroleum diesel to biodiesel, there is a drastic drop in the carbon monoxide emissions as most organic compounds are converted into carbon dioxide instead. Biodiesel virtually eliminates the notorious black soot emissions associated with diesel engines and the total particulate matter emissions are also significantly lower. Usage of biodiesel needs, however, requires a strict balance to be sought between feedstock farming and economic development (Kalam and Masjuki 2002). Reduced emissions make biodiesel suitable for use in major cities where air pollution is a problematic. In addition, its lower emissions make biodiesel suitable for use in confined areas such as mines where ventilation is a concern. Pure biodiesel has low aquatic toxicity and more than 90% biodiesel can be biodegraded within 21 days (Mudge and Pereira 1999; Speidel et al. 2000). This property substantially reduces the impact of accidental spills and makes this liquid fuel an ideal candidate for use in environmentally sensitive areas, such as inland water ways. As a useful by-product, oxygen is produced by the photoautotrophic feedstock from which biofuels are made. Oxygen makes up almost 10% (w/w) of biodiesel, making it a naturally "oxygenated" fuel. As an oxygenated hydrocarbon, biodiesel itself burns cleanly but it also improves the efficiency of combustion in blends with petroleum fuel (Murayama 1994). Biodiesel hydrocarbon chains are generally 16–20 carbons in chain length, and they are all oxygenated at one end, making the product an excellent fuel, which is safe for transport due to its high flash point (130°C), which is significantly higher than that of petrodiesel (64°C) (Knothe 2006). At present two common standards of biodiesel fuel exists: American ASTM D6751 and European EN 14214 (Demirbas 2009).

Bioethanol properties

Bioethanol is one of the most commonly used biofuel worldwide, particularly in Brazil and the USA. It is a clear colourless, food-grade liquid that burns to produce carbon dioxide and water; it is less harmful to the atmosphere than fossil fuel, readily biodegradable, and its use produces fewer air-borne pollutants than that of petroleum fuel if spilt. Bioethanol is a high-octane fuel and it can also be used to replace octane enhancers such as methyl-*tert*-butyl ether (MTBE), methyl-cyclopentadienyl-manganese tricarbonyl (MMT), and aromatic hydrocarbons in petrol (Champagne 2007). Blending bioethanol with petrodiesel, can also oxygenate the fuel mixture, burning it more completely, and thereby reducing polluting emissions. Ethanol fuel blends are widely sold in the United States. The most common blend is E10 (10:90; ethanol and petrol). Automobile engines do not require any modifications to run on E10 and vehicle warranties are also unaffected. Only flexible fuel vehicles can run on up to E85 (15:85; petrol and ethanol) with conventional fuel without the need of engine modifications. Bioethanol has the advantage of being a renewable resource, contributing to the mitigation of greenhouse gas emissions by preventing further burning of fossil fuels. In addition, using bioethanol in older engines can help reduce the amount of carbon monoxide produced by the vehicle, thus improving air quality. Another advantage of bioethanol is the ease with which it can be integrated into the existing road transport fuel system. Bioethanol can also be used in rocket engines or internal combustion engines in its pure form (Pfromm et al. 2010).

Biogas properties

Biogas is a mixture of methane (55–75%) and CO_2 (25–45%) (Harun et al. 2010b; Jasvinder and Gu 2010) including minor amounts of hydrogen sulphide (H_2S), nitrogen, siloxanes, and moisture. Biogas may be not suitable to be used as fuel gas for transportation/or machinery, due to the high amount of carbon dioxide, which can be upgraded into a mixture of biomethane and carbon dioxide with a more favourable ratio: 97% biomethane and 2% CO_2. Biomethane is environmentally friendly and CO_2 neutral, due to the life cycle of it. There are both emissions and mitigations that balance and produce a net reduction in GHG (i.e., CO_2) when compared to natural gas (i.e., fossil fuel). Biomethane is being used as vehicular fuel gas and to generate electricity as well as for cooking and heating purposes (Holm et al. 2009).

Feedstocks for biofuel manufacture

There is increasing interest in alternative fossil fuels from various sources including food, municipal and agricultural wastes, vegetable oils, animal fat, as well as microalgal biomass as feedstocks for the production of biofuel worldwide. Biofuels such as biodiesel, bioethanol, and biogas differ from fossil fuels in their chemical nature, as they are derived from first generation feedstocks such as food-based crops. Biodiesel is typically produced from vegetable oils such as rapeseed, canola, soybean, palm, sunflower, mustard oil, and Karanja oil (Issariyakul and Dalai 2014). However, the use of food crops for biodiesel has raised much debate and criticism involving food security concerns, for vegetable oils compete with their use as food and require considerable use of arable lands and fresh water. The higher prices of biodiesel compared to those of petrodiesel and the possible use of vegetable oils as food are great limiting factors for their application in large scale ventures. Therefore, scientists and researchers have focussed on waste materials as, for instance, animal fats (Meher et al. 2006), waste cooking or non-edible oils (Issariyakul and Dalai 2014) and free fatty acids containing oils (e.g., those from rice bran; Srilatha et al. 2012) to prepare biodiesel. Nonetheless, these oils might not be appropriate as biodiesel due to their high amount of saturated fatty acids. In addition, as most of them solidify at room temperature, they cannot be used as fuel in a diesel engine in their original form (Leung et al. 2010).

Bioethanol, which is derived from edible biomass, primarily corn, soybeans, and sugarcane raised concerns about the impact of first generation biofuel on food prices and increased deforestation (Cassman and Liska 2007; Fargione et al. 2008). Therefore, attention has been diverted into second generation feedstocks like lignocellulosic and *Miscanthus* biomass as well as agricultural and municipal waste (Hattori and Morita 2010; Daroch and Mos 2011). Despite lignocellulosic feedstocks being cheaper than first-generation feedstocks, they are more difficult to break down/or convert into small molecules like ethanol than starch, sugar, or oils, and the technology to convert them into liquid fuels is at present not cost-effective due to its resistance to saccharification (i.e., hydrolysis) usually caused by their high lignin content. Consequently, there is currently a great effort to find alternative third generation feedstocks. Fortunately, microalgae have been identified as a very good candidate for biofuel production (Schenk et al. 2008), being already cultivated as source of food-based and pharmaceuticals applications (Apt and Behrens 1999; Guedes et al. 2011; Virginie et al. 2012). Therefore, microalgae feedstocks are currently receiving a lot of attention due to a multiplicity of reasons such as higher photosynthetic efficiencies and higher growth rates, as compared to first and second-generation crops, ability of being cultivated on marginal lands, continuous biomass production (i.e., day and night), the fact that they do not compete with food or other crops, and the possibility of using saline and wastewater streams for biomass production (Schenk et al. 2008). The other main advantages of microalgae-derived biofuels over the first and second-generation biofuels are to produce considerably greater amounts of biomass including lipids, carbohydrates, and proteins per hectare than any kind of terrestrial photoautotrophs (Chisti 2007). Therefore, the combined potential of biofuel production, CO_2 fixation, and bio-treatment of wastewater emphasises the utilization of microalgae as a promising feedstock for biodiesel, bioethanol, and biomethane, including bio-oil, synthetic gas, as well as other valuable pharmaceutical and nutraceutical products.

Strengths and opportunities encompassing marine microalgae

Microalgae are a large and diverse group, ranging from unicellular to multi-cellular microorganisms, including both prokaryotic microalgae (e.g., cyanobacteria) and eukaryotic microalgae such as diatoms (Bacillariophyta), green (Chlorophyta), and red (Rhodophyta) algae. Microalgae are very important from an ecological point of view, which are the food source for many animals and belonging to the bottom of the food chain, being thus the main oxygen producers on earth. They were once considered to be aquatic plants but are now classified separately because they lack true roots, stems, leaves, and embryos. The simpler morphology of microalgae is perhaps one of their main advantages, as compared to terrestrial crops, allowing them to grow faster. Photosynthetic growth of microalgae requires light, water, and CO_2 to synthesise lipids, carbohydrates, nucleic acids, proteins, and other metabolites. The use of a wide range of microalgae such as *Phormidium, Arthrospira, Chlorella, Scenedesmus, Botryococcus,* and *Chlamydomonas* for treating domestic wastewater has been reported and efficacy of this method is promising (Olguın 2003; Schulze et al. 2017a; Chinnasamy et al. 2010; Kong et al. 2010; Wang et al. 2010). According to reported studies, to produce 1000 Kg of microalgal biomass, about 1800 Kg CO_2, 70 Kg N, 10 Kg P, and 8 Kg K are required (Chisti 2007; Wijffels and Barbosa 2010; Collet et al. 2011). Hence, microalgal biomass can contribute to the bio-fixation of atmospheric CO_2, improving air quality. Microalgae also exhibit other advantages when compared with terrestrial plants. Apart from higher growth rates, microalgae do not require arable land and can be grown all year round (Borowitzka 1999) without the use of polluting, potentially toxic chemicals such as herbicides or pesticides (Rodolfi et al. 2008). They can produce lipids, carbohydrates, and proteins in large amounts over short periods of time that can be further processed into both biofuels and useful chemicals (Gangadhar et al. 2016a; Pereira et al. 2016). Moreover, some microalgal species can be grown in the absence of fresh water and perform very well in brackish or salt water (Christenson and Sims 2011). It is estimated that the biomass productivity of microalgae could be 50 times higher than that of switchgrass (i.e., *Miscanthus*), which is the fastest growing terrestrial plant (Demirbas 2006; Nakamura 2006). The exponential growth rates can double their biomass in periods as short as 3.5 h (Metting 1996; Spolaore et al. 2006; Chisti 2007). Secondly, in spite of their growth in aqueous media, algae need less freshwater than vegetable oil crops (Dismukes et al. 2008). Furthermore, as microalgae can be cultivated in brackish water on non-arable land, associated environmental impacts of land use are minimised (Searchinger et al. 2008), without compromising the production of food, fodder, and other products derived from terrestrial crops (Chisti 2007). The dual application of microalgae for phycoremediation of organic effluents from the agrochemical industry (Cantrell et al. 2008) turns microalgal cultivation into an eco-friendly process with no secondary pollution, allowing also an efficient recycling of nutrients (e.g., nitrogen and phosphorus) present in industrial and/or municipal wastewaters (Munoz and Guieysse 2006; Pizarro et al. 2006; Mulbry et al. 2008; Giorgos and Dimitris 2011; Rawat et al. 2011; Schulze et al. 2017a). Other studies have reported that lipid content in some microalgae increases under different cultivation conditions such as nitrogen deprivation (Iillman et al. 2000; Hsieh and Wu 2009), high light intensity (Khotimchenko and Yakovleva 2005), and high salt (Araujo et al. 2011). Moreover, phosphorus depletion found to be more efficient inducer of TAG as compared to that of nitrate depletion (Wu et al. 2015; Schüler et al. 2017 and references therein). Taken together, the use of microalgae is thus highly desirable since they are able to serve a dual role of bioremediation of wastewater as well as generating biomass for biofuel production with resultant carbon dioxide sequestration, even if temporary (Olguin 2003; Mulbry et al. 2008; Jasvinder and Gu 2010; Mata et al. 2010; Amaro et al. 2011).

Nutraceutical applications of microalgae

Biofuels from microalgae biomass cannot be commercially feasible unless value-added products are optimally utilized. Microalgae are a potential source of various applications such as nutraceutical (i.e., polyunsaturated fatty acids, PUFA), vitamins (Schmid 2009; Pereira et al. 2012), anti-oxidants, and metal chelators (Custódio et al. 2012; Gangadhar et al. 2016b) as well as bioactive value-added products as healthy ingredients for functional food (Matos et al. 2017). PUFA such as eicosapentaenoic (EPA) and docosahexaenoic (DHA) acids are *n*-3 fatty acids (Pereira et al. 2012), which are generally obtained from fish oil for human consumption (Luiten et al. 2003), providing a high-value food supplement (Harun et al. 2010a). Contrary to fish and mammals, microalgae produce their own *n*-3 fatty acids (Belarbi et al. 2000). EPA and DHA have their individual medical applications, including treatment of cardiovascular disease, migraine headache, and psoriasis (Singh et al. 2005); prevention and cure of cancer, AIDS, and hypercholesterolemia; as well as stimulation of the immune system, and body detoxification (Patil et al. 2007). Chlorophyll is another pharmaceutically important compound and it has been reported as source of a natural ingredient in processed food (Humphrey 2004) and cosmetics industries (Erica 1996) and as a chelating agent in ointment. It has also been used in pharmaceutical applications, especially liver recovery and ulcer treatment; it appears to repair cells, increase haemoglobin levels in blood, and cell growth (Puotinen 1999). Moreover, the cyanobacterium *Arthrospira* sp. is being used in food supplements due to its excellent nutrient composition and digestibility (Kumar et al. 2005), being a rich source of vitamins, particularly vitamin B_{12} and β-carotene and minerals (Thajuddin and Subramanian 2005). *Chlorella* sp. has been reported as potential food, principally because it contains most nutrients required for human nourishment (Spolaore et al. 2006). In addition to green algae, red algae, mainly *Porphyra,* and brown microalgae were deemed as safe for human consumption as food (Besada et al. 2009). Microalgae also play a key role in high-grade animal nutrition food (Dhargalkar and Verlecar 2009), due to their low caloric content, high concentration of minerals, vitamins and proteins, as well as a low fat content. The cultivation of *Arthrospira* sp. is possible at high salinity and high pH. Other studies have reported the use of microalgae (e.g., *Arthrospira*) in animal feed (Belay et al. 1996), *Arthrospira* as food additive (Erica 1996) and *Porphyridium* sp. as feed supplement for poultry (Ginzberg et al. 2000). Moreover, cyanobacteria have been used as feed for aquaculture based on their nutritional and non-toxic performance (Thajuddin and Subramanian 2005). *Chlorella* sp. has a high amount of chlorophyll among various species of microalgae (Dring 1991; Nakanishi 2001; Deng et al. 2008). Microalgae such as cyanobacteria, are also native producers of poly-β-hydroxybutyrate (PHB), providing a viable potential alternative to plastic carry bags made from petroleum-derived polymers (Brandl et al. 1995).

Cultivation and harvesting

Cultivation

The main way to produce microalgae biomass is by cultivation, which can be carried out in two different systems: open (e.g., raceway ponds) and closed (e.g., photobioreactors) systems. Several methods have been developed for the optimization of cultivation (Li et al. 2008; Wang and Lan 2011), and harvesting process for the production of biomass (Singh et al. 2011). Currently, industrial microalgae cultivation system is mostly carried out in open ponds due to their low construction and capital costs. However, it has some drawbacks such as lack of control of operational conditions, leading to water evaporation and contaminations with unwanted species, inefficient exposure of microalgal cells to sunlight and CO_2, due to poor mixing, thus sustaining low biomass yields (Suh and Lee 2003). Although sunlight in outdoor systems is free and abundant, seasonal variations of sunlight can significantly limit productivity (Borowitzka 1999). On the other hand, the main advantages of using photobioreactors (i.e., closed systems) are linked to a better control of the microalgal culture and its environment, large surface to volume ratios, less water evaporation, better isolation from outside contaminations, and higher biomass productivity.

Microalgae can be grown in various types of water like fresh, salt, waste, or brackish water and with as well as without an organic carbon source. Depending upon their carbon metabolism, microalgae can be divided into three types: (i) phototrophic, (ii) heterotrophic, and (iii) mixotrophic. Phototrophic microalgae utilize light as energy source and CO_2 as an inorganic carbon source from atmosphere and flue gases, whereas heterotrophic growth is directly independent of light, utilizing organic compounds (i.e., glucose, acetate, glycerol, fructose, sucrose, lactose, galactose, mannose, and corn powder hydrolysate instead of sugars) as both energy and carbon sources (Mata et al. 2010). Mixotrophic microalgae are capable of growing both photo- or heterotrophically. Most of the microalgal strains can shift from phototrophic to heterotrophic growth, and some can grow mixotrophically (Carlsson et al. 2007). To date, only phototrophic cultivation is technically and economically viable to cultivate microalgae in large scale, preferentially using seawater (Matsunaga et al. 2005). High salinity helps prevent the contamination of culture media, allowing seawater to be directly used, instead of depleting freshwater resources, simultaneously mitigating CO_2 from flue gases and removing/or re-using nutrients from wastewater and water returned to the growth facility. Several parameters are important to consider for microalgal growth, namely light, nutrient availability, temperature, pH, and culture mixing.

Light

Light is a source of energy for microalgal growth to produce lipids, carbohydrates, and proteins from CO_2 and H_2O. Due to its simpler morphology, metabolism, and development, microalgae achieve higher photosynthetic efficiency (PE) than terrestrial plants (Vasudevan and Briggs 2008). The use of natural light sources for commercial microalgal production has the advantage of using sunlight freely (Janssen et al. 2003). Light intensity plays an important role in microalgal photosynthesis, as exposure of cells to long periods of high light intensity causes photoinhibition, which leads to the formation of free radicals, causing photo oxidative damage (Rubio et al. 2003; Torzillo et al. 2003). Saturation light intensity determines the light utilization efficiency and overall photosynthetic efficiency (Torzillo et al. 2003). Sunlight within the range of 400–700 nm is photosynthetic active radiation (PAR), which accounts for 40–50% of total sunlight (Akkerman et al. 2002; Suh and Lee 2003). *Cyanobacterium anabaena* was studied (Gao et al. 2007) for growth responses under different solar radiation experiments. It was found that exposure to natural levels of PAR and UV light increased and inhibited cell growth, respectively (Gao et al. 2007). Another important parameter is the effects of the diurnal cycle (Pulz and Scheinbenbogan 1998) for outdoor cultivation, which have remarkable effects on the overall efficiency of solar energy capture. However, this may be limited by available sunlight due to diurnal cycles and the seasonal variations, thereby limiting the viability of commercial production to areas with high solar radiation. Both photo-inhibition and low light stress of photosynthesis causes decrease in microalgal biomass production (Barbosa et al. 2003; Kuda et al. 2005). As a result, up to 42% of biomass produced during daytime could be lost throughout the subsequent night (Tredici et al. 1991; Jacob et al. 2009). To address the limitations of natural sunlight growth conditions, artificial light in the form of fluorescent lamps are usually used for the cultivation of phototrophic microalgae at large scale (Muller et al. 1998), which allows for continuous production, but at significantly higher energy input. Light limitation due to high volume to surface ratios may also induce reduced biomass productivity. However, enhanced light supply is possible by reducing layer thickness, using thin layer photobioreactors with optimised culture mixing (Pulz 2001; Chisti 2007; Ugwu et al. 2008). More recently, flashing light-emitting diodes (LEDs) have been proposed as an alternative to fluorescent lamps to illuminate photobioreactors in order to decrease light attenuation and improve microalgal productivity (Schulze et al. 2014, 2017b).

Carbon dioxide (CO_2)

Carbon is an essential nutrient for microalgae growth; nearly 50% of the microalgae biomass is made up of carbon (Becker 1994), as it is a major nutrient for cell growth. Approximately 1.8 Kg of CO_2 are needed to produce 1 Kg of microalgae biomass (Amaro et al. 2011). Microalgae can be used to mitigate CO_2 from various sources, such as organic compounds (Lodi et al. 2005), inorganic carbonates (Emma et al. 2000; Wang et al. 2008), atmosphere (i.e., CO_2), as well as flue gas from industrial and power plants (Sydney et al. 2010). Some microalgae species such as *Arthrospira platensis* and *Chlamydomonas reinhardtii* (Chen et al. 1996) and *Arthrospira* sp. (Chojnacka and Noworyta 2004) are capable of growing in darkness and of using organic carbons (such as acetate or glucose) as energy and carbon sources (Ogbonna et al. 2000); however, using organic carbons for microalgal growth is expensive. Thus, a cheap source of CO_2 for photosynthetic production of biofuels is needed (Wang et al. 2008). Mitigation of atmospheric CO_2 is probably the most basic method to mitigate carbon, and relies on the mass transfer from the air to the microalgae in their aquatic growth environments during photosynthesis (Wang et al. 2008). However, atmospheric CO_2 (0.03%) is not sufficient to support the microalgal growth rates and productivities needed for full-scale biofuel production, which makes it economically infeasible (Stepan et al. 2002). Combustion of fossil fuel is the largest source of CO_2 emissions globally and CO_2 mitigation from flue gas emissions may be a way to achieve higher biomass productivities due to the higher CO_2 concentration (Bilanovic et al. 2009; Chiu et al. 2009). Consequently, when CO_2-enriched air is employed, light usually becomes the limiting factor.

Temperature

Temperature is another limiting factor for microalgae cultivation, which influences oxygen evolution and production efficiency (Ras et al. 2013). The optimal growth temperature for most species of microalgae is between 20 to 25°C (Ras et al. 2013 and references therein). A few microalgae can tolerate temperatures below 16°C, but this will result in reduced production of biomass and slower growth rates. Temperatures higher than 35°C are normally harmful (Hanagata et al. 1992; Andersen and Andersen 2006; Graham et al. 2008). However, some cyanobacteria such as *Anabaena variabilis* (Fontes et al. 1987) and *Arthrospira* sp. (Rafiqul et al. 2003; Vonshak and Tomaselli 2003; Ogbonda et al. 2007) can grow optimally at temperatures between 30–38°C. In addition, thermophilic cyanobacteria belonging to the *Synechococcus* genus (Murata 1989) can tolerate temperatures as high as 60°C (Miyairi 1995). The optimum temperature range required to support microalgae growth is strain dependent (Ras et al. 2013). Temperature has also been found to have a major effect on the fatty acid synthesis of microalgae (Guschina and Harwood 2006; Dean et al. 2010; Sharma et al. 2012). It has been reported that as temperature increases there is an increase in saturated fatty acids (Sushchik et al. 2003), whereas when temperature are lower the accumulation of unsaturated fatty acids is induced (Hu et al. 2008). Some studies have also reported that carbohydrate content increases with temperature (Oliveira et al. 1999; Sharma et al. 2012).

pH

Most of microalgae are sensitive to pH changes, its control being essential for achieving high growth rates. As microalgae are able to capture CO_2, there is a trend for pH to increase. Most microalgae (e.g., *Nannochloropsis salina*) usually grow faster at pH values between 8.0 and 9.0 (Bartley et al. 2014). Nonetheless, some species withstand and thrive at lower (*Chlorococcum littorale*) and higher pH (*Arthrospira platensis*) values. For example, *A. platensis* grows well at 11.5 but not at 7.0; however, productivity decreases significantly at pH higher than 9 (Fontes et al. 1987; Kodama et al. 1993; Vieira et al. 2004). In the case of high-density microalgal cultures in controlled systems using air enriched with carbon dioxide (CO_2 from flue gas or in pure form), the concentration of dissolved CO_2 may be the dominant factor determining the culture pH (Sanchez et al. 2003). On the other hand, the dissolved CO_2 is a result of balance between the mass transfer of CO_2 from gas to liquid phase and the consumption of CO_2 by culture cells. In contrast, microalgae have shown to cause a rise at pH from 10 to 11 in open ponds because of CO_2 mitigation (Oswald 1988). This increase in pH can be beneficial for inactivation of pathogens in microalgal wastewater treatment, but can also inhibit microalgal growth. With elevated CO_2 concentrations, pH drops down to pH 5, and with higher SOx concentrations pH values as low as 2.6 have been reported (Maeda et al. 1995), which can cause growth inhibition. With buffered medium, the pH drop can be avoided, also preventing changes in the growth rate caused by SOx (Maeda et al. 1995). Therefore, the pH control mechanisms need to be integrated with the aeration system by adding buffer to the culture (Qiu et al. 2017).

Mixing

Mixing is an important factor that (i) promotes gas exchange in a microalgal culture, (ii) uniforms CO_2 distribution, (iii) prevents sedimentation of culture cells, (iv) optimises exposure to light and nutrients, and (v) helps remove excess oxygen (Pulz 2001). A high concentration of dissolved oxygen is toxic to microalgal cells, and continuous exposure to light generates oxygen radicals, which can lead to cell damage and growth inhibition.

Harvesting

The term harvesting refers to the concentration of diluted algae suspensions to a thick paste and it may account for up to 20–30% (Borodyanski and Konstantinov 2002; Mata et al. 2010) of the total production costs. Microalgae with cell/filament sizes around 200 μm appear to be suitable candidates for cultivation and harvesting, compared to other smaller microalgae with sizes ranging between 0.5 to 30 μm (Molina et al. 2003; Waterbury 2006; Pereira et al. 2016). Microalgae with large cells or filaments can help reduce the harvesting problem, as they may be harvested with relative ease by filtration. For example, the cyanobacterium *Arthrospira* forms along spiral shape filament (trichome) with lengths varying, in most cases, between 20–100 μm, naturally allowing the use of more cost-efficient, energy-efficient harvesting methods (Benemann and Oswald 1996). However, the main drawback of using oleaginous microalgae for biomass production is that they are generally unicellular and are in suspension, thus being very difficult to harvest (Moreno 2008). Generally, a two-stage harvesting process, involving dewatering followed by thickening, can be quite costly if an additional drying step is needed (Barros et al. 2015). As a universal harvesting technique does not exist, intense research has been carried out to find suitable processes for specific microalgae.

Settling

Settling is a harvesting technique for microalgae biomass separation from suspension because of low value of the biomass generated from large volumes (Nurdogan and Oswald 1996). However, it is only suitable for species with large cells/filaments such as *Tetraselmis* sp. CTP4 (Pereira et al. 2016) and *Arthrospira* sp. (Munoz and Guieysse 2006), which minimizes the dewatering cost of biomass. Therefore, the settling velocity of microalgal cells could be enhanced by increasing the dimension of the suspended cells, that is, by cell aggregation, which can be achieved with chemical coagulants, resulting in large microalgal flocs that settle rapidly to the bottom (Salim et al. 2011).

Centrifugation

Centrifugation is a common method to harvest microalgae, as no additional chemicals are needed. It is preferred for harvesting biomass for high value-added metabolites and extended shelf-life concentrates (Heasman et al. 2000). Biomass recovery depends on the settling characteristics of the cells, slurry residence time in the centrifuge, and settling depth (Molina et al. 2003). Harvesting efficiency of more than 95% (Heasman et al. 2000), and increase in slurry concentration by up to 150 times for 15% total suspended solids are technically feasible (Mohn 1980). However, the main drawbacks of the technique include high energy costs and potentially higher maintenance requirements due to freely moving parts (Bosma et al. 2003).

Flocculation

Microalgal cells carry a negative charge that prevents natural aggregation of cells in suspension. Flocculants block these surface charges on the microalgal cells, stimulating cell-cell adhesion and generating flocs. These agents are widely used in harvesting processes as they increase the effective particle size and facilitate aggregation, which is a step prior to other harvesting techniques such as filtration, flotation, and/or sedimentation by centrifugation (Molina et al. 2003; Munoz and Guieysse 2006). Microalgal biomass is typically separated from large volumes of water more easily by flocculation and sedimentation than filtration (Divakaran and Pillai 2002). Several methods have been reported using flocculants such as chitosan, aluminium salts, polyacrylamide polymers, alum, lime, alkali and ferric chlorides, and aluminium and ferric sulphates (Semerjian and Ayoub 2003; Knuckey et al. 2006). Aluminium salts are also widely used as flocculants, but are unsuitable for animal feed unless the aluminium is removed. Microalgal biomass also can be separated by bio-flocculation (Salim et al. 2011) and anionic polyacrylamide (Bukenik et al. 1985; Knuckey et al. 2006) followed by pH adjustment using alkaline solutions. This process was successfully employed to a range of species with flocculation efficiencies approximately 80%. The use of sedimentation in combination with flocculation is reported to be cost effective due to minimal power consumption and use of gravity for biomass settling. Recently, calcium hydroxide has been reported as a less expensive non-toxic flocculant. Calcium is a cation that can and should be present in animal feed, and it typically precipitates more cells than other basic flocculants (Ami et al. 2012).

Flotation

Flotation method is based on the trapping of microalgae cells using dispersed micro-air bubbles (Wang et al. 2008). Harvesting of microalgae biomass needs a high overflow rate, which promotes flotation during which microalgae move upward instead of settling down. It does not require any addition of chemicals. It is a simple method by which algae can be made to float on the surface of the suspension, being removed as scum. Flotation is generally carried out with a combination of flocculation for microalgae harvesting. Some strains naturally float on the surface when their lipid contents increase (Bruton et al. 2009; Hamed 2016). Froth flotation is a technique of separating microalgae from the suspension by manipulating pH and bubbling air through a column to create algal froth that accumulates on the surface of a liquid, which can be removed by suction. Froth flotation and drying are currently considered very expensive for commercial use. Flotation has advantages of flexible operation and needs a small footprint as compared with sedimentation and coagulation-flocculation (Liu et al. 1999). With added cationic surfactants, dissolved air flotation (DAF) effectively separated microalgae from water (Liu et al. 1999). Although flotation has been reported as a potential harvesting method, there is very limited evidence of its technical or economic feasibility. *Chlorella vulgaris* has been harvested using a dispersed ozone flotation process (Ya et al. 2010), whereas *Microcystis* cyanobacteria have been eliminated by using an ozoflotation method (Benoufella et al. 1994). Pure oxygen aeration failed to yield algal flotation, while ozone produced flotation efficiently. The ozone dose needed to harvest microalgae ranged between 0.005 and 0.03 mg/mg biomass, making the ozone flotation a promising option for algal harvesting. However, ozone is a strong oxidant that normally oxidizes unsaturated compounds and aromatic rings into carbonyls and carboxylic acids, respectively (Von 2003; Beltran 2004; Li et al. 2008). Production of biofuel from microalgae should minimize the addition of chemicals for minimizing contamination.

Conversion to biofuels

There are several ways to convert biofuels from microalgae biomass including: (i) direct combustion (ii) thermochemical (iii) and chemical reactions, as well as (iv) biochemical conversions. Direct combustion corresponds to burning biomass in the presence of air at temperatures above 800°C to produce heat and electricity. These produced gases cannot be stored and must be used immediately. The conversion efficiency of biomass to energy is more favourable than that of direct combustion of coal. The main disadvantage of biomass is that it contains more water; thus, prior to combustion, biomass needs to be dried, and undergo other pre-treatments that may affect the energy balance (Brennan and Owende 2010). Gasification also generates a syngas (i.e., carbon dioxide, hydrogen gas, nitrogen, and methane) and bio-oil that can be combusted directly at high temperature (800–1000°C) by partial oxidation, producing heat, or used in turbines for electricity generation (Brennan and Owende 2010; Mutanda et al. 2011). Thermochemical liquefaction of biomass is used for conversion of wet biomass to bio-crude oil and into small molecules in the presence of a catalyst at 300–350°C and high pressure (50–200 bar) (Ross et al. 2010). Pyrolysis is another alternative method of bio-oil production, as it is more cost effective for microalgal biomass as feedstock than lignocellulosic biomass. This is due to the fact that microalgae contain higher amounts of cellular lipids, carbohydrates, and proteins, which are more easily pyrolysed and result in higher quality bio-oil production (Huang et al. 2010). Both gasification and pyrolysis require dried biomass as feedstock, and the processes operate at a temperature higher than 800°C. Several methods reported on the pyrolysis of *Chlorella prototheocoides* and *Oscillatoria tenuis* (Wu et al. 1996), *Arthrospira platensis* and *Chlorella protothecoides* (Peng et al. 2001), *Microcystis aeruginosa* (Miao et al. 2004; Miao and Wu 2006), *Chlorella muelleri* and *Synechococcus* (Grierson et al. 2009), and *Dunaliella tertiolecta* (Shuping et al. 2010) have exhibited competitive potential for biofuel production from microalgal biomass (Xiaoling et al. 2004). Pyrolysis of microalgal biomass has yielded promising results and shown to produce higher quality bio-oil than lignocellulosic compounds (Brennan and Owende 2010). Nonetheless, there are still some problems in the process of producing biofuels from microalgae by pyrolysis (Miao et al. 2004) due to longer residence time, which can cause secondary cracking of the primary products, reducing yield, and adversely affecting bio-oil properties. Other investigations have been carried out regarding the suitability of microalgal biomass for bio-oil production (Miao and Wu 2004; Miao et al. 2004). It was shown that microalgal bio-oils are of higher quality than bio-oil from wood (Demirbas 2006). All of these are not attractive for commercial application of liquid fuel production like biodiesel and bioethanol.

Biodiesel

Biodiesel is considered a promising alternative to fossil fuel. It is a monoalkyl ester of long-chain fatty acids derived from first generation renewable feedstocks, such as vegetable oil or animal fats (Meher et al. 2006), and waste cooking oils (Issariyakul and Dalai 2014). Microalgae are also identified as potential sources for biodiesel production, as they have the ability to synthesize triacylglycerols (TAG) and other neutral lipids that can be converted to biodiesel. The rapid growth potential and numerous species such as *Chlorella, Dunaliella, Isochrysis, Nannochloris, Nannochloropsis, Neochloris, Nitzschia, Tetraselmis* sp., *Tetraselmis* sp. CTP4, and *Phaeodactylum* sp. with oil content in the range of 20–50% biomass dry weight is another advantage for their choice as a potential biomass (Huerlimann et al. 2010; Gangadhar et al. 2016a; Pereira et al. 2016). *Chlorella* appears indeed a particularly good option for biodiesel. The biochemical composition of the microalgae biomass can be altered by varying growth conditions, and thus significantly stimulating oil yield (Qin 2005; Schüler et al. 2017 and references are therein). Eukaryotic microalgae are preferred to prokaryotes as they have been shown to produce more lipids (Williams and Laurens 2010). Microalgal biomass is attracting interest not only as a possible source of biodiesel (Amaro et al. 2011), but also as a source of other types of biofuels: bioethanol, biomethane, biohydrogen, and biobutanol (Brennan and Owende 2010; Jasvinder and Gu 2010; Parmar et al. 2011; Varfolomeev and Wasserman 2011). In addition, microalgae can also produce value-added co-products such as proteins and residual biomass after oil extraction, which may be used as feed or fertilizers (Spolaore et al. 2006) and fermented to produce bioethanol or biomethane (Hirano et al. 1997). Moreover, a few microalgae are capable of photo-biological production of biohydrogen (Ghirardi et al. 2000).

Oil extraction

Storage of lipids in microalgae differs from strain to strain and even within a single culture under different growth conditions (Huerlimann et al. 2010). Lipid identification is essential as its composition determines the properties of the biodiesel produced. Lipid qualification and quantification are be determined by several methods such as Nile red fluorescence microscopy, Nile red spectrofluorometric, and BODIPY staining (Medina et al. 1998; Mutanda et al. 2011; Pereira et al. 2011; Pereira et al. 2016). These methods have primarily been employed to identify the presence of lipid bodies within cells as an initial screen for lipid accumulation and as a semi-quantitative method for lipid storage.

Various methods for extraction of lipids from microalgae have been reported in literature, but most common methods are expeller/oil press, ultrasound, microwave, osmotic, solvent extraction, supercritical fluid extraction, and enzymatic extractions. The extraction of microalgal lipids from microalgal biomass is generally more expensive and technologically more challenging than that of terrestrial plant seeds. However, commercial scale microalgae biodiesel production is restricted by unfavourable downstream costs of lipid extraction and availability of water, CO_2, and nutrients (Pate et al. 2011). A possible solution would be to integrate microalgal cultivation with existing biogas plants, where algae could be cultivated using discharges of CO_2 and digestate as nutrient input, and then the attained biomass could be converted directly to biomethane by existing infrastructures. The downstream processes of biomass drying and lipid extraction would take up 50–90% of the overall energy consumption (Lardon et al. 2009, Stephenson et al. 2010). It is therefore worth to explore simple and robust processes for the energy utilization of microalgal biomass (Collet et al. 2011).

Mechanical extraction methods minimize the contamination from external sources (Greenwell et al. 2010), while maintaining the chemicals originally contained within the biomass. These methods are usually used in combination with some kind of solvent extraction. Expellers are the common method for extraction of oil from biomass such as oil seeds (Popoola and Yangomodou 2006) and microalgae. To ensure the efficacy of this process, the microalgae need to be dried first. These methods are able to extract almost 75% of oil, with no special skill required. However, this conventional method has been reported to be less effective due to relatively longer extraction times (Popoola and Yangomodou 2006). Microwave extraction directly affects solvents and biomass, even though trace amount of moisture content in cells are affected. It had been reported to be a useful method for extraction of oils from renewable materials (Angelis et al. 2005). Microwave-assisted extraction to recover oil has proved the most efficient method for *Botryococcus* sp. (Geciova et al. 2002) and *Scenedesmus obliquus* (Balasubramanian et al. 2010). Microwave-assisted hexane extractions were found to result in higher oil yields compared to conventionally water-heated hexane extraction. It can be easily scaled-up in commercial scale production (Sahena et al. 2009). However, microwave-assisted extraction presents some drawbacks, as it has the potential for causing oxidative damage to value-added products (Sahena et al. 2009).

Ultrasonic extraction methods were most effective at disrupting cell walls (Cravotto et al. 2008; Wei et al. 2008), increasing oil production from *Crypthecodinium cohnii* and *S. obliquus* using Soxhlet extraction in hexane (Cravotto et al. 2008; Balasubramanian et al. 2010). Both ultrasonication and microwave-assisted methods improve oil extraction of microalgae significantly, with higher efficiency, reduced extraction times and increased yields, as well as low to moderate costs and negligible added toxicity. This technology, however, may negatively impact oil quality and/or stability of polyunsaturated fatty acid-rich oils, as it is difficult to scale up.

Solvent extraction entails extracting oil from microalgae by repeated washing or percolation with an organic solvent. Extraction methods used should be fast, effective, and non-damaging to oils extracted and easily scaled up (Medina et al. 1998). The choice of solvent for oil extraction, as with the harvesting process, will depend on the type of the microalgae selected. Lipids have different types of associations which need to be disrupted for effective extraction. Pre-treatment of samples may be required for oil extraction of certain types of biomass. This is generally not necessary for extraction from wet biomass, as solvents generally rupture cells by disassembling microalgal cell membranes and cell walls as well (Mercer and Armenta 2011; Gangadhar et al. 2016a). In addition, a suitable organic solvent should be commercially available, inexpensive, and insoluble in water, have a low boiling point to facilitate its removal after extraction, and have a considerably different density than water, enabling its re-use. Hexane is typically the solvent of choice for large scale extractions, as it is inexpensive, has high extraction efficiency, and is less dense than water (Benerjee et al. 2002). However, if the solvent is not harmful to the cells it is also possible to isolate the oils from microalgae such as *Botryococcus braunii* without breaking the cell walls (Benerjee et al. 2002). To optimise efficiency, cell mechanical rupture is, however, usually needed before exposure to the organic solvent (Cooney et al. 2009). Solvent extraction is most effective at recovering the oil from *Scenedesmus* sp. (Shen et al. 2009) and *Nannocloropsis* sp. using Soxhlet extraction (Wiyarno et al. 2011). With Soxhlet extraction, higher solvent concentration led to higher FFA levels and saponification. Normally, lipids are extracted from microalgal biomass using a mixture of solvents like chloroform, methanol and water (Bligh and Dyer 1959) and a modified Bligh and Dyer (Mutanda et al. 2011) method is most commonly used. Disadvantages of this method is that it produces waste solvent that is costly to recycle at large scale, raising safety concerns due to handling of large amounts of organic solvents (Sahena et al. 2009). However, the Bligh and Dyer method has been proven moderately effective when used on wet material. In addition, organic solvents can lead to contamination in the form of solvent residues being present in the final product. A mixture of hexane and isopropanol was used to extract oil from *Chlorococcum* sp., because is an alternative set of solvents with lower toxicity as compared to a mixture of chloroform and methanol (Halim et al. 2011).

Enzymatic treatment of microbial biomass has the potential to partially or fully disrupt cells with minimal damage to the desired biomaterials (i.e., oil). It has proven to be successful extracting oil from plant seeds using sonication and cold pressing (Shah et al. 2004; Soto et al. 2007). One distinct challenge with enzymes is that in order to design an effective enzymatic procedure for hydrolysing microbial cells, it is necessary to determine their composition. Therefore, the most appropriate enzymes can be chosen to optimize extraction conditions.

Supercritical CO_2 extraction is a promising green technology that can potentially be used for large-scale microalgal lipid extraction (Ronald et al. 2012). It is rapid, non-toxic, has high selectivity towards triacylglycerols, and produces solvent-free lipids. Since CO_2 is a gas at room temperature, it is safe for food applications, and it is thus being favoured with respect to other solvents due to its relatively low critical temperature and pressure, and the possibility of being recycled upon extraction completion (Sahena et al. 2009; Wiyarno et al. 2011). Supercritical carbon dioxide extraction is regarded with interest as an industrial process, due to its shorter extraction time and the ability of obtaining pure compounds without solvent contamination, being also deemed as safe for thermally sensitive products (Mendes et al. 2006; Sahena et al. 2009). Moreover, isolation of important compounds such as anti-oxidants (e.g., canthaxanthin and astaxanthin) from *Chlorella vulgaris* and *Haematococcus pluvial* and β-carotene from *Dunaliella salina* are feasible, which may reduce separation costs, as well as possibly counteracting greenhouse gas effects by using CO_2 waste from industry (e.g., Cyanotech and Aquasearch) (Perrut 2001; Guerin et al. 2003; Herrero et al. 2006; Mendiola et al. 2007). In addition, the reported methods have been compared to the traditional solvent extraction method with supercritical fluid extraction (Mendes et al. 2005; Andrich et al. 2006; Xu et al. 2008) of oils from microalgal strains, such as *A. maxima, A. platensis, B. braunii, C. vulgaris, Ochronomas danica, Skeletonema costatum,* and *Isochrysis galbana* (Santos et al. 1997; Mendes et al. 1999; Perretti et al. 2003). Supercritical fluid extraction efficiency is affected by four main factors: pressure, temperature, flow rate, and extraction time (Andrich et al. 2006; Xu et al. 2008). These factors, along with the use of a co-solvent (i.e., ethanol), can be altered and adjusted to optimize extractions. When ethanol is used as a co-solvent, the polarity of the extracting solvent is increased and the viscosity of the fluid is subsequently altered. The resulting effect is an increase of solvating power of CO_2, and the extraction requires lower temperature and pressure, making it more efficient (Wiyarno et al. 2011). High moisture content can reduce contact time between the solvent and sample. Hence samples are dried prior to supercritical fluid extraction (Sahena et al. 2009). Its main disadvantages are associated with the high capital cost and the high energy requirement for supercritical fluid compression.

Transesterification for biodiesel production

The main disadvantage of vegetable and microalgae oils is that they usually have lower volatility and higher viscosity than that of petroleum diesel (Fuls et al. 1984; Ma and Hanna 1999; Zuhair 2007; Rawat et al. 2011). For instance, the fuel properties of biodiesel, such as cetane number, heat of combustion, melting point and viscosity, increase with increasing carbon number, and saturation degree (Pinto et al. 2005). Therefore, these oils cannot be used directly due to engine coking, carbon depositing, and gelling of the lubricating oil (Ma and Hanna 1999; Meher et al. 2006; Akoh et al. 2007), requiring conversion to lower molecular weight constituents with higher volatility and lower viscosity (Rawat et al. 2011). Hence, the properties of biodiesel can be improved by several methods, namely (i) genetic engineering of the cells producing the parent oils in order to enrich the fuel with specific fatty acids (Narasimharao et al. 2007), (ii) dilution, (iii) pyrolysis, (iv) cracking, and (v) transesterification (Fukuda et al. 2001; Helwani et al. 2009). Therefore, these oils need to be modified or transesterified to biodiesel using catalysts.

Transesterification (i.e., alcoholysis) is the process of treating a triacylglycerol (TAG) molecule with alcohol in the presence of a catalyst to produce alkyl esters and glycerol. These alkyl esters are known as biodiesel. As the reaction is reversible, an excess of alcohol must be used. Methanol, ethanol, and propanol are the most commonly used alcohols, but often methanol is used due to its low price and high availability. Stoichiometrically, for each mole of TAG, three moles of alcohol are required; however, in industrial processes this is usually upgraded to a 6:1 molar ratio to increase biodiesel yield (Fukuda et al. 2001). Transesterification is often carried out by two methods, namely chemical transesterification and enzymatic esterification (Fangrui et al. 1999; Shimada et al. 2002).

Chemical catalysed transesterification: Chemically, the reaction is catalysed by either an acid or an alkali. Biodiesel production from low quality oils (i.e., waste cooking oils) is challenging due to the presence of free fatty acids (FFA) and water (Schuchardt et al. 1998). Usage of homogeneous base catalyst for transesterification of such feedstock suffers from soap formation, which creates serious problems of product separation and lowers the biodiesel yield substantially (Sharma et al. 2008). In this case, biodiesel production catalysed by acids (Georgogianni et al. 2008; Zhang et al. 2008) can replace base catalysts since they do not show measurable susceptibility to FFA and can catalyse esterification (i.e., by acid catalyst) and transesterification (i.e., by acid or base catalyst) simultaneously (Fig. 1). In current commercial processes; however, excess alkali is added to remove all FFA from crude feedstock (Fukuda et al. 2001). For alkali-catalysed transesterification, as water and FFA do not favour transesterification, anhydrous triacylglycerol and alcohol are necessary to minimize the production of soap. Soap production decreases the amount of biodiesel and the separation of glycerol and esters becomes difficult. In spite of this limitation, as alkaline catalysts are often less corrosive than their acidic counterparts and alkali-catalysed transesterification is typically faster than acid-catalysed reactions (Freedman et al. 1984), industrial processes usually favour metal hydroxides (Wimmer 1995) and metal alkoxides (Freedman et al. 1984; Schwab et al. 1987) as well as its carbonates (Filip et al. 1992). Therefore, alkaline catalysts such as sodium and potassium hydroxides are commonly used as commercial catalysts at a concentration of about 1% by weight of the oil. Sodium alkoxides are the most active catalysts, since they give excellent

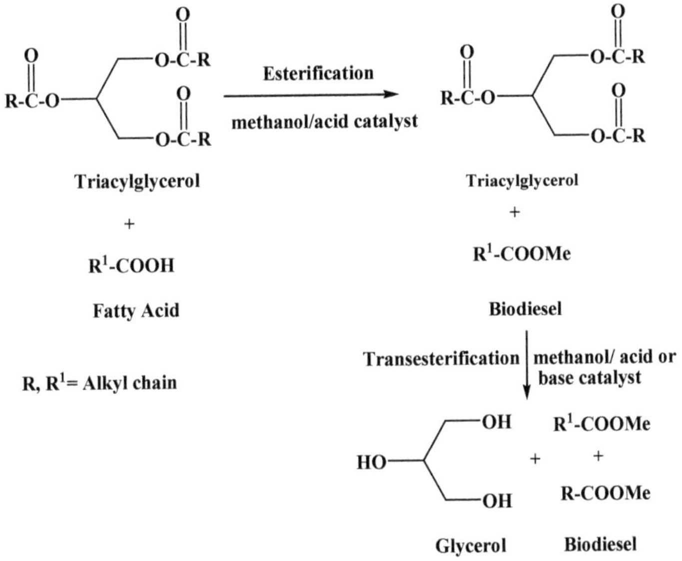

Fig. 1. Preparation of biodiesel from fatty acids and triacylglycerols (Oil).

yields in a short reaction time, even if they are applied at low molar concentrations. Metal hydroxides are cheaper than metal alkoxides, but less active. However, they are a good alternative since they can give the same high conversions of oils just by increasing the catalyst concentration to 1 or 2 mol%. However, homogeneous transesterification conversions involve high-energy consumption, a difficult separation of the catalyst from the homogenous reaction mixtures, and are expensive as well as chemically wasteful (Lotero et al. 2005). Re-usable heterogeneous catalysts have gained attention as replacement for homogenous catalysts, due to the ease of product separation and reusability.

Several reports about the use of heterogeneous acid catalysts to produce biodiesel, including MCM–41 (Pariente et al. 2003), Nafion (Ngaosuwan et al. 2007), lanthanum zeolite-β (Shu et al. 2007), metal carbonates or oxides (Demirbas 2008), zeolites (Ramos et al. 2008), Amberlyst-15 (Talukder et al. 2009), heteropolyacid catalyst (Zhang et al. 2010), sulphonated amorphous carbon (Toda et al. 2005; Shu et al. 2010), and metal oxides (Joana et al. 2012) have been published in the last few years. However, the aforementioned catalysts have also some drawbacks, such as tedious preparation, elevated price of starting materials, and possible leaching into the reaction medium, which makes them non-reusable (Ngaosuwan et al. 2007; Park et al. 2009). Therefore, carbon-based heterogeneous acid (Devi et al. 2009) and base (Devi et al. 2017) catalysts are the key to new developments in the production of biodiesel, combined with their reusable application in a continuous process and easy separation from the reaction mixture. These catalysts have therefore the potential of substantially decreasing the costs of biodiesel production.

Enzymatic-catalysed transesterification: There is increasing interest in using biocatalysts in TAG conversion to biodiesel. Biocatalysts allow easy substrate-catalyst separation, the use of mild, simple reaction conditions, attaining high product purity. Biocatalysts can be prepared from renewable sources, being re-usable, and favour stereospecific environment-friendly reactions of esterification and transesterification. Contrary to chemical catalysts, enzymes do not form soaps and catalyse esterification of FFA and TAG in one step without any need for the washing step. A number of studies have reported that lipases yield promising results as alternative catalysts (Nelson et al. 1996; Iso et al. 2001; Kose et al. 2002; Shimada et al. 2002). Nonetheless, the enzymatic process for the production of biodiesel is still not commercially feasible, due to the relatively high cost of the biocatalyst (Fukuda et al. 2001) and enzyme inactivation by alcohols such as methanol, ethanol, and glycerol (Samukawa et al. 2000). It was reported that the activity of the lipase inhibited by methanol can be restored to a certain extent by washing the enzyme with secondary and tertiary alcohols such as isopropanol, 2-butanol, and tertiary butanol (Chen and Wu 2003). Isopropanol showed virtually no negative effect on lipase activity and also the fatty acid isopropyl esters improved the cold weather performance of the biofuel (Lee et al. 1995). Producing ethyl esters rather than methyl esters is of considerable interest, as the extra carbon atom increases the heat content and cetane number (Encinar et al. 2002).

Bioethanol production

Bioethanol is a biofuel that is obtained from starch-based (corn, wheat, barley) and sugar-based (sugar beets, sugarcane) first generation feedstocks and has partially replaced petrol in some parts of the world, for instance, Brazil and United States (Bai et al. 2008). Although bioethanol is easily produced via fermentation, the use of food crops for its production impacts on food security and agricultural land availability (Sun and Cheng 2002). Second generation feedstock such as lignocellulosic (i.e., agriculture waste, forest residues, and others sources) biomass contains a complex mixture of carbohydrate polymers from plant cell walls such as cellulose, hemicellulose, and lignin. In order to produce sugars, lignocellulosic biomass must be pre-treated with acids or enzymes in order to breakdown these polymers into smaller sized compounds, thus increasing the cost of conversion into bioethanol. Lignin could be used as starting material for bioethanol production. However, it is a polymer very difficult to degrade or ferment biologically (Lynd 1996), thereby decreasing overall biomass to biofuel yields, leading also to higher waste treatment costs. Conversely, algae have been purported as alternative feedstocks for bioethanol production (Harun et al. 2010b), due to their fast growth, efficient carbon dioxide fixation, and potentially accumulating high amounts of carbohydrates, apart from lipids and proteins, when compared to biomass derived from food crops such as sugarcane and maize. As microalgae can contain high levels of starch and cellulose, having no lignin and low hemicellulose content, their saccharification for bioethanol production appears to be a more straightforward process (Hamelinck et al. 2005; Harun et al. 2010b). For instance, *Chlamydomonas reinhardtii* UTEX 90 (Choi et al. 2010) and *Chlorella vulgaris* (Branyikova et al. 2011) accumulate their energy reserves in form of 55 to 60% (w/w) starch, which can be easily hydrolysed into glucose via enzymatic and/or chemical methods. Starch contents of other microalgae ranging between 20 to 50% (w/w) have also been reported (Matsumoto et al. 2003; Rodjaroen et al. 2007; Rojan et al. 2011; Chun et al. 2013).

Bioethanol production from microalgae such as *Chlorococcum* sp. (Harun et al. 2010b) and *Chlamydomonas perigranulata* was fermented to produce bioethanol, butanediol, acetic acid, and CO_2 (Hon 2006), which supports the suitability of microalgae as promising substrates for bioethanol production. It was found that hydrogen and carbon recovery from that fermentation was about 139 and 105%, respectively.

There are several methods of extracting sugars from biomass by using chemical and enzymatic saccharification (Daroch et al. 2013). Chemical saccharification of biomass is typically carried out using dilute and concentrated sulphuric acids and alkaline solutions (Van de Vyver et al. 2010). Enzymatic saccharification (i.e., hydrolysis) methods, involving the use of cellulases, amylases, and glucoamylases, are widely employed to hydrolyse microalgal biomass to sucrose (Rui et al. 2012; Chun et al. 2013). Sucrose can be hydrolysed into glucose and fructose by invertase and then converted to ethanol using zymase. Both enzymes are produced by yeasts during fermentation for bioethanol production. Cellulosic biomass can be converted into any type of fuel including ethanol, gasoline, diesel, and jet fuel (Huber and Dale 2009). This has resulted in considerable attention towards the application of biomethane fermentation of microalgae to produce value-added by-products such as biogas (Daroch et al. 2013). Fermentation of the microalgal biomass can be carried out by bacteria, yeast, and filamentous fungi, resulting in by-products such as CO_2 and water. CO_2 produced in fermentation processes can be re-used as carbon sources for microalgae cultivation, as well as reducing the greenhouse gases emissions. After bioethanol production by fermentation, spent biomass is also useful for anaerobic digestion process for production of biogas (Ueda et al. 1996; Bush and Hall 2006), making all organic matter useful for biofuel production (Jasvinder and Gu 2010; Harun et al. 2010b). As an example, production of bioethanol, biobutanol, and biohydrogen from pre-treated microalgae biomass by anaerobic fermentation with immobilized *Clostridium acetobutylicum* cells has been studied (Efremenko et al. 2012), which offers promising multiple applications using immobilized biocatalysts for converting waste/microalgal biomass into biofuel. Recently, *Mychonastes afer* PKUAC9 and *Scenedesmus abundans* PKUAC12 were tested for saccharification followed by fermentative bioethanol production. *S. abundans* exhibited the highest production of total glucose (5.730 g/L) and sugars hydrolysates (10.752 g/L) as compared to *M. afer*. Chemo-enzymatic treated *S. abundans* was the best feedstock for fermentative production (0.103 g/g dry weight) of bioethanol (Hui et al. 2013).

Biogas production

Biogas is the biofuel substitute for natural gas, being produced by anaerobic (i.e., in the absence of oxygen/or in an oxygen-deficient environment) digestion or fermentation of biodegradable materials such as cow manure, municipal waste, plant residues, and microalgal biomass. For example, as large-scale production of microalgal-based biodiesel is still limited by the downstream costs of lipid extraction (Collet et al. 2011), there is a need to explore other energy uses of spent microalgae biomass. The absence of lignin makes microalgae ideal candidates for efficient biomethane production by fermentation in biogas manufactures. Thus, microalgae are not only a possible source of biodiesel (Amaro et al. 2011), but also as a source of other types of biofuels: bioethanol, biomethane, hydrogen, and butanol (Brennan and Owende 2010; Varfolomeev et al. 2010; Parmar et al. 2011; Varfolomeev and Wasserman 2011). The coupled process of microalgae cultivation and biogas production may be a better option than biodiesel production alone (Collet et al. 2011). Upon oil extraction, the spent biomass can be used to produce biogas and/or biofertilisers (Jasvinder and Gu 2010). In addition to being renewable, this would encourage sustainable farming procedures, providing higher efficiencies and reduce production costs (Vergara et al. 2008). Generally, anaerobic digestion can be divided into four types: (i) hydrolysis (ii) acidogenesis, (iii) acetogenesis, and (iv) methanogenesis. Methanogenic digestion is widely used in digestion of organic waste. First, in hydrolysis, the complex materials are degraded into small particle such as soluble sugars, which are later fermented to alcohols and acetic acid and/or volatile fatty acids and other by-products like hydrogen and carbon dioxide, which are metabolized to biogas by methanogens. The biogas production from this anaerobic digestion process is primarily affected by organic loadings, pH, temperature, and retention time in reactors (Harun et al. 2010). Mainly long solid retention time and high organic loading rate give significant results in terms of high methane yield (Chynoweth 2005). In addition, anaerobic digestion can operate in either mesophilic (35°C) or thermophilic (55°C) conditions (Otsuka and Yoshino 2004). Biomethane production by anaerobic digestion of microalgal biomass has been reported for freshwater and marine microalgae in various combinations (Aino et al. 2013). Net energy analysis of two production systems for producing biodiesel and biogas from *Nannochloropsis* and *Haematococcus pluvialis*, grown in a raceway in salt and freshwater, respectively, were studied (Luis and Tan 2011). Life cycle assessment of the system has shown that biomethane production without greenhouse gas heating would have a net energy ratio of 1.54, which is slightly lower than that of biomethane from ley crop (1.78) (Aino et al. 2013). Biogas production from microalgae change due to variation in cellular lipid, carbohydrate and protein content, cell wall structure, cultivation, and digestion temperature (Aino et al. 2013). Alternatively, biomethane, as well as bio-oil, can be produced from *Emiliania huxleyi via* direct pyrolysis (Wu et al. 1999), thus, making this coccolithophore a promising candidate for biofuel production. Methanation of syngas produced from gasification of microalgal biomass is another route to produce biofuels. Although microalgae offer a good potential for biogas production, commercial ventures exploiting this possibility have not yet been implemented (Jasvinder and Gu 2010).

Acknowledgements

The authors would like to thanks financial assistance for a postdoctoral fellowship (ref: SFRH/BPD/81882/2011) by Foundation for Science and Technology (FCT), Portugal. We are grateful for financial support through project DINOSSAUR—PTDC/BBB-EBB/1374/2014 - POCI-01-0145-

FEDER-016640, coordinated by author F.X.M., and funded by FEDER funds through COMPETE2020—Programa Operacional Competitividade e Internacionalização (POCI), and by national funds through FCT—Fundação para a Ciência e a Tecnologia, I.P.

References

Abreu, F.R., M.B. Alves and C.C.S. Macedo. 2005. New multi-phase catalytic systems based on tin compounds active for vegetable oil transesterification reaction. J. Mol. Catal. A: Chem. 227: 263–267.

Aino, M.L., O.H. Tuovinen and J.A. Puhakka. 2013. Anaerobic conversion of microalgal biomass to sustainable energy carriers—a review. Bioresour. Technol. 135: 222–231.

Akkerman, I., M. Janssen, J. Rocha and R.H. Wijffels. 2002. Photobiological hydrogen production: photochemical efficiency and bioreactor design. Int. J. Hydrogen Energy 27: 1195–1208.

Akoh, C.C., S.W. Chang, G.C. Lee and J.F. Shaw. 2007. Enzymatic approach to biodiesel production. J. Agric. Food Chem. 55: 8995–9005.

Amaro, H.M., A.C. Guedes and F.X. Malcata. 2011. Advances and perspectives in using microalgae to produce biodiesel. Appl. Energ. 88: 3402–3410.

Ami, S., E. Doron, B.G. Amicam, C. Hilla, E. Shai and G. Jonathan. 2012. Inexpensive non-toxic flocculation of microalgae contradicts theories; overcoming a major hurdle to bulk algal production. Biotechnol. Adv. 30: 1023–1030.

Andersen, T. and F.O. Andersen. 2006. Effects of CO_2 concentration on growth of filamentous algae and *Littorellauniflora* in a Danish softwater lake. Aquat. Bot. 84: 267–271.

Andrich, G., A. Zinnai, U. Nesti, F. Venturi and R. Fiorentini. 2006. Supercritical fluid extraction of oil from microalgae *Spirulina (Arthrospira) platensis*. Acta Aliment. 35: 195–203.

Angelis, L.D., P. Rise, F. Giavarini, C. Galli, C.L. Bolis and M.L. Colombo. 2005. Marine microalgae analyzed by mass spectrometry are rich sources of polyunsaturated fatty acids. Journal of Mass Spectroscopy 40: 1605–1608.

Antolin, G., F.V. Tinaut, Y. Briceno, V. Castano, C. Perez and A.I. Ramiez. 2002. Optimisation of biodiesel production by sunflower oil transesterification. Bioresour. Technol. 83: 111–114.

Apt, K.E. and P.W. Behrens. 1999. Commercial developments in microalgal biotechnology. J. Phycol. 35: 215–226.

Araujo, G.S., L.J.B.L. Matos, L.R.B. Goncalves, F.A.N. Fernandes and V.R.L. Farias. 2011. Bioprospecting for oil producing microalgal strains: evaluation of oil and biomass production for ten microalgal strains. Bioresour. Technol. 102: 5248–5250.

Bai, F.W., W.A. Anderson and M.M. Young. 2008. Ethanol fermentation technologies from sugar and starch feedstocks. Biotechnol. Adv. 26: 89–105.

Balasubramanian, S., J.D. Allen, A. Kanitkar and D. Boldor. 2010. Oil extraction from *Scenedesmus obliquus* using a continuous microwave system-design, optimization, and quality characterization. Bioresour. Technol. 102: 3396–3403.

Barbosa, M.J., M. Janssen, N. Ham, J. Tramper and R.H. Wijffels. 2003. Microalgae cultivation in air-lift reactors: modeling biomass yield and growth rate as a function of mixing frequency. Biotechnlo. Bioeng. 82: 170–179.

Barnwal, B.K. and M.P. Sharma. 2005. Prospects of biodiesel production from vegetable oils in India. Renew. Sust. Energ. Rev. 9: 363–78.

Barros, A.I., A.L. Gonçalves, M. Simões and J.C.M. Pires. 2015. Harvesting techniques applied to microalgae: a review. Renew. Sust. Energ. Rev. 41: 1489–1500.

Bartley, M.L., W.J. Boeing, H.N. Dungan, F.O. Holguin and T. Schaub. 2014. pH effects on growth and lipid accumulation of the biofuel microalgae *Nannochloropsis* salina and invading organisms. J. Appl. Phycol. 26: 1431–1437.

Becker, E.W. 1994. Large-scale cultivation. pp. 63–171. *In*: E.W. Becker (ed.). Microalgae: Biotechnology and Microbiology. Cambridge University Press, New York.

Belarbi, E.H., E. Molina and Y. Chisti. 2000. A process for high yield and scaleable recovery of high purity eicosapentaenoic acid esters from microalgae and fish oil. Enzyme Microb. Technol. 26: 516–529.

Belay, A., T. Kato and Y. Ota. 1996. *Spirulina. (Arthrospira)*: potential application as an animal feed supplement. J. Appl. Phycol. 8: 303–311.

Beltran, F.J. 2004. Ozone reaction kinetics for water and wastewater systems. CRC Press, Boca Raton, FL, USA.

Benemann, J.R. and W.J. Oswald. 1996. Systems and Economic Analysis of Microalgae Ponds for Conversion of CO_2 to Biomass. US Department of Energy, Pittsburgh Energy Technology Centre.

Benerjee, A., R. Sharma, Y. Chisti and U.C. Banerjee. 2002. *Botryococcus braunii*: a renewable source of hydrocarbons and other chemicals. Crit. Rev. Biotechnol. 22: 245–279.

Benoufella, F., A. Laplanche, V. Boisdon and M.M. Bourbigot. 1994. Elimination of *Microcystis* cyanobacteria (blue-green-algae) by an ozoflotation process a pilot plant study. Water Sci. Technol. 30: 245–257.

Besada, V., J.M. Andrade, F. Schultze and J.J. Gonzalez. 2009. Heavy metals in edible seaweeds commercialised for human consumption. J. Marine Syst. 75: 305–313.

Bhatti, H.N., M.A. Hanif, M. Qasim and A. Rehman. 2008. Biodiesel production from waste tallow. Fuel 87: 2961–2966.

Bilanovic, D., A. Andargatchew, T. Kroeger and G. Shelef. 2009. Freshwater and marine microalgae sequestering of CO_2 at different C and N concentrations response surface methodology analysis. Energy Convers. Manage. 50: 262–267.

Bligh, E.G. and W.J. Dyer. 1959. A rapid method for total lipid extraction and purification. Can. J. Biochem. Physiol. 37: 911–917.

Borodyanski, G. and I. Konstantinov. 2002. Microalgae separator apparatus and method. US Patent US 2002/0079270A1.

Borowitzka, M.A. 1999. Commercial production of microalgae: ponds, tanks, tubes and fermenters. J. Biotechnol. 70: 313–321.

Bosma, R., W.A. van Spronsen, J. Tramper and R.H. Wijffels. 2003. Ultrasound, a new separation technique to harvest microalgae. J. Appl. Phycol. 15: 143–153.

BP, 2010. BP statistical review of world energy June 2010. UK, British Petroleum.

Brandl, H., R. Bachofen, J. Mayer and E. Wintermantel. 1995. Degradation and applications of polyhydoxyalkanoates. Can. J. Microbiol. 41: 143–153.

Branyikova, I., B. Marsalkova, J. Doucha, T. Branyik, K. Bisova, V. Zachleder and M. Vitova. 2011. Microalgae-novel highly efficient starch producers. Biotechnol. Bioeng. 108: 766–776.

Brennan, L. and P. Owende. 2010. Biofuels from microalgae a review of technologies for production, processing, and extraction of biofuels and co-products. Renew. Sust. Energ. Rev. 14: 557–577.

Briens, C., J. Piskorz and F. Berruti. 2008. Biomass valorization for fuel and chemicals production—a review. International Int. J. Chem. React. Eng. 6: 1–49.

Bruton, T., H. Lyons, Y. Lerat, M. Stanley and M.B. Rasmussen. 2009. A review of the potential of marine algae as a source of biofuel in Ireland. Sust. Energ. Ireland 1–88.

Bush, R.A. and K.M. Hall. 2006. Process for the production of ethanol from algae. U.S. Patent 7,135,308.

Cantrell, K.B., T. Ducey, K.S. Ro and P.G. Hunt. 2008. Livestock waste-to-bioenergy generation opportunities. Bioresour. Technol. 99: 7941–7953.

Carlsson, A.S., J.V. Beilen, R. Moller and D. Clayton. 2007. Macro and Micro-algae: Utility for Industrial Applications, University of York, UK.

Cassman, K.G. and A.J. Liska. 2007. Food and fuel for all: realistic or foolish? Biofuel Bioprod. Bior. 1: 18–23.

Champagne, P. 2007. Feasibility of producing bio-ethanol from waste residues: a Canadian perspective: feasibility of producing bio-ethanol from waste residues in Canada. Resour. Conserv. Recycl. 50: 211–230.

Chen, C.Y., K.L. Yeh, R. Aisyah, D.J. Lee and J.S. Chang. 2011. Cultivation, photobioreactor design and harvesting of microalgae for biodiesel production: a critical review. Bioresour. Technol. 102: 71–81.

Chen, F., Y. Zhang and S. Guo. 1996. Growth and phycocyanin formation of *Spirulina platensis* in photoheterotrophic culture. Biotechnol. Lett. 18: 603–608.

Chen, J.W. and W.T. Wu. 2003. Regeneration of immobilized *Candida antarctica* lipase for transesterification. J. Biosci. Bioeng. 95: 466–469.

Chinnasamy, S., A. Bhatnagar, R.W. Hunt and K.C. Das. 2010. Microalgae cultivation in a wastewater dominated by carpet mill effluents for biofuel applications. Bioresour. Technol. 101: 3097–4105.

Chisti, Y. 2007. Biodiesel from microalgae. Biotechnol. Adv. 25: 294–306.

Chiu, S.Y., C.Y. Kao, M.T. Tsai, S.C. Ong, C.H. Chen and C.S. Lin. 2009. Lipid accumulation and CO_2 utilization of *Nanochloropsis oculata* in response to CO_2 aeration. Bioresour. Technol. 100: 833–838.

Choi, S.P., M.T. Nguyen and S.J. Sim. 2010. Enzymatic pretreatment of *Chlamydomonas reinhardtii* biomass for ethanol production. Bioresour. Technol. 101: 5330–5336.

Chojnacka, K. and A. Noworyta. 2004. Evaluation of *Spirulina* sp. growth in photoautotrophic, heterotrophic and mixotrophic cultures. Enzyme Microb. Technol. 34: 461–465.

Christenson, L. and R. Sims. 2011. Production and harvesting of microalgae for wastewater treatment, biofuels, and bioproducts. Biotechnol. Adv. 29: 686–702.

Chun, Y.C., X.Q. Zhao, H.W. Yen, S.H. Ho, C.L. Cheng, D.J. Lee, F.W. Bai and J.S. Chang. 2013. Microalgae based carbohydrates for biofuel production. Biochem. Eng. J. 78: 1–10.

Chynoweth, D.P. 2005. Renewable biomethane from land and ocean energy crops and organic wastes. Hortic. Sci. 40: 283–286.

Collet, P., A. Helias, L. Lardon, M. Ras, R.A. Goy and J.P. Steyer. 2011. Life-cycle assessment of microalgae culture coupled to biogas production. Bioresour. Technol. 102: 207–214.

Cooney, M., G. Young and N. Nagle. 2009. Extraction of bio-oils from microalgae. Sep. Purif. Rev. 38: 291–325.

Cravotto, G., L. Boffa, S. Mantegna and P. Perego. 2008. Improved extraction of vegetable oils under high-intensity ultrasound and/or microwaves. Ultrason. Sonochem. 15: 898–902.

Custódio L., T. Justo, L. Silvestre, A. Barradas, C.V. Duarte, H. Pereira, L. Barreira, A.P. Rauter, F. Albericio and J. Varela. 2012. Microalgae of different phyla display antioxidant, metal chelating and acetylcholinesterase inhibitory activities. Food Chem. 131: 134–140.

Daroch, M. and M. Mos. 2011. Role of biotechnology in developing the bioenergy sector in Poland. BioTechnologia 92: 23–32.

Daroch, M., S. Geng and G. Wang. 2013. Recent advances in liquid biofuel production from algal feedstocks. Appl. Energ. 102: 1371–1381.

Dean, A.P., D.C. Sigee, B. Estrada and J.K. Pittman. 2010. Using FTIR spectroscopy for rapid determination of lipid accumulation in response to nitrogen limitation in freshwater microalgae. Bioresour. Technol. 101: 4499–4507.

Demirbas, A. 2006. Oily products from mosses and algae via pyrolysis. Energ. Source Part A-Recovery Utilization and Environmental Effects 28: 933–940.

Demirbas, A. 2008. Biofuels sources, biofuel policy, biofuel economy and global biofuel projections. Energy Convers. Manag. 49: 2106–2116.

Demirbas, A. 2009. Biodiesel from waste cooking oil via base catalytic and supercritical methanol transesterification. Energy Convers. Manag. 50: 923–927.

Demirbas, A. 2010. Social, economic, environmental and policy aspects of biofuels. Energy Educ. Sci. Tech. 2: 75–109.

Deng, Z., Q. Hu, F. Lu, G. Liu and Z. Hu. 2008. Colony development and physiological characterization of the edible blue-green alga, *Nostoc sphaeroides* (Nostocaceae, Cyanophyta). Prog. Nat. Sci. 18: 475–484.

Devi, B.L.A.P., K.N. Gangadhar, P.S.S. Prasad, B. Jagannadh and R.B.N. Prasad. 2009. A glycerol-based carbon catalyst for the preparation of biodiesel. Chem. Sustain. Chem. 2: 617–620.

Devi, B.L.A.P., K.V. Lakshmi, K.N. Gangadhar, R.B.N. Prasad, P.S.S. Prasad, B. Jagannadh, P.P. Kundu, G. Kumari and C. Narayana. 2017. Novel heterogeneous SO_3Na-carbon transesterification catalyst for the production of biodiesel. ChemistrySelect 2: 1925–1931.

Dhargalkar, V.K. and X.N. Verlecar. 2009. Southern Ocean seaweeds: a resource for exploration in food and drugs. Aquacult. 287: 229–242.

Dismukes, G.C., D. Carrieri, N. Bennette, G.M. Ananyev and M.C. Posewitz. 2008. Aquatic phototrophs: efficient alternatives to land-based crops for biofuels. Curr. Opin. Biotechnol. 19: 235–240.

Divakaran, R. and V.N.S. Pillai. 2002. Flocculation of algae using chitosan. J. Appl. Phycol. 14: 419–422.

Dring, J.M. 1991. The Biology of Marine Plants, Cambridge University Press, USA: 46.

Efremenko, E.N., A.B. Nikolskaya, I.V. Lyagin, O.V. Senko, T.A. Makhlis, N.A. Stepanov, O.V. Maslova, F. Mamedova and S.D. Varfolomeev. 2012. Production of biofuels from pretreated microalgae biomass by anaerobic fermentation with immobilized *Clostridium acetobutylicum* cells. Bioresour. Technol. 114: 342–348.

Emma, H.I., B. Colman, G.S. Espie and L.M. Lubian. 2000. Active transport of CO_2 by three species of marine microalgae. J. Phycol. 36: 314–320.

Emma, S. and S. Rosalam. 2012. Conversion of microalgae to biofuel. Renew. Sust. Energ. Rev. 16: 4316–4342.

Encinar, J.M., J.F. Gonzalez, J.J. Rodriguez and A. Tejedor. 2002. Biodiesel fuels from vegetable oils: transesterification of *Cynara cardunculus* L. oils with ethanol. Energy Fuels 16: 443–450.

Erica, M.T. 1996. SalonOvations' day spa techniques. Cengage Learning. 137.

Fangrui, M. and M.A. Hanna. 1999. Biodiesel production: a review. Bioresour. Technol. 70: 1–15.

Fargione, J., J. Hill, D. Tilman, S. Polasky and P. Hawthorne. 2008. Land clearing and the biofuel carbon debt. Science 319: 1235–1238.

Filip, V., V. Zajic and J. Smidrkal. 1992. Methanolysis of rapeseed oil triglyceride. Revue Francaise des Corps Gras 39: 91–94.

Fontes, A.G., V.M. Angeles, J. Moreno, M.G. Guerrero and M. Losada. 1987. Factors affecting the production of biomass by a nitrogen-fixing blue-green alga in outdoor culture. Biomass 13: 33–43.

Freedman, B., E.H. Pryde and T.L. Mounts. 1984. Variables affecting the yields of fatty esters from transesterified vegetable oils. J. Am. Oil Chem. Soc. 61: 1638–1643.

Fukuda, H., A. Kondo and H. Noda. 2001. Biodiesel fuel production by transesterification of oils. J. Biosci. Bioeng. 92: 405–416.

Fuls, J., C.S. Hawkins and F.J.C. Hugo. 1984. Tractor engine performance on sunflower oil fuel. J. Agr. Eng. Res. 30: 29–35.

Gangadhar, K.N., H. Pereira, H.P. Diogo, R.M.B. dos Santos, B.L.A.P. Devi, R.B.N. Prasad, L. Custodio, F.X. Malcata, J. Varela and L. Barreira. 2016a. Assessment and comparison of the properties of biodiesel synthesized from three different microalgae biomass. J. Appl. Phycol. 28: 1571–1578.

Gangadhar, K.N., H. Pereira, M.J. Rodrigues, L. Custódio, L. Barreira, F.X. Malcata and J. Varela. 2016b. Microalgae-based unsaponifiable matter as source of natural antioxidants and metal chelators to enhance the value of wet *Tetraselmis chuii* biomass. Open Chem. 14: 299–307.

Gao, K., Yu, H and M.T. Brown. 2007. Solar PAR and UV radiation affects the physiology and morphology of the cyanobacterium *Anabaena sp.* PCC 7120. J. Photochem. Photobiol. B 14: 117–124.

Geciova, J., D. Bury and P. Jelen. 2002. Methods for disruption of microbial cells for potential use in the dairy industry—a review. Int. Dairy J. 12: 541–553.

Georgogianni, K.G., M.G. Kontominas, P.J. Pomonis, D. Avlonitis and V. Gergis. 2008. Conventional and *in situ* transesterification of sunflower seed oil for the production of biodiesel. Fuel Process. Technol. 89: 503–509.

Gerpen, J.V. 2005. Biodiesel processing and production. Fuel Process. Technol. 86: 1097–1107.

Ghirardi, M.L., L. Zhang, J.W. Lee, T. Flynn, M. Seibert, E. Greenbaum and A. Melis. 2000. Microalgae: a green source of renewable H_2. Trends Biotechnol. 18: 506–511.

Ginzberg, A., M. Cohen, U.A. Sod-Moriah, S. Shany, A. Rosenshtrauch and S. Arad. 2000. Chickens fed with biomass of the red microalga *Porphyridium* sp. have reduced blood cholesterol level and modified fatty acid composition in egg yolk. J. Appl. Phycol. 12: 325–330.

Giorgos, M. and G. Dimitris. 2011. Cultivation of filamentous cyanobacteria (blue-green algae) in agro-industrial wastes and wastewaters. a review. Appl. Energ. 88: 3389–3401.

Graham, L.E., J.E. Graham and L.W. Wilcox. 2008. Algae, Benjamin-Cummings Publishing, Menlo Park, CA, USA.

Greenwell, H.C., L.M.L. Laurens, R.J. Shields, R.W. Lovitt and K.J. Flynn. 2010. Placing microalgae on the biofuels priority list: a review of the technological challenges. J. R. Soc. Interface 7: 703–726.

Grierson, S., V. Strezov, G. Ellem, R. Mcgregor and J. Herbertson. 2009. Thermal characterisation of microalgae under slow pyrolysis conditions. J. Anal. Appl. Pyrolysis 85: 118–123.

Guedes, A.C., H.M. Amaro and F.X. Malcata. 2011. Microalgae as sources of high added-value compounds, a brief review of recent work. Biotechnol. Progr. 27: 597–613.

Guerin, M., M.E. Huntley and M. Olaizola. 2003. *Haematococcus* astaxanthin: applications for human health and nutrition. Trends Biotechnol. 21: 210–216.

Guschina, I.A. and J.L. Harwood. 2006. Lipids and lipid metabolism in eukaryotic algae. Prog. Lipid. Res. 45: 160–186.

Halim, R., Gladman, M.K. Danquah and P.A. Webley. 2011. Oil extraction from microalgae for biodiesel production. Bioresour. Technol. 102: 178–185.

Hamed, I. 2016. The evolution and versatility of microalgal biotechnology: a review. Compr. Rev. Food Sci. Food Saf. 15: 1104–1123.

Hamelinck, C.N., G. van Hooijdonk and A.P.C. Faaij. 2005. Ethanol from lignocellulosic biomass: techno-economic performance in short-, middle- and long-term. Biomass Bioenergy 28: 384–410.

Hanagata, N., T. Takeuchi, Y. Fukuju, D.J. Barnes and I. Karube. 1992. Tolerance of microalgae to high CO_2 and high temperature. Phytochem. 31: 3345–3348.

Harun, R., M. Singh, G.M. Forde and M.K. Danquah. 2010a. Bioprocess engineering of microalgae to produce a variety of consumer products. Renew. Sust. Energ. Rev. 14: 1037–1047.

Harun, R., M.K. Danquah and G.M. Forde. 2010b. Microalgal biomass as a fermentation feedstock for bioethanol production. J. Chem. Technol. Biotechnol. 85: 199–203.

Hattori, T. and S. Morita. 2010. Energy crops for sustainable bioethanol production; which, where and how? Plant Prod. Sci. 13: 221–234.

Heasman, M., J. Diemar, W.O. Connor, T. Sushames and L. Foulkes. 2000. Development of extended shelf-life microalgae concentrate diets harvested by centrifugation for bivalve mollusks a summary. Aquacult. Res. 31: 637–659.

Helwani, Z., M.R. Othman, N. Aziz, W.J.N. Fernando and J. Kim. 2009. Technologies for production of biodiesel focusing on green catalytic techniques: a review. Fuel Process. Technol. 90: 1502–1514.

Herrero, M., A. Cifuentes and E. Ibanez. 2006. Sub- and supercritical fluid extraction of functional ingredients from different natural sources: plants, food-by-products, algae and microalgae: a review. Food Chem. 98: 136–148.

Hirano, A., R. Ueda, S. Hirayama and Y. Ogushi. 1997. CO_2 fixation and ethanol production with microalgal photosynthesis and intracellular anaerobic fermentation. Energy 22: 137–142.

Holm, N.J.B., T. al Seadi and P.O. Popiel. 2009. The future of anaerobic digestion and biogas utilization. Bioresour. Technol. 100: 5478–5484.

Hon, N.K. 2006. A unique feature of hydrogen recovery in endogenous starch to alcohol fermentation of the marine microalga, *Chlamydomonas perigranulata*. Appl. Biochem. Biotechnol. 131: 808–828.

Hsieh, C.H.H. and W.T. Wu. 2009. Cultivation of microalgae for oil production with a cultivation strategy of urea limitation. Bioresour. Technol. 100: 3921–3926.

Hu, C., M. Li, J. Li, Q. Zhu and Z. Liu. 2008. Variation of lipid and fatty acid compositions of the marine microalga *Pavlova viridis* (Prymnesiophyceae) under laboratory and outdoor culture conditions. World J. Microbiol. Biotechnol. 24: 1209–1214.

Huang, G., F. Chen, D. Wei, X. Zhang and G. Chen. 2010. Biodiesel production by microalgal biotechnology. Appl. Energ. 87: 38–46.

Huber, G.W. and B.E. Dale. 2009. Biofuels: grassoline at the pump. Scientific American 301: 52–59.

Huerlimann, R., R. de Nys and K. Heimann. 2010. Growth, lipid content, productivity, and fatty acid composition of tropical microalgae for scale-up production. Biotechnol. Bioeng. 107: 245–257.

Hui, G., M. Daroch, L. Liu, G. Qiu, S. Geng and G. Wang. 2013. Biochemical features and bioethanol production of microalgae from coastal waters of Pearl River Delta. Bioresour. Technol. 127: 422–428.

Humphrey, A.M. 2004. Chlorophyll as a colour and functional ingredient. J. Food Sci. 69: 422–425.

IEA. 2010. Key world energy statistics. 2010. International Energy Agency.

Iillman, A.M., A.H. Scragg and S.W. Shales. 2000. Increase in *Chlorella* strains calorific values when grown in low nitrogen medium. Enzyme Microb. Technol. 2: 631–635.

Iso, M., B. Chen, M. Eguchi, T. Kudo and S. Shrestha. 2001. Production of biodiesel fuel from triglycerides and alcohol using immobilized lipase. J. Mol. Catal. B: Enzym. 16: 53–58.

Issariyakul, T. and A.K. Dalai. 2014. Biodiesel from vegetable oils. Renew. Sust. Energ. Rev. 31: 446–471.

Jacob, L.E., C.H.G. Scoparo, L.M.C.F. Lacerda and T.T. Franco. 2009. Effect of light cycles (night/day) on CO_2 fixation and biomass production by microalgae in photobioreactors. Chem. Eng. Process. 48: 306–310.

Janssen, M., J. Tramper, L.R. Mur and R.H. Wijffels. 2003. Enclosed outdoor photobioreactors: light regime, photosynthetic efficiency, scale-up, and future prospects. Biotechnol. Bioeng. 81: 193–210.

Jasvinder, S. and S. Gu. 2010. Commercialization potential of microalgae for biofuels production. Renew. Sust. Energ. Rev. 14: 2596–2610.

Joana, M.D., M.C.M. Alvim-Ferraz, M.F. Almeida, J.D.M. Diaz, M.S. Polo and J.R. Utrilla. 2012. Selection of heterogeneous catalysts for biodiesel production from animal fat. Fuel 94: 418–425.

Kalam, M.A. and H.H. Masjuki. 2002. Biodiesel from palm oil-an analysis of its properties and potential. Biomass Bioenergy 23: 471–479.

Khotimchenko, S.V and I.M. Yakovleva. 2005. Lipid composition of the red alga *Tichocarpuscrinitus* exposed to different levels of photon irradiance. Phytochem. 66: 73–79.

Knothe, G., R.O. Dunn and M.O. Bagby. 1996. Technical aspects of biodiesel standard. Inform 7: 827–829.

Knothe, G. and K.R. Steidley. 2005. Kinematic viscosity of biodiesel fuel components and related compounds. Influence of compound structure and comparison to petrodiesel fuel components. Fuel 84: 1059–1065.

Knothe, G. 2006. Analyzing biodiesel: standards and other methods. J. Am. Oil Chem. Soc. 83: 823–833.

Knuckey, R.M., M.R. Brown, R. Robert and D.M.F. Frampton. 2006. Production of microalgal concentrates by flocculation and their assessment as aquaculture feeds. Aquacult. Eng. 35: 300–313.

Kodama, M., H. Ikemoto and S. Miyachi. 1993. A new species of highly CO_2-tolerant, fast-growing marine microalgae suitable for high-density culture. J. Mar. Biotechnol. 1: 21–25.

Kong, Q.X., L. Li, B. Martinez, P. Chen and R. Ruan. 2010. Culture of microalgae *Chlamydomonas reinhardtii* in wastewater for biomass feedstock production. Appl. Biochem. Biotechnol. 160: 9–18.

Kose, O., M. Tuter and A.H. Aksoy. 2002. Immobilized *Candida antarctica* lipase-mediated alcoholysis of cotton seed oil in a solvent-free medium. Bioresour. Technol. 83: 125–129.

Kuda, T., M. Tsunekawa, H. Goto and Y. Araki. 2005. Antioxidant properties of four edible algae harvested in the Noto Peninsula, Japan. J. Food Compost. Anal. 18: 625–633.

Kumar, A., S. Ergas, X. Yuan, A. Sahu, Q. Zhang, J. Dewulf, F.X. Malcata and H. van Langenhov. 2010. Enhanced CO_2 fixation and biofuel production via microalgae: recent developments and future directions. Trends Biotechnol. 28: 371–380.

Kumar, M., M.K. Sharma and A. Kumar. 2005. *Spirulina fusiformis*: a food supplement against mercury induced hepatic toxicity. J. Health Sci. 51: 424–430.

Lam, M.K. and K.T. Lee. 2012. Microalgae biofuels: a critical review of issues, problems and the way forward. Biotechnol. Adv. 30: 673–690.

Lardon, L., A. Helias, B. Sialve, J.P. Steyer and O. Bernard. 2009. Life-cycle assessment of biodiesel production from microalgae. Environ. Sci. Technol. 43: 6475–6481.

Lee, I., L.A. Johnson and E.G. Hammond. 1995. Use of branched chain esters to reduce the crystallization temperature of biodiesel. J. Am. Oil Chem. Soc. 72: 1155–1160.

Leung, D.Y.C., X. Wu and M.K.H. Leung. 2010. A review on biodiesel production using catalyzed transesterification. Appl. Energ. 87: 1083–1095.

Li, L., W. Du, D. Liu, L. Wang and Z. Li. 2006. Lipase-catalyzed transesterification of rapeseed oils for biodiesel production with a novel organic solvent as the reaction medium. J. Mol. Catal. B: Enzym. 43: 58–62.

Li, Y.Q., M. Horsman, B. Wang, N. Wu and C.Q. Lan. 2008. Effects of nitrogen sources on cell growth and lipid accumulation of green alga *Neochloris oleoabundans*. Appl. Microbiol. Biotechnol. 81: 629–636.

Liu, J.C., M.Y. Chen and Y.H. Ju. 1999. Separation of algal cells from water by column flotation. Sep. Sci. Technol. 34: 2259–2272.

Lodi, A., L. Binaghi, D.D. Faveri, J.C.M. Carvalho and A. Converti. 2005. Fed batch mixotrophic cultivation of *Arthrospira* (*Spirulina*) *platensis* (Cyanophycea) with carbon source pulse feeding. Ann. Microbiol. 55: 181–185.

Lotero, E., Y. Liu, D.E. Lopez, K. Suwannakaran, D.A. Bruce and J.G. Goodwin. 2005. Synthesis of biodiesel via acid catalysis. Ind. Eng. Chem. Res. 44: 5353–5363.

Luis, F.R. and R.R. Tan. 2011. Net energy analysis of the production of biodiesel and biogas from the microalgae *Haematococcus pluvialis* and *Nannochloropsis*. Appl. Energ. 88: 3507–3514.

Luiten, E.E.M., I. Akkerman, A. Koulman, P. Kamermans, H. Reith, M.J. Barbosa, D. Sipkema and R.H. Wijffels. 2003. Realizing the promises of marine biotechnology. Biomol. Eng. 20: 429–439.

Lynd, L.R. 1996. Overview and evaluation of fuel ethanol from cellulosic biomass: technology, economics, the environment and policy. Annu. Rev. Energy Environ. 21: 100–165.

Ma, F. and M.A. Hanna. 1999. Biodiesel production: a review. Bioresour. Technol. 70: 1–15.

Maeda, K., M. Owada, N. Kimura, K. Omata and I. Karube. 1995. CO_2 fixation from flue gas on coal fired thermal power plant by microalgae. Energy Convers. Manag. 36: 717–720.

Mata, T.M., A.A. Martins and N.S. Caetano. 2010. Microalgae for biodiesel production and other applications: a review. Renew. Sust. Energ. Rev. 14: 217–232.

Matos, J., C. Cardoso, N.M. Bandarra and C. Afonso. 2017. Microalgae as healthy ingredients for functional food: a review. Food Funct. 8: 2672–2685.

Matsumoto, M., H. Yokouchi, N. Suzuki, H. Ohata and T. Matsunaga. 2003. Saccharification of marine microalgae using marine bacteria for ethanol production. Appl. Biochem. Biotechnol. 105: 247–254.

Matsunaga, T., H. Takeyama, H. Miyashita and H. Yokouchi. 2005. Marine microalgae. Adv. Biochem. Eng./Biotechnol. 96: 165–188.

Medina, A.R., E.M. Grima, A.G. Gimenez and M.J.I. Gonzalez. 1998. Downstream processing of algal polyunsaturated fatty acids. Biotechnol. Adv. 3: 517–580.

Meher, L.C., D.V. Sagar and S.N. Naik. 2006. Technical aspects of biodiesel production by transesterification: a review. Renew. Sust. Energ. Rev. 10: 248–268.

Mendes, R.L., A. Reis, H.L. Fernandes, J.M. Novais and A.F. Palavra. 1999. Supercritical CO_2 extraction of lipids from a GLA-rich *Arthrospira* (*Spirulina*) *maxima* biomass. *In*: Proc. 5th Int. Conference on Supercritical Fluids and their Applications. Verona, Italy, 13–16 June: 209–216.

Mendes, R.L., A.D. Reis, A.P. Pereira. M.T. Cardoso, A.F. Palavra and J.P. Coelho. 2005. Supercritical CO_2 extraction of γ-linolenic acid (GLA) from the *Cyanobacterium Arthrospira* (*Spirulina*) *maxima*: experiments and modelling. Chem. Eng. J. 105: 147–152.

Mendes, R.L., A.D. Reis and A.F. Palavra. 2006. Supercritical CO_2 extraction of γ-linolenic acid and other lipids from *Arthrospira* (*Spirulina*) *maxima*: comparison with organic solvent extraction. Food Chem. 99: 57–63.

Mendiola, J.A., L. Jaime, S. Santoyo, G. Reglero, A. Cifuentes, E. Ibanez and F.J. Senoeans. 2007. Screening of functional compounds in supercritical fluid extracts from *Spirulina platensis*. Food Chem. 102: 1357–1367.

Mercer, P. and R.E. Armenta. 2011. Developments in oil extraction from microalgae. Eur. J. Lipid Sci. Technol. 113: 539–547.

Metting, F.B. 1996. Biodiversity and application of microalgae. J. Ind. Microbiol. 17: 477–489.

Miao, X. and Q. Wu. 2004. High yield bio-oil production from fast pyrolysis by metabolic controlling of *Chlorella protothecoides*. J. Biotechnol. 110: 85–93.

Miao, X., Q. Wu and C. Yang. 2004. Fast pyrolysis of microalgae to produce renewable fuels. J. Anal. Appl. Pyrolysis 71: 855–863.

Miao, X. and Q. Wu. 2006. Biodiesel production from heterotrophic microalgal oil. Bioresour. Technol. 97: 841–846.

Miyairi, S. 1995. CO_2 assimilation in a thermophilic cyanobacterium. Energy Convers. Manag. 36: 763–766.

Mohn, F.H. 1980. Experiences and strategies in the recovery of biomass in mass culture of microalgae. pp. 547–571. *In*: G. Shelef and C.J. Soeder (eds.). Algal Biomass. Elsevier Amsterdam.

Molina, G.E., E.H. Belarbi, F.F.G. Acien, M.A. Robles and Y. Chisti. 2003. Recovery of microalgal biomass and metabolites: process options and economics. Biotechnol. Adv. 20: 491–515.

Moreno, G.I. 2008. Microalgae immobilization: current techniques and uses. Bioresour. Technol. 99: 3949–3964.

Mudge, S.M. and G. Pereira. 1999. Stimulating the biodegradation of crude oil with biodiesel preliminary results. Spill Science and Technology Bulletin 5: 353–355.

Mulbry, W., S. Kondrad, C. Pizarro and K.E. Westhead. 2008. Treatment of dairy manure effluent using freshwater algae: algal productivity and recovery of manure nutrients using pilot-scale algal turf scrubbers. Bioresour. Technol. 99: 8137–8142.

Muller, F.A., G.R. Le, A. Herve and P. Durand. 1998. Comparison of artificial light photobioreactors and other production systems using *Porphyridium cruentum*. J. Appl. Phycol. 10: 83–90.

Munoz, R. and B. Guieysse. 2006. Algal-bacterial processes for the treatment of hazardous contaminants: a review. Water Res. 40: 2799–2815.

Murata, N. 1989. Low-temperature effects on cyanobacterial membranes. J. Bioenerg. Biomembr. 21: 61–75.

Murayama, T. 1994. Evaluating vegetable oils as a diesel fuel. Inform. 5: 1138–1145.

Mussgnug, J.H., V. Klassen, A. Schluter and O. Kruse. 2010. Microalgae as substrates for fermentative biogas production in a combined biorefinery concept. J. Biotechnol. 150: 51–56.

Mutanda, T., D. Ramesh, S. Karthikeyan, S. Kumari, A. Anandraj and F. Bux. 2011. Bioprospecting for hyper-lipid producing microalgal strains for sustainable biofuel production. Bioresour. Technol. 102: 57–70.

Nakamura, D.N. 2006. Journally speaking: the mass appeal of biomass. Oil Gas J. 104: 15.

Nakanishi, K. 2001. Chlorophyll rich and salt resistant *Chlorella*. European Patent 1,142,985.

Narasimharao, K., A. Lee and K. Wilson. 2007. Catalysis in production of biodiesel: a review. J. Biobased. Mater. Bio. 1: 19–30.

Nelson, L.A., T.A. Foglia and W.N. Marmer. 1996. Lipase-catalyzed production of biodiesel. J. Am. Oil Chem. Soc. 73: 1191–1195.

Ngaosuwan, K., E. Lotero, K. Suwannakarn, J.G. Goodwin and P. Praserthdam. 2009. Hydrolysis of triglycerides using solid acid catalysts. Ind. Eng. Chem. Res. 48: 4757–4767.

Nurdogan, Y. and W.J. Oswald. 1996. Tube settling rate of high-rate pond algae. Water Sci. Technol. 33: 229–234.

Ogbonda, K.H., R.E. Aminigo and G.O. Abu. 2007. Influence of temperature and pH on biomass production and protein biosynthesis in a putative *Spirulina* sp. Bioresour. Technol. 98: 2207–2211.

Ogbonna, J.C., H. Yoshizawa and H. Tanaka. 2000. Treatment of high strength organic wastewater by a mixed culture of photosynthetic microorganisms. J. Appl. Phycol. 12: 277–284.

Olguin, E.J. 2003. Phycoremediation: key issues for cost-effective nutrient removal processes. Biotechnol. Adv. 22: 81–91.

Oliveira, M.A.C.L., M.P.C. Monteiro, P.G. Robbs and S.G.F. Leite. 1999. Growth and chemical composition of *Spirulina maxima* and *Spirulina platensis* biomass at different temperatures. Aquacult. Int. 7: 261–275.

Oswald, W.J. 1988. Large-scale algal culture systems (engineering aspects). pp. 357–394. *In*: M.A. Borowitzka and L.J. Borowitzka (eds.). Micro-Algal Biotechnology. Cambridge University Press: New York:

Otsuka, K. and A. Yoshino. 2004. A fundamental study on anaerobic digestion of sea lettuce Ocean'04-MTS/IEEE Techno-Ocean'04: bridges across the oceans. Conference Proc. 1770–1773.

Pariente, J.P., I. Diaz, F. Mohino and E. Sastre. 2003. Selective synthesis of fatty monoglycerides by using functionalised mesoporous catalysts. Appl. Catal., A: General 254: 173–188.

Park, J.Y., D.K. Kim and J.S. Lee. 2009. Esterification of free fatty acids using water tolerable Amberlyst as a heterogeneous catalyst. Bioresour. Technol. 101: S62–S65.

Parmar, A., N.K. Singh, A. Pandey, E. Gnansounou and D. Madamwar. 2011. Cyanobacteria and microalgae: a positive prospect for biofuels. Bioresour. Technol. 102: 10163–10172.

Pate, R., G. Klise and B. Wu. 2011. Resource demand implications for US algae biofuels production scale-up. Appl. Energ. 88: 3377–3388.

Patil, V., T. Kallqvist, E. Olsen, G. Vogt and H.R. Gislerod. 2007. Fatty acid composition of 12 microalgae for possible use in aquaculture feed. Aquacult. Int. 15: 1–9.

Peng, W., Q. Wu, P. Tu and N. Zhao. 2001. Pyrolytic characteristics of microalgae as renewable energy source determined by thermogravimetric analysis. Bioresour. Technol. 77: 1–7.

Pereira, H., L. Barreira, A. Mozes, C. Florindo, C. Polo, C. Vizetto-Duarte, L. Custodio and J. Varela. 2011. Microplate-based high throughput screening procedure for the isolation of lipid-rich marine microalgae. Biotechnol. Biofuels 4: 61–72.

Pereira, H., L. Barreira, F. Figueiredo, L. Custódio, C. Vizetto-Duarte, C. Polo, E. Rešek, A. Engelen and J. Varela. 2012. Polyunsaturated fatty acids of marine macroalgae: potential for nutritional and pharmaceutical applications. Mar. Drugs 10: 1920–1935.

Pereira, H., K.N. Gangadhar, P.S.C. Schulze, T. Santos, C.B. de Sousa, L.M. Schueler, L. Custódio, F.X. Malcata, L. Gouveia, J.C.S. Varela and L. Barreira. 2016. Isolation of a euryhaline microalgal strain, *Tetraselmis* sp. CTP4, as a robust feedstock for biodiesel production. Scientific Reports 6: 35663.

Perretti, G., E. Bravi, L. Montanari and P. Fantozzi. 2003. Extraction of PUFAs rich oil from algae with supercritical carbon dioxide. *In*: Proc. 6th Int. Symposium on Supercritical Fluids, Versailles, France 29–34.

Perrut, M. 2001. Supercritical fluid application: industrial developments and economical issues. pp. 1–8. *In*: Proc. 5th Int. Proc. Conference on Supercritical Fluids and their Applications. Maiori, Italy.

Pfromm, P.H., V.A. Boadu, R. Nelson, P. Vadlani and R. Madl. 2010. Bio-butanol vs. bio-ethanol: a technical and economic assessment for corn and switchgrass fermented by yeast or *Clostridium acetobutylicum*. Biomass Bioenergy 34: 515–524.

Pinto, A.C., L.L.N. Guarieiro, M.J.C. Rezende, N.M. Ribeiro, E.A. Torres, W.A. Lopes, P.A. de P Pereira and J.B. de Andrade. 2005. Biodiesel: an overview. J. Brazil Chem. Soc. 16: 1313–1330.

Pizarro, C., W. Mulbry, D. Blersch and P. Kangas. 2006. An economic assessment of algal turf scrubber technology for treatment of dairy manure effluent. Ecol. Eng. 26: 321–327.

Popoola, T.O.S. and O.D. Yangomodou. 2006. Extraction, properties and utilization potentials of cassava seed oil. Biotechnol. 5: 38–41.

Pulz, O. and K. Scheinbenbogan. 1998. Photobioreactors: design and performance with respect to light energy input. Adv. Biochem. Eng./Biotechnol. 59: 123–152.

Pulz, O. 2001. Photobioreactors: production systems for phototrophic microorganisms. Appl. Microbiol. Biotechnol. 57: 287–293.

Puotinen, C.J. 1999. Herbs for detoxification. McGraw-Hill, Nevada, USA: 25.

Qin, J. 2005. Bio-hydrocarbons from algae-impacts of temperature, light and salinity on algae growth. Rural Industries Research and Development Corporation. Barton, Australia.

Qiu, R., S. Gao, P.A. Lopez and K.L. Ogden. 2017. Effects of pH on cell growth, lipid production and CO_2 addition of microalgae *Chlorella sorokiniana*. Algal Res. 28: 192–199.

Rafiqul, I.M., A. Hassan, G. Sulebele, C. Orosco and P. Roustaian. 2003. Influence of temperature on growth and biochemical composition of *Spirulina platensis, S. Fusiformis*. Iranian Int J. Sci. 4: 97–106.

Ramos, M.J., A. Casas, L. Rodriguez, R. Romero and A. Perez. 2008. Transesterification of sunflower oil over zeolites using different metal loading: a case of leaching and agglomeration studies. Appl. Catal., A: 346: 79–85.

Ras, M., J.P. Steyer and O. Bernard. 2013. Temperature effect on microalgae: a crucial factor for outdoor production. Rev. Environ. Sci. Biotechnol. 12: 153–164.

Rawat, I., R.R. Kumar, T. Mutanda and F. Bux. 2011. Dual role of microalgae: phycoremediation of domestic wastewater and biomass production for sustainable biofuels production. Appl. Energ. 88: 3411–3424.

Rodjaroen, S., N. Juntawong, A. Mahakhant and K. Miyamoto. 2007. High biomass production and starch accumulation in native green algal strains and cyanobacterial strains of Thailand Kasetsart. J. Nat. Sci. 41: 570–575.

Rodolfi, L., G.C. Zittelli, N. Bassi, G. Padovani, N. Biondi, G. Bonini and M.R. Tredici. 2008. Microalgae for oil: strain selection, induction of lipid synthesis and outdoor mass cultivation in a low-cost photobioreactor. Biotechnol. Bioeng 102: 100–112.

Rojan, P.J, G.S. Anisha, K.M. Nampoothiri and A. Pandey. 2011. Micro and macroalgal biomass: a renewable source for bioethanol. Bioresour. Technol. 102: 186–193.

Rojas-Downing, M.M., A.P. Nejadhashemi, T. Harrigan and S.A. Woznicki. 2017. Climate change and livestock: Impacts, adaptation, and mitigation. Climate Risk Management 16: 145–163.

Ronald, H., M.K. Danquah and P.A. Webley. 2012. Extraction of oil from microalgae for biodiesel production: a review. Biotechnol. Adv. 30: 709–732.

Ross, A.B., P. Biller, M.L. Kubacki, H. Li, A.L. Langton and J.M. Jones. 2010. Hydrothermal processing of microalgae using alkali and organic acids. Fuel 89: 2234–2243.

Rubio, F.C., F.G. Camacho, J.M.F. Sevilla, Y. Chisti and E.M. Grima. 2003. A mechanistic model of photosynthesis in microalgae. Biotechnol. Bioeng. 81: 459–473.

Rui, C., Z. Yue, L. Deitz, Y. Liu, W. Mulbry and W. Liao. 2012. Use of an algal hydrolysate to improve enzymatic hydrolysis of lignocelluloses. Bioresour. Technol. 108: 149–154.

Sahena, F., I.S.M. Zaidul, S. Jinap, A.A. Karim, K.A. Abbas, N.A.N. Norulaini and A.K.M. Omar. 2009. Application of supercritical CO_2 in lipid extraction-a review. J. Food Eng. 95: 240–253.

Salim, S., R. Bosma, M.H. Vermuë and R.H. Wijffels. 2011. Harvesting of microalgae by bio-flocculation. J. Appl. Phycol. 23: 849–855.

Samukawa, T., M. Kaieda, T. Matsumoto, K. Ban, A. Kondo, Y. Shimada, H. Noda and H. Fukuda. 2000. Pretreatment of immobilized *Candida antarctica* lipase for biodiesel fuel production from plant oil. J. Biosci. Bioeng. 90: 180–183.

Sanchez, J.L.G., M. Berenguel, F. Rodriguez, J.M.F. Sevilla, C.B. Alias and F.G.A. Fernandez. 2003. Minimization of carbon losses in pilot-scale outdoor photobioreactors by model-based predictive control. Biotechnol. Bioeng. 84: 533–543.

Santos, R., T. Lu, M.B. King and J.A. Empis. 1997. Extraction of valuable components from micro-alga *Spirulina platensis*. pp. 217–224. *In*: Proc. 4th Int. Conference on Supercritical Fluids and their Applications, Capri, Italy.

Schenk, P.M., S.R.T. Hall, E. Stephens, U.C. Marx, J.H. Mussgnug, C. Posten, O. Kruse and B. Hankamer. 2008. Second generation biofuels: high-efficiency microalgae for biodiesel production. BioEnergy Res. 1: 20–43.

Schmid, S.U. 2009. Algae biorefinery-concept. National German Workshop on Biorefineries, 15 September, Worms, Germany.

Schuchardt, U., R. Sercheli and R.M. Vargas. 1998. Transesterification of vegetable oils: a review. J. Brazil Chem. Soc. 9: 199–210.

Schüler, L.M., P.S.C. Schulze, H. Pereira, L. Barreira, R. León and J. Varela. 2017. Trends and strategies to enhance triacylglycerols and high-value compounds in microalgae. Algal Res. 25: 263–273.

Schulze, P.S.C., L.A. Barreira, H.G.C. Pereira, J.A. Perales and J.C.S. Varela. 2014. Light emitting diodes (LEDs) applied to microalgal production. Trends Biotechnol. 32: 422–430.

Schulze, P.S.C., F.M.C. Carolina, H. Pereira, K.N. Gangadhar, L.M. Schüler, T.F. Santos, J.C.S. Varela and L. Barreira. 2017a. Urban wastewater treatment by *Tetraselmis* sp. CTP4 (Chlorophyta). Bioresour. Technol. 223: 175–183.

Schulze, P.S.C., R. Guerra, H. Pereira, L.M. Schüler and J.C.S. Varela. 2017b. Flashing LEDs for microalgal production. Trends Biotechnol. 35: 1088–1101.

Schwab, A.W., M.O. Baghy and B. Freedman. 1987. Preparation and properties of diesel fuels from vegetable oils. Fuel 66: 1372–1378.

Searchinger, T., R. Heimlich, R.A. Houghton, F. Dong, A. Elobeid, J. Fabiosa, S. Tokgoz, D. Hayes and T.H. Yu. 2008. Use of U.S. croplands for biofuels increases greenhouse gases through emissions from land-use change. Science 319: 1238–1240.

Semerjian, L. and G.M. Ayoub. 2003. High-pH-magnesium coagulation-flocculation in wastewater treatment. Adv. Environ. Res. 7: 389–403.

Shah, S., A. Sharma and M.N. Gupta. 2004. Extraction of oil from *Jatropha curcas* L. seed kernels by enzyme assisted three phase partitioning. Ind. Crops Prod. 20: 275–279.

Shakeel, A.K., Rashmi, M.Z. Hussain, S. Prasad and U.C. Banerjee. 2009. Prospects of biodiesel production from microalgae in India. Renew. Sust. Energ. Rev. 13: 2361–2372.

Sharma, K.K., H. Schuhmann and P.M. Schenk. 2012. High lipid induction in microalgae for biodiesel production. Energy 5: 1532–1553.

Sharma, Y.C., B. Singh and S.N. Upadhyay. 2008. Advancements in development and characterization of biodiesel, a review. Fuel 87: 2355–2373.

Shen, Y., Z. Pei, W. Yuan and E. Mao. 2009. Effect of nitrogen and extraction method on algae lipid yield. Int. J. Agric. Biol. Eng. 2: 51–57.

Shimada, Y., Y. Watanabe, A. Sugihara and Y. Tominaga. 2002. Enzymatic alcoholysis for biodiesel fuel production and application of the reaction to oil processing. J. Mol. Catal. B: Enzym. 17: 133–142.

Shu, Q., B.L. Yang, H. Yuan, S. Qing and G.L. Zhu. 2007. Synthesis of biodiesel from soybean oil and methanol catalyzed by zeolite β-modified with La^{3+}. Catal. Commun. 8: 2159–2165.

Shu, Q., Z. Nawaz, J. Gao, Y. Liao, Q. Zhang, D. Wang and J. Wang. 2010. Synthesis of biodiesel from a model waste oil feedstock using a carbon-based solid acid catalyst: reaction and separation. Bioresour. Technol. 101: 5374–5384.

Shuping, Z., W. Yulong, Y. Mingde, L. Chun and T. Junmao. 2010. Pyrolysis characteristics and kinetics of the marine microalgae *Dunaliella tertiolecta* using thermogravimetric analyzer. Bioresour. Technol. 101: 359–365.

Singh, A., P.S. Nigam and J.D. Murphy. 2011. Mechanism and challenges in commercialisation of algal biofuels. Bioresour. Technol. 102: 26–34.

Singh, S., B.N. Kate and U.C. Banerjee. 2005. Bioactive compounds from cyanobacteria and microalgae: an overview. Crit. Rev. Biotechnol. 25: 73–95.

Soto, C., R. Chamy and M.E. Zuniga. 2007. Enzymatic hydrolysis and pressing conditions effect on borage oil extraction by cold pressing. Food Chem. 102: 834–840.

Speidel, H.K., R.L. Lightner and I. Ahmed. 2000. Biodegradability of new engineered fuels compared to conventional petroleum fuels and alternative fuels in current use. Appl. Biochem. Biotechnol. 84: 879–897.

Spolaore, P., C.J. Cassan, E. Duran and A. Isambert. 2006. Commercial applications of microalgae. J. Biosci. Bioeng. 101: 87–96.

Srilatha, K., R. Sree, B.L.A.P. Devi, P.S.S. Prasad, R.B.N. Prasad and N. Lingaiah. 2012. Preparation of biodiesel from rice bran fatty acids catalyzed by heterogeneous cesium-exchanged 12-tungstophosphoric acids. Bioresour. Technol. 116: 53–57.

Stepan, D.J., R.E. Shockey, T.A. Moe and R. Dorn. 2002. Carbon dioxide sequestering using microalgae systems. U.S. Department of Energy. Pittsburgh, PA, USA.

Stephenson, A.L., E. Kazamia, J.S. Dennis, C.J. Howe, S.A. Scott and A.G. Smith. 2010. Life-cycle assessment of potential algal biodiesel production in the United Kingdom: a comparison of raceways and air-lift tubular bioreactors. Energy Fuels 24: 4062–4077.

Suh, I.S. and C.G. Lee. 2003. Photobioreactor engineering: design and performance. Biotechnol. Bioprocess Eng. 8: 313–321.

Sukenik, A., W. Schroder, J. Lauer, G. Shelef and C.J. Soeder. 1985. Co-precipitation of microlagal biomass with calcium and phosphate ions. Water Res. 19: 127–129.

Sun, Y. and J. Cheng. 2002. Hydrolysis of lignocellulosic materials for ethanol production: a review. Bioresour. Technol. 83: 1–11.

Sushchik, N.N., G.S. Kalacheva, N.O. Zhila, M.I. Gladyshev and T.G. Volova. 2003. A temperature dependence of the intra- and extracellular fatty-acid composition of green algae and cyanobacterium. Russ. J. Plant Physiol. 50: 374–380.

Sydney, E.B., W. Sturm, J.C. de Carvalho, V.T. Soccol, C. Larroche, A. Pandey and C.R. Soccol. 2010. Potential carbon dioxide fixation by industrially important microalgae. Bioresour. Technol. 101: 5892–5896.

Talukder, M.M.R., J.C. Wu, S.K. Lau, L.C. Cui, G. Shimin and A. Lim. 2009. Comparison of Novozyme 435 and Amberlyst 15 as heterogeneous catalyst for production of biodiesel from palm fatty acid distillate. Energy Fuels 23: 1–4.

Thajuddin, N. and G. Subramanian. 2005. Cyanobacterial biodiversity and potential applications in biotechnology. Curr. Sci. 89: 47–57.

Toda, M., A. Takagaki, M. Okamura, J.N. Kondo, S. Hayashi, K. Domen and M. Hara. 2005. Biodiesel made with sugar catalyst. Nature 438: 178.

Torzillo, G., B. Pushparaj, J. Masojidek and A. Vonshak. 2003. Biological constraints in algal biotechnology. Biotechnol. Bioprocess Eng. 8: 338–348.

Tredici, M.R., P. Carlozzi, G.C. Zittelli and R. Materassi. 1991. A vertical alveolar panel (VAP) for outdoor mass cultivation of microalgae and cyanobacteria. Bioresour. Technol. 38: 153–159.

Ueda, R., S. Hirayama, K. Sugata and H. Nakayama. 1996. Process for the production of ethanol from microalgae. U.S. Patent 5,578,472.

Ugwu, C.U., H. Aoyagi and H. Uchiyama. 2008. Photobioreactors for mass cultivation of algae. Bioresour. Technol. 99: 4021–4028.

van de Vyver, S., L. Peng, J. Geboers, H. Schepers, F. de Clippel, C.J. Gommes, B. Goderis, P.A. Jacobs and B.F. Sels. 2010. Sulfonated silica/carbon nanocomposites as novel catalysts for hydrolysis of cellulose to glucose. Green Chem. 12: 1560–1563.

Varfolomeev, S.D., E.N. Efremenko and L.P. Krylova. 2010. Biofuels. Russ. Chem. Rev. 79: 544–564.

Varfolomeev, S.D. and L.A. Wasserman. 2011. Microalgae as source of biofuel, food, fodder, and medicines. Appl. Biochem. Microbiol. 47: 789–807.

Vasudevan, P. and M. Briggs. 2008. Biodiesel production-current state of the art and challenges. J. Ind. Microbiol. Biotechnol. 35: 421–430.

Vergara, F.A., G. Vargas, N. Alarcon and A. Velasco. 2008. Evaluation of marine algae as a source of biogas in a two-stage anaerobic reactor system. Biomass Bioenergy 32: 338–344.

Vieira, C.J.A., L.M. Colla and F.P.F. Duarte. 2004. Improving *Spirulina platensis* biomass yield using a fed-batch process. Bioresour. Technol. 92: 237–241.

Virginie, M., L. Ulmann, V. Pasquet, M. Mathieu, L. Picot, G. Bougaran, J.P. Cadoret, A.M. Manceau and B. Schoefs. 2012. The potential of microalgae for the production of bioactive molecules of pharmaceutical interest. Curr. Pharm. Biotechnol. 13: 2733–2750.

Von, G.U. 2003. Ozonation of drinking water: part I. Oxidation kinetics and product formation. Water Res. 37: 1443–1467.

Vonshak, A. and L. Tomaselli. 2003. *Arthrospira (Spirulina)*: systematics and ecophysiology biochemistry. pp. 505–522. *In*: A. Vonshak (ed.). *Spirulina platensis (Arthrospira)*: Physiology, Cell-biology and Biotechnology. London: Taylor & Francis.

Wang, B. and C.Q. Lan. 2011. Optimising the lipid production of the green alga *Neochloris oleoabundans* using Box-Behnken experimental design. Can. J. Chem. Eng. 89: 932–939.

Wang, B., Y. Li, N. Wu and C. Lan. 2008. CO_2 bio-mitigation using microalgae. Appl. Microbiol. Biotechnol. 79: 707–718.

Wang, L., M. Min, Y. Li, P. Chen, Y. Chen, Y. Liu, Y. Wang and R. Ruan. 2010. Cultivation of green algae *Chlorella* sp. in different wastewaters from municipal wastewater treatment plant. Appl. Biochem. Biotechnol. 162: 1174–1186.

Waterbury, J.B. 2006. The cyanobacteria: isolation, purification, and identification. *In*: M. Dworkin, S. Falkow, E. Rosenberg, K.-H. Schleifer and E. Stackebrandt (eds.). The Prokaryotes: A Handbook on the Biology of Bacteria, Springer, New York 3: 1053–1073.

Wei, F., G.Z. Gao, X.F. Wang and X.Y. Dong. 2008. Quantitative determination of oil content in small quantity of oilseed rape by ultrasound-assisted extraction combined with gas chromatography. Ultrason. Sonochem. 15: 938–942.

Wijffels, R.H. and M.J. Barbosa. 2010. An outlook on microalgal biofuels. Science 329: 796–799.

Williams, P.J.B. and L.M.L. Laurens. 2010. Microalgae as biodiesel and biomass feedstocks: review and analysis of the biochemistry, energetic and economics. Energ. Environ. Sci. 3: 554–590.

Wimmer, T. 1995. Process for the production of fatty acid esters of lower alcohols. US Patent US5,399,731.

Wiyarno, B., R.M. Yunus and M. Mel. 2011. Extraction of algae oil from *Nannocloropsis* sp.: a study of soxhlet and ultrasonic-assisted extractions. J. Appl. Sci. 11: 3607–3612.

Wu, Q., B. Zhang and N.G. Grant. 1996. High yield of hydrocarbon gases resulting from pyrolysis of yellow heterotrophic and bacterially degraded *Chlorella protothecoides*. J. Appl. Phycol. 8: 181–184.

Wu, Q., J. Dai, Y. Shiraiwa, G. Sheng and J. Fu. 1999. A renewable energy source - hydrocarbon gases resulting from pyrolysis of the marine nanoplanktonic alga *Emiliania huxleyi*. J. Appl. Phycol. 11: 137–142.

Wu, Y.H., Y. Yu and H.Y. Hu. 2015. Microalgal growth with intracellular phosphorus for achieving high biomass growth rate and high lipid/triacylglycerol content simultaneously. Bioresour. Technol. 192: 374–381.

Xiaoling, M., Q. Wu and C. Yang. 2004. Fast pyrolysis of microalgae to produce renewable fuels. J. Anal. Appl. Pyrolysis 71: 855–863.

Xu, X., Y. Gao, G. Liu, Q. Wang and J. Zhao. 2008. Optimization of supercritical carbon dioxide extraction of sea buckthorn (*Hippophaethamnoides* L.) oil using response surface methodology. Food Sci. Technol. 41: 1223–1231.

Ya, L.C., C.J. Yu, Y.L. Guan, H.H. Shih, L.Y. Kuei, Y.C. Chun, S.C. Jo, C.L. Jhy and J.L. Duu. 2010. Dispersed ozone flotation of *Chlorella vulgaris*. Bioresour. Technol. 101: 9092–9096.

Yanan, H., S. Wang and K.K. Lai. 2010. Global economic activity and crude oil prices: a cointegration analysis. Bioresour. Technol. 32: 868–876.

Zhang, J.J. and L.F. Jiang. 2008. Acid-catalyzed esterification of *Zanthoxylum bungeanum* seed oil with high free fatty acids for biodiesel production. Bioresour. Technol. 99: 8995–8998.

Zhang, S., Y.G. Zu, Y.J. Fu, M. Luo, D.Y. Zhang and T. Efferth. 2010. Rapid microwave-assisted transesterification of yellow horn oil to biodiesel using heteropolyacid solid catalyst. Bioresour. Technol. 101: 931–936.

Zhao, B. 2017. Why will dominant alternative transportation fuels be liquid fuels, not electricity or hydrogen? Energy Policy 108: 712–714.

Zuhair, S.A. 2007. Production of biodiesel: possibilities and challenges. Biofuel Bioprod. Biorefin. 1: 57–66.

Genetic Engineering Approaches for the Exploitation of Marine Algae as Potential Sources of Organic and Inorganic Biofuels

Maria Gloria Esquível,[1,*] Julia Marín-Navarro,[2] Teresa S. Pinto[1] and Joaquín Moreno[2]

Introduction

Marine algae include cyanobacteria, eukaryotic microalgae, and seaweed. They are critical to the health of our planet and are dominant in the oceans, which produce approximately half of the oxygen that we breathe (Brodie et al, 2017). Unicellular algae (which sometimes associate in colonies) are known as microalgae. Examples of prokaryotic microalgae are cyanobacteria (*Cyanophyceae*), and of eukaryotic microalgae are green algae (*Chlorophyceae*) and diatoms (*Mediophyceae and Bacillariophyceae*). Due to the favourable natural features of microalgae, such as fast growth and high photosynthetic efficiency (compared to higher plants), these organisms appear as ideal for high biomass production. Algal biomass is a potential source of feed/foodstuff, added-value products, or biofuel. The latter possibility is especially attractive because algal metabolism naturally produces substances that can be directly used as (or easily converted to) fuel. Among these are hydrogen (Melis et al. 2000; Ghirardi et al. 2007; Esquível et al. 2011), alcohols (Spolaore et al. 2006), and lipids (Mata et al. 2010). Furthermore, biofuels can be produced simultaneously with value-added by-products for the chemical industry and human health, including oils (e.g., triglycerides), polysaccharides (e.g., alginate, agar), and pigments (e.g., phycobiliproteins, carotenoids). Current protocols allow the genetic manipulation of the algal strains to redesign the cellular

[1] Landscape, Environment, Agriculture and Food (LEAF), Instituto Superior de Agronomia (ISA), Universidade de Lisboa, Lisbon, Portugal.
Email: tpinto@isa.utl.pt
[2] Dept. Biochemistry & Molecular Biology, Faculty of Biology, University of Valencia C Dr. Moliner 50, Burjassot E-46100 (Spain).
Emails: juvicma@uv.es; joaquin.moreno@uv.es
* Corresponding author: gesquivel@isa.utl.pt

metabolism to promote synthesis of the target compound, and this approach appears the most promising to support further developments in this field.

Transformation in marine algae

Eukaryotic microalgae include cells resulting from a primary endosymbiotic event (as land plants) and cells derived from subsequent secondary and/or tertiary endosymbiotic events. Consequences of multiple endosymbiotic events include (1) the endosymbiotic gene transfer (EGT) which enriched the nuclear genome. Most EGT-derived gene products have complex targeting mechanisms to be imported back into the plastid and (2) coordination of enzymes, from "symbiogenetic" genes, derived from different originator cells (host or various prokaryotes), to form complex metabolic pathways that include responses to stress, phototropism, and adaptation to nutrients limitations (Parker et al. 2008; Brodie et al. 2017). Due to the complex and unique gene organization of algae, genetic engineering approaches must frequently consider the methodologies already in use for both prokaryotic and green lineage organisms.

The number of marine algae successful transformed has been increasing in recent times. These include members of the chlorophytes, such as *Chlorella vulgaris* (Dawson et al. 1997; Hawkins and Nakamura 1999; Cha et al. 2012), *Chlorella ellipsoidea* (Chen et al. 2001; Liu et al. 2012), *Ulva lactuca* (Huang 1996), *Dunaliella salina* (Tan et al. 2005; Feng et al. 2009), and *Ostreococcus tauri* (van Ooijen et al. 2012); from the phaeophytes, such as *Laminaria japonica* (Qin et al. 1999) and *Undaria pinnatifada* (Qin et al. 2003); from the rhodophytes, such as *Porphyra yezoensis* (Cheney et al. 2001; Hirata et al. 2011; Uji et al. 2013), *Porphyratenera* (Son et al. 2012), *Gracilaria changii* (Gan et al. 2003), and *Porphyridium* sp. (Lapidot et al. 2002); and from diatoms, such as *Thalassiosira pseudonana* (Poulsen et al. 2006), *Cylindrotheca fusiformis* (Poulsen and Kröger 2005), *Fistulifera* sp. (Muto et al. 2013), and *Phaeodactylum tricornutum* (Zaslavskaia et al. 2001; Niu et al. 2012). Special attention must be paid to the industrially relevant oleaginous marine green alga *Nannochloropsis* sp. Different methods of transformation are available for *Nannochloropsis*, a species which is easily transformed (Cha et al. 2011; Kilian et al. 2011; Radakovits et al. 2012). In the future, the marine algae *Nannochloropsis* may emerge as significant new model algal system due to their naturally high production of biomass and lipids that could be used as feedstock for biofuel. However, the green alga from soil and fresh water *Chlamydomonas reinhardtii* is still the most significant experimental model, and the most comprehensively studied (Harris et al. 2009)—thus remaining as a reference point in all works of genetic transformation in algae.

In general, the transformation efficiencies of the marine algae are low, which may be due to the composition and thickness of the cell wall of these species. For example, the efficiencies for marine *Synechococcus* strains were ten times lower than those for freshwater *Synechococcus* strains. Unique polysaccharides, which surround the cell wall, may prevent DNA uptake (Qin et al. 2012).

Currently, the main model species of algae – the green alga *C. reinhardtii*, the diatom *Phaeodactylum tricornutum* and the oleaginous algae *Nannochloropsis* sp., had their genome sequenced (Merchant et al. 2007; Bowler et al. 2008; Radakovits et al. 2012), each one has well established transformation methods (Zaslavskaia et al. 2001; Harris et al. 2009; Radakovits et al. 2012) and the CRISPR/Cas9 gene editing is available for the three organisms (Nymark et al. 2016; Shin et al. 2016; Wang et al. 2016).

The biolistic method (micro-particle bombardment) proved to be an efficient and highly reproducible method for delivering exogenous DNA into algal chloroplasts. *C. reinhardtii* chloroplasts were the first to be transformed by high velocity microprojectiles, 30 years ago (Boynton et al. 1988). Biolistic procedures were used to transform other chlorophytes, such as the genus *Chlorella* (Dawson et al. 1997) and *Dunaliella salina* (Tan et al. 2005); rhodophytes such as *Gracilaria changii* (Gan et al. 2003) and *Porphyra tenera* (Son et al. 2012); phaeophytes, such as *Laminaria japonica* (Qin et al. 1999); and diatoms, such as *Phaeodactylum tricornutum* (Zaslavskaia et al. 2001), *Cylindrotheca fusiformis* (Poulsen and Kröger 2005), and *Fistulifera* sp. (Muto et al. 2013).

Glass beads continue to be the simplest and most convenient method for algal nuclear transformation (Harris et al. 2009). This method was first used in *Chlamydomonas reinhardtii*, and achieved a high-frequency of nuclear transformation (Kindle 1990). *Dunaliella salina* was also successfully transformed with glass beads (Feng et al. 2009).

Electroporation is a highly efficient method for transferring small amounts of DNA. This transformation methodology was first utilized with a marine cyanobacterium *Synechococcus* sp. (Matsunaga et al. 1990). In eukaryotic algae, efficient transformation mediated by electroporation was firstly achieved with *C. reinhardtti* (Shimogawara et al. 1998). The maximum transformation frequency was two orders of magnitude higher than with glass beads (Shimogawara et al. 1998). Recently, transformation by electroporation was also successful with the diatoms *Nannochloropsis* sp. (Kilian et al. 2011), *Ostreococcus tauri* (van Ooijen et al. 2012), and *Phaeodactylum tricornutum* (Niu et al. 2012).

Agrobacterium tumefaciens-mediated genetic transformation is a very popular method for introducing gene in higher plants. The first report of an effective genetic transformation by *A. tumefaciens* in algae was conducted in the marine red seaweed *Porphyrayezoensis* (Cheney et al. 2001). Only in 2004 transformation of *C. reinhardtii* by co-cultivation with *A. tumefaciens* was described holding promising results—as the efficiency of transformation was higher than that of glass bead transformation (Kumar et al. 2004), but this method has not yet been widely adopted for this alga. Recently, *Agrobacterium*-mediated transformations of *Nannochloropsis* sp. (Cha et al. 2011) and *Chlorella vulgaris* (Cha et al. 2012) have been described. The microalgae commonly used as bioenergy sources and the most important methods for their genetic transformation are shown in Table 1.

Achieving high biomass for biofuels production

Biofuels or their precursors are composed entirely of carbon, hydrogen, and oxygen – and their synthesis results from the photosynthetic assimilation of CO_2 and water, not needing a net investment of other nutrients into the final product. Hence, biofuel synthesis could in principle be uncoupled from the nutrient expenses associated to cell proliferation. For these and other reasons, great hopes have been placed on microalga cultures as a sustainable source of biofuel. However, it is recognized that, in order to achieve an economically competitive status, the natural efficiency of the process has to be increased. Light into chemical free energy conversion, under optimal conditions, is estimated as ca. 5% efficient in microalgae, which compares favourably with 3% in higher plants but is still below 10% of current silicon-based artificial solar cells producing hydrogen through water electrolysis (Blankenship et al. 2011). However, algal biomass may still be a preferred source because of a higher versatility of extractable products, including liquid fuels. Moreover, it is perceived that the energy gain of the natural photosynthetic process taking place in microalgae could be further improved at several steps by biotechnological means. In this regard, biotechnologically oriented genome manipulation has been encouraged by recent developments in genetic engineering, including a specific adaptation of the CRISPR/Cas9 technology for editing genes of marine algae (Nymark et al. 2016) Prospects, current limitations, and the research agenda for the technological improvement of biomass and biofuel production have been recurrently reviewed in the past years (Anemaet et al. 2010; Kumar et al. 2010; Stephens et al. 2010; Stephenson et al. 2011; Day et al. 2012; Larkum et al. 2012; Rosgaard et al. 2012; Work et al. 2012; Ho et al. 2014; Hedge et al. 2015; Gomaa et al. 2016; Banerjee et al. 2016; Ng et al. 2017). This section will focus on organic fuels resulting from photosynthetic assimilation, while hydrogen production will be considered below in a separate section.

Improving light capture

Sunlight delivers an average power of 170 W/m^2 on the Earth surface (Blankenship et al. 2011). However, only a reduced range of wavelengths (roughly between 400 and 700 nm, with a narrow "green" window of transmission around 555 nm) is photosynthetically active (the so-called PAR). Therefore, only a fraction of light power is profitable through natural photosynthesis. The efficiency of radiation energy capture may be improved through different strategies of artificial illumination (reviewed by Ramanna et al. 2017), including wavelength selection and flashing. However, this section will focus on genetic engineering approaches to enhance radiation capture under natural sunlight.

PAR matches the overlapping absorption spectra of the different pigments harboured by the light-harvesting complexes that deliver the excitation energy to the photosynthetic reaction center.

Table 1. Examples of marine microalgae used in biofuel production and in biotechnology.

Algal class	Algal species	Biotechnological applications (Value compounds)	Transformation techniques	Advantages or drawbacks
Chlorophytes	*Chlorella* sp.	Rich in chlorophyll, essential amino acids, high-quality protein, fatty acids, and bioactive compounds. High biomass production. One of the best candidate for biofuel and H_2 production.	Biolistic procedures (Dawson et al. 1997) PEG-CaCl2 and Electroporation (Hawkins and Nakamura 1999) *Agrobacterium tumefaciens*-mediated (Cha et al. 2012)	Whole genome available, transformable. Widespread, but with rigid cell wall makes cell disruption difficult
	Dunaliella salina	Cells accumulate abundant quantities of β-carotene and glycerol. High production of valuable polypeptides and proteins. Potential as a feedstock for biofuel production.	Biolistic procedures (Tan et al. 2005) Glass beads (Feng et al. 2009)	Needs high salinity in the culture media. Very little is known about the genome
	Ostreococcus tauri	Rich in oils. Good candidate for biodiesel production.	Electroporation (van Ooijen et al. 2012)	Smallest free-living eukaryote. Whole genome available. Easy to culture
Phaeophytes	*Laminaria japonica*	Consumed as a subsidiary food, and used as a source of low value products, such as iodine, mannitol and alginate.	Biolistic procedures (Qin et al. 1999)	Large size. Little is known about the genome
Rhodophytes	*Porphyra yezoensis*	Traditionally used as ingredient in foods, food additives, and medicines. Good biomass production.	Biolistic procedures (Hirata et al. 2011; Uji et al. 2012)	Little is known about the genome
	Porphyridium sp.	Attractive for biotechnological purposes because they synthesize unique cell wall sulfated polysaccharides, accessory photosynthetic pigments (phycobilins and carotenoids), and unsaturated fatty acids.	Biolistic procedures (Lapidot et al. 2002)	Little is known about the genome
Diatoms	*Thalassiosira* sp	Potential source of biodiesel and specialty chemicals, such as omega-3 fatty acids.	Biolistic procedur (Poulsen et al. 2006)	Whole genome available.
	Fistulifera sp.	Rich in oils, Biodiesel production.	Biolistic procedures (Muto et al. 2012)	Chloroplast genome sequenced
	Phaeodactylum tricornutum	Abundant sources of long chained polyunsaturated omega-3 and omega-6 fatty acids.	Electroporation (Niu et al. 2012) Biolistic procedures (Zaslavskaia et al. 2001)	Whole genome available
Eustigmato-phyceae	*Nannochloropis* sp.	Stores relatively large amounts of lipids, in the form of triacylglycerols, even during logarithmic growth. Could be used as feedstock for biofuel production.	Electroporation (Kilian et al. 2011)	Whole genome available, easily transformed
Cyanobacteria	*Anabaena* sp.	Possible candidate for H_2 production.	Natural transformation-plasmid transfer of DNA (Happe et al. 2000; Masukawa et al. 2010)	Whole genome available. Capable of nitrogen fixation. Filamentous morphology
	Synechocystis sp. and *Synecococcus* sp.	Showing oils and terpens suitable for biofuel production. H_2 producer.	Natural transformation-plasmid transfer of DNA (Baebprasert et al. 2011)	Whole genome available, genetically transformable

A first approach to increase the light conversion efficiency would be to extend the range of radiation wavelengths that can be captured by the photosynthetic machinery. Due to different algae having slightly shifted ranges of PAR, mixed cultures of different species may cover the solar spectrum better than axenic ones. Alternatively, algae that are susceptible of genetic transformation may be manipulated for extended absorption – and this may be achieved in several ways. The discovery and characterization of the cyanobacterial chlorophylls *d* (Larkum and Kühl 2005) and *f* (Chen et al. 2010), which absorb light beyond the visible limit into the near infrared (Li et al. 2012), prompted suggestions for introducing such pigments in selected algae, thereby expanding the spectrum of PAR (Chen and Blankenship 2011). Cyanobacteria using almost exclusively chlorophyll *d* are known to carry out oxygenic photosynthesis with an energy conversion efficiency that is comparable to, or higher than, other species typically relying on chlorophyll *a* (Mielke et al. 2011). This is possible because chlorophyll *d* substitutes chlorophyll *a* also at the photosynthetic reaction center, thus shifting its absorption peak to longer wavelengths and letting the excitation energy collected by the antennae at the far red to run downhill to the reaction center. Chlorophyll *d* differs from the more common chlorophyll *a* in that the C3 vinyl group at ring A has been oxidized to a formyl group. Oxidation of chlorophyll *a* is carried out by molecular oxygen (Schliep et al. 2010) in a reaction that is apparently catalyzed by a P450 oxygenase. Chlorophyll *f* contains also a formyl group substitution at the C2 position of the same ring, but details of its biosynthesis are still unknown. In any case, the intended goal would be to transform algae with genes coding for the enzymes needed to synthesize these red-shifted chlorophylls. This would extend the span of light harvested by the antenna complexes, and would potentially increase the photosynthetic efficiency—although this is yet to be experimentally demonstrated.

A different strategy has been followed by introducing fluorescent proteins. These are variants of the well known green fluorescent protein (GFP) from the jellyfish *Aequorea victoria*, which acquires fluorescence through a spontaneous condensation of three vicinal amino acid residues into a high quantum yield fluorophore upon folding (reviewed by Stepanenko et al. 2011). Therefore, fluorescence can be introduced directly by transformation with a gene encoding GFP (or any of its mutants with shifted excitation and/or emission wavelengths). The aim is to express proteins that can harvest sunlight at the near-ultraviolet region (the so-called UV-A) and re-emit it in the visible range to be subsequently captured by the antenna complexes (and its energy funnelled into the reaction center). This approach, protected by a US patent (Gressel et al. 2010), may be currently under research. Nevertheless, the introduction of fluorescent dyes in solution (with the same aim of converting unusable radiation into photosynthetically profitable wavelengths) has been shown to increase growth and lipid accumulation in microalgal cultures (Seo et al. 2015). In addition, the expression of light-harvesting phycobilisomes of cyanobacteria in algae has been proposed also to improve absorption at the "green" window of transmission of natural photosynthetic pigments (Stephenson et al. 2011).

Under relatively high light intensity, the photosynthetic centers of exposed cells may become saturated and a significant part of the absorbed energy may be lost to heat or fluorescence. Even worse, intense illumination in bioreactors may result in photosynthetic center damage and photoinhibition in cells of the most exposed layer (Simionato et al. 2013). This is unfortunate because other cells located deeper inside the culture (i.e., receiving less intense light) may not be saturated and could potentially channel this energy in a productive fashion. This problem may be alleviated in shallow or constantly mixed liquid cultures, but there is always a certain degree of unavoidable self-shading – especially in the case of dense cultures as intended for high biomass production. In order to reduce this effect, different algal strains have been equipped with defective antenna complexes, which capture less light and thus allow photons to penetrate deeper in the culture. This has been achieved in several species by different strategies (reviewed in Stephens et al. 2010; Simionato et al. 2013), including targeted interference-RNA repression and several forms of mutagenesis to disrupt light harvesting components. Under such conditions, which distribute radiation more evenly among cells of dense cultures, a significant increase in photosynthetic efficiency and productivity has been reported for *C. reinhardtii* grown in mass culture (Polle et al. 2003; Mussgnug et al. 2007).

Improving CO₂ fixation

Carbon assimilation takes place in all algae, as well as in cyanobacteria, through the metabolic route known as Calvin (or Calvin-Benson-Bassham) cycle. The initial step in this pathway is the fixation of CO_2 mediated by ribulose 1,5-bisphosphate carboxylase/oxygenase (Rubisco). Cyanobacterial and most eukaryotic Rubiscos are composed of eight large and eight small subunits, assembled in a hexadecameric holoenzyme. Sequence comparison has revealed two evolutive branches among the hexadecameric Rubiscos (Tabita et al. 2008). One of them comprises the enzymes of cyanobacteria, green algae, and higher plants, encoding the large subunit in the chloroplast genome and the small subunit in the nucleus. These enzymes are called "green-like" Rubiscos. On the other hand, red and brown algae encode both subunits in the chloroplast, and the corresponding enzymes show a significant divergence (at the sequence level) from the green-like forms, termed "red-like" Rubiscos. In all cases, the catalytic site resides at the contact surface between large subunits. Rubisco catalyzes the carboxylation of ribulose 1,5-bisphosphate (RuBP) to produce two molecules of 3-phosphoglycerate, which are further metabolized through the Calvin cycle. However, Rubisco also promotes the oxygenation of RuBP, thus rendering only one molecule of phosphoglycerate and another of 2-phosphoglycolate. This latter reaction initiates the photorespiratory pathway, which metabolizes 2-phosphoglycolate to recover two thirds of its carbon as 3-phosphoglycerate, while the other third is released as CO_2. Therefore, loss of CO_2 through photorespiration opposes CO_2 fixation through the Calvin cycle, and the net balance between the two outcomes determines the rate of carbon biomass accumulation. Rubisco oxygenase activity is thought to be an unavoidable escape of the catalytic mechanism, while the photorespiratory pathway appears to be just a metabolic solution to avoid the accumulation of 2-phosphoglycolate (which inhibits Calvin cycle enzymes) and to salvage part of its carbon. Despite photorespiration having acquired some secondary functions (such as protection against oxidative stress), the experimental fact is that favouring carbon fixation over photorespiration—by raising the environmental CO_2/O_2 ratio—substantially increases the growth rate of plants. Therefore, biotechnological suppression or reduction of the oxygenase activity of Rubisco would, in principle, enhance biomass production significantly.

The ratio of the carboxylase (Vc) to oxygenase (Vo) activity of Rubisco is known to be proportional to the corresponding ratio of substrate concentrations:

$$Vc/Vo = \Omega \cdot ([CO_2]/[O_2])$$

The proportionality constant Ω (called the specificity factor) represents the intrinsic preference of the enzyme for one or the other substrate. The specificity factor is known to vary among enzymes of different species, and is thought to reflect environmental adaptation. Accordingly, photosynthetic organisms living in anoxic conditions or possessing mechanisms to increase the natural CO_2/O_2 ratio around Rubisco, have comparatively low specificity factors (reviewed in Whitney et al. 2011). In spite of extended mutagenesis work, the critical residues and molecular determinants that sustain the variability of specific factor among enzymes of different species have remained elusive. Several structural domains of green-like Rubiscos have been identified as relevant for substrate partitioning (Spreitzer and Salvucci 2002). However, directed mutagenesis yielded only limited advances in specificity, and those were counterbalanced by other kinetic drawbacks—perhaps because most enzymes appear to be (nearly) optimized under their corresponding intracellular and environmental conditions (Tcherkez et al. 2006; Savir et al. 2010). The natural variability of the catalytic properties of Rubisco has been explored (see, for example, Orr et al. 2016), and the screening for desirable variants of the enzyme has been recently extended to uncultivated species from natural microbial communities through a metagenomic approach (Varaljay et al. 2016). Highest specificity factors found for green-like Rubiscos are of the order of 100. In contrast, significantly higher values have been reported for some red-like enzymes (e.g., 238 for the thermophilic red alga *Galdieria partita*) (Uemura et al. 1997), suggesting that better solutions for substrate specificity could have evolved from alternative sequences. A direct approach to boost photosynthetic carbon fixation by biotechnological means would be to express the genes encoding the subunits of the superior red-like Rubiscos in the algal species selected for biomass production. Similar attempts carried out in higher plants have been unsuccessful because of failure of the foreign holoenzyme to assemble correctly (which appears to require very specific chaperones) in the hosting chloroplast (Whitney et al. 2001). Nevertheless, full expression in algae which are phylogenetically nearer the donor species might be plausible, and will surely be attempted as new transformation procedures become suitable for an extended number of microalgae.

Artificial directed evolution of Rubisco is also possible, as demonstrated by a clever scheme designed to improve Rubisco features (Parikh et al. 2006). The Calvin cycle was partially reconstructed in an engineered strain of *E. coli*, which still required Rubisco and phosphoribulokinase for growth in a medium with a pentose as the sole source of carbon supplemented with a pentose. After random mutation of the gene encoding the large subunit of the *Synechococcus* Rubisco, the resulting library was co-expressed together with the small subunit of Rubisco and the phosphoribulokinase from *Synechococcus* in the engineered *E. coli*, and screened for hypermorphic variants. After three rounds of random mutagenesis the selected strains exhibited 5-fold improvement in Rubisco specific activity relative to the wild type enzyme. The selected Rubisco displayed higher catalytic efficiency and folding capacity (i.e., better interaction with the chaperones from *E. coli*), but showed no improvement in the specificity factor (Greene et al. 2007). Further attempts of evolving a better Rubisco have been carried out in *E. coli* starting from enzymes of cyanobacteria (*Synechococcus*) (Mueller-Cajar and Whitney 2008a) or archea (*Methanococcoides burtoni*) (Wilson et al. 2016). These approaches yielded enzyme variants with higher turnover number (Durao et al. 2015), but no significant increase of substrate specificity. While directed evolution of Rubisco appears as highly promising strategy, it is possible that an increase of the natural specificity factor (without compromising the catalytic activity itself during the change) requires an array of simultaneous mutations that cannot be explored, even using high-throughput techniques, without a better understanding of the complex structure-function relationships and the catalytic chemistry of the enzyme (Mueller-Cajar and Whitney 2008b). Regardless the moderate specificity factor may be tolerated in organisms that possess mechanisms to concentrate CO_2 at the Rubisco location, thereby achieving a high carboxylation to oxygenation rate. C4 metabolism is a well-known strategy used by land plants. Evidence suggesting a C4-like metabolism has been also reported in diatoms and other algae (Reinfelder 2011), although its possible physiological significance is still under discussion (Raven 2010; Haimovich-Dayan et al. 2013; Raven and Giordano 2017). However, the most common device found in aquatic organism to favour carboxylation is the active pumping of bicarbonate into cellular compartments containing Rubisco, followed by the local release of CO_2 from bicarbonate by carbonic anhydrase. Many marine microalgae employ this procedure or variants thereof (Spalding 2008; Reinfelder 2011). All these strategies are collectively known as carbon concentrating mechanisms (CCM). The existence of a CCM seems to be more the rule than the exception, since the alternative (i.e., diffusive entry of CO_2) appears to be present only in a small minority of eukaryotic algae (Raven 2010). As CCMs from different species display variable efficiency, it has been suggested that the identification of genes coding for high-activity versions of carbonic anhydrases and bicarbonate transporters would be of interest in order to introduce them into suitable hosts as a mean to improve Rubisco carboxylation to oxygenation ratio (Work et al. 2012).

Besides, cyanobacteria possess their own CCM in the form of carboxysomes, which are self-assembling particles containing mostly Rubisco and carbonic anhydrase inside a proteinaceous shell (Rae et al. 2013). A hypothetical assembly of carboxysomes inside the chloroplast could potentially raise the carboxylation/oxygenation ratio of Rubisco (McGrath and Long 2014; Hanson et al. 2016). The correct assembly of functional carboxysomes in *E. coli* (Bonacci et al. 2012) and the expression of a functional cyanobacterial Rubisco in the chloroplast of higher plants (Lin et al. 2014) have been hailed as first steps towards the introduction of carboxysomes in photosynthetic eukaryotes.

Redesigning the carbon assimilation metabolism

A different approach to enhance biomass production is to modify the natural metabolic routes of algae to permit a more efficient processing of fixed carbon to fuels or fuel precursors. Several attempts have been reported along these lines.

Accepting that the Rubisco oxygenase activity is inescapable; the 2-phosphoglycolate metabolism may be redesigned to diminish its cost in terms of nutrients (since the conventional photorespiratory pathway releases nitrogen in the form of ammonia) and energy (Peterhansel and Maurino 2011). Toward that end, the bacterial glycolate pathway—which converts glycolate to glycerate in three steps, was engineered into the chloroplast of a higher plant (*Arabidopsis thaliana*) (Kebeish et al. 2007). The final product of this pathway (glycerate) can enter the Calvin cycle after phosphorylation. With a similar aim but following a different strategy, a new pathway was created also in *Arabidopsis* to achieve full oxidation of glycolate to CO_2. This approach was based on the conversion of glycolate to glyoxylate, followed by two sequential decarboxylations carried out by malate synthase and the NADP malic enzyme to yield pyruvate (Peterhansel and Maurino 2011). In both cases, the new routes skipped ammonia loss (and subsequent expense of re-fixation energy), released CO_2 in the chloroplast (raising the local CO_2/O_2 ratio that governs Rubisco partitioning), and produced additional NADH and/or NADPH. Consequently, the engineered plants showed improved growth and produced higher leaf biomass (up to 30% more in case of overexpression of the glycolate pathway) (Kebeish et al. 2007). In principle, this strategy could work equally well in the case of algae, although possession of a strong CCM may leave less room for metabolic improvement.

Under intense light, radiant energy to biomass conversion can be limited by the speed of RuBP regeneration through the Calvin cycle. Theoretical analyses have shown that this pathway still allows for a potential increase of photosynthetic rate, highlighting sedoheptulose 1,7-bisphosphatase (SBPase) as the key enzyme controlling RuBP regeneration and main target for biotechnological manipulation of the metabolic flux through the cycle (Raines 2003; Zhu et al. 2007). In fact, overexpression of a bifunctional fructose 1,6-bisphosphatase/SBPase from *Synechococcus* in tobacco chloroplasts resulted an increase in photosynthesis and growth (Miyagawa et al. 2001). A further comparison of the effect of expressing either fructose 1,6-bisphosphatase or SBPase alone showed that both of them could separately contribute to increase photosynthetic rate, with SBPase being the most important for RuBP regeneration (Tamoi et al. 2006). Similarly, overexpression of an *Arabidopsis* cDNA coding SBPase in transgenic tobacco promoted sucrose and starch accumulation and increased biomass up to 30% (Lefebvre et al. 2005). These results demonstrate that further improvements in photosynthetic efficiency for selected algal species could result from metabolic control analysis and engineering of Calvin cycle enzymatic activities. Indeed, significant enhancements of photosynthetic activity and biomass production have been reported through transgenic expression of a fructose 1,6-bisphosphate aldolase in *C. reinhardtii* (Yang et al. 2017) and a bifunctional fructose 1,6-bisphosphatase/SBPase in *Euglena gracilis* (Ogawa et al. 2015). In both cases, the cyanobacterial enzymes were fused to plastid transit peptides in order to ensure expression inside the chloroplast.

Finally, a radical solution for eliminating photorespiratory losses has been proposed (but still not successfully achieved) by replacing the Calvin cycle altogether by another different carbon fixation pathway, as found in bacteria and archaea (Blankenship et al. 2011). Most of these routes cannot be easily transferred to photosynthetic organisms performing oxygenic photosynthesis because the relevant enzymes (which are to be expressed in the host) are strongly oxygen-sensitive. However, the 3-hydroxypropionate/malyl-CoA cycle, present in the green nonsulfur bacterium *Chlorofexus*, has been found not to be inhibited by oxygen (Thauer 2007) and could, in principle, be used (Blankenship et al. 2011). Another possibility would be to design a new ("synthetic") fixation pathway from existing enzymatic activities collected from different organisms. In this regard, a new pathway converting 2-phosphoglycolate into pyruvate (bypassing photorespiration) has been implemented in the cyanobacterium *Synechococcus elongatus* by expressing enzymes from *Accumulibacter phosphatis* and from the 3-hydroxypropionate/malyl-CoA cycle from *Chlorofexus*. This new route has been demonstrated to work although it did not result in faster growth, probably because *S. elongatus* possesses already a highly active CCM that curtails the metabolic benefit (Shih et al. 2014). Alternative synthetic routes for carbon fixation, employing the superior phosphoenolpyruvate carboxylase instead of Rubisco, have also been proposed (Bar-Even et al. 2010). Nevertheless, it should be kept in mind that all carbon fixation pathways face similar thermodynamic constraints, which determine the free energy and cellular resources that should be invested (Bar-Even et al. 2012). Therefore, a detailed analysis of the expected costs and gains should be carried out before embarking in such an ambitious project, which would probably need an extensive re-design of the whole metabolism.

Besides all such possibilities to increase the efficiency of carbon fixation, the metabolism of photosynthates can also be steered to boost the synthesis of end products that can be used as fuel. Accordingly, the production of alcohols, sugars starch, fatty acids, and other metabolites has been enhanced in different photosynthetic organisms (especially cyanobacteria) either by overexpression of key enzymes in their corresponding biosynthetic pathways, or by avoiding them (bypassing unnecessary intermediaries and flux branching points), or by engineering suitable secretion routes (reviewed by Rosgaard et al. 2012). Advances in the specific production of oils and ethanol will be considered in a separate section below. Besides improving the existing metabolic pathways, another strategy consists of creating new ones that lead to the desired final product. In a pioneer work back in 1999, genes coding for pyruvate decarboxylase and alcohol dehydrogenase from the bacterium *Zymomonas mobilis* were introduced in *Synechococcus* in order to deviate pyruvate to the synthesis of ethanol (Deng and Coleman 1999). An extension of this idea, which has been termed the "photanol" approach (Hellingwerf and Teixeira de Mattos 2009), may be used to produce a variety of fuels by linking metabolites generated by the carbon fixation pathway (particularly glyceraldehyde 3-phosphate and pyruvate) to the fermentative pathways of chemoautotrophic bacteria (Anemaet et al. 2010). The authors propose cyanobacteria (in particular, *Synechococcus*) to host the bacterial genes coding the enzymes required to implement the fermentative route. However, this approach may in principle be also extended to those algae that can be transformed. In a different strategy carried out in *E. coli*, 2-ketoacid decarboxylase and alcohol dehydrogenase were introduced in the bacterium in order to synthesize several branched-chain higher alcohols (which are superior to ethanol as biofuel) from 2-ketoacids that are naturally produced from glucose through the amino acid biosynthetic pathways (Atsumi et al. 2008). A similar approach may again be attempted in photosynthetic organisms like cyanobacteria, and perhaps eukaryotic algae; these are, most likely, just the first examples of promising developments derived from the synthetic biology approach to biofuel production.

Production of specific organic biofuels

Generation of oils, storage lipids, and triacylglycerides for biodiesel production

One of the most abundant forms of reduced carbon chains on Earth is fatty acids—plant seeds and algae are the largest sources of these compounds.

Biodiesel is made by a mixture of fatty acid alkyl esters obtained by transesterification (ester exchange reaction) of triacylglycerides (TAGs) from oils or fats. Transesterification comprises a few reactions where triacylglycerides and a short chain alcohol react in the presence of a strong base catalyst (usually NaOH) to form a mixture of fatty acid alkyl esters (Mata et al. 2010). This conversion of fatty acids to biodiesel is a rapid and technically straightforward process that renders triacylglycerides attractive for bioenergy generation.

Lipid biosynthesis and catabolism was first studied in detail in oil seed plants. Because homologues can be found in algal genomes for the majority of genes related to lipid metabolism in land plants, the strategies used to improve lipid production in higher plants are expected to also be effective with microalgae. In higher plants and eukaryotic algae, the reactions for *de novo* fatty acid synthesis are located in plastids and, consequently, the synthase machinery is similar to prokaryotes in that the enzymatic components are separable polypeptides rather than large multifunctional proteins as found in animals and fungi. This sequence of reactions is called the "prokaryotic pathway". The priming reaction, the conversion of acetyl-coenzyme A (CoA) to malonyl-CoA, is catalyzed by the plastidic acetyl-CoA carboxylase (ACCase), which is considered the first control step in fatty acid biosynthesis in many organisms. However, several attempts to utilize ACCase overexpression to increase lipid content in various systems have failed. The overexpression of ACCase in two diatom species increased the activity of ACCase, but enhancement of lipid production was not observed (Dunahay et al. 1995). After the priming reaction, the elongation of nascent acyl chains requires the sequential entrance of malonyl CoA until reaching a 16-carbon acyl-CoA. The first desaturation step for fatty acids is catalyzed by a plastidial stearoyl-acyl carrier protein (ACP) desaturase. Termination of fatty acid chain elongation in the plastids is catalyzed by (acyl-ACP) thioesterases (TE), which hydrolyze acyl chains from ACP. After termination, free fatty acids are activated to CoA esters, exported from the plastid, and assembled into glycerolipids at the endoplasmic reticulum (ER) (Thelen and Ohlrogge 2002). Genetic modification of genes involved in TAG assembly in the ER produced better results toward the increase of the amount of lipids than genetic manipulation of ACCase in the chloroplasts. For example, the overexpression of genes involved in TAG assembly, such as glycerol-3-phosphate dehydrogenase (G3PDH) and diacylglycerol acyltransferase (DGT), results in significant increases in plant lipid production (Lardizabal et al. 2008). An increase of TAGs accumulation was also accomplished in *C. reinhardtii,* using the type-2 diacylglycerol acyltransferase (DGTT2) with a phosphorus starvation–inducible promoter, which was up-regulated during P starvation (Iwai et al. 2014). Further lipid modifications (desaturation, hydroxylation, and elongation) occur in the ER (Thelen and Ohlrogge 2002). The low polarity of triacylglycerols (TAG), which are neutral lipids, causes accumulation of these lipids between bilayer leaflets (termed oil bodies or oleosomes) in the cytosol that is the general lipid storage form.

It has been known for a long time that nitrogen starvation, which stops protein (and starch) synthesis, can induce lipid synthesis. Based on this, a possible approach to improve cell lipid contents is via blocking metabolic pathways that lead to the accumulation of storage compounds such as starch (Radakovits et al. 2010a). *Chlamydomonas reinhardtii* has a mutant collection defective for starch biosynthesis—*sta1-1, sta6,* and *sta7* (see Ball and Deschamps 2009), which can be good candidates to improve lipid production. When the sta6 strain is nitrogen starved in acetate and then was "boosted" with additional acetate, the cells become "obese". The genes of G3PDH, DGTT2, and lipases were selectively upregulated in this condition (Goodenough et al. 2014). Within the genus *Chlorella, C. pyrenoidosa* has fewer lipid contents than the other species (Mata et al. 2010)—which may be due to the presence of pyrenoid, a structure rich in starch and Rubisco. It should be noted that a starch-less mutant of *C. pyrenoidosa* accumulated much higher polyunsaturated fatty acids per dry weight than the wild type (Ramazanov and Ramazanov 2006). Rubisco could be a potential target because modifications in the enzyme cause changes in carbon allocation in terms of membrane fatty acid composition and storage lipid accumulation (Esquivel et al. 2017). On the other hand, the identification of genes that are up-regulated upon nitrogen depletion resulted in the elucidation of a putative transcription factor that triggers lipid accumulation in *C. reinhardtii,* as well as the identification of three acyltransferases that are implicated in nitrogen starvation-induced TAG accumulation (Boyle et al. 2012). Those genes are potential targets for manipulating TAG hyperaccumulation in marine microalgae. Another strategy was patented to increase lipid production using homologous recombination in the alga *Nannochloropsis* to knock out the alternative oxidase (AOX). With mitochondrial respiration inhibited, a high concentration of lipids inside the cells was achieved (Bailey et al. 2011).

Engineering microalgae is also desirable to improve the quality of algal lipids to be used as biodiesel. The carbon chain length of fatty acids affects the cold flow properties of biodiesel, with shorter chain lengths being preferred. By resorting to genetic engineering approaches, a shorter fatty acid chain was attained with the diatom *Phaeodactylum tricornutum.* Two medium-chain acyl-ACP thioesterases from *Umbellularia californica* and *Cinnamomum camphora* were overexpressed in this diatom, resulting in increased production of lauric acid (12:0) and myristic acid (14:0) (Radakovits et al. 2010b). Interestingly, the overexpression of these thioesterases also resulted in an increased production of total fatty acids on a per cell basis. Another approach to improve the production of shorter fatty acid chain was achieved with a transgenic strain of *Chlamydomonas* overexpressing thioesterase (TE) in the algal chloroplast. This effect is thought to be due to protein-protein interactions between the fatty acid acyl carrier protein (ACP) and the TE, resulting in short-circuiting of fatty acid chain elongation—which leads to increasing myristic acid content by 2.5fold, compared with the wild-type (Blatti et al. 2012). These results unfold a new tool to manipulate chain lengths in fatty acid biosynthesis through protein-protein interactions.

Bioethanol production

Ethanol for biofuels is currently produced from the fermentation of starches or cellulose-derived sugars. The predominant energy storage polysaccharide in Chlorophyta (green algae), Dinophyta (dinoflagellates), Glaucophyta and Rhodophyta (red algae) is starch, while Phaeophyceae (brown algae) and Bacillariophyceae (diatoms) store glucans in laminarian and chrysolaminarin, respectively. In industry, the algal polysaccharides are firstly hydrolyzed and then fermented to ethanol by other organisms such as yeast. However, an approach that would couple ethanol production directly to photosynthetic carbon fixation *in situ* may be preferred. Numerous microalgae have fermentative metabolic pathways to ethanol, but the coupling of ethanol production to photoautotrophic metabolism would require changes in regulatory pathways or the insertion of new metabolic pathways (Radakovits et al. 2010a). Increasing carbohydrate production in algae would help biofuel production. Proposed strategies to that end include overexpression of key enzymes in starch biosynthesis (e.g., ADP-glucose pyrophosphorylase or isoamylase), knockout of starch degrading enzymes (e.g., glucan-water dikinases and amylases), and secretion of soluble carbohydrates (Work et al. 2012).

As brown macroalgae does not contain lignin, the sugars can be released by such simple operations as milling or crushing. This bio-architectural feature gives macroalgae a distinct advantage over lignocellulosic biomass, by facilitating higher yields and avoiding the need for energy-intensive pretreatment and hydrolytic processes before fermentation (Wargacki et al. 2012). However, seaweeds have been ignored as a source of renewable fuel because their primary sugar component is not easily fermented. A microbe able to extract sugars from brown seaweeds and convert them into low-carbon, renewable fuels, and chemicals was recently engineered (Wargacki et al. 2012).

Production of hydrogen

The biosynthesis of hydrogen has attracted an outstanding interest in the last decades due to its potential application as non-polluting and renewable biofuel. Hydrogen can indeed, be converted to electricity thus liberating high amounts of energy (122 kJ/g) and releasing only water as a collateral product (Ballat 2008). Reviews about hydrogen production, facts, and potentials, in green microalgae and cyanobacteria include Burgess et al. (2011); Eroglu and Melis (2011); Esquivel et al. (2011); Srirangan et al. (2011); Masukawa et al. (2012); Antal et al. (2015); Dubini and Ghirardi (2015); Khanna and Lindblad (2015); Oey et al. (2016); Khetkorn et al. (2017); Nagarajan et al. (2017) and Martin and Frymier (2017).

Enzymes involved in hydrogen production in green microalgae and cyanobacteria

Cyanobacteria and eukaryotic microalgae have evolved different systems to produce hydrogen via specific enzymes. Green microalgae possess [FeFe]-hydrogenases, whereas cyanobacteria may produce hydrogen by a [NiFe]-hydrogenase or, in the case of nitrogen-fixing cyanobacteria, by a [MoFe]-nitrogenase.

Green algae [FeFe]-hydrogenases are located in the chloroplast and act in a unidirectional way reducing free protons into hydrogen. The enzyme is ferredoxin-dependent, and the reducing power may be supplied by the photosynthetic electron transport chain in the light, or by fermentative metabolism through a pyruvate-ferredoxin oxidoreductase (PFOR) (Burgess et al. 2011) (Fig. 1A). The active site of [FeFe]-hydrogenases contains a six-iron complex, termed the H-cluster, formed by a [4Fe4S] cubane connected to a di-iron subcluster [2Fe] through a cysteine thiolate (Mulder et al. 2010). *C. reinhardtii* contains two [FeFe]-hydrogenase-encoding genes, namely *HydA1* and *HydA2*, of which the former contributes to 75% of total hydrogen-production (Meuser et al. 2012). The formation of an active enzyme requires expression of a maturation cassette composed of genes *HydE*, *HydF* and *HydG*. Transcription, maturation, and activity of [FeFe]-hydrogenase are oxygen-sensitive (Srirangan et al. 2011). Oxygen reacts in a stepwise manner with the di-iron subcluster, releasing reactive oxygen-species that subsequently attack the [4Fe4S] cubane leading to complete and irreversible destruction of the H-cluster (Stripp et al. 2009; Lambertz et al. 2011). This property of the enzyme is the main bottleneck for efficient hydrogen-production in green microalgae, as will be discussed below.

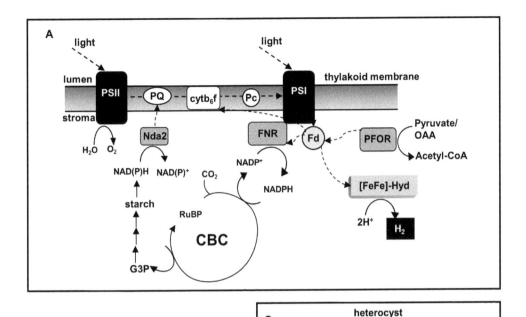

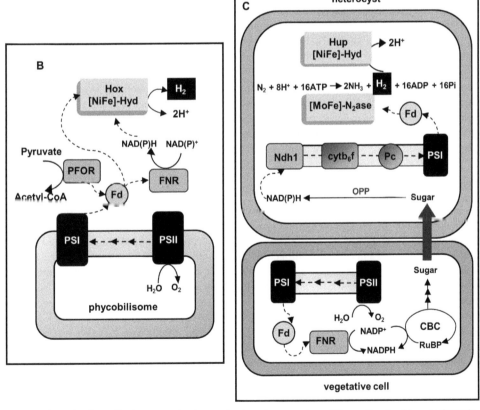

Fig. 1. (A) Schematic view of hydrogen production mechanisms in green algae, (B) non-diazotrophic cyanobacteria, and (C) and heterocyst-forming cyanobacteria. The electron flux is indicated with dashed arrows. PSI, photosystem I; PSII, photosystem II; PQ, plastoquinone pool; cytb6f, cytochrome b6f; Pc, plastocyanine; Fd, ferredoxin; PFOR, pyruvate-ferredoxin oxidoreductase; OAA, oxaloacetate; FNR, ferredoxin-NADP oxidoreductase; RuBP, ribulose-1,5-bisphosphate; Nda2, type II NAD(P)H dehydrogenase; CBC, Calvin-Benson cycle; Ndh1, NADPH dehydrogenase (complex I); OPP, oxidative pentose phosphate pathway; [FeFe]-Hyd, [FeFe] hydrogenase; Hup [FeNi]-Hyd, uptake [FeNi]-hydrogenase; Hox [FeNi]-Hyd, bidirectional [FeNi] hydrogenase; [MoFe]-N2ase [MoFe], nitrogenase. (See text for further details.)

Cyanobacterial [NiFe]-hydrogenases contain a nickel atom at their active site, linked to a $Fe(CN)_2CO$ molecule (Eroglu and Melis 2011). Two types of [NiFe]-hydrogenases are present in cyanobacteria: an uptake [NiFe]-hydrogenase, encoded by the *hup* genes, which recycles hydrogen recovering energy-rich electrons and is present in almost all nitrogen-fixing cyanobacteria (Khetkorn et al. 2017), and a multi-subunit bidirectional [NiFe]-hydrogenase, encoded by the *hoxFUYH* genes (Fig. 1B). The bidirectional enzyme was thought to be peripherally associated with the cytoplasmatic membrane and to accept electrons from NAD(P)H to produce hydrogen (Eroglu and Melis 2011; Srirangan et al. 2011). However, further research indicated that Hox subunits are associated to thylakoids and both ferredoxin and flavodoxin have been suggested as the main direct electron donors for this hydrogenase (Khanna et al. 2015). Hydrogen synthesis by this enzyme seems to be sustained in the dark, under fermentative conditions, when pyruvate may be used as electron source through pyruvate-ferredoxin oxidoreductase (PFOR). In contrast, only a short bust of hydrogen is detected upon illumination, followed by hydrogen consumption. These enzymes are also inhibited by oxygen, but in a reversible manner (McIntosh et al. 2011).

Nitrogen-fixing, filamentous cyanobacteria are able to synthesize hydrogen through a [MoFe]-nitrogenase, localized in specialized cells named heterocysts. This enzyme contains two metal clusters, the [8Fe-7S] P-cluster and the [1Mo-7Fe-9S-1X-homocitrate] FeMo cofactor, where the homocitrate molecule is required for efficient nitrogen fixation (Mayer et al. 2002). The nitrogenase catalyzes the following reaction in a unidirectional way: $N_2 + 8e^- + 8H^+ + 16ATP \rightarrow 2NH_3 + H_2 + 16ADP + 16P_i$. In the absence of molecular nitrogen, the enzyme exclusively catalyzes hydrogen production with a higher yield. Although the nitrogenase is also oxygen-sensitive, this is not a problem for hydrogen production since heterocysts keep an internal microoxic environment. These cells do not produce oxygen because they lack PSII and are surrounded by a thick cell wall that restricts oxygen diffusion (Eroglu and Melis 2011; Masukawa et al. 2012). The reaction catalyzed by the nitrogenase requires a high energy expenditure in the form of ATP, which is achieved via cyclic photophosphorylation. The electrons are provided by ferredoxin, which in turn requires PSI illumination. The source of both reducing power and energy depends ultimately on the photosynthetic activity of neighbouring vegetative cells, which synthesize carbohydrates via the photosynthetic electron transport chain and the Calvin cycle. These carbohydrates are exported to heteterocysts and used as a source of NAD(P)H and ATP. NAD(P)H feeds electrons into the plastoquinone pool via NADH dehydrogenase (Ndh1), the reducing equivalents being subsequently directed to PSI, and eventually to nitrogenase (Skizim et al. 2012) (Fig. 1C).

Hydrogen production mechanisms

Green algae and cyanobacteria from different genus have shown diverse hydrogen production yields (Meuser et al. 2009; Eroglu and Melis 2011; Khetkorn et al. 2012a; Kothari et al. 2012). Hydrogen bioproduction by *C. reinhardtii* has been the most studied example among all photosynthetic organisms. This unicellular alga has a fairly high hydrogen synthesis yield compared to other green algae, in its wild-type version (Timmins et al. 2009a); it exhibits also clear advantages for genetic-engineering studies, as previously mentioned. When *C. reinhardtii* cultures are sealed to induce anaerobiosis and exposed to light, only a transient hydrogen pulse (60–90 s) is detected, since the oxygen release concomitant to water oxidation at PSII immediately poisons the [FeFe]-hydrogenase (Ghirardi et al. 1997). A sustained (four days) hydrogen production was achieved by Melis et al. (2000) with *C. reinhardtii* under sulfur deprivation conditions. The changes occurring after the onset of the sulfur-starvation phase have been thoroughly studied from a transcriptomic and metabolomic perspective (Timmins et al. 2009b; Doebbe et al. 2010; Nguyen et al. 2011). The sequence of events begins with an aerobic phase, characterized by the accumulation of starch and triacylglycerides (Zhang et al. 2002; Timmins et al. 2009b). This is followed by a rapid degradation of ribulose 1,5-bisphosphate carboxylase/oxygenase (Rubisco) that impairs Calvin-cycle function, and thus, eliminates one of the main electron sinks of the photosynthetic electron-transport chain. In parallel, the nutrient deficiency stress hampers new protein synthesis, which is crucial for repair of the PSII protein D1, so, the number of active PSII centers decreases to 5–10% of normal levels (Zhang et al. 2002). Since the mitochondria activity is essentially unaffected, the photosynthesis/respiration rate drops dramatically, the cells turn to anaerobiosis and switch on the synthesis of [FeFe]-hydrogenase. During this period, a low activity of PSII keeps electron flow from water through the photosynthetic chain, and ultimately into ferredoxin and the hydrogenase. In addition, limited oxygen availability reduces respiratory rates at the mitochondria, with glycolysis and anaerobic fermentation becoming important pathways for ATP formation (Timmins et al. 2009b). On the other hand, the mobilization of starch reserves generates reducing power in the form of NAD(P)H, feeding electrons to plastoquinone through the enzyme Nda2 (Fig. 1A). Moreover, pyruvate derived from glycolysis may also contribute to reduce ferredoxin through pyruvate:ferredoxin oxidoreductase. Both sources of reducing power (water and starch) are important for hydrogen production under these conditions (Hemschemeier et al. 2008; Chochois et al. 2009). Eventually, hydrogen production comes to an end, as a result of the generalized cell damage induced by the extended sulfur shortage. Therefore, hydrogen production through sulfur deprivation requires not only a previous lag-phase for anaerobic conditions to be established, but also a regeneration phase in a non-deficient medium for cell recovery after nutrient stress. Anyhow, this induction system may compromise cell viability for extended periods, and maximum light-conversion efficiencies (less than 0.5%) are anyway below the values recommended for technical and economic feasibility (i.e., above 10%) (Posewitz et al. 2008). Subsequent studies have evaluated hydrogen production by *C. reinhardtii* subjected to different nutrient deprivation regimes, including N-, P-, and Mg-deficient media (reviewed in González-Ballester et al. 2015). Despite hydrogen production rates were comparable (in the same order of magnitude) to those achieved with S-deprived cultures, the underlying physiological mechanisms differed, specially, in the relative contribution of PSII or in the routes to feed electrons to ferredoxin. Alternative electron sources arise from starch mobilization (in P- and Mg-deficient cultures), as previously indicated, or protein degradation (in N-deficient cultures). Furthermore, the finding that pyruvate:ferredoxin oxidoreductase from *C. reinhardtii* was able to transfer electrons from oxaloacetate to hydrogenase through ferredoxin (Fig. 1A) indicated that acetate fermentation via the glyoxylate pathway may also be coupled to hydrogen production (Noth et al. 2013). This route has been shown to operate *in vivo* with mixotrophic, non-stressed cultures of *C. reinhardtii*, grown with acetate under low light intensity, which were able to produce hydrogen to similar yields than S-deprived cultures (Jurado-Oller et al. 2015). In this case, acetate consumption rather than starch mobilization was used as the source of reducing power by hydrogenase. Low light intensities were required to restrict oxygen evolution through PSII. Interestingly, slight aeration that kept low oxygen levels improved hydrogen production, by enhancing acetate uptake and releasing inhibitory hydrogen partial pressure. Altogether, these results indicate that manipulation of growth conditions is a key aspect to optimize hydrogen yields. On the other hand, genetic engineering approaches have also aimed at increasing hydrogen production, mainly directed to alleviate the problem of the oxygen-sensitivity of the [FeFe]-hydrogenase or to enhance the electron flow from ferredoxin into hydrogenase, as will be discussed later.

In cyanobacteria, a two-stage strategy has been also addressed to sustain hydrogen production (reviewed by Srirangan et al. 2011). In the case of heterocyst-forming cyanobacteria (*Anabaena cylindrica, Nostoc muscorum, Anabaena variabilis*), nitrogen starvation is required to induce the differentiation of vegetative cells into heterocysts. Therefore, cell growth under aerobic conditions in light is promoted at a first stage; and hydrogen production is induced at a second stage, by imposing a metabolic stress (nitrogen deficiency) under an argon atmosphere supplemented in some cases with CO_2. The slower turnover rate of nitrogenases (6.4 s^{-1}) compared to bidirectional [NiFe]-hydrogenases (98 s^{-1}), together with the high energetic cost of the nitrogenase catalyzed reaction have encouraged the study of hydrogen production in non-diazotrophic cyanobacteria (*Synechoccoccus, Synechocystis, Gloebacter violaceus*). In this case, a temporal separation is established between oxygen-evolving photosynthesis (generating energy and reducing power) and anoxic hydrogen evolution—as happens in green algae and in contrast to heterocyst-forming cyanobacteria, where a physical separation exists between the two processes. The two-stage operational mode consists of an initial phase in a nutrient-enriched medium in the light to accumulate cell biomass, which is subsequently transferred to a nutrient-limited medium (deficient in nitrogen or sulfur). In this second stage, anaerobic conditions are kept via argon bubbling, and glycogen is further accumulated. As previously mentioned, hydrogen production by the bidirectional [NiFe]-hydrogenase is promoted in the dark (which is opposed to the heterocysts system, requiring light for hydrogen production, as shown in Fig. 1) and begins at the onset of anaerobiosis supported by glycogen catabolism. Efficiency of hydrogen production is limited in cyanobacteria by specific constraints of these systems (such as the presence of an uptake [NiFe]-hydrogenase, or the low frequency of heterocyst formation), as well as common restrictions to the green algae mechanism (such as the oxygen-sensitivity of the hydrogenase or the competition for electrons from non-hydrogenase pathways) (Eroglu and Melis 2011; Srirangan et al. 2011).

Targets to improve hydrogen production

Based on the different pathways leading to hydrogen synthesis (Fig. 1) and the regulation mechanisms described above, different strategies have been undertaken to obtain new strains capable of producing hydrogen to higher yields and/or with an improved process for industrial scale up. The final goal of some of these designs is to bypass the problem of the hydrogenase inhibition by oxygen. This may be accomplished by engineering the enzyme to decrease its oxygen sensitivity (reviewed in Ghirardi et al. 2015). In the case of [FeFe]-hydrogenase, rational design of hydrogenase mutants may be directed to narrow the putative oxygen diffusion channels into the active site (Ghirardi et al. 2005; Lambertz et al. 2011) or to change the redox potential of the [4Fe4S] cubane (Lambertz et al. 2011). Other strategies include random mutagenesis (Nagy et al. 2007; Stapleton and Swartz 2010) or metagenomic analysis to explore the natural diversity of these enzymes (Rusch et al. 2007; Warnecke et al. 2007). However, no oxygen-insensitive [FeFe]-hydrogenase has been found so far. In contrast, oxygen-tolerant [NiFe]-hydrogenases have been discovered in non-photosynthetic microorganisms, although their expression in cyanobacteria was not successful until they were co-translated with their corresponding maturation machinery. This strategy has permitted the expression of active oxygen-tolerant [NiFe]-hydrogenases from *Alteromonas macleodii* and *Thiocapsa roseopersicina* in *Synechococcus elongatus* (Weyman et al. 2011).

As an alternative to alleviate the oxygen-sensitivity problem, another plausible approach is to decrease the oxygen concentration in the environment of the enzyme. This has been accomplished in *C. reinhardtii* by the overexpression of oxygen-sequestering proteins, such as ferrochelatase or leghemoglobin, which caused an increase in hydrogen production up to 4.5-fold compared to the wild-type strain (Wu et al. 2010, 2011). A different strategy may be to decrease oxygen evolution through PSII. Constitutive and full inhibition of the complex by knocking-out any of its subunits resulted in detrimental hydrogen production (Makarova et al. 2007; Hemschemeier et al. 2008)—probably because photosynthetic activity is required for starch accumulation. However, a regulated partial inhibition of PSII could be more effective, as suggested by the CyC6Nac2.49 *C. reinhardtii* transgenic strain, where the expression of the PSII D2 protein is down regulated in the presence of Cu^{2+} and up-regulated under copper depletion or anaerobiosis. When this strain was grown aerobically in a medium lacking copper, anaerobiosis, and hydrogen synthesis could be transiently induced upon copper addition (Surzycki et al. 2007). Furthermore, because in a copper-repleted medium PSII expression is only induced under anaerobic conditions, a system based on this strain may operate with a negative feedback mechanism. Thus, excessive oxygen evolution from PSII (i.e., further surpassing oxygen consumption by mitochondrial respiration) would prompt down-regulation of PSII expression, therefore extending the anaerobiosis period and hydrogen synthesis. Indeed, this strain supported sustained hydrogen evolution under a light/dark regime with higher efficiency than that achieved with the wild-type strain under the same conditions (Batyrova and Hallenbeck 2017). However, hydrogen production yields were lower than those reported with the nutrient-deprivation protocols. In a similar line of action, reduced oxygen concentrations may be achieved by the generation of mutants with a decreased photosynthesis to respiration ratio. The *apr1* mutant, with dramatically reduced photosynthetic rates and a slightly increased respiration, turned into anaerobiosis when placed in a sealed container upon illumination. However, hydrogen production required the simultaneous inhibition of the Calvin cycle by addition of glycolaldehyde (Ruhle et al. 2008), which eliminates one of the main competitors with the hydrogenase for the electrons derived from ferredoxin. This result is indicative that efficient hydrogen production probably requires multi-targeting approaches. Another strategy, also based in decreasing photosynthetic oxygen evolution, employed temperature sensitive-PSII mutants (Bayro-

Kaiser and Nelson 2016). Such strains showed a photosynthetic performance comparable to wild-type *C. reinhardtii* when grown at 25°C, but a seriously reduced PSII activity when shifted to 37°C, which allowed on-going hydrogen production for three days. Moreover, after this period, mutants returned to the wild-type phenotype upon incubation at 25°C. This behaviour may elicit a cyclic hydrogen production system with a biomass-generating phase at 25°C and a hydrogen-production phase at 37°C.

According to what has been discussed before, an important goal to optimize hydrogen production is to enhance the electron flow from ferredoxin into hydrogenase. The identification of the potential surfaces of interaction between ferredoxin and hydrogenase (Long et al. 2008) raises the possibility of designing hydrogenase mutants to form strengthened complexes with ferredoxin, thus improving electron transfer. In this regard, the HydA2 double mutant D102K/T99K has shown increased association rate with ferredoxin (Long et al. 2008, 2009). Alternatively, electrons can be redirected to ferredoxin by decreasing their flow to the competitive acceptor ferredoxin:NADP reductase. This was achieved with a ferredoxin mutant with lower affinity for the competitor but a similar interaction with hydrogenase, resulting in higher hydrogen synthesis *in vitro* (Rumpel et al. 2014). A more straightforward approach has been the replacement of hydrogenase by a ferredoxin-hydrogenase fusion (Fd-HydA), in *C. reinhardtii*. The chimeric protein accumulated to lower amounts than the wild-type hydrogenase but showed a 4.5-fold increased hydrogen photoproduction rate per mol of enzyme, indicative of a more efficient electron transfer from PSI to hydrogenase (Eilenberg et al. 2016). Importantly, the ferredoxin module in the fusion protein is not active as an electron intermediate between PSI and hydrogenase, as evidenced by *in vitro* experiments carried out with thylakoid preparations, where addition of wild-type ferredoxin was required for hydrogen release (Yacoby et al. 2011). The higher activity of the chimeric hydrogenase can be explained taking in account that ferredoxin:NADP reductase is physically attached to PSI. Thus, the ferredoxin module in the hybrid protein may direct the hydrogenase to a closer proximity to PSI, facilitating electron transfer. A direct electron transfer from PSI to hydrogenase, avoiding the ferredoxin intermediate, was achieved by fusing PSI subunit PsaE from *Thermosynechococcus elongatus* with the oxygen-tolerant [NiFe]-hydrogenase from *Ralstonia eutropha*. The hydrogenase-PsaE fusion was assembled together with a PsaE-depleted PSI isolated from a *Synechocystis* sp. PCC6803 mutant. The resulting hydrogenase-PSI complex was able to sustain light-driven hydrogen production *in vitro* (Ihara et al. 2006). Alternatively, the electron current from ferredoxin may be diverted to hydrogen synthesis by hampering the activity of competing pathways. The use of specific inhibitors redirecting the electron flow towards nitrogenase and bidirectional [NiFe]-hydrogenase has indeed enhanced hydrogen production in the cyanobacterium *Anabaena siamensis* TISTR 8012 (Khetkorn et al. 2012a). *Synechocystis* mutant strains disrupted in the nitrate assimilatory pathway (potentially competing with hydrogenase for reducing potential) also showed increased hydrogen production (Baebprasert et al. 2011). A *C. reinhardtii* strain (CC-2803) lacking Rubisco, the enzyme catalyzing the first step in CO_2 fixation through the Calvin cycle, is able to produce hydrogen without submitting the cells to sulfur-starvation (Hemschemeier et al. 2008). Due to CC-2803 being light sensitive and grows only in the dark, a similar strategy was attempted with *C. reinhardtii* Rubisco mutants, carrying a single substitution at the small subunit which decreased the structural stability of the enzyme. These strains were characterized by Rubisco low levels and enhanced proteolysis (Esquivel et al. 2006; Pinto et al. 2013). Mutant Y67A showed particularly high rates of Rubisco degradation, even in sulfur-repleted cultures, and this was accompanied by PSII inactivation, low photosynthetic rate, high expression of Fe-hydrogenases, and high levels of hydrogen production (around 9-fold compared to the wild-type strain under the same conditions). Rubisco manipulation may be a target not only to down-regulate the Calvin-cycle but also to increase oxygen consumption through photorespiration, since this enzyme is located at a metabolic crossroad between these two pathways (Marin-Navarro et al. 2010). Cyclic-electron transport from ferredoxin back into plastoquinone is another electron waste, in terms of hydrogen productivity. Photosynthetic organisms promote transitions between linear (state 1) and cyclic (state 2) transport to balance the energy absorbed by PSII and PSI. The *C. reinhardtii* mutant *stm6*, which is permanently blocked at state 1 (Kruse et al. 2005; Volgusheva et al. 2013) showed a 9-fold enhanced hydrogen production compared to wild-type. This is another example of the efficacy of multi-targeting strategies, since this mutant presents other relevant phenotypes, such as increased respiration rate (accelerating the transition into anoxia) and increased starch reserves.

Manipulation of carbohydrate metabolism routes is another interesting target, since these may serve as a reducing power source for ferredoxin reduction, and ultimately for feeding electrons into hydrogenase. The hydrogen production yield of the *C. reinhardtii stm6* mutant has been improved a further 1.5-fold by coexpression of a hexose symporter (Doebbe et al. 2007). A *C. reinhardtii* mutant possessing delayed starch catabolism (*sda6*) showed lower initial hydrogen release rates, but an overall increase (1.5-fold) in hydrogen production compared to a wild-type strain (Chochois et al. 2010). A prolonged synthesis of carbohydrates upon sulfur-deprivation was possible also in the *C. reinhardtii* L159I/N230Y D1 mutant, as a result of a higher amount of D1 protein that slowed-down PSII down-regulation (Scoma et al. 2012). In this mutant, anaerobiosis was sustained because the higher PSII activity was counter-balanced by a higher respiratory activity. Because of this, and probably other accompanying phenotypes (i.e., lower chlorophyll content, accumulation of carotenoids), this mutant showed a 10-fold increase in hydrogen production rate compared to the corresponding wild-type strain (Torzillo et al. 2009). Other strategies, undertaken also in *C. reinhardtii*, attempted to enhance the injection of reducing power into ferredoxin via the anaerobic fermentative pathways. The approach in some of these cases was to focus the electron flow up to pyruvate-ferredoxin oxidoreductase (PFOR), by knocking down competing pathways. Knocking-out pyruvate formate lyase (PFL1) increased hydrogen production in the dark (Philipps et al. 2011), although hydrogen evolution in the light was not significantly changed (Burgess et al. 2012) or negatively affected (Philipps et al. 2011). Alternatively, overexpression of type II NAD(P)H dehydrogenase (Nda2) in *C. reinhardtii* also enhanced hydrogen photoproduction under nutrient-deprivation conditions by increasing the reducing state of plastoquinones (Baltz et al. 2014).

Sulfur deprivation is itself a multi-targeting strategy to induce hydrogen production in green algae and (in some cases) non-heterocysts forming cyanobacteria (Srirangan et al. 2011), and this effect may be mimicked by restricting sulfur uptake. Antisense technology was used in *C. reinhardtii* to dampen the expression of sulphate permease (SulP) (Chen et al. 2005). The resulting strains were able to produce hydrogen under limited amounts of sulphate during seven days after sealing the cultures, with a 5-fold increased yield compared to a wild-type strain.

Specific strategies have been designed in the case of cyanobacteria to improve hydrogen production. As previously mentioned, nitrogen-fixing cyanobacteria contain an uptake-hydrogenase encoded by the *hup* genes, which catalyzes the unidirectional oxidation of hydrogen and therefore decreases hydrogen yield. Elimination of the Hup activity in several *Anabaena* and *Nostoc* cyanobacteria strains has enhanced hydrogen production compared to the corresponding wild-type strains (Happe et al. 2000; Lindbgerg et al. 2002; Masukawa et al. 2002; Yoshino et al. 2007; Khetkorn et al. 2012b). In the case of diazotrophic cyanobacteria, an interesting goal would be to increase heterocyst frequency upon nitrogen starvation. However, heterocyst formation is a complex process involving activation of 600–1000 genes (Lynn et al. 1986), and the critical regulatory steps are not yet completely understood. Candidate genes for manipulation are *hetR* causing an increase in heterocyst frequency when overexpressed, *patS* and *patN* that suppress heterocyst formation, and *hglK* that fortifies the glycolipid layer of the hetorocyst when upregulated (reviewed in Srirangan et al. 2011). Indeed, *patN* disruption in an *Anabaena* strain deficient in Hup hydrogenase, resulted in increased heterocysts formation and higher hydrogen production compared to the parental strain (Masukawa et al. 2017). Finally, another improvement of the two-stage process commonly used for hydrogen production with nitrogen-fixing cyanobacteria would be to avoid the replacement of nitrogen by argon, which increases the operational cost at industrial scale (Masukawa et al. 2012). Interestingly, a [MoFe]-nitrogenase containing citrate instead of homocitrate in the FeMo cofactor catalyzed nitrogen reduction poorly but was still able to reduce protons effectively in a nitrogen atmosphere (Mayer et al. 2002). In agreement with this, gene disruption of one of the genes encoding homocitrate synthase (*nifV1*) in a ΔHup strain of *Nostoc* sp. PCC 7120 increased hydrogen production in the presence of nitrogen, compared to the parental ΔHup strain (Masukawa et al. 2007). Another approach has been site-directed mutagenesis of *Anabaena* [MoFe]-nitrogenase in residues located in the vicinity of the active site to direct electron flow selectively to proton reduction in the presence of nitrogen. Several *Anabaena* mutants increased their *in vivo* rates of hydrogen production in a nitrogen atmosphere. In particular, the R284H mutant was able to accumulate up to 87% hydrogen after 1 wk under nitrogen, as compared to the reference strain under argon (Masukawa et al. 2010).

Conclusion

Microalgae use sunlight to produce TAGs and starch, and their biomass productivity greatly exceeds the best producing crops. In addition, some of the microalgae are also capable of producing hydrogen—a fuel that liberates a large amount of energy per unit mass. Microalgae biofuel can be produced on non-arable land, using for this purpose saline and wastewater streams. In this way the production of biofuel by algae will allow arable land to remain available for cultivation of crops, thus avoiding the adverse impacts on food supplies. For these features, microalgae have been considered a promising platform to produce of biofuels. However, it is of critical economic importance—for the practical use of microalgae as source of renewable energy, to achieve a significant improvement in the efficiency of biofuel production. Nowadays, with genome sequence data and a wide variety of genetic tools and powerful analytical techniques available, new genetic engineering modifications can be accomplished to exploit marine algae as efficient sources of biofuels if protocols for chloroplast and nuclear transformations, and stable expression of genes are developed for species of interest. Such modifications might include accurate reprogramming and manipulation of metabolic pathways, combined with strategies to overcome the limitations of CO_2 fixation and light capture, which are crucial to optimize biomass productivity.

Acknowledgements

This study was funded by Fundação para a Ciência e Tecnologia (FCT-Portugal), Units UID/AGR/04129/2013 and by a grant from the University of Valencia (UV-INV-AE14-269247).

References

Anemaet, I.G., M. Bekker and K.J. Hellingwerf. 2010. Algal photosynthesis as the primary driver for a sustainable development in energy, feed and food production. Mar. Biotechnol. 12: 619–629.

Antal, T.K., T.E. Krendeleva and E. Tyystjärvi. 2015. Multiple regulatory mechanisms in the chloroplast of green algae: relation to hydrogen production. Photosynth. Res. 125: 357–381.

Atsumi, S., T. Hanai and J.C. Liao. 2008. Non-fermentative pathways for synthesis of branched-chain higher alcohols as biofuels. Nature 451: 86–90.

Baebprasert, W., S. Jantaro, W. Khetkorn, P. Lindblad and A. Incharoensakdi. 2011. Increased H_2 production in the cyanobacterium *Synechocystis* sp. strain PCC 6803 by redirecting the electron supply via genetic engineering of the nitrate assimilation pathway. Metab. Eng. 13: 610–616.

Bailey, S., B. Vick and J. Moseley. 2011. Manipulation of an alternative respiratory pathway in photoautotrophs. US Patent US8709765 B2.

Ball, S.G. and P. Deschamps. 2009. Starch metabolism. The *Chlamydomonas* Sourcebook. Elsevier Science and Technology, New York, 2nd Ed. Vol. 2: 1–40.

Ballat, M. 2008. Potential importance of hydrogen as a future solution to environmental and transportation problems. Int. J. Hydrogen Energy 33: 4013–4029.

Baltz, A., K.V. Dang, A. Beyly, P. Auroy, P. Richaud, L. Cournac and G. Peltier. 2014. Plastidial expression of type II NAD(P)H dehydrogenase increases the reducing dtate of plastoquinones and hydrogen photoproduction rate by the indirect pathway in *Chlamydomonas reinhardtii*. Plant Physiol. 165: 1344–1352.

Banerjee, C., K.K. Dubey and P. Shukla. 2016. Metabolic engineering of microalgal based biofuel production: prospects and challenges. Front. Microbiol. 7: 432.

Bar-Even, A., E. Noor, N.E. Lewis and R. Milo. 2010. Design and analysis of synthetic carbon fixation pathways. Proc. Natl. Acad. Sci. USA 107: 8889–8894.

Bar-Even, A., A. Flamholz, E. Noor and R. Milo. 2012. Thermodynamic constraints shape the structure of carbon fixation pathways. Biochim. Biophys. Acta 1817: 1646–1659.

Batyrova, K. and P.C. Hallenbeck. 2017. Hydrogen production by a *Chlamydomonas reinhardtii* strain with inducible expression of Photosystem II. Int J. Mol. Sci. 18: E647.

Bayro-Kaiser, V. and N. Nelson. 2016. Temperature-sensitive PSII: a novel approach for sustained photosynthetic hydrogen production. Photosynth. Res. 130: 113–121.

Blankenship, R.E., D.M. Tiede, J. Barber, G.W. Brudvig, G. Fleming, M. Ghirardi, M.R. Gunner, W. Junge, D.M. Kramer, A. Melis, T.A. Moore, C.C. Moser, D.G. Nocera, A.J. Nozik, D.R. Ort, W.W. Parson, R.C. Prince and R.T. Sayre. 2011. Comparing photosynthetic and photovoltaic efficiencies and recognizing the potential for improvement. Science 332: 805–809.

Blatti, J.L., J. Beld, C.A. Behnke, M. Mendez, S.P. Mayfield and M.D. Burkart. 2012. Manipulating fatty acid biosynthesis in microalgae for biofuel through protein-protein interactions. PLoS ONE 7(9): e42949. doi:10.1371/journal.pone.0042949.

Bonacci, W., P.K. Teng, B. Afonso, H. Niederholtmeyer, P. Grob, P.A. Silver and D.F. Savage. 2012. Modularity of a carbon-fixing protein organelle. Proc. Natl. Acad. Sci. USA 109: 478–483.

Boyle, N.R., M.D. Page, B. Liu, I.K. Blaby, D. Casero, J. Kropat, S.J. Cokus, A. Hong-Hermesdorf, J. Shaw, S.J. Karpowicz, S.D. Gallaher, S. Johnson, C. Benning, M. Pellegrini, A. Grossman and S.S. Merchant. 2012. Three acyltransferases and nitrogen-responsive regulator are implicated in nitrogen starvation-induced triacylglycerol accumulation in *Chlamydomonas*. J. Biol. Chem. 287: 15811–15825.

Boynton, J.E., N.W. Gillham, E.H. Harris, J.P. Hosler, A.M. Johnson, A.R. Jones, B.L. Randolph-Anderson, D. Robertson, T.M. Klein, K.B. Shark and J.C. Sanford. 1988. Chloroplast transformation in *Chlamydomonas* with high velocity microprojectiles. Science 240: 1534–1538.

Bowler, C., A.E. Allen, J.H. Badger, J. Grimwood, K. Jabbari, A. Kuo, U. Maheswari, C. Martens, F. Maumus, R.P. Otillar, E. Rayko, A. Salamov, K. Vandepoele, B. Beszteri, A. Gruber, M. Heijde, M. Katinka, T. Mock, K. Valentin, F. Verret, J.A. Berges, C. Brownlee, J.P. Cadoret, A. Chiovitti, C.J. Choi, S. Coesel, A. De Martino, J.C. Detter, C. Durkin, A. Falciatore, J. Fournet, M. Haruta, M.J. Huysman, B.D. Jenkins, K. Jiroutova, R.E. Jorgensen, Y. Joubert, A. Kaplan, N. Kröger, P.G. Kroth, J. La Roche, E. Lindquist, M. Lommer, V. Martin-Jézéquel, P.J. Lopez, S. Lucas, M. Mangogna, K. McGinnis, L.K. Medlin, A. Montsant, M.P. Oudot-Le Secq, C. Napoli, M. Obornik, M.S. Parker, J.L. Petit, B.M. Porcel, N. Poulsen, M. Robison, L. Rychlewski, T.A. Rynearson, J. Schmutz, H. Shapiro, M. Siaut, M. Stanley, M.R. Sussman, A.R. Taylor, A. Vardi, P. von Dassow, W. Vyverman, A. Willis, L.S. Wyrwicz, D.S. Rokhsar, J. Weissenbach, E.V. Armbrust, B.R. Green, Y. Van de Peer and I.V. Grigoriev. 2008. The *Phaeodactylum* genome reveals the evolutionary history of diatom genomes. Nature 456: 239–244.

Brodie, J., C.X. Chan, O. De Clerck, J.M. Cock, S.M. Coelho, C. Gachon, A.R. Grossman, T. Mock, J.A. Raven, A.G. Smith, H.S. Yoon and D. Bhattacharya. 2017. The algal revolution. Trends Plant Sci. 22: 726–738.

Burgess, S.J., B. Tamburic, F. Zemichael, K. Hellgardt and P.J. Nixon. 2011. Solar-driven hydrogen production in green algae. Adv. Appl. Microbiol. 75: 71–110.

Burgess, S.J., G. Tredwell, A. Molnar, J.G. Bundy and P.J. Nixon. 2012. Artificial microRNA-mediated knockdown of pyruvate formate lyase (PFL1) provides evidence for an active 3-hydroxybutyrate production pathway in the green alga *Chlamydomonas reinhardtii*. J. Biotechnol. 162: 57–66.

Cha, T.S., C.F. Chen, W. Yee, A. Aziz and S.H. Loh. 2011. Cinnamic acid, coumarin and vanillin: alternative phenolic compounds for efficient Agrobacterium-mediated transformation of the unicellular green alga, *Nannochloropsis* sp. J. Microbiol. Methods 84: 430–434.

Cha, T.S., W. Yee and A. Aziz. 2012. Assessment of factors affecting *Agrobacterium*-mediated genetic transformation of the unicellular green alga, *Chlorella vulgaris*. World J. Microbiol. Biotechnol. 28: 1771–1779.

Chen, H.C., A.J. Newton and A. Melis. 2005. Role of SulP, a nuclear-encoded chloroplast sulfate permease, in sulfate transport and H_2 evolution in *Chlamydomonas reinhardtii*. Photosynth. Res. 84: 289–296.

Chen, M., M. Schliep, R.D. Wllows, Z.L. Cai, B.A. Neilan and H. Scheer. 2010. A red-shifted chlorophyll. Science 329: 1318–1319.

Chen, M. and R.E. Blankenship. 2011. Expanding the solar spectrum used by photosynthesis. Trends Plant. Sci. 16: 427–431.

Chen, Y., Y. Wang, Y. Sun, L. Zhang and W. Li. 2001. Highly efficient expression of rabbit neutrophil peptide-1 gene in *Chlorella ellipsoidea* cells. Curr. Genet. 39: 365–370.

Cheney, D., B. Metz and J. Stiller. 2001. Agrobacterium-mediated genetic transformation in the microscopic marine red alga *Porphyra yezoensis*. J. Phycol. 37(Suppl.). 11.

Chochois, V., D. Dauvillee, A. Beyly, D. Tolleter, S. Cuine, H. Timpano, S. Ball, L. Cournac and G. Peltier. 2009. Hydrogen production in *Chlamydomonas*: photosystem II-dependent and -independent pathways differ in their requirement for starch metabolism. Plant Physiol. 151: 631–640.

Chochois, V., S. Constans, D. Dauvillee, A. Beyly, M. Soliveres, S. Ball, G. Peltier and L. Cournac. 2010. Relationships between PSII-independent hydrogen bioproduction and starch metabolism as evidenced from isolation of starch catabolism mutants in the green alga *Chlamydomonas reinhardtii*. Int. J. Hydrogen Energy 35: 10731–10740.

Day, J.G., S.P. Slocombe and M.S. Stanley. 2012. Overcoming biological constraints to enable the exploitation of microalgae for biofuels. Biores. Technol. 109: 245–251.

Dawson, H.N., R. Burlingame and A.C. Cannons. 1997. Stable transformation of *Chlorella*: rescue of nitrate reductase-deficient mutants with the nitrate reductase gene. Curr. Microbiol. 35: 356–362.

Deng, M.D. and J.R. Coleman. 1999. Ethanol synthesis by genetic engineering in cyanobacteria. Appl. Environ. Microbiol. 65: 523–528.

Doebbe, A., J. Rupprecht, J. Beckmann, J.H. Mussgnug, A. Hallmann, B. Hankamer and O. Kruse. 2007. Functional integration of the HUP1 hexose symporter gene into the genome of *C. reinhardtii*: Impacts on biological H_2 production. J. Biotechnol. 131: 27–33.

Doebbe, A., M. Keck, R.M. La, J.H. Mussgnug, B. Hankamer, E. Tekce, K. Niehaus and O. Kruse. 2010. The interplay of proton, electron, and metabolite supply for photosynthetic H_2 production in *Chlamydomonas reinhardtii*. J. Biol. Chem. 285: 30247–30260.

Dubini, A. and M.L. Ghirardi. 2015. Engineering photosynthetic organisms for the production of biohydrogen. Photosynth. Res. 123: 241–253.

Dunahay, T.G., E.E. Jarvis and P.G. Roessler. 1995. Genetic transformation of the diatoms *Cyclotella cryptica* and *Navicula saprophila*. J. Phycol. 31: 1004–1011.

Durao, P., H. Aigner, P. Nagy, O. Mueller-Cajar, F.U. Hayer-Hartl and M. Hartl. 2015. Opposing effects of folding and assembly chaperones on evolvability of Rubisco. Nat. Chem. Biol. 11: 148–155.

Eilenberg, H., I. Weiner, O. Ben-Zvi, C. Pundak, A. Marmari, O. Liran, M.S. Wecker, Y. Milrad and I. Yacoby. 2016. The dual effect of a ferredoxin-hydrogenase fusion protein *in vivo*: successful divergence of the photosynthetic electron flux towards hydrogen production and elevated oxygen tolerance. Biotechnol. Biofuels 9: 182.

Eroglu, E. and A. Melis. 2011. Photobiological hydrogen production: Recent advances and state of the art. Bioresour. Technol. 102: 8403–8413.

Esquível, M.G., T.S. Pinto, J. Marín-Navarro and J. Moreno. 2006. Substitution of tyrosine residues at the aromatic cluster around the betaA-betaB loop of rubisco small subunit affects the structural stability of the enzyme and the *in vivo* degradation under stress conditions. Biochemistry 45: 5745–5753.

Esquível, M.G., H.M. Amaro, T.S. Pinto, P.S. Fevereiro and F.X. Malcata. 2011. Efficient H_2 production via *Chlamydomonas reinhardtii*. Trends Biotechnol. 29: 595–600.

Esquível, M.G., A.R. Matos and J. Marques Silva. 2017. Rubisco mutants of *Chlamydomonas reinhardtii* display divergente photosynthetic parameters and lipid allocation. Appl. Microbiol. Biotechnol. 101: 5569–5580.

Feng, S.Y., L.X. Xue, H.T. Liu and P.J. Lu. 2009. Improvement of efficiency of genetic transformation for *Dunaliella salina* by glass beads method. Mol. Biol. Rep. 36: 1433–1439.

Gan, S.Y., S. Qin, R.Y. Othman, D. Yu and S.M. Phang. 2003. Transient expression of lacZ in particle bombarded *Gracilaria changii* (Gracilariales, Rhodophyta). J. Appl. Phycol. 15: 345–349.

Ghirardi, M.L., M.C. Posewitz, P.C. Maness, A. Dubini, J. Yu and M. Seibert. 2007. Hydrogenases and hydrogen photoproduction in oxygenic photosynthetic organisms. Annu Rev. Plant Biol. 58: 71–91.

Ghirardi, M.L., R.K. Togasaki and M. Seibert. 1997. Oxygen sensitivity of algal H_2-production. Appl. Biochem. Biotechnol. 63: 141–151.

Ghirardi, M.L., P.W. King, M.C. Posewitz, P.C. Maness, A. Fedorov, K. Kim, J. Cohen, K. Schulten and M. Seibert. 2005. Approaches to developing biological H_2-photoproducing organisms and processes. Biochem. Soc. Trans. 33: 70–72.

Ghirardi, M.L. 2015. Implementation of photobiological H_2 production: the O_2 sensitivity of hydrogenases. Photosynth. Res. 125: 383–393.

González-Ballester, D., J.L. Jurado-Oller and E. Fernandez. 2015. Relevance of nutrient media composition for hydrogen production in *Chlamydomonas*. Photosynth. Res. 125: 395–406.

Gomaa, M.A., L. Al-Haj and R.M.M. Abed. 2016. Metabolic engineering of cyanobacteria and microalgae for enhanced production of biofuels and high-value products. J. Appl. Microbiol. 121: 919–931.

Goodenough, U., I. Blaby, D. Casero, S.D. Gallaher, C. Goodson, S. Johnson J.H. Lee, S.S. Merchant, M. Pellegrini, R. Roth, J. Rusch, M. Singh, J.G. Umen, T.L. Weiss and T. Wulan. 2014. The path to triacylglyceride obesity in the sta6 strain of *Chlamydomonas reinhardtii*. Eukaryot Cell. 13: 591–613.

Greene, D.N., S.M. Whitney and I. Matsumura. 2007. Artificially evolved *Synechococcus* PCC6301 Rubisco variants exhibit improvements in folding and catalytic efficiency. Biochem. J. 404: 517–524.

Gressel, J., D. Eisenstadt, D. Schatz, S. Einbinder and S. Ufaz. 2010. Use of fluorescent protein in cyanobacteria and algae for improving photosynthesis and preventing cell damage. US Patent US20100087006.

Haimovich-Dayan, M., N. Garfinkel and D. Ewe. 2013. The role of C4 metabolism in the marine diatom *Phaeodactylum tricornutum*. New Phytol. 197: 177–185.

Hanson, M.R., M.T. Lin, A.E. Carmo-Silva and M.A.J. Parry. 2016. Towards engineering carboxysomes in C_3 plants. Plant J. 87: 38–50.

Happe, T., K. Schutz and H. Bohme. 2000. Transcriptional and mutational analysis of the uptake hydrogenase of the filamentous cyanobacterium *Anabaena variabilis* ATCC 29413. J. Bacteriol. 182: 1624–1631.

Harris, E.H., D.B. Stern and G.B. Witman. 2009. The *Chlamydomonas* Sourcebook. Elsevier Science and Technology, New York, 2nd Ed.

Hawkins, R.L. and M. Nakamura. 1999. Expression of human growth hormone by the eukaryotic alga, *Chlorella*. Curr. Microbiol. 38: 335–341.

Hedge, K., N. Chandra, S.J. Sarma, S.K. Brar and V.D. Veeranki. 2015. Genetic engineering strategies for enhanced biodiesel production. Mol. Biotechnol. 57: 606–624.

Hellingwerf, K.J. and M.J. Teixeira de Mattos. 2009. Alternative routes to biofuels: Light-driven biofuel formation from CO_2 and water based on the "photanol" approach. J. Biotechnol. 142: 87–90.

Hemschemeier, A., S. Fouchard, L. Cournac, G. Peltier and T. Happe. 2008. Hydrogen production by *Chlamydomonas reinhardtii*: an elaborate interplay of electron sources and sinks. Planta 227: 397–407.

Hirata, R., M. Takahashi, N. Saga and K. Mikami. 2011. Transient gene expression system established in *Porphyra yezoensis* is widely applicable in *Bangiophycean* algae. Mar. Biotechnol. 13: 1038–1047.

Ho, S.H., X. Ye, T. Hasunuma, J.S. Chang and A. Kondo. 2014. Perspectives on engineering strategies for improving biofuel production from microalgae—A critical review. Biotechnol. Adv. 32: 1448–1459.

Huang, X., J.C. Weber, T.K. Hinson, A.C. Mathieson and S.C. Minocha. 1996. Transient expression of the GUS reporter gene in the protoplasts and partially digested cells of *Ulva lactuca* L. (*Chlorophyta*). Bot. Mar. 39: 467–474.

Ihara, M., H. Nishihara, K.S. Yoon, O. Lenz, B. Friedrich, H. Nakamoto, K. Kojima, D. Honma, T. Kamachi and I. Okura. 2006. Light-driven hydrogen production by a hybrid complex of a [NiFe]-hydrogenase and the cyanobacterial photosystem I. Photochem. Photobiol. 82: 676–682.

Iwai, M., K. Ikeda, M. Shimojima and H. Ohta. 2014. Enhancement of extraplastidic oil synthesis in *Chlamydomonas reinhardtii* using a type-2 diacylglycerol acyltransferase with a phosphorus starvation–inducible promoter. Plant Biotechnol. J. 12: 808–819.

Jurado-Oller, J.L., A. Dubini, A. Galván, E. Fernández and D. González-Ballester. 2015. Low oxygen levels contribute to improve photohydrogen production in mixotrophic non-stressed Chlamydomonas cultures. Biotechnol. Biofuels, 8: 149.

Kebeish, R., M. Niessen, K. Thiruveedhi, R. Bari, H.J. Hirsch, R. Rosenkranz, N. Stäbler, B. Schönfeld, F. Kreuzaler and C. Peterhansel. 2007. Chloroplastic respiratory bypass increases photosynthesis and biomass production in *Arabidopsis thaliana*. Nat. Biotechnol. 25: 593–599.

Khanna, N. and P. Lindblad. 2015. Cyanobacterial hydrogenases and hydrogen metabolism revisited: recent progress and future prospects. Int. J. Mol. Sci. 16: 10537–10561.

Khetkorn, W., W. Baebprasert, P. Lindblad and A. Incharoensakdi. 2012a. Redirecting the electron flow towards the nitrogenase and bidirectional Hox-hydrogenase by using specific inhibitors results in enhanced H2 production in the cyanobacterium *Anabaena siamensis* TISTR 8012. Bioresour. Technol. 118: 265–271.

Khetkorn, W., P. Lindblad and A. Incharoensakdi. 2012b. Inactivation of uptake hydrogenase leads to enhanced and sustained hydrogen production with high nitrogenase activity under high light exposure in the cyanobacterium *Anabaena siamensis* TISTR 8012. J. Biol. Eng. 6: 19. doi:10.1186/1754-1611-6-19.

Khetkorn, W., R.P. Rastogi, A. Incharoensakdi, P. Lindblad, D. Madamwar, A. Pandey and C. Larroche. 2017. Microalgal hydrogen production—a review. Bioresour. Technol. 243: 1194–1206.

Kilian, O., C.S.E. Benemann, K.K. Niyogi and B. Vick. 2011. High efficiency homologous recombination in the oil-producing alga *Nannochloropsis* sp. Proc. Natl. Acad. Sci. USA 108: 21265–21269.

Kindle, K.L. 1990. High-frequency nuclear transformation of *Chlamydomonas reinhardtii*. Proc. Natl. Acad. Sci. USA 87: 1228–1232.

Kothari, A., R. Potrafka and F. Garcia-Pichel. 2012. Diversity in hydrogen evolution from bidirectional hydrogenases in cyanobacteria from terrestrial, freshwater and marine intertidal environments. J. Biotechnol. 162: 105–114.

Kruse, O., J. Rupprecht, K.P. Bader, S. Thomas-Hall, P.M. Schenk, G. Finazzi and B. Hankamer. 2005. Improved photobiological H2 production in engineered green algal cells. J. Biol. Chem. 280: 34170–34177.

Kumar, A., S. Ergas, X. Yuan, A. Sahu, Q. Zhang, J. Dewulf, F.X. Malcata and H. van Langenhove. 2010. Enhanced CO_2 fixation and biofuel production via microalgae: recent developments and future directions. Trends Biotechnol. 28: 371–380.

Kumar, S.V., R.W. Misquitta, V.S. Reddy, B.J. Rao and M.V. Rajam. 2004. Genetic transformation of the green alga - *Chlamydomonas reinhardtii* by *Agrobacterium tumefaciens*. Plant Sci. 166: 731–738.

Lambertz, C., N. Leidel, K.G. Havelius, J. Noth, P. Chernev, M. Winkler, T. Happe and M. Haumann. 2011. O_2 reactions at the six-iron active site (H-cluster) in [FeFe]-hydrogenase. J. Biol. Chem. 286: 40614–40623.

Lapidot, M., D. Raveh, A. Sivan, S.M. Arad and M. Shapira. 2002. Stable chloroplast transformation of the unicellular red alga *Porphyridium* species. Plant Physiol. 129: 7–12.

Lardizabal, K., R. Effertz, C. Levering, J. Mai, M.C. Pedroso, T. Jury, E. Aasen, K. Gruys and K. Bennett. 2008. Expression of *Umbelopsis ramanniana* DGAT2A in seed increases oil in soybean. Plant Physiol. 148: 89–96.

Larkum, A.W. and M. Kühl. 2005. Chlorophyll *d*: the puzzle resolved. Trends Plant Sci. 10: 355–357.

Larkum, A.W., I.L. Ross, O. Kruse and B. Hankamer. 2012. Selection, breeding and engineering of microalgae for bioenergy and biofuel production. Trends Biotechnol. 30: 198–205.

Lefebvre, S., T. Lawson, O.V. Zakhleniuk, J.C. Lloyd and C.A. Raines. 2005. Increased sedoheptulose-1,7-bisphosphatase activity in transgenic tobacco plants stimulates photosynthesis and growth from an early stage in development. Plant Physiol. 138: 451–460.

Li, Y., N. Scales, R.E. Blankenship, R.D. Willows and M. Chen. 2012. Extinction coefficients for red-shifted chlorophylls: Chlorophyll *d* and chlorophyll *f*. Biochim. Biophys. Acta 1817: 1292–1298.

Lin, M.T., A. Occhialini, P.J. Andralojc, M.A.J. Parry and M.R. Hanson. 2014. A faster Rubisco with potential to increase photosynthesis in crops. Nature 513: 545–550.

Lindberg, P., K. Schütz, T. Happe and P. Lindblad. 2002. A hydrogen producing, hydrogenase-free mutant strain of *Nostoc punctiforme* ATCC 29133. Int. J. Hydrogen Energy 27: 1291–1296.

Liu, L., Y. Wang, Y. Zhang, X. Chen, P. Zhang and S. Ma. 2012. Development of a new method for genetic transformation of the green alga *Chlorella ellipsoidea*. Mol. Biotechnol. 54: 211–219.

Long, H., C.H. Chang, P.W. King, M.L. Ghirardi and K. Kim. 2008. Brownian dynamics and molecular dynamics study of the association between hydrogenase and ferredoxin from *Chlamydomonas reinhardtii*. Biophys. J. 95: 3753–3766.

Long, H., P.W. King, M.L. Ghirardi and K. Kim. 2009. Hydrogenase/ferredoxin charge-transfer complexes: effect of hydrogenase mutations on the complex association. J. Phys. Chem. A 113: 4060–4067.

Lynn, M.E., J.A. Bantle and J.D. Ownby. 1986. Estimation of gene expression in heterocysts of *Anabaena variabilis* by using DNA-RNA hybridization. J. Bacteriol. 167: 940–946.

Makarova, V.V., S. Kosourov, T.E. Krendeleva, B.K. Semin, G.P. Kukarskikh, A.B. Rubin, R.T. Sayre, M.L. Ghirardi and M. Seibert. 2007. Photoproduction of hydrogen by sulfur-deprived *C. reinhardtii* mutants with impaired photosystem II photochemical activity. Photosynth. Res. 94: 79–89.

Marin-Navarro, J., M.G. Esquivel and J. Moreno. 2010. Hydrogen production by *Chlamydomonas reinhardtii* revisited: Rubisco as a biotechnological target. World J. Microbiol. Biotechnol. 26: 1785–1793.

Martin, B.A. and P.D. Frymier. 2017. A review of hydrogen production by photosynthetic organisms using whole-cell and cell-free systems. Appl. Biochem. Biotechnol. 183: 503–519.

Masukawa, H., K. Inoue and H. Sakurai. 2007. Effects of disruption of homocitrate synthase genes on *Nostoc* sp. strain PCC 7120 photobiological hydrogen production and nitrogenase. Appl. Environ. Microbiol. 73: 7562–7570.

Masukawa, H., K. Inoue, H. Sakurai, C.P. Wolk and R.P. Hausinger. 2010. Site-directed mutagenesis of the *Anabaena* sp. strain PCC 7120 nitrogenase active site to increase photobiological hydrogen production. Appl. Environ. Microbiol. 76: 6741–6750.

Masukawa, H., M. Mochimaru and H. Sakurai. 2002. Disruption of the uptake hydrogenase gene, but not of the bidirectional hydrogenase gene, leads to enhanced photobiological hydrogen production by the nitrogen-fixing cyanobacterium *Anabaena* sp. PCC 7120. Appl. Microbiol. Biotechnol. 58: 618–624.

Masukawa, H., M. Kitashima, K. Inoue, H. Sakurai and R.P. Hausinger. 2012. Genetic engineering of cyanobacteria to enhance biohydrogen production from sunlight and water. Ambio. 41 Suppl 2: 169–173.

Masukawa, H., H. Sakurai, R.P. Hausinger and K. Inoue. 2017. Increased heterocyst frequency by patN disruption in *Anabaena* leads to enhanced photobiological hydrogen production at high light intensity and high cell density. Appl. Microbiol. Biotechnol. 101: 2177–2188.

Mata, T.M., A.A. Martins and N.S. Caetano. 2010. Microalgae for biodiesel production and other applications: a review. Renew. Sustain. Energy Rev. 14: 217–232.

Matsunaga, T., H. Takeyama and N. Nakamura. 1990. Characterization of cryptic plasmids from marine cyanobacteria and construction of a hybrid plasmid potentially capable of transformation of marine cyanobacterium, *Synechococcus* sp. and its transformation. Appl. Biochem. Biotechnol. 24: 151–160.

Mayer, S.M., C.A. Gormal, B.E. Smith and D.M. Lawson. 2002. Crystallographic analysis of the MoFe protein of nitrogenase from a nifV mutant of *Klebsiella pneumoniae* identifies citrate as a ligand to the molybdenum of iron molybdenum cofactor (FeMoco). J. Biol. Chem. 277: 35263–35266.

McGrath, J.M. and S.P. Long. 2014. Can the cyanobacterial carbon-concentrating mechanism increase photosynthesis in crop species? A theoretical analysis. Plant Physiol. 164: 2247–2261.

McIntosh, C.L., F. Germer, R. Schulz, J. Appel and A.K. Jones. 2011. The [NiFe]-hydrogenase of the cyanobacterium *Synechocystis* sp. PCC 6803 works bidirectionally with a bias to H$_2$ production. J. Am. Chem. Soc. 133: 11308–11319.

Melis, A., L. Zhang, M. Forestier, M.L. Ghirardi and M. Seibert. 2000. Sustained photobiological hydrogen gas production upon reversible inactivation of oxygen evolution in the green alga *Chlamydomonas reinhardtii*. Plant Physiol. 122: 127–136.

Merchant, S.S., S.E. Prochnik, O. Vallon, E.H. Harris, S.J. Karpowicz, G.B. Witman, A. Terry, A. Salamov, L.K Fritz-Laylin, L. Maréchal-Drouard, W.F. Marshall, L.H. Qu, D.R. Nelson, A.A. Sanderfoot, M.H. Spalding, V.V. Kapitonov, Q. Ren, P. Ferris, E. Lindquist, H. Shapiro, S.M. Lucas, J. Grimwood, J. Schmutz, P. Cardol, H. Cerutti, G. Chanfreau, C.L. Chen, V. Cognat, M.T. Croft, R. Dent, S. Dutcher, E. Fernández, H. Fukuzawa, D. González-Ballester, D. González-Halphen, A. Hallmann, M. Hanikenne, M. Hippler, W. Inwood, K. Jabbari, M. Kalanon, R. Kuras, P.A. Lefebvre, S.D. Lemaire, A.V. Lobanov, M. Lohr, A. Manuell, I. Meier, L. Mets, M. Mittag, T. Mittelmeier, J.V. Moroney, J. Moseley, C. Napoli, A.M. Nedelcu, K. Niyogi, S.V. Novoselov, I.T. Paulsen, G. Pazour, S. Purton, J.P. Ral, D.M. Riaño-Pachón, W. Riekhof, L. Rymarquis, M. Schroda, D. Stern, J. Umen, R. Willows, N. Wilson, S.L. Zimmer, J. Allmer, J. Balk, K. Bisova, C.J. Chen, M. Elias, K. Gendler, C. Hauser, M.R. Lamb, H. Ledford, J.C. Long, J. Minagawa, M.D. Page, J. Pan, W. Pooltham, S. Roje, A. Rose, E. Stahlberg, A.M. Terauchi, P. Yang, S. Ball, C. Bowler, C.L. Dieckmann, V.N. Gladyshev, P. Green, R. Jorgensen, S. Mayfield, B. Mueller-Roeber, S. Rajamani, R.T. Sayre, P. Brokstein, I. Dubchak, D. Goodstein, L. Hornick, Y.W. Huang, J. Jhaveri, Y. Luo, D. Martínez, W.C. Ngau, B. Otillar, A. Poliakov, A. Porter, L. Szajkowski, G. Werner, K. Zhou, I.V. Grigoriev, D.S. Rokhsar and A.R. Grossman. 2007. The *Chlamydomonas* genome reveals the evolution of key animal and plant functions. Science 318: 245–250.

Meuser, J.E., G. Ananyev, L.E. Wittig, S. Kosourov, M.L. Ghirardi, M. Seibert, G.C. Dismukes and M.C. Posewitz. 2009. Phenotypic diversity of hydrogen production in chlorophycean algae reflects distinct anaerobic metabolisms. J. Biotechnol. 142: 21–30.

Meuser, J.E., S. D'Adamo, R.E. Jinkerson, F. Mus, W. Yang, M.L. Ghirardi, M. Seibert, A.R. Grossman and M.C. Posewitz. 2012. Genetic disruption of both *Chlamydomonas reinhardtii* [FeFe]-hydrogenases: Insight into the role of HYDA2 in H$_2$ production. Biochem. Biophys. Res. Commun. 417: 704–709.

Miyagawa, Y., M. Tamoi and S. Shigeoka. 2001. Overexpression of a cyanobacterial fructose-1,6-/sedoheptulose-1,7-bisphosphatase in tobacco enhances photosynthesis and growth. Nat. Biotechnol. 19: 965–969.

Mielke, S.P., N.Y. Kiang, R.E. Blankenship, M.R. Gunner and D. Mauzerall. 2011. Efficiency of photosynthesis in a Chl d-utilizing cyanobacterium is comparable to or higher than in Chl a-utilizing oxygenic species. Biochim. Biophys. Acta 1807: 1231–1236.

Mueller-Cajar, O. and S.M. Whitney. 2008a. Evolving improved *Synechococcus* Rubisco functional expression in *Escherichia coli*. Biochem. J. 414: 205–214.

Mueller-Cajar, O. and S.M. Whitney. 2008b. Directing the evolution of Rubisco and Rubisco activase: first impressions of a new tool for photosynthesis research. Photosynth. Res. 98: 667–675.

Mulder, D.W., E.S. Boyd, R. Sarma, R.K. Lange, J.A. Endrizzi, J.B. Broderick and J.W. Peters. 2010. Stepwise [FeFe]-hydrogenase H-cluster assembly revealed in the structure of HydA(DeltaEFG). Nature 465: 248–251.

Mussgnug, J.H., S. Thomas-Hall, J. Rupprecht, A. Foo, V. Klassen, A. McDowall, P.M. Schenk, O. Kruse and B. Hankamer. 2007. Engineering photosynthetic light capture: impacts on improved solar energy to biomass conversion. Plant Biotechnol. J. 5: 802–814.

Muto, M., Y. Fukuda, M. Nemoto, T. Yoshino, T. Matsunaga and T. Tanaka. 2013. Establishment of a genetic transformation system for the marine pennate diatom *Fistulifera* sp. strain JPCC DA0580 a high triglyceride producer. Mar Biotechnol. (NY) 15: 48–55.

Nagarajan, D., D.J. Lee, A. Kondo and J.S. Chang. 2017. Recent insights into biohydrogen production by microalgae - From biophotolysis to dark fermentation. Bioresour. Technol. 227: 373–387.

Nagy, L.E., J.E. Meuser, S. Plummer, M. Seibert, M.L. Ghirardi, P.W. King, D. Ahmann and M.C. Posewitz. 2007. Application of gene-shuffling for the rapid generation of novel [FeFe]-hydrogenase libraries. Biotechnol. Lett. 29: 421–430.

Ng, I.S., S.I. Tan, P.H. Kao, Y.K. Chang and J.S. Chang. 2017. Recent developments on genetic engineering of microalgae for biofuels and bio-based chemicals. Biotechnol. J. 12: 1600644. doi:10.1002/biot.201600644.

Nguyen, A.V., J. Toepel, S. Burgess, A. Uhmeyer, O. Blifernez, A. Doebbe, B. Hankamer, P. Nixon, L. Wobbe and O. Kruse. 2011. Time-course global expression profiles of *Chlamydomonas reinhardtii* during photo-biological H$_2$ production. PLoS One 6: e29364.

Niu, Y-F., Z-K. Yang, M-H. Zhang, C-C. Zhu, W-D. Yang, J-S. Liu and H-Y. Li. 2012. Transformation of diatom *Phaeodactylum tricornutum* by electroporation and establishment of inducible selection marker. Biotechniques. 52: 1–3. doi: 10.2144/000113881.

Noth, J., D. Krawietz, A. Hemschemeier and T. Happe. 2013. Pyruvate:ferredoxin oxidoreductase is coupled to light-independent hydrogen production in *Chlamydomonas reinhardtii*. J. Biol. Chem. 288: 4368–4377.

Nymark, M., A.K. Sharma, T. Sparstad, A.M. Bones and P. Winge. 2016. A CRISPR/Cas9 system adapted for gene editing in marine algae. Sci. Rep. 6: 24951. doi: 10.1038/srep24951

Oey, M., A.L. Sawyer, I.L. Ross and B. Hankamer B1. 2016. Challenges and opportunities for hydrogen production from microalgae. Plant Biotechnol. J. 14: 1487–1499.

Ogawa, T., M. Tamoi, A. Kimura, A. Mine, H. Sakuyama, E. Yoshida, T. Maruta, K. Suzuki, T. Ishikawa and S. Shigeoka. 2015. Enhancement of photosynthetic capacity in *Euglena gracilis* by expression of cyanobacterial fructose-1,6-/sedoheptulose-1,7-bisphosphatase leads to increases in biomass and wax ester production. Biotechnol. Biofuels. 8: 80, doi: 10.1186/s13068-015-0264-5.

Orr, D., A. Alcantara, M.V. Kapralov, P.J. Andralojc, E. Carmo-Silva and M.A.J. Parry. 2016. Surveying Rubisco diversity and temperatura response to improve crop photosynthetic efficiency. Plant Physiol. 172: 707–717.

Parker, M.S., T. Mock and E.V. Armbrust. 2008. Genomic insights into marine microalgae. Annu. Rev. Genet. 42: 619–645.

Parikh, M.R., D.N. Greene, K.K. Woods and I. Matsumura. 2006. Directed evolution of Rubisco hypermorphs through genetic selection in engineered *E. coli*. Protein Eng. Des. Sel. 19: 113–119.

Peterhansel, C. and V.G. Maurino. 2011. Photorespiration redesigned. Plant Physiol. 155: 49–55.

Philipps, G., D. Krawietz, A. Hemschemeier and T. Happe. 2011. A pyruvate formate lyase-deficient *Chlamydomonas reinhardtii* strain provides evidence for a link between fermentation and hydrogen production in green algae. Plant J. 66: 330–340.

Pinto, T.S., F.X. Malcata, J.D. Arrabaça, J.M. Silva, R.J. Spreitzer and M.G. Esquível. 2013. Rubisco mutants of *Chlamydomonas reinhardtii* enhance photosynthetic hydrogen production. Appl. Microbiol. Biotechnol. 97: 5635–5643

Polle, J.E.W., S.D. Kanakagiri and A. Melis. 2003. tla1, a DNA insertional transformant of the green alga *Chlamydomonas reinhardtii* with a truncated light harvesting chlorophyll antenna size. Planta 217. 49–59.

Posewitz, M.C., A. Dubini, J.E. Meuser, M. Seibert and M.L. Ghirardi. 2008. Hydrogenases hydrogen production and anoxia. pp. 217–255. *In*: D. Stern (ed.). The *Chlamydomonas* Sourcebook, vol. 2. Academic Press. Oxford.

Poulsen, N., P.M. Chesley and N. Kröger. 2006. Molecular genetic manipulation of the diatom *Thalassiosira pseudonana* (Bacillariophyceae). J. Phycol. 42: 1059–1065.

Poulsen, N. and N. Kröger. 2005. A new molecular tool for transgenic diatoms. FEBS J. 272: 3413–3423.

Qin, S., G.Q. Sun, P. Jiang, L.H. Zou, Y. Wu and C.K. Tseng. 1999. Review of genetic engineering of *Laminaria japonica* (Laminariales, Phaeophyta) in China. Hydrobiologia 398: 469–472.

Qin, S., D.Z. Yu, P. Jiang, C.Y. Teng and C.K. Zeng. 2003. Stable expression of lacZ reporter gene in seaweed *Undaria pinnatifida*. High Technol. Lett. 13: 87–89.

Qin, S., H. Lin and P. Jiang. 2012. Advances in genetic engineering of marine algae. Biotechnol. Adv. 30: 1602–1613.

Radakovits, R., R.E. Jinkerson, A.l. Darzins and M.C. Posewitz. 2010a. Genetic engineering of algae for enhanced biofuel production. Eukaryotic Cell 9: 486–501.

Radakovits, R., P.M. Eduafo and M.C. Posewitz. 2010b. Genetic engineering of fatty acid chain length in *Phaeodactylum tricornutum*. Metabolic Engineer. 13: 89–95.

Radakovits, R., R.E. Jinkerson, S.I. Fuerstenberg, H. Tae, R.E. Settlage, J.L. Boore and M.C. Posewitz. 2012. Draft genome sequence and genetic transformation of the oleaginous alga *Nannochloropsis gaditana*. Nat. Commun. 3: 686. doi: 10.1038/ncomms1688.

Rae, B.D., B.M. Long, M.R. Badger and G.D. Price. 2013. Functions, compositions, and evolution of the two types of carboxysomes: polyhedral microcompartments that facilitate CO$_2$ fixation in cyanobacteria and some proteobacteria. Microbiol. Mol. Biol. Rev. 77: 357–379.

Raines, C.A. 2003. The Calvin cycle revisited. Photosynth. Res. 75: 1–10.

Ramanna, L., I. Rawat and F. Bux. 2017. Light enhancement strategies improve microalgal biomass productivity. Renew. Sust. Energ. Rev. 80: 765–773.

Ramazanov, A. and Z. Ramazanov. 2006. Isolation and characterization of a starchless mutant of *Chlorella pyrenoidosa* STL-PI with a high growth rate, and high protein and polyunsaturated fatty acid content. Phycol. Res. 54: 255–259.

Raven, J.A. 2010. Inorganic carbon acquisition by eukaryotic algae: four current questions. Photosynth. Res. 106: 123–134.

Raven, J.A. and M. Giordano. 2017. Acquisition and metabolism of carbon in the Ocrophyta other than diatoms. Phil. Trans. Roy. Soc. B 372: 20160400. doi:10.1098/rstb.2016.0400.

Reinfelder, J.R. 2011. Carbon concentrating mechanisms in eukaryotic marine phytoplankton. Annu. Rev. Mar. Sci. 3: 291–315.

Rosgaard, L., A.J. de Porcellinis, J.H. Jacobsen, N.U. Frigaard and Y. Sakuragi. 2012. Bioengineering of carbon fixation, biofuels, and biochemicals in cyanobacteria and plants. J. Biotechnol. 162: 134–147.

Ruhle, T., A. Hemschemeier, A. Melis and T. Happe. 2008. A novel screening protocol for the isolation of hydrogen producing *Chlamydomonas reinhardtii* strains. BMC. Plant Biol. 8: 107.

Rumpel, S., J. Siebel, C, Far, J. Duan, E. Reijerse, T. Happe, W. Lubitza and M. Winkler. 2014. Enhancing hydrogen production of microalgae by redirecting electrons from photosystem I to hydrogenase. Energy Environ. Sci. 7: 3296–3301.

Rusch, D.B., A.L. Halpern, G. Sutton, K.B. Heidelberg, S. Williamson, S. Yooseph, D. Wu, J.A. Eisen, J.M. Hoffman, K. Remington, K. Beeson, B. Tran, H. Smith, H. Baden-Tillson, C. Stewart, J. Thorpe, J. Freeman, C. ndrews-Pfannkoch, J.E. Venter, K. Li, S. Kravitz, J.F. Heidelberg, T. Utterback, Y.H. Rogers, L.I. Falcon, V. Souza, G. Bonilla-Rosso, L.E. Eguiarte, D.M. Karl, S. Sathyendranath, T. Platt, E. Bermingham, V. Gallardo, G. Tamayo-Castillo, M.R. Ferrari, R.L. Strausberg, K. Nealson, R. Friedman, M. Frazier and J.C. Venter. 2007. The sorcerer II global ocean sampling expedition: northwest Atlantic through eastern tropical Pacific. PLoS. Biol. 5: e77.

Savir, Y., E. Noor, R. Milo and T. Tlusty. 2010. Cross-species analysis traces adaptation of Rubisco toward optimality in a low-dimensional landscape. Proc. Natl. Acad. Sci. USA 107: 3475–3480.

Schliep, M., B. Crossett, R.D. Willows and M. Chen. 2010. ¹⁸O Labeling of chlorophyll *d* in *Acaryochloris marina* reveals that chlorophyll *a* and molecular oxygen are precursors. J. Biol. Chem. 285: 28450–28456.

Scoma, A., D. Krawietz, C. Faraloni, L. Giannelli, T. Happe and G. Torzillo. 2012. Sustained H$_2$ production in a *Chlamydomonas reinhardtii* D1 protein mutant. J. Biotechnol. 157: 613–619.

Seo, Y.H., Y. Lee, D.Y. Jeon and J.I. Han. 2015. Enhancing the light utilization efficiency of microalgae using organic dyes. Bioresour. Technol. 181: 355–359.

Shih, P.M., J. Zarzycki, K.K. Niyogi and C.A. Kerfeld. 2014. Introduction of a synthetic CO$_2$-fixing photorespiratory bypass into a cyanobacterium. J. Biol. Chem. 289: 9493–9500.

Shin, S.E., J.M. Lim, H.G. Koh, E.K. Kim, N.K. Kang, S. Jeon, S. Kwon, W.S. Shin, B. Lee, K. Hwangbo, J. Kim, S.H. Ye, J.Y. Yun, H. Seo, H.M. Oh, K.J. Kim, J.S. Kim, W.J. Jeong, Y.K. Chang and B.R. Jeong. 2016. CRISPR/Cas9-induced knockout and knock-in mutations in *Chlamydomonas reinhardtii*. Sci Rep. 6: 27810. doi:10.1038/srep27810.

Shimogawara, H., S. Fujiwara, A. Grossman and H. Usuda. 1998. High-Efficiency Transformation of *Chlamydomonas reinhardtii* by Electroporation. Genetics 148: 1821–1828.

Simionato, D., S. Basso, G.M. Giacometti and T. Morosinotto. 2013. Optimization of light use efficiency for biofuel production in algae. Biophys. Chem. 182: 71–78

Skizim N.J., G.M. Ananyev, A. Krishnan and G.C. Dismukes. 2012. Metabolic pathways for photobiological hydrogen production by nitrogenase and hydrogenase-containing unicellular cyanobacteria *Cyanothece*. J. Biol. Chem. 287: 2777–2786.

Son, S.H., J-W. Ahn, T. Uji, D-W. Choi, E-J. Park, M.S. Hwang, J.R. Liu, D. Choi, K. Mikami and W-J Jeong. 2012. Development of an expression system using the heat shock protein 70 promoter in the red macroalga, *Porphyra tenera*. J. Appl. Phycol. 24: 79–87.

Spalding, M.H. 2008. Microalgal carbon-dioxide-concentrating mechanisms: *Chlamydomonas* inorganic carbon transporters. J. Exp. Bot. 59: 1463–1473.

Spolaore, P., C. Joannis-Cassan, E. Duran and A. Isambert. 2006. Commercial applications of microalgae. J. Biosci. Bioeng. 101: 87–96.

Spreitzer, R.J. and M.E. Salvucci. 2002. Rubisco, Structure, regulatory interactions, and possibilities for a better enzyme. Annu. Rev. Plant. Biol. 53: 449–475.

Srirangan, K., M.E. Pyne and C.C. Perry. 2011. Biochemical and genetic engineering strategies to enhance hydrogen production in photosynthetic algae and cyanobacteria. Bioresour. Technol. 102: 8589–8604.

Stapleton, J.A. and J.R. Swartz. 2010. A cell-free microtiter plate screen for improved [FeFe] hydrogenases. PLoS One 5: e10554.

Stepanenko, O.V., O.V. Stepanenko, D.M. Shcherbakova, I.M. Kuznetsova, K.K. Turoverov and V.V. Verkhusha. 2011. Modern fluorescent proteins: from cromophore formation to novel intracellular applications. BioTechniques 51: 313–327.

Stephens, E., I.L. Ross, J.H. Mussgnug, L.D. Wagner, M.A. Borowitzka, C. Posten, O. Kruse and B. Hankamer. 2010. Future prospects of microalgal biofuel production systems. Trends Plant Sci. 15: 554–564.

Stephenson, P.G., C.M. Moore, M.J. Terry, M.V. Zubkov and T.S. Bibby. 2011. Improving photosynthesis for algal biofuels: toward a green revolution. Trends Biotechnol. 29: 615–623.

Stripp, S.T., G. Goldet, C. Brandmayr, O. Sanganas, K.A. Vincent, M. Haumann, F.A. Armstrong and T. Happe. 2009. How oxygen attacks [FeFe] hydrogenases from photosynthetic organisms. Proc. Natl. Acad. Sci USA 106: 17331–17336.

Surzycki, R., L. Cournac, G. Peltier and J.D. Rochaix. 2007. Potential for hydrogen production with inducible chloroplast gene expression in *Chlamydomonas*. Proc. Natl. Acad. Sci. USA 104: 17548–17553.

Tabita, F.R., S. Satagopan, T.E. Hanson, N.E. Kreel and S.S. Scott. 2008. Distinct form I, II, III, and IV Rubisco proteins from the three kingdoms of life provide clues about Rubisco evolution and structure/function relationships. J. Exp. Bot. 59: 1515–1524.

Tamoi, M., M. Nagaoka, Y. Miyagawa and S. Shigeoka. 2006. Contribution of fructose-1,6-bisphosphatae and sedoheptulose-1,7-bisphosphatase to the photosynthetic rate and carbon flow in the Calvin cycle in transgenic plants. Plant Cell Physiol. 47: 380–390.

Tan, C., S. Qin, Q. Zhang, P. Jiang and F. Zhao. 2005. Establishment of a micro-particle bombardment transformation system for *Dunaliella salina*. J. Microbiol. 43: 361–365.

Tcherkez, G.G., G.D. Farquhar and T.J. Andrews. 2006. Despite slow catalysis and confused substrate specificity, all ribulose bisphosphate carboxylases may be nearly perfectly optimized. Proc Natl. Acad. Sci. USA 103: 7246–7251.

Thauer, R.K. 2007. A fifth pathway for carbon fixation. Science 318: 1732–1733.

Thelen, J.J. and J.B. Ohlrogge. 2002. Metabolic engineering of fatty acid biosynthesis in plants. Metabolic Engineering 4: 12–21.

Timmins, M., S.R. Thomas-Hall, A. Darling, E. Zhang, B. Hankamer, U.C. Marx and P.M. Schenk. 2009a. Phylogenetic and molecular analysis of hydrogen-producing green algae. J. Exp. Bot. 60: 1691–1702.

Timmins, M., W. Zhou, J. Rupprecht, L. Lim, S.R. Thomas-Hall, A. Doebbe, O. Kruse, B. Hankamer, U.C. Marx, S.M. Smith and P.M. Schenk. 2009b. The metabolome of *Chlamydomonas reinhardtii* following induction of anaerobic H_2 production by sulfur depletion. J. Biol. Chem. 284: 23415–23425.

Torzillo, G., A. Scoma, C. Faraloni, A. Ena and U. Johanningmeier. 2009. Increased hydrogen photoproduction by means of a sulfur-deprived *Chlamydomonas reinhardtii* D1 protein mutant. Int. J. Hydrogen Energy 34: 4529–4536.

Uemura, K., M. Anwaruzzaman, S. Miyachi and A. Yokota. 1997. Ribulose-1,5-bisphosphate carboxylase/oxygenase from thermophilic red algae with a strong specificity for CO_2 fixation. Biochem. Biophys. Res. Commun. 233: 568–571.

Uji, T., H. Mizuta and N. Saga. 2013. Characterization of the sporophyte-preferential gene promoter from the red alga *Porphyra yezoensis* using transient gene expression. Mar. Biotechnol. (NY) 15: 188–196.

van Ooijen, G., K. Knox, K. Kis, F.Y. Bouget and A.J. Millar. 2012. Genomic transformation of the picoeukaryote *Ostreococcus tauri*. J. Vis. Exp. (65): e4074, doi:10.3791/4074.

Varaljay, V.A., S. Satagopan, J.A. North, B. Witte, M.N. Dourado, K. Anantharaman, M.A. Arbing, S.H. McCann, R.S. Oremland, J.F. Banfield, K.C. Wrighton and F.R. Tabita. 2016. Functional metagenomic selection of ribulose 1,5-bisphosphate carboxylase/oxygenase from uncultivated bacteria. Environ. Microbiol. 18: 1187–1199.

Volgusheva, A., S. Styring and F. Mamedov. 2013. Increased photosystem II stability promotes H_2 production in sulfur-deprived *Chlamydomonas reinhardtii*. Proc. Natl. Acad. Sci. USA 110: 7223–7228.

Wang, Q., Y. Lu, Y. Xin, L. Wei, S. Huang and J. Xu. 2016. Genome editing of model oleaginous microalgae *Nannochloropsis* spp. by CRISPR/Cas9. Plant J. 88: 1071–1081.

Wargacki, A.J., E. Leonard, M.N. Win, D.D. Regitsky, C.N. Santos, P.B. Kim, S.R. Cooper, R.M. Raisner, A. Herman, A.B. Sivitz, A. Lakshmanaswamy, Y. Kashiyama, D. Baker and Y. Yoshikuni. 2012. An engineered microbial platform for direct biofuel production from brown macroalgae. Science 335: 308–313.

Warnecke, F., P. Luginbuhl, N. Ivanova, M. Ghassemian, T.H. Richardson, J.T. Stege, M. Cayouette, A.C. McHardy, G. Djordjevic, N. Aboushadi, R. Sorek, S.G. Tringe, M. Podar, H.G. Martin, V. Kunin, D. Dalevi, J. Madejska, E. Kirton, D. Platt, E. Szeto, A. Salamov, K. Barry, N. Mikhailova, N.C. Kyrpides, E.G. Matson, E.A. Ottesen, X. Zhang, M. Hernandez, C. Murillo, L.G. Acosta, I. Rigoutsos, G. Tamayo, B.D. Green, C. Chang, E.M. Rubin, E.J. Mathur, D.E. Robertson, P. Hugenholtz and J.R. Leadbetter. 2007. Metagenomic and functional analysis of hindgut microbiota of a wood-feeding higher termite. Nature 450: 560–565.

Weyman, P.D., W.A. Vargas, Y. Tong, J. Yu, P.C. Maness, H.O. Smith and Q. Xu. 2011. Heterologous expression of *Alteromonas macleodii* and Thiocapsa roseopersicina [NiFe] hydrogenases in *Synechococcus elongatus*. PLoS. One. 6: e20126.

Wilson, R.H., H. Alonso and S.M. Whitney. 2016. Evolving *Methanococcoides burtonii* archaeal Rubisco for improved photosynthesis and plant growth. Sci. Rep. 6: 22284, doi:10.1038/srep22284.

Whitney, S.M., P. Baldet, G.S. Hudson and T.J. Andrews. 2001. Form I Rubiscos from non-green algae are expressed abundantly but not assembled in tobacco chloroplasts. Plant J. 26: 535–547.

Whitney, S.M., R.L. Houtz and H. Alonso. 2011. Advancing our understanding and capacity to engineer nature's CO_2-sequestering enzyme, Rubisco. Plant Physiol. 155: 27–35.

Work, V.H., S. D'Adamo, R. Radakovits, R.E. Jinkerson and M.C. Posewitz. 2012. Improving photosynthesis and metabolic networks for the competitive production of phototroph-derived biofuels. Curr. Op. Biotechnol. 23: 290–297.

Wu, S., R. Huang, L. Xu, G. Yan and Q. Wang. 2010. Improved hydrogen production with expression of hemH and lba genes in chloroplast of *Chlamydomonas reinhardtii*. J. Biotechnol. 146: 120–125.

Wu, S., L. Xu, R. Huang and Q. Wang. 2011. Improved biohydrogen production with an expression of codon-optimized hemH and lba genes in the chloroplast of *Chlamydomonas reinhardtii*. Bioresour. Technol. 102: 2610–2616.

Yacoby, I., S. Pochekailov, H. Toporik, M.L. Ghirardi, P.W. King and S. Zhang. 2011. Photosynthetic electron partitioning between [FeFe]-hydrogenase and ferredoxin:NADP+-oxidoreductase (FNR) enzymes *in vitro*. Proc. Natl. Acad. Sci. USA 108: 9396–9401.

Yang, B., J. Liu, X. Ma, B. Guo, B. Liu, T. Wu, Y. Jiang and F. Cheng. 2017. Genetic engineering of the Calvin cycle towards enhanced photosynthetic fixation in microalgae. Biotechno. Biofuels 10: 229. doi. 10.1186/s13068-017-0916-8.

Yoshino, F., H. Ikeda, H. Masukawa and H. Sakurai. 2007. High photobiological hydrogen production activity of a *Nostoc* sp. PCC 7422 uptake hydrogenase-deficient mutant with high nitrogenase activity. Mar. Biotechnol. 9: 101–112.

Zaslavskaia, L.A., J.C. Lippmeier, C. Shih, D. Ehrhardt, A.R. Grossman and K.E. Apt. 2001. Trophic conversion of an obligate photoautotrophic organism through metabolic engineering. Science 292: 2073–2075.

Zhang, L., T. Happe and A. Melis. 2002. Biochemical and morphological characterization of sulfur-deprived and H_2-producing *Chlamydomonas reinhardtii* (green alga). Planta 214: 552–561.

Zhu, X.G., E. de Sturler and S.P. Long. 2007. Optimizing the distribution of resources between enzymes of carbon metabolism can dramatically increase photosynthetic rate: a numerical simulation using an evolutionary algorithm. Plant Physiol. 145: 513–526.

16

Present and Future Economic and Environmental Impacts of Microalgal Technology#

Miguel Olaizola,[2,*] *Rob C. Brown*[1] *and Elizabeth D. Orchard*[1]

Introduction

In this chapter, we review some of the economic and environmental impacts, current and expected, from commercialization of algal technology. This is not an exhaustive review; we will limit ourselves to the fast growing number of commercial applications and products of microalgae already accomplished and those expected to materialize within a relatively short time horizon (less than a decade). We will also limit ourselves to microalgal products and applications driven by phototrophic processes (light driven; non-dependent on organic carbon resources).

Here, we make the case that, on balance, the economic and environmental impacts of microalgal technology are positive but not without a few caveats and that the economics of microalgal technologies are very much dependent on the type of product and the scale at which it is being produced. Further, we argue that without large improvements in productivity, and concomitant decrease in costs, some anticipated products will have difficulty finding markets. We propose that improvements in strains (over wild type; both genetically modified and classically improved) will significantly aid in increasing both the positive economic and environmental impacts of microalgal technology. Finally, we also discuss what some of those improvements will be such as improvements in photosynthetic efficiency and carbon partitioning.

Present scale of commercial microalgal cultivation

The promise of novel products and applications as well as increased demand for established products and applications has resulted in renewed interest in microalgae cultivation. In 2010, it was estimated that the

---
[1] Synthetic Genomics, Inc., 11149 N Torrey Pines Rd., La Jolla, CA 92037, USA.
[2] OLAS - All Things Algae, LLC, Snowflake, AZ 85937.
* Corresponding author: miguel@olasalgae.com

We dedicate this contribution to Professor Guillermo Garcia-Blairsy Reina (1958–2012), innovating and visionary algal technologist and educator who will be sorely missed.

amount of microalgae produced commercially is around 10×10^3 metric tons worldwide most of which was *Arthrospira* sp. (Grewe and Pulz 2012). More recent estimates (as of 2016) indicate about 15×10^3 and 10×10^3 metric tons of *Arthrospira* and *Chlorella* respectively (Slocombe and Benemann 2016). Note that the estimate for *Chlorella* includes heterotrophic production.

Much of this biomass is sold as high value food supplements that can command prices well above \$10/kg (Benemann 2010). In 2004, Pulz and Gross (2004) estimated the size of the market at 5000 metric tons/yr with a value of US\$1.25 Bn/yr. Assuming that the tonnage has already increased significantly as noted above, the value of the market today could is expected to be significantly larger. If one is to believe the projections made by researchers and venture capitalists, future commercial microalgal production is certain to increase substantially.

Interestingly, the total tonnage of all microalgae utilized worldwide is much higher; it has been calculated as high as 240 million tons. Most of this biomass is not produced in photobioreactors or open ponds but as "green water" to support the nutrition of mostly herbivorous fish and shrimp for aquaculture (Neori 2011). That author (that is, Neori 2011), estimates that the cost of producing such "green water" algal biomass must be on the order of less than US\$0.1/kg to support the economic production of fish species with value as low as US\$0.5/kg. Note that 240 million tons at US\$0.1/kg is equivalent to 24 Bn US\$/yr, one order of magnitude higher than that of commercial microalgae production.

The current scale of commercial microalgal production is small; the largest operations include substantially less than 50 ha of cultivation area made up of mostly open raceway reactors (Cyanotech, Earthrise, Parry in Fig. 1). Commercial operations using enclosed photobioreactor (PBRs) are substantially smaller (for example, Algatechnologies' plant in Kibbutz Ketura, Israel). Plants under construction for commercial purposes such as Necton's new PBR plant in Portugal (Verdelho 2012) fit within these parameters. Using open raceways, Sapphire developed a new plant in Columbus (New Mexico, USA), which was expected to reach a size of some 100 ha of cultivation (Fig. 1). Therefore, in spite of multiple

Fig. 1. Scale of large microalgal farms. From left to right and top to bottom: Cyanotech aerial photograph provided by Dr. Gerry Cysewski (Cyanotech), Earthrise satellite photograph obtained from Google Earth, Sapphire aerial photograph obtained from their website, and Parry Nutraceuticals/Valensa photograph courtesy of Mr. Umadsuhan.

optimistic projections over the last decades, the scale of microalgal commercial cultivation is still quite small (compared, for example, to agriculture). The economic and environmental impacts of the microalgae industry have been, therefore, quite small. However, as the scale of activities increase in the future, fueled by increases in productivity and market size and lowering of costs, we expect that these impacts will increase in significance.

Microalgal commercial products at present

Apart from "green water" production, touched upon in the previous section, most photoautotrophic commercial microalgal production consists of relatively high value products such as astaxanthin and β-carotene (from *Haematococcus* and *Dunaliella*) and whole cell biomass (such as *Arthrospira*, *Nannochloropsis*, and *Chlorella*). These are, at this point in time, specialty products, not commodities. Out of the myriad of proposed microalgal products, many will continue to be specialty products. Two families of products, however, have the potential to truly change the scale of microalgal cultivation: feeds and fuels. As these new products come into being, we expect that the industry's environmental and economic impacts will become significant.

The cost of microalgae and microalgal products on the present commercial scale

Producers of commercial microalgal products are not likely to share their true production costs for competitive reasons. However, based on pricing and availability of commercial microalgal products educated guesses can be made. Depending on the process used, microalgal biomass production costs range over three orders of magnitude. Here we consider two established commercial microalgal products, Spirulina biomass (*Arthrospira*) and astaxanthin (*Haematococcus*), and we compare them to the estimated costs of producing "green water" (see above, Neori 2011).

First, we consider the production costs of Spirulina biomass. Spirulina has been cultivated commercially for over 40 years in the US and production has been expanding in Asia, especially China (Yun-Ming et al. 2011). Spirulina has many applications (see later sections) but is mainly being sold as a nutritional supplement. Because of its growth and physical characteristics (high alkalinity medium, large size), it is a relatively inexpensive biomass to produce that needs relatively little processing. Several authors have estimated the cost of producing Spirulina biomass at about US$3–5/kg depending on the locale (e.g., Grewe and Pulz 2012).

Second, we consider the production costs of significantly more costly microalgal products, such as astaxanthin from *Haematococcus*, *Haematococcus* astaxanthin has been produced commercially for over 15 years (Olaizola 2000). *Haematococcus* is much costlier than Spirulina to produce: it requires an enclosed PBR growth phase, has a higher probability of culture crashes than *Arthrospira* cultures and requires more complex processing, including cell cracking and extraction. Estimates of production costs range up to US$100/kg dry biomass (see for example, Carlsson et al. 2007).

Taking those biomass production costs, we can loosely estimate the cost of microalgal products which may accumulate in microalgal biomass at different concentrations (Table 1). Clearly, production of microalgal biomass is quite expensive if one needs to control the population composition (as opposed to "green water") and if one does not use an extremophile strain (like *Arthrospira*). The cost data may help explain why some products have not yet made it to market. For example, algal biofuel precursors from a non-robust strain (even if highly lipid accumulating: 20–50% is commonly claimed) can be expected to cost > US$200 per kg. From a protected strain, but less lipid enriched, it might cost as little as US$20 per kg. This explains why the products that have reached the market tend to be very valuable, that is, can demand a high price (i.e., higher than the cost of production).

Economic impacts

The development of the microalgal technology industry has spurred growing economic activities. As indicated earlier, several US$Bn per year would appear to be a reasonable size for the worldwide market

Table 1. Estimated ranges of costs for microalgal biomass and microalgal products (authors estimates based on generalized production cost model and product specific flowsheets).

	Green water	***Arthospira***	***Haematococcus***
US$/kg dry biomass	$0.10	$5.00	$100.00
Production and Processing	None	Minimal	Significant
Scale of production	1,000's ha	100's ha	10's ha
Cost of 1 kg of			
70% component	$0.14	$7.14	$142.86
30% component	$0.33	$16.67	$333.33
3% component	$3.33	$167	$3,333
1% component	$10	$500	$10,000

in produced microalgal products today. Although not an economic activity per se, we might consider the value of green water production as a near term goal for the industry's output (US$ 24 Bn per year). The value and economic impact of village-level microalgal production (see for example Habib et al. 2008) is much more difficult to estimate although Abdulqader et al. (2000) have estimated the value of dihé (a local Spirulina product) naturally harvested from Lake Kossorom in Chad at more than US$100,000 per year (about 40 ton biomass). In this section, we will explore some of the economic impacts that microalgal technology is having on two geographically but economically different areas in Southern California (USA).

Microalgal technologists claim that the economic impacts of their industry are positive. Here we explore two case studies where microalgal economic activity has had positive effects: San Diego and Imperial Counties. Many economic activities can be considered when one evaluates the economic impacts of the microalgal industry such as

- public money awarded to universities to conduct research,
- public money given to private companies to spur research and innovation,
- venture capital raised to found new companies,
- money raised from product sales (i.e., revenue) and
- employment created both directly and indirectly from those above.

It is also true that money spent by the microalgal industry do not always bear fruit. In some cases companies have ceased operations resulting in loss of investment and public money and creating unemployment. We also briefly touch upon some of the challenges that starting companies face in the microalgal space.

Positive economic impacts: 2 case studies in the US

Southern California in the US, where Synthetic Genomics is located, has been a hotbed of microalgal research and production for many years. Here we explore the economic impact that microalgal technology activities have had in two very different areas: San Diego County, a coastal metropolitan area that hosts several large public research institutions as well as established and new private enterprises, and Imperial County, a largely traditional agricultural area in the southern California desert about 200 km inland from San Diego (Table 2).

Southern California's San Diego County

San Diego County is the second most populous county in the State of California (after Los Angeles County). It has a varied economy including manufacturing, military, services, education, and agriculture. Several public educational entities are located here engaged in microalgal research (e.g., University of California-San Diego, San Diego State University, University of San Diego, Scripps Institution of

Table 2. Some characteristics of San Diego and Imperial Counties (2010 US Census data available at http://www.census.gov/).

	San Diego County	**Imperial County**
Population	3,095,313	174,528
Total payroll	US$ 52 Bn	US$ 0.9 Bn
Persons per square mile	736	42
Population employed[†] (persons)	1,592,037	77,531
Unemployed[†] (%)	9.4%	28.9%

[†] Data from United States Bureau of Labor Statistics covering the period November 2011–December 2012 (http://www.bls.gov/lau/laucntycur14.txt).

Oceanography, and Scripps Research Institute). Several microalgal technology companies are also based in San Diego. Some are relatively new and focused on microalgal applications such as Synthetic Genomics Inc. and Sapphire Energy. There are also more established companies that have recently started programs in microalgal technology such as SAIC and General Atomics. Public/private collaborations, such as the San Diego Center for Algal Biotechnology (SD-CAB), have been established to help identify microalgae technology applications that may result in new products, processes, and economic activity. Although a few pilot scale facilities exist (e.g., at General Atomics, Synthetic Genomics, and SD-CAB, Fig. 2) within the County, most of the algal technology research occurs in laboratories at the bench scale and much of it has been fueled by the heightened interest in algal biofuels.

Photo of GA

Photos of SGI

Fig. 2. Pilot scale facilities for microalgal technology research in San Diego County. SD-CAB (top-left and right photographs courtesy of Dr. Steve Mayfield), General Atomics (middle photograph courtesy of Dr. Aga Pinowska) and Synthetic Genomics (bottom left and right).

Over the last decade and in response to the perceived increase in microalgal research and commercial activities in San Diego County, SD-CAB has commissioned several studies on the economic impact of algal biofuel research in the area (Mayfield, S. personal communication). Three studies have been released (April 2009, June 2010 and January 2012) which indicate an increasing trend in positive economic impact in the County. The economic impacts were quantified as follows: direct impacts (reflecting jobs and expenditures that are directly related to research in algal biofuels), indirect impacts (reflecting the numerous business products, materials, and services required and supplied locally to support the direct activities of the research) and induced impacts (reflecting the local household expenditures of employees involved in the research). Figure 3 summarizes the results of those studies and reflects that by 2011 algal biofuels directly employed 466 workers in San Diego which resulted in US$41 million in direct payroll and US$81 million in economic activity for the region. Furthermore, these studies concluded that this activity had generated (indirect impacts, see above) an additional US$12.5 million in annual payroll and US$33 million in additional economic activity at other local companies. Additionally, the study reports additional induced (see above) economic impacts (economic activity and further employment) based on the expenditures of employees resulting in a grand total of 1,005 employees and US$157 million total economic output for the year that can be traced to the algal biofuels effort in San Diego County.

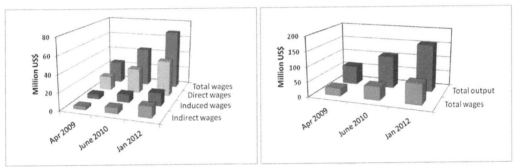

Fig. 3. Economic activity credited to the algal biofuel research expenditures in San Diego County, California. Studies were carried out by the San Diego Association of Governments (www.sandag.org/servicebureau).

Southern California's Imperial County

The Imperial County in southern California is a largely agricultural area which is much less populated than San Diego County with which it shares a border. It is fast becoming a renewable energy center: companies here are focused on hydrothermal, wind, solar and, more recently, algal biofuels. Imperial County is also one the highest unemployment areas in the US (Table 2). There are two institutions of higher education (university level) in the County: Imperial Valley College and San Diego State University.

This area is conducive to microalgal production: the area is very flat and enjoys high solar average radiation during the year of 5.6 kWh/m²/d: Winter lows average about 3.1 kWh/m²/d which more than doubles during the summer to peak at above 8 kWh/m²/d (NREL 2012). There are large seasonal variations in air temperature but conditions within seasons are fairly predictable: the average (22 year) winter daily minimum is 5.3°C in December; the average summer daily maximum is 41.8°C (Fig. 4) although microalgal raceway temperatures rarely reach above 35°C. Of special note, freezing winter conditions are nearly non-existent. Due to the high temperatures, evaporation rates can reach over 32 cm/month (US Bureau of Reclamation 2004). Due to the climatic conditions and access to water from the Colorado River, the area produces nearly US$ 4 Bn of agricultural commodities per year. Of these, aquatic products (combined fish and algae) represent US$ 8.1 million per year (Valenzuela 2011).

There is a long history of microalgal production in the County. Proteus Corporation (later, Earthrise Nutritionals), one of the pioneers in Spirulina production, started activities in this desert area in the late 1970s and it is one of the largest Spirulina producers in the world today. In 1984, another microalgal company, Microbio Resources, was reported to be the first company to successfully produce and market

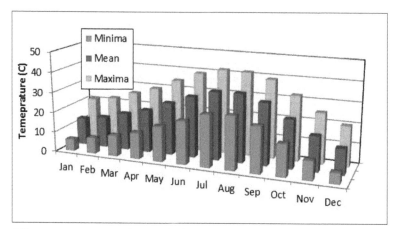

Fig. 4. 22 year daily temperature averages (per month) for daily minimum, mean, and maximum in Calipatria, CA. Data obtained from: http://www.wunderground.com/about/data.asp.

β-carotene from *Dunaliella* (Spencer 1989). This activity continued with Amway/Nutrilite during the early 1990s. Over the years, other companies have taken advantage of the facilities created by Microbio Resources including Carbon Capture Corporation (interested in microalgal-based CO_2 capture and sequestration) and, since 2012, Synthetic Genomics.

We are not aware of any specific studies commissioned to analyze the economic impacts of microalgal technology activities in Imperial County. However, we can make some estimates based on parallels with San Diego County. We estimate that microalgae technology provides approximately 60 jobs directly in Imperial County (employment at Earthrise plus SGI and research and education conducted at Imperial Valley College and San Diego State University's satellite campus). Using similar multipliers as those used in the SANDAG report (above) for San Diego County, we estimate total employment in Imperial County (direct plus indirect plus induced) of about 113 that can be traced to microalgae technology efforts as well as some US$ 14 million in total economic output per year. While this is much smaller than the total value of agricultural commodities produced in the County (US$4 Bn, Valenzuela 2011) the area is attracting much interest in the microalgae community and is expected to increase.

Its combination of large tracks of inexpensive flat land, sunshine, temperature, water, high unemployment, and alternative sources of microalgal nutrients (e.g., cattle food lot waste) present Imperial County with a tremendous opportunity to conduct R&D and commercial microalgal production at scales yet unmatched anywhere else. We expect that the economic impact of microalgal technology in the area will increase significantly in the future.

The challenges of sustaining positive economic impacts

The type of positive economic impacts described above tend to be quite localized because of the scale of the industry at present. The industry is in "start-up" mode. Some of us have experienced first hand the up and downs of small companies trying to make their mark in this space. The problems that these small companies are presented with are those common to other industries. In some cases, companies are not well capitalized; they may be too dependent on public money (which can dry up) or the capital that they do have may not be well spent. Furthermore, these companies may take too long to develop products or processes that can be used to raise revenue and they may not have the resources to properly market their products and services. In some cases, these problems have been compounded because of making exaggerated promises (including overly optimistic cost projections which some consider necessary to raise investment) and a "if we make it they will buy it" attitude (Olaizola 2003a). In some cases, the result has been layoffs and even bankruptcy.

As the industry has been affected by growing pains, we believe that lessons have been learned. The industry is entering a maturing phase as a true understanding of the challenges and costs of microalgal production is appreciated. We predict that from this base of understanding the industry will enter a phase of steady, if not explosive, growth.

Environmental impacts

Like any other economic activity, algal technology can be expected to have environmental impacts. Large scale micro-algae cultivation is dependent on water sources (either fresh or salt water), inorganic nutrients (including nitrogen, phosphorus, iron, potassium, and other trace metals), and CO_2 and energy. In this section, we will explore the pros and cons of various nutrient and water sources, and how algae compare to terrestrial sources. For the most part, estimates of water and nutrient usage used here are based on current algae biomass productivities and lipid content. Improving either of these parameters has direct effects on improving the environmental and economics of microalgae production. Possible methods for improving productivity and lipid content will be discussed later in this chapter.

Water

Fresh water footprint

Successful cultivation of algae is dependent on consistent supply of quality water at a predictable cost. Algae biofuel production is often touted for the fact that they can grow in seawater or brackish water, and therefore do not have to compete with agricultural uses. While some current algae production does require fresh water, we will only consider the case of marine algae here, as that is the most pragmatic approach at commodity scales. However, depending on the cultivation system, and location of production, there will be significant water demand to compensate for harvest, biomass processing, and evaporative related losses (Table 4). At a biofuel production level of 10 billion gallons/year in the southwestern United States, Pate et al. (2011) estimated that algae production would consume 24% percent of all water used for irrigation in this region. This is a significant concern in the water deficient southwest. More efficient water recycling, and partnering with waste water streams seem the best way to reduce some of these requirements. Yang et al. (2011) estimated that as much as 90% of the freshwater usage could be eliminated by using seawater or waste water sources. Furthermore, thoughtful location, and cultivation system demand will be critical to reduce the draw and the water supply. Even without taking this into account, algae water usage is on par with estimates of the water foot print for other biofuel feedstocks (Table 3). Therefore with respect to water utilization, algae production appears to be the most environmental favorable option for biofuel production if recycling is part of the process. Still, water and nutrient recycle likely will be required for any large scale microalgal cultivation operation.

Table 3. Water footprint for microalgae versus other biofuels feedstocks. Adapted from Yang et al. 2011.

Feedstock	Water footprint (kg water/kg biofuel)
Maize	4,015
Sugarcane	3,931
Potatoes	3,748
Soybean	13,676
Switchgrass	2,189
Microalgae	591-3,650[†]

[†] Range due to variations in recycle rate.

Replacing farm animal feeds

According to UNESCO (2009) total global freshwater use is about 4,000 km^3/yr. About 70% of this amount is used in irrigated agriculture (20% goes to industry and energy and 10% for domestic use). Agriculture also benefits from about 6,400 km^3 of rain water (responsible for about 60% of crop production). A recent report (Mekonnen and Hoekstra 2012) has compiled data on the footprint of animal production. The authors note that animal production and consumption play a significant role in depleting (and polluting) the world's freshwaters. They estimate that 98% of global animal production footprint arises from the feed the animals consume (about 2374 km³). From these numbers, one can estimate a global average of 26% of the fresh and rain water being used for animal feed production. As the world population increases and the standard of living improves, it is expected that feed production (and its water footprint) will also increase. In this section, we explore the environmental benefits of replacing animal feed with microalgal products, especially those from marine species.

For over 60 years, studies have shown that farm animal feed can be partially replaced with microalgal biomass (e.g., Fisher and Burlew 1953; Hintz et al. 1966; Lipinsky et al. 1970). We propose that replacement of farm animal feed with microalgae produced in seawater would have a significant impact on freshwater use and that this would have a positive impact on the amount of fresh water available for direct human consumption: replacing 10% of farm animal feed with marine microalgal biomass would save nearly 240 km³ per year of freshwater which could be redirected to other uses.

Nutrients

Algae require the addition of inorganic nutrients to grow optimally. The most significant of these components in algal biomass are nitrogen (N), phosphorus (P), potassium (K), and iron (Fe). These nutrients can be provided from fertilizers, waste water (municipal or industrial), or recycled (from digested biomass). In this section, we explore nutrient demand, possible nutrient sources, and their relative environmental and economic impacts.

Nutrient requirements

Relative to terrestrial autotrophs, algae biomass is relatively rich in nutrients (Elser et al. 2000). Unlike more traditional agriculture systems where nutrients may be lost or gained from the soil, in a well-run production system, algae can be grown with little to no nutrient run off down-stream of "the algae farm". Additionally, unlike land plants where nutrients can come from the soil, all nutrients have to be supplemented to the system. A nutrient replete algae culture is typically ~ 10% nitrogen (N) and ~ 1% phosphorus (P), while an algae culture which has been induced to make lipids can be ~ 6% N. However, this means that a well-managed pond can be run such that there is little to no excess nutrients (i.e., run off) that are lost from the system. This is in direct contrast to land plants where eutrophication of water sheds is often a major issue.

Eutrophication of bodies of water results in high rates of accumulation of organic matter, mainly from aquatic plants and algae. This in turn can cause an anoxic condition, which results in fish kills and decreased biodiversity. On small scales, this can result in decreased recreational and beneficial use of affected waters. On large scales, eutrophication induced hypoxic events have resulted in mass mortality in coastal and estuarine areas such as the "dead zone" in the Gulf of Mexico which has been linked to high nutrient run off from the Mississippi river (Diaz and Rosenberg 2008). Therefore, there is a major environmental interest in reducing anthropogenic fertilization of water sheds. Clarens et al. (2010) estimates a 2- to 8-fold reduction in eutrophication potential (as estimated by phosphate) by using algae for biofuel production relative to switchgrass or corn respectively.

Nutrient usage is a major issue when considering the feasibility of any biofuel feedstock. We will focus on N and P as those are by far the most important nutrients in terms of consumption. Using fertilizers is the easiest method of nutrient delivery and allows for the most controlled scenario, and has been effective at small scale for high value products. On larger scales, nutrient consumption from fertilizers becomes a major issue (Table 4). Nitrogen and phosphorus requirements to produce 10 billion gallons/year of biofuel from algae (roughly 15% of US diesel fuel demand) without recycling nutrients requires roughly 44 and 20% of the total US use of these nutrients respectively (Pate et al. 2011), but this issue is not unique to microalgae. Corn grain ethanol, for example, requires even more N and P to produce ethanol per unit energy than biodesiel from microalgal based on current estimates from Hill et al.

(2006) and Pate et al. (2011) (Table 5). Soybeans, on the other hand, require less N to produce biodiesel than corn and microalgae because of their association with nitrogen fixers. When the biomass is recycled, microalgae-based fuel production consumes about as much P as soybeans and much less than corn.

Microalgae do have the advantage over more tranditional crops in that they are able to grow on waste water sources, and this can represent a major source of nutrients in microalgae production. Therefore, pairing microalgae production with waste waters sources and/or recycling water is critical not only for improving the environmental impacts of algal biofuel but for the economic feasibility of commodity-scale microalgal production. From a waste water treatment stand point, there is significant interest in using algae to consume nutrients in waste water to further reduce sources of eutrophication. Indeed the more algae biofuel production can limit the use of chemical fertilizers, the more advantages to the environment. Furthermore, 60–80% of energy consumption at waste water treatment plants comes from nutrient removal (Clarens et al. 2010). Therefore, there is a significant advantage for waste water treatment plants in partnering with algae cultivation.

Table 4. Estimated nutrient consumption for algal biofuel production of 10 billion gallons/year (BGY) of biofuel in the Southwestern United States assuming open cultivation systems. Adapted from Pate et al. (2011).

Resource	Harvest water not recycled	Harvest water recycled[†]
Nitrogen (Million Mt/year)	6.1	1.8
% of national usage	44	13
Phosphorus (Million Mt/year)	0.8	0.24
% of national usage	20	6

[†] Assuming 70% recycle efficiency.

Table 5. Estimated nutrient usage in g/MJ to produce biodiesel or ethanol from various biofeedstocks.

Nutrient usage for biofuel production (g/MJ)				
Nutrient	Corn ethanol	Soy biodiesel	Algae biodiesel	Algae biodiesel (recycling)
N	7	0.1	5	1.5
P	2.6	0.2	0.7	0.2

[1]Hill et al. 2006
[2]Pate et al. 2011
[3]Assuming a 70% recovery efficiency.

Remediation, using waste water and other wastes

Humans produce very large quantities of waste water through agriculture, animal husbandry, and industrial and domestic processes (UNESCO 2017). Using algae as a means to treat waste water has already been long utilized (Hoffmann 1998), and it has been shown that algae can be more efficient at nutrient removal than the activated sludge process (Tam and Wong 1989). However, using waste water as a nutrient source for algae biofuel production does not come without its technical hurdles, particularly if the target is to growth a single species of algae. Depending on the source, waste water (e.g., animal waste versus municipal waste) is not always uniform in it's nutrient content and contains bacteria and organic carbon which would allow heterotrophically growing organisms a growth advantage. Despite these challenges, there are several examples from the literature of lab-scale or pilot scale experiments where microalgae were utilized in conjunction with waste water treatment and resulted in high biomass productivities (Gonzalez et al. 1997; Samori et al. 2013). While these are very useful in assessing the feasibility of waste water-grown algae, there is still a need to demonstrate high lipid and biomass productivities of waste water grown algae at pond scale.

If algae production is to be paired with waste water treatment, then complete nutrient removal is desirable, particularly N and P. Algae are very efficient at scavenging dilute nutrients. Lab scale studies have shown near complete removal of N and P from waste water media (Picot et al. 1991; Garcia et al. 2000; Woertz et al. 2009). Although at pilot scale nutrient removal efficiencies, at least in one study, have been less efficient (65% N removal and 2% P removal), these cultures appeared to be carbon limited, so growth optimization is required (Craggs et al. 2012). Assuming nutrient uptake is optimized, using microalgae to scavenge nutrients from waste water has the added advantage of allowing for recovery of phosphate. While nitrogen can be produced from dinitrogen gas (via the Haber-Bosch process), phosphorus is a finite resource. We are currently at peak production levels for phosphorus, and it has been estimated that the global supply of phosphorus may be depleted in 50–100 years (Cordell et al. 2009). Therefore, efficient recovery of phosphorus from waste waters is likely to become increasingly critical, not only from a cost basis, but also to recycle this limiting global resource.

Some waste water sources, however, may contain high levels of heavy metals, such as cadmium, lead, copper, nickel, and silver. Algae have been shown to be effective at removing some of these metals (Khoshmanesh et al. 1996). This is quite advantageous from a waste water treatment standpoint, where heavy metal removal can be expensive but may be a major issue for algae production depending on the products produced. Heavy metals may have minimal effect on algae biofuel production, but may have significant consequences in algae food or feed replacement stocks. Therefore, significant care must be given with respect to waste water streams for microalgae production.

Energy

Microalgal cultivation and the production of microalgal products require energy for cultivation (e.g., pumps and paddlewheels), harvesting, and processing. This energy can be provided via fossil fuel combustion (most common) but can also be resourced from renewable sources such as solar and wind. Environmental impacts from energy use will be dependent on the type of energy used. The products of the microalgae industry also contain energy. Of special interest, from the energy point of view, is the use of microalgae to produce energy crops such as biofuels.

Much has been written on the use of microalgae for the production of biofuels (see Chapter 14 in this volume for a recent review) and the environmental impact of such processes, especially since the completion of the US Department of Energy's Aquatic Species program (Sheehan et al. 1998). The conclusions of that program were, essentially, that microalgae can indeed be used to produce biofuels but at a very high cost. The concept is simple: microalgae produce oils and starches that can be converted to biodiesel and bioalcohol, respectively, utilizing existing technologies. Companies have demonstrated the process to completion producing liquid fuels that have actually been used in existing transportation equipment: for example, Sapphire Energy produced jet fuel that was used in a blend in a Continental Airlines Boeing 737 jet in 2009 (http://www.sapphireenergy.com/documents/Biofuel_News_Release_1–7.pdf). However, at the present time, most research is centered on the production of bioethanol and biodiesel.

Most bioethanol today is produced from agricultural sources such as grain and sugarcane but microalgae heve been gaining attention (Mussatto et al. 2010). As with other products, microalgae have the advantages that they can thrive in salt water on land not useful for agriculture thus avoiding the fuel vs. food debate. Microalgae, however, have the disadvantage that production is expensive mainly because of the energy required to harvest and concentrate the algae before fermentation. At least one company is exploring selecting for microalgae that can secrete bioethanol directly (Algenol: www.Algenolbiofuels.com) which would avoid some of these expenses (although the relatively dilute bioethanol still needs to be concentrated at some cost).

Microalgal biodiesel has received the most attention since the result would be a drop-in fuel. Some studies based on model predictions indicate that this activity would result in (modest) energy and GHG emissions gains (Batan et al. 2010) while others predict that for the balance to be positive, renewable sources of energy would be required to power the microalgal process (Sander and Murthy 2010). See also Chapter 14 for a more extensive review.

At the present scale, it is not possible to determine whether microalgal products (fuels, feeds, foodstuffs) will have a more favorable energy balance than the products they replace (such as petroleum-derived fuels, fish meal, and other agricultural ingredients). However, there does not seem to be any indication that they will be worse.

CO₂

Microalgae consume CO_2 via photosynthesis and carbon represents about 50% of the biomass by weight. Thus, many researchers and policy makers have suggested that microalgae could be used to capture CO_2 and alleviate emissions.

A multiyear project funded by the US Department of Energy (Nakamura et al. 2005) demonstrated production of microalgal biomass using simulated flue gases (representing those emitted from power plants combusting bituminous and sub-bituminous coal, natural gas, and diesel fuel oil) and real flue gases produced by a coal combustor and a propane combustor (a household water heater). Specifically it was shown that the algae did not show any negative effects caused by exposure to the different gas mixtures (and up to 100% CO_2) as long as the pH of the cultures was properly managed (Olaizola 2003b). Several research installations have demonstrated the ability of microalgae to use combustion gases directly (Fig. 5).

Thus, algae-based processes for carbon capture offer several desirable characteristics such as:

- Microalgae do not require high purity CO_2. Flue gas of different CO_2 content can be fed directly into an algal culture which would alleviate the need to separate and concentrate CO_2 from the flue gas,
- Microalgae can use other combustion products present in the flue gas, such as SO_x and NO_x as nutrients which would simplify flue gas scrubbing, and
- The costs of the CO_2 capture would be minimal since the product of the microalgal-based process is expected to have been conducted as an economic activity designed to produce revenue.

Global CO_2 emissions from fuel combustion alone were about 30.3 Gt in 2010 (which is equivalent to 8.2 Gt C) and about 41% of CO_2 emissions are from electricity and heat generation point sources (IEA 2012). We noted in the introduction that global commercial microalgae production might total about 20,000 metric tons equivalent to about 10,000 tons of C. For commercial microalgae production to capture 1% of such emissions (82 million tons C) the output of the microalgal industry would need to

Fig. 5. Examples of microalgal cultures grown on flue gases and waste heat. Top left: Quarter acre Greenfuel PBR (lighted for night-time maintenance) grown with natural gas combustion gases at the Red Hawk power plant in Arizona in 2007 (Photo by MO). Top right: Test bag PBRs grown on coal combustion gases at the NRG power plant in Dunkirk, New York, in 2007 (Photo by MO). Bottom left: Proviron PBR during a snow storm in Belgium kept warm by water heated with natural gas recovered from a nearby landfill and obtaining CO_2 from the gas combustion in 2012 (photograph provided by Dr. Mark Michiels, Proviron). Bottom right: Subitec PBRs at the GMB coal power plant in Senftenberg, Germany, growing on coal combustion flue gases in 2012 (Photo by MO, used with permission, Dr. Peter Ripplinger, Subitec).

increase by a factor of over 8,000 over what it is today. Therefore, microalgae production today has little potential to reduce carbon emissions.

Commercial production of microalgae, however, also produces CO_2. Energy is required for cultivation, harvesting, processing (including drying), and formulation (whether for feed, food, or fuel). A recent study (Taylor et al. 2013) suggests that for microalgal-based fuel production to be carbon negative (consume more CO_2 than it produces) two conditions must be met: (1) the energy inputs into the plant must be produced via carbon neutral renewable energy generation such as concentrated solar power and (2) carbon containing coproducts must be produced that would not be later burnt (such as glycerol). One would also take credit for the CO_2 releases avoided by not burning non-renewable fossil fuel.

A similar situation can be assumed to exist for feed and food applications: one may expect that the use of renewable, carbon neutral, energy plus the credits of CO_2 avoided by not producing food and feed via standard processes could result in CO_2 savings. One might imagine that once renewable energy is available at the same cost as fossil energy the transition will occur. In the meantime, only the CO_2 credits generated via emissions avoidance can be counted on. In the interim, we expect that CO_2 capture via microalgal photosynthesis will not significantly slowdown the increase in global atmospheric CO_2 but it could be useful to decrease emissions from point sources in places where regulations or costs would encourage capture of produced CO_2.

Other ecological and human impacts

Besides those impacts explored above, there are others of a more ecological and human nature that microalgae can have a large positive effect on. First, the decimation of wild fisheries to supply feed ingredients to the aquaculture industry can be limited and even reversed with the use of feed ingredients of microalgal origin. Second, microalgae can supply foodstuffs to bolster the nutritional and economic wellbeing of disadvantaged human populations.

Aquaculture feed replacements

Aquaculture feeds is the fastest growing sector of the animal feeds industry worldwide increasing at 6–8% yearly (Rust et al. 2011; Tacon et al. 2011). It is expected that aquaculture production will surpass wild catch within a few years (Fig. 6). Fishmeal and fish oil have been traditionally used to supply balanced protein and essential long chain polyunsaturated omega-3 fatty acids (PUFA's). Demand for fishmeal and fish oil is increasing and the supply is not, causing a sustained increase in price (e.g., fishmeal price data is available at http://www.indexmundi.com/commodities/?commodity=fish-meal&months=180). Further increases in supply from this source are not feasible because most forage fisheries are at or near maximum exploitation levels; further harvesting will have undesirable ecological impacts on the populations and marine food webs that depend on them. Therefore, there is a strong push to replace fishmeal and fish oil with other sources of nutrition.

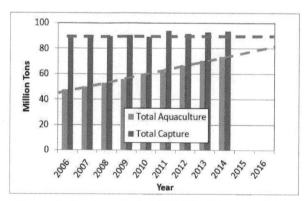

Fig. 6. Fisheries vs. aquaculture production of aquatic food 2006–2014. It is expected that aquaculture production will surpass fisheries within the next five years. Data from FAO 2016.

Aquafeed manufacturers have been successful in sourcing partial replacements for fishmeal using protein from vegetable origin (Tacon and Metian 2008; Rust et al. 2011), although digestibility of some vegetable materials is limited by low levels and imbalances in essential amino acids, anti-nutrient factors, and high level of indigestible starch (Naylor et al. 2009). The latter not only reduces the nutritional quality of the feed, but also increases waste and excretion that contributes to water quality problems (Cho and Bureau 2001). Vegetable sources also do not supply essential long chain omega-3 fatty acids. Given that body composition and nutritional requirements are linked (Pike et al. 1990), preferred alternative ingredients would ideally have similar composition as fishmeal, assuming similar digestibility levels. As we have noted above, agricultural products used in feed applications also have a large freshwater footprint. Microalgae are receiving growing attention as aquafeed ingredients. Habib et al. (2008) summarize some of the uses that Spirulina has been put to including as a protein, color, and nutritional enhancer for fish, shrimp, poultry, cattle, and domestic animals. However, as with vegetable sources, cyanobacteria do not supply the essential long chain omega-3 fatty acids necessary for optimal fish health (Sargent et al. 1999; Pettersen et al. 2010). Therefore, a fishmeal and fish oil replacement with high levels of long-chain omega-3 fatty acids is necessary to expand the industry and alleviate pressure on the wild fisheries.

Marine eukaryotic algae are a natural source of balanced protein (Becker 2007) and oil for fish; they represent the primary source of essential long chain omega-3 fatty acids for wild marine fish. Work carried out at SGI has shown that microalgae protein provides the closest match to that provided by menhaden, an important source of fishmeal and fish oils (Fig. 7) as well as omega-3 fatty acids. A preferred target feed composition for carnivorous fish such as trout and salmon is 65% protein, 12% oil, and < 10% moisture with the amino acid and fatty acid profile of menhaden meal and menhaden oil, respectively. Previous studies indicated that many marine algae accumulate protein at levels close to this target, and many exceed the target for oil content and composition (Becker 2007); this has been confirmed by SGI's in-house evaluations. Selected examples of strains from the SGI culture collection (*Pavlova* sp. and

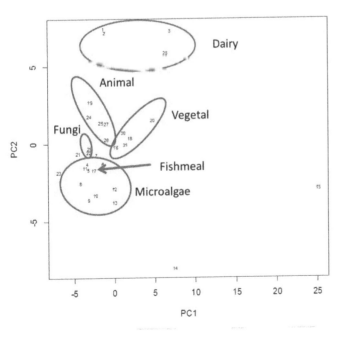

Labels	Source	Data source
1	Whey isolate	SGI
2	Whey concentrate	SGI
3	Rennet casein	SGI
4	SGIC0739	SGI
5	SGIC0609	SGI
6	SGIC0537	SGI
7	SGIC0463	SGI
8	SGIC0675	SGI
9	SGIC1328	SGI
10	SGIC0907	SGI
11	SGICU	SGI
12	SGIS0250	SGI
13	SGIS0573	SGI
14	SGIS0285	SGI
15	SGIS0886	SGI
16	Schizochytrium	Pyle et al., 2008
17	Fishmeal	IAFMM, 1970
18	Soymeal	Miller, 1970
19	Egg	Becker, 2007
20	Soy bean	Becker, 2007
21	Chlorella	Becker, 2007
22	Arthrospira	Becker, 2007
23	Spirulina	Becker, 2007
24	Eggwhite	Miller, 1970
25	Tuna	Miller, 1970
26	Beef	Miller, 1970
27	Chicken	Miller, 1970
28	Casein	Miller, 1970
29	Yeast	Miller, 1970
30	Greenpea	Iqbal et al., 2006
31	Chickpea	Iqbal et al., 2006

Fig. 7. Principal components analysis (PCA) of amino acids in proteins from fishmeal and other sources. Each number represents a single protein source. Circles indicate clusters of organisms from similar biological groups. Unpublished data provided by Dr. G. Toledo (SGI).

Cyclotella sp.) show that the fatty acid composition closely matches the profile of menhaden oil and their protein composition has an amino acid composition similar to fishmeal (Fig. 7). Additionally, eukaryotic microalgae are a rich source of many vitamins; levels in four microalgae from different genera are higher than that found in a commercial menhaden meal (Fabregas and Herrero 1990).

Studies evaluating the use of biomass from marine eukaryotic algae as a fishmeal replacement for valuable species such as Atlantic salmon have shown promising results at levels of replacement around 10%. For example, Cellana showed that algal protein from two marine microalgae strains that were selected for biofuel (not aquafeed) production was suitable as a fishmeal replacement for salmon at levels up to 10% (Kiron et al. 2012). Replacement of fishmeal in diets of Atlantic cod with a mixture of 30% microalgae (*Nannochloropsis* and *Isochrysis*) resulted in a reduction in feeding, possibly because of reduced palatability (Walker and Berlinsky 2011). Additional research, therefore, is required in order to identify microalgae strains with the optimum biomass composition coupled with digestibility and palatability levels that will allow an eventual complete replacement of fishmeal and fish oil. For example, the ProAlgae 2012 initiative (Kleivdal 2012) is supporting research to identify preferred microalgae strains suitable for cold water cultivation for Norwegian aquaculture.

The acceptance by manufacturers of microalgae biomass as an alternative feed ingredient will depend on having data for available nutrient utilization and efficiency in growing fish. The NOAA/USDA report "Future of Aquafeeds" (Rust et al. 2011) emphasizes that research is required on the suitability of algae biomass (either whole cells or secondary streams resulting from extraction of high-value products) so that manufacturers can properly formulate feeds. Success also depends on creating a reliable supply at the scale and cost needed for aquafeed manufacturing.

In summary, because of needed replacements for wild-caught fish meal and microalgae's nutritional characteristics, expansion in the use of microalgal meal in aquafeeds will result in environmental benefits including the recovery of wild fisheries that are currently under pressure.

Nutrition for humans

Spirulina has been part of the human diet for, at least, centuries. Diaz del Castillo (in Ciferri 1983) described the sale of dried Spirulina in the Tenochtitlan (today's Mexico City) market some 500 yr ago as human food. Dangeard (in Ciferri 1983) also described the use of Spirulina as human food in the area around Lake Chad in Africa early in the 20th century. Today, Spirulina products (*Arthrospira* sp.) are widely available. The biomass of the *Arthrospira* cyanobacterium has long been recognized as an excellent source of protein for humans based on its amino acid composition (Habib et al. 2008). Furthermore, on a land utilization basis, it is probably the most efficient phototrophic source of proteins for humans (Alisan et al. 2008). Commercial Spirulina plants (Fig. 1) have been operating in the US since the early 1980s. Smaller plants have been built since across the world (see, for example, Yun-Ming et al. 2011). Spirulina capacity is believed to represent the largest share of the global microalgal market (Benemann 2010).

Habib et al. (2008) summarize a number of reports and conclude that Spirulina has the potential to produce protein with a much smaller environmental footprint than many other food crops. Spirulina production from commercial farms ranges from 10 to 40 tons per hectare per year. Due to its high protein content it yields 20, 40, and over 200 times more protein per unit area than soy, corn, or beef respectively. Because of its ability to use non-potable non-fresh water it would not require use of water otherwise destined for people or agricultural crops. Additionally, it uses only about 2100 L of water per kg of protein produced which is about 25%, 17%, and 2% of the water used for soy, corn, and beef respectively. Further, cultivation can be carried out on land not suitable to agriculture. Thus, Spirulina represents an excellent protein source.

Other microalgae, although with a somewhat lower protein concentration, would be expected to have similar environmental advantages over conventional protein sources (Chacón-Lee and González-Mariño 2010). *Chlorella, Dunaliella, Haematococcus, Schizochytrium, Scenedesmus, Aphanizomenon, Odontella,* and *Porphyridium* are noted not only for their protein content but an array of other nutritional benefits such as polyunsaturated fatty acids, polysaccharides, and carotenoid pigments. In spite of the many nutritional benefits, microalgal biomass has usually been available as a nutritional supplement in capsules or powders. Algae biomass tends to have strong smell and taste that is hard to mask when formulated with ordinary foods (e.g., breads, pasta, sauces). The authors suggest possible solutions including formulation of exotic (to western palates) snacks and foods together with familiar Asian and Indian spices, encapsulation of the algal biomass or removal of odor-causing compounds.

In summary, microalgae represent a valid source of protein for human nutrition with a smaller fresh water footprint than protein obtained from traditional agriculture (vegetable and, especially, animal). Some strains, like the relatively easy to grow, harvest and process *Arthrospira*, have been used by human populations for centuries and hold the promise of providing nutritional and economic benefits to disadvantaged populations.

The future of microalgal technology

As has been mentioned earlier in this chapter and many other places, microalgae hold the promise of environmentally benign production of many useful chemicals, environmental remediation, and economic opportunities. However, this promise will only be realized in the costs make these diverse applications economically feasible. In this section we will briefly consider:

1. The different scales at which different types of microalgal products are expected to have an impact,
2. The critical need to select the best suited microalgal strains,
3. The genetic resources or products available from microalgae that have, so far, garnered little attention,
4. The use of biotechnology to transform microalgal production, and
5. The future impacts of a more efficient and much larger microalgal technology industry.

Products versus scale

Microalgae produce chemicals and services of varied value. For example, see the chapters on biofuels, chemicals, feeds, nutrition products, and cosmetics in this volume. Some of these are high value (per mass produced) such as specialty chemicals and nutraceuticals. Others, such as biofuels, have less value but presumably much larger markets (Fig. 8). At the present time, we find that the higher value products such as nutraceuticals are produced at relatively small scale and that lower value products such as biofuels are not commercially produced (because they are cost prohibited nowadays) but are expected to be produced at very large scale when economically feasible. In between, we find products such as specialty microalgal feeds for hatcheries or the cheaper nutritional products (e.g., *Spirulina*). Thus there is an inverse relationship between scale and value.

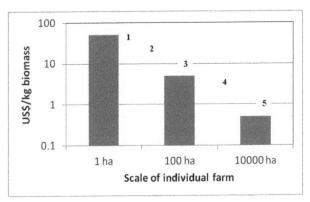

Fig. 8. Predicted relationship between crop value and farm size based on present knowledge. The numerals represent possible products such as, for example, (1) specialty chemicals/nutraceuticals, (2) specialty feeds and nutritional products, (3) Spirulina, (4) protein ingredients, and (5) future biofuels. X-axis represents the scale of real or predicted microalgal production facilities and parallels the progression in scale expected for the different products in the industry as a whole.

The biorefinery concept attempts to illustrate how, by finding a use for every component of microalgal biomass, one producer could, in principle, serve markets of different size and value: lipids and carbohydrates for fuel, protein for feeds, and other small components (e.g., pigments, specialty lipids, vitamins) for specialty markets like nutrition and cosmetics. Furthermore, those components of high value could be used to raise revenue that would bring the net cost of microalgal biofuels production within range of fossil fuels. However, it is not clear how the biorefinery concept will actually work: if one produces microalgal biomass at a scale to be significant in the biofuels market, any secondary component will be produced in quantities that will surely flood their respective markets and destroy any price support. We suspect that one would have to be very careful with the microalgal strain chosen and any secondary products that are accumulated. In the meantime, we predict continuation of the reverse relationship between scale and value.

Strain selection and improvement

The success of microalgal technology will be highly dependent on selecting the proper strains. Specifically strains need to be robust to withstand outdoor conditions such as temperature swings, changes in salinity as water evaporates, and exposure to high oxygen concentrations and to contamination. Synthetic Genomics has identified high-productivity photosynthetic eukaryotic algal strains by a multi-step process that begins with the collection of novel environmental isolates followed by evaluation in advanced screening tests. This broadly-sourced native algae culture collection contains about 2,000 strains selected for robustness and biomass and oil productivity under production conditions. To create this culture collection, Synthetic Genomics obtained permits and legal access to all sampling areas, primarily in North and Central America, continental US and Pacific Islands, prior to collecting samples from environments with conditions similar to those predicted for photobioreactors (PBRs) and open raceway ponds.

Our strategy for strain selection has resulted in the identification of several production strains (for lipids, starch, and high value products) for which we are now investigating the best process options: type and size of growth reactor, contamination mitigation, nutrient management, harvest, and downstream processing. The identification of these production strains also permits us to concentrate efforts aimed at further increasing biomass, starch, and lipid productivity of high-productivity strains through strain engineering. These efforts are essential because a key factor impacting the cost of algal products is the productivity (grams of biomass per square meter per day, Olaizola 2003a; Davis et al. 2011). Higher productivity ultimately will reduce the cost of algal-based food, feed, and fuel components and improve their environmental footprint. State of the art tools that SGI has developed include the use of novel continuous culture methods to select for improved strains and analysis of changes to the strain via a systems biology analysis (genomics, proteomics, transcriptomics, and lipidomics) performed using an SGI proprietary software platform (Archetype™).

We believe that this multi-faceted approach encompassing the latest synthetic biology, photosynthetic improvements, bioinformatics (transcriptomics, proteomics, metabolomics, lipidomics, and fluxomics) and scale down physiology provides a sound basis for strain improvement for commercial scale production.

Genetic resources of microalgae

Besides the well known products and processes that result from microalgal technology, microalgae can also provide us with valuable genetic resources. Here we discuss the availability of microalgal genomes, wild population genes, and the ability of transforming microalgae with other species' genes into microbial factories.

Microalgal genomes

Progress in genome sequencing is proceeding at an exponential pace, and new algal genomes are becoming available every year. One of the challenges facing the community is the association of protein sequences encoded in the genomes with biological function (Lopez et al. 2011), a process known as annotation or functional genomics. While most genome assembly projects generate annotations for predicted protein sequences, they are usually limited and integrate functional terms from a small number of databases. There are around 20 algal eukaryotic genomes in draft form that are publically available. Several commercial entities and research institutes have proprietary genomes in-house and are considered potential intellectual property upon specific utilities of these strains. Synthetic Genomics has many algal genomes, metagenomes, and transcriptomes that are considered almost complete when compared to the assembly and annotation of public counterparts. This level of assigning biological function comes from a legacy of decades of genomic and bioinfomatic knowledge accrued during the human genome project, at The Institute of Genome Research (TIGR) and J. Craig Venter Institute (JCVI), and a drive to commercial utilization of strains, genes, RNA, and proteins to meet the needs of pharmaceutical or industrial innovation to address the unmet needs of disease, sustainable food, fuels, and chemicals. The culmination of this heritage is a dynamic bioinformatics proprietary software platform; Archetype™ developed at Synthetic Genomics. The ability to assemble the Gigabases of genetic data, annotate, and re-annotate genomes' genes, RNAs, and proteins based on literature and biochemical functionality. This capacity provides an unparalleled technology advantage in the field of synthetic biology where the assembly of functional genes and other genetic elements (regulatory and structural) is key to the programming of new pathways and cells.

The value in this type of gene discovery is that new genes, or different versions of genes, can be found that might have advantages over others found in, for example, commercial crop plants. Companies are taking advantage of this concept. For example, Sapphire Energy has a collaboration with Monsanto to leverage algal expertise and research tools, which can be used to screen for promising traits in algae that might have applicability in modern agriculture. The use of algal traits may be two fold; impact on traditional agricultural crops (corn, wheat, cotton, and soy) and accelerate commercializing algae as a renewable energy crop. Sapphire plans to focus on identifying genes that positively affect growth in algae, which might increase crop yields as well. The idea is that a discovery platform in algae is a more rapid approach than the conventional plant approach undertaken by companies like Pioneer Hybrid, Syngenta, and Monsanto because of the much faster generation time. Syngenta had a successful program with Diversa Corp. using a similar approach in heterotrophic microbes such as *E. coli* and yeast, which lead to the discovery of an alpha-amylase (Enogen®) and corn amylase to improve ethanol production.

Genetically engineered plants which produce essential omega-3 fish oils could offer a new way of improving people's diets. Presently, the only sources of EPA and DHA are marine microalgae and the fish that move these algal fats up the food chain. As the availability of fish decreases, other sources for omega-3 fatty acids are drawing interest. Monsanto has developed a GMO soy that expressed genes for two enzymes – one derived from a flower (*Primula juliae*), the other from a red bread mold (*Neurospora crassa*) effectively turning the legume's oil into an omega-3 rich oil (Harris et al. 2008). In 2011, the FDA granted GRAS status to stearidonic acid-enriched soy, but the FDA has not yet granted Monsanto permission to grow this genetically modified line of soybeans in open fields.

Many algal strains are proficient at producing such oils and therefore represent genetic resources that can be transferred into plants. In 2004, Bayer CropScience presented preliminary results from transgenic flax equipped with algal genes, which can produce omega-3, omega-4, and other polyunsaturated fatty acids (Breithaupt 2004). In recent years, work has intensified to move these genes and pathways into a number of plant species. The eventual aim is to feed the GM-enhanced plants to animals such as chicken and cattle so as to produce omega-3 enriched meat, milks, and eggs.

Wild population genes

Although genomes from microalgae are becoming more available, the large majority of microorganisms are believed to be unculturable. Since 2003, scientists at JCVI have been on a quest to uncover novel genes by sampling, sequencing, and analyzing the DNA of the microorganisms living in oceanic waters. While this world is invisible to us, its importance is immeasurable: the microbes in the sea, land, and air sustain our life on Earth. In the initial Sargasso Sea study, 200 liters of filtered surface seawater were used to isolate microorganisms for metagenomic analysis. DNA was isolated from the collected organisms, and genome shotgun sequencing methods were used to identify more than 1.2 million new genes (Venter et al. 2004).

After that successful pilot project in the Sargasso Sea in 2003, Venter and the expedition team set out to evaluate the microbial diversity in the world's oceans using the tools and techniques developed to sequence the human and other genomes. With a better understanding of marine microbial biodiversity, scientists will be able to understand how ecosystems function and to discover new genes of ecological and evolutionary importance. The results of the first phase of the expedition have been published (Kannan et al. 2007; Rusch et al. 2007; Yooseph et al. 2007).

Since then JCVI has conducted sampling in waters off California and the west coast of the United States, and carried out sampling in extreme conditions such as Antarctica and deep sea ocean vents. The research vessel Sorcerer II also sampled the waters of the Baltic, Mediterranean, and Black Seas. These are scientifically important because they are among the world's largest seas isolated from the major oceans. Differences in gene content between samples can identify functions that reflect the lifestyles of the community in the context of its local environment. The results highlight the astounding diversity contained within microbial communities, as revealed through whole genome shotgun sequencing carried out on a global scale. Much of this microbial diversity is organized around phylogenetically related, geographically dispersed populations.

These efforts have produced a very large amount of genetic information much of which is presumably from the so-called unculturable microorganisms. We expect that this collection of information will be fruitful in discovery of new genes, many of microalgal origin, with different characteristics than those of well know strains opening the possibility of, for example, greater metabolic efficiencies. Transfer of those genetic codes to production strains will result in higher productivities of desired chemicals which will improve the economic and environmental out look of microalgal technology. We can also envision transfer of these novel genes to higher plants that could result in more efficient agricultural crops that would also result in environmental and economic benefits.

Microalgae as factories

The genomes of microalgae can also be modified to convert them into factories for products that are now produced in plant or mammalian platforms. For example, several research efforts have highlighted that algae are an attractive platform for producing subunit vaccines because of the low cost of production, genetic tractability, scalability, and short generation time. Furthermore, inflammatory issues, viral, or prion contaminants have been not been observed to date. The Mayfield team at UCSD tested whether algal chloroplasts can produce malaria transmission blocking vaccine candidates (Gregory et al. 2012). Mayfield and his colleagues showed they could produce a mammalian serum amyloid protein and human antibody protein in *Chlamydomonas*. Other therapies include delivery of drug or protein molecules to specific cells. The chloroplast of the green alga *Chlamydomonas reinhardtii* has been shown to contain the machinery necessary to fold and assemble complex eukaryotic proteins including eukaryotic toxins that otherwise would kill a eukaryotic host. The expression and accumulation of immunotoxin proteins in algal chloroplasts was demonstrated; these fusion proteins contain an antibody domain targeting CD22, a B-cell surface epitope, and the enzymatic domain of exotoxin A from *Pseudomonas aeruginosa*. These algal-produced immunotoxins bind target B cells and efficiently kill them *in vitro* and significantly prolong the survival of mice with implanted human B-cell tumors (Tran et al. 2013).

Advances in this field may widen the usefulness of microalgae as manufacturing platforms for many other products that would have applications in nutrition, chemistry, and environmental industries.

Use of biotechnology tools to transform algal production

Microalgal production needs to be transformed if promise of commodity-level products and services is to be achieved; otherwise, production costs (Table 1) will result in the inability of expanding into those markets. To lower costs we need to develop cultivation systems and strategies that result in higher biomass productivity and titers: we need to improve the conversion of light energy into fixed carbon and the channeling of that fixed carbon into the desired product. We believe that engineering of cultivation systems (raceways and PBRs) may result in gains in productivity and development of cultivation strategies (for example continuous vs. batch, with or without nutrient limitation) may be used to selectively accumulate certain products. However, we believe that these gains will be modest. We need to develop systems and strategies that will result in order of magnitude improvements in economic efficiency. We propose that these processes and strategies will necessarily include advanced microalgal strains generated through selection, mutation, and genetic modification. In this section, we explore three approaches that may result in significant gains in productivity.

Photosynthetic efficiency

Microalgae have adapted to survive under conditions of varying irradiance induced by vertical movement through the water column, changes with cloud cover, and time of day and season. In response, microalgal photosystems are quite flexible (Falkowski and Chen 2003): microalgae can, for example, change the size of their photosynthetic antenna (useful but slow) and the relative importance of wasteful reactions designed to dissipate excess light energy to avoid damage (useful and fast but wasteful). The microalga strategy of survival is to adapt to low light conditions by increasing the size of their antennae and, when exposed to high light (for example at midday), open up safety valves to dissipate excessive absorbed light energy such as non-photochemical quenching (dissipation of absorbed light energy as heat), reduction of O_2 to H_2O via the water-water cycle and associated Mehler reaction (photorespiration) or the use of oxydases (e.g., plastid terminal oxidase or PTOX). It has been suggested that up to 80% of absorbed light on a sunny day may be wasted by photosynthetic organisms through NPQ and photorespiration (Melis 2009).

We have used a simple light-limited production model where photosynthetic efficiency of a culture can be adjusted to estimate the resulting changes in culture productivity. The model maintains constant the physical characteristics of the reactor (such as depth at 0.3 m), amount of incoming energy (5.5 kWh/m²/d), and respiration losses equal to 10% of the biomass in the system. We calculate the gross biomass produced by considering the incoming light energy, the photosynthetic efficiency, and the energy equivalence of biomass. Here we run the model to calculate daily productivity when the bulk photosynthetic efficiency (PE) of a culture (e.g., in a raceway) is 2% versus when it is 50% higher at 3% (Fig. 9). In the figure we estimate the changes in gross and net productivity as well as respiration in units of g/m²/d as a new culture is inoculated and allowed to grow for ten days in batch fashion. The model predicts that the culture at 2% PE will become light limited (all available light in the culture is intercepted) on day 3 (vs. day 4 if PE = 3%) and maximum net productivity reaches near 20 g/m²/d (vs. nearly 30 when PE = 3%). Clearly, increases in PE will result in much improved productivity and, therefore, economics of microalgal production.

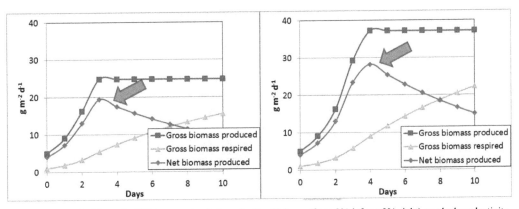

Fig. 9. Top panel: Effect of a 50% increase in photosynthetic efficiency (from 2%-left- to 3%-right) on algal productivity.

It has been proposed (e.g., Benemann 1989) that smaller antennae in microalgae would result in higher light utilization efficiency by eliminating the need to dissipate absorbed light energy thus eliminating waste. It is also generally expected that cells with smaller antenna would allow more light into the culture thus possibly increasing the carrying capacity of such a system. Some, but not all, published studies have supported the concept that reduction in antenna size can lead to improved photosynthetic efficiency. Nakajima and Ueda (1997, 1999) showed that productivity of *Synechocystis* PCC6714 and *Chlorella pyrenoidosa* strains with smaller antenna sizes each increased of 30 to 50% in dense cell cultures. Antenna size was controlled by light intensity for *Chlorella* and by the use of a phycocyanin-deficient mutant for *Synechocystis*. These researchers observed similar properties for a *Chlamydomonas* mutant strain lacking an antenna complex protein (Nakajima and Ueda 2000; Nakajima et al. 2001). Polle et al. (2003) screened 129 low chlorophyll fluorescence mutants and isolated one stable mutant from an insertional knockout library in *Chlamydomonas* that had a functional chlorophyll antenna size for PSI and PSII of 50% and 65% relative to wild type, respectively. This mutant required a higher light intensity for the saturation of photosynthesis (Melis 2009; Mitra and Melis 2010). It also exhibited a greater photosynthetic efficiency and higher cell density in greenhouse grown cultures (Polle et al. 2003). Beckmann et al. (2009) engineered a small antenna strain of *Chlamydomonas* by permanently expressing an LHC repressor; this mutation led to a 17% decrease in antenna size, resulting in a 50% increase in photosynthetic efficiency in saturating light. They also observed significant increases in biomass productivity compared to wild type cells in laboratory photobioreactors. Note, however, that certain attempts to increase photosynthetic efficiency by antenna reduction have been unsuccessful. For example, Huesemann et al. (2009) examined chemically- and UV-induced mutants of *Cyclotella* that had smaller antenna sizes; although these strains required a higher light intensity to reach saturation, no improvements in biomass productivities were observed in either semi-continuous laboratory cultures or outdoor ponds. This may be due to the decrease in antenna size occurring out of balance with other aspects of the acclimation response, leading to aberrant photosynthetic control; an appropriate antenna composition must occur at each reaction center in order to ensure that photosynthetic electron transport (PET) is poised to adequately match Calvin cycle activity. During photosynthetic acclimation, most algae alter antenna size and composition along with Calvin cycle activity and PET to ensure that the saturation point for photosynthesis is reached in a balanced manner. Therefore, obtaining decreased antenna strains, with higher light saturation points, that retain all aspects of the high light acclimation response, should ensure that the light and dark reactions saturate in balance.

Although some success has been achieved some hurdles remain to understand the actual changes in the cell's photosystem and their significance;

- How stable are these transformed strains?
- Are we decreasing the amount of chl-a in the cell and the number of reaction centers in parallel? Or are we decreasing the size of the antenna complex per reaction center?
- How is cellular respiration changing?
- How competitive are these small antenna strains in an outdoor competitive environment?

We believe that answering these questions will help realize large gains in microalgal photosynthetic efficiency which will result in concomitant increases in economic and environmental efficiency.

Redirect photosynthate towards desired products

In photosynthetic algae, "carbon partitioning" refers to the proportion of photosynthetically fixed carbon that is directed into the various biochemical components of a cell (i.e., carbohydrates, proteins, and lipids). In general, carbon partitioning is determined by the relative activities of the enzymatic pathways that are responsible for biosynthesis of the major cell components. This is particularly important at the branch points of metabolism (Fig. 10).

The flux of carbon through these pathways is controlled both at the local enzyme level and by means of a global regulatory network that responds to the sensed environment. Global regulatory systems utilize specialized regulatory proteins and cascades that usually perceive the environment through the

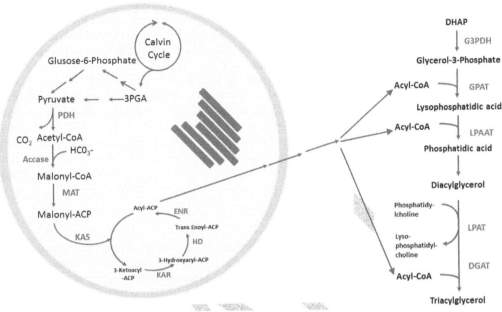

Fig. 10. Areas of targeted modifications for improved carbon partitioning to lipid. Free fatty acids are synthesized in the chloroplast, while TAGs may be assembled at the ER. ACCase, acetyl-CoA carboxylase; ACP, acyl carrier protein; CoA, coenzyme A; DAGAT, diacylglycerol acyltransferase; DHAP, dihydroxyacetone phosphate; ENR, enoyl-ACP reductase; FAT, fatty acyl-ACP thioesterase; G3PDH, gycerol-3-phosphate dehydrogenase; GPAT, glycerol-3-phosphate acyltransferase; HD, 3-hydroxyacyl-ACP dehydratase; KAR, 3-ketoacyl-ACP reductase; KAS, 3-ketoacyl-ACP synthase; LPAAT, lyso-phosphatidic acid acyltransferase; LPAT, lyso-phosphatidylcholine acyltransferase; MAT, malonyl-CoA:ACP transacylase; PDH, pyruvate dehydrogenase complex; TAG, triacylglycerols (Radakovits et al. 2010).

absolute or relative levels of intercellular or intracellular metabolites; this regulation can occur at the transcriptional or translational levels. The tools for modifying carbon partitioning have evolved over time from natural strain selection, through mutagenesis and screening, through the use of recombinant methods for metabolic engineering, through "industrialized" molecular biology in the form of the various high throughput "omics" platforms (genomics, transcriptomics, proteomics, and metabolomics). These tools can be utilized to accelerate targeted metabolic engineering of algal strains through synthetic biology as demonstrated in Ajjawi et al. (2017). In this work a combination of transcriptomics and a CRISPR/cas9 genome editing pipeline was used to identify a transcription factor (ZnCys) that negatively regulates lipid accumulation in *Nannochloropsis gaditana*. By attenuating and fine-tuning the expression of this transcription factor, lipid productivity was doubled with minimal impact on biomass productivity.

Of special interest is the partitioning of fixed carbon into lipids. Lipids are high energy molecules that are desirable for the production of biofuels. In algae, the chloroplast is the location not only for the photosynthetic light reactions and the Calvin cycle, but for three enzyme complexes required for fatty acid synthesis: pyruvate dehydrogenase, acetyl-CoA carboxylase, and fatty acid synthase. Fatty acids are released from the "acyl carrier protein" component of the fatty acid synthase, presumably by a thioesterase or some intermediary, and then transported into the cytoplasm. After conversion to an acyl-CoA, the acyl groups are moved to the endoplasmic reticulum and transferred serially to glycerol-3-phosphate, with a dephosphorylation to diacylglycerol as the penultimate step, resulting finally in a triacylglycerol (TAG). The TAG accumulates in lipid bodies which bud off from the endoplasmic reticulum with the TAG accumulating inside the fragments of inner and outer ER membrane. The factors determining the number and size of lipid bodies are not yet known (Radakovits et al. 2010).

Although fatty acid synthesis occurs in the chloroplast, many of the genes encoding proteins involved in the essential complexes are located in the nuclear genome. These genes are transcribed and processed before the mRNA is translated to protein in the cytoplasm. Translocation of the cytoplasmically-translated chloroplast-localized proteins requires targeting peptide sequences across membranes. In certain classes of algae, that is, the Eustigmatophytes, this can be across four membranes (i.e., two ER membranes and two chloroplast membranes, Ott and Oldham-Ott 2003). Therefore, metabolic engineering of proteins to complement or supplement chloroplastic functions requires identification of not only nuclear promoters and terminators, but appropriate targeting sequences.

In addition to enhancing the biochemical pathway to lipids, partitioning in that direction can be increased by eliminating alternate routes of carbon flux. In algae that accumulate storage carbohydrates (such as starch) in place of, or in addition to lipids, the synthesis of the carbohydrates can be attenuated or blocked by targeted inactivation or down-regulation of appropriate biosynthetic enzymes. Likewise, carbon flux to metabolic energy generation, other biosynthetic pathways and waste carbon products can be interrupted, with the carbon channeled to additional lipid. The limitation on these manipulations is the impact they can have on strain robustness. There must be sufficient energy reserves available for dark metabolism to make it through the night.

The artificial chromosome

The two previous sections describe some of the attempts that different groups have carried out to either increase the PE of microalgae or the proportion of fixed carbon that is destined to, for example, lipid metabolism. A common outcome of bioengineering molecular pathways one enzyme at a time is that as one bottleneck is resolved, another one downstream is revealed. A possible solution to this problem is the introduction of multiple changes in a biochemical pathway at once.

It is anticipated that multiple metabolic modifications will be required to generate algae capable of cost effective production of biofuels. Genes for improved photosynthetic efficiency, fatty acid synthesis, triacylglycerol synthesis, attenuation of alternative carbon sinks, and to improve agronomic properties suggest that we may be targeting as many as fifty modifications. This is not far from the exemplary twenty six genetic modifications utilized by the Genencor/DuPont team to engineer production of 1,3-propanediol in *Escherichia coli* (Nakamura and Whited 2003).

Current strategies for introduction of genetic modifications into eukaryotes algae rely on random integration of the exogenous DNA into the chromosome (Coll 2006; Hallmann 2007) and the number of genetic changes attempted is typically one gene at a time. Random integration results in an unpredictable number of gene copies due to the tendency of the cellular machinery to concatenate exogenous DNA. Also, genetic marker availability limits the number of genes that can be introduced. This results in the need to screen many colonies to identify the few with stable and desired levels of gene expression. Synthetic chromosomes are preferred when multiple genetic elements are required for metabolic modification. The synthetic chromosome provides a single copy number and stable, autonomous inheritance with the ability to accommodate a large number of genes (Zieler et al. 2009).

Early eukaryotic synthetic chromosome constructs in yeast suggested that there are three required elements for maintenance and stability during replication; a centromere, replication origins, and telomeres (Murray and Szostak 1983). Significant work has since been directed at the construction of synthetic chromosomes in mammals, plants, and in some algae strains (Carlson et al. 2007; Houben et al. 2008; Zieler et al. 2009; O'Neill et al. 2011). Strategies for establishing synthetic chromosomes are described as *Top-down or Bottom-up*. Top-down strategies involve reducing functional chromosomes to the minimal number of elements that are required for synthetic genomic applications. Although it has been reported that irradiation of *Chlorella* with electron beams results in minichromosomes (Yamada et al. 2003), a more typical approach for chromosome reduction is based on the observation that introduction of a telomere sequence in an ectopic location results in truncation of the distal portion of the chromosome (Farr et al. 1991). Bottom-up strategies require the isolation of a centromere, origins of replication, and for linear constructs, telomeres which when combined function as a synthetic chromosome scaffold. Based on the diversity of reported performance, it appears that the nature of the sequences required to establish centromere function and the ability of circular constructs to function as synthetic chromosomes are species-specific.

Introduction of genes of interest for metabolism into a synthetic chromosome scaffold has been reported by several methods, for example, inclusion of recombination sites, transposition, or homologous recombination. Additional genetic elements to control expression of genes of interest will be required, such as promoters and terminators, other regulatory elements, and chromatin organizing regions. The other regulatory elements may contain synthetic genes expressing RNAi or sequence specific nucleases such as Meganucleases, TALENs, zinc-finger nucleases or CRISPR/Cas Systems (Cong et al. 2013) in order to attenuate endogenous gene function. To allow facile construction and manipulation of the synthetic chromosome, elements enabling replication and selection in an alternative host, for example, *Escherichia coli* and *Saccharomyces cerevisiae*, can be included.

Synthetic chromosomes provide a scaffold for genetic engineering, enabling the ability to regulate the target genes in a consistent context, free of the positional effects of random integration, and to stack genetic traits. The implications of synthetic chromosome assembly extend beyond the limited pathway and gene engineering of the past to include the engineering or whole metabolisms, regulatory networks, and even ecosystems (Montague et al. 2012). While DNA modification and assembly are becoming routine, regulatory RNA, DNA, and repetitive genetic elements can still provide challenges in genetic engineering (Treangen and Salzberg 2012). However, in order for those potentials to be met, certain limitations and barriers must be overcome. Once the genome of an organism has been sequenced and annotated, the subsequent manipulation of the organism or it synthetic progeny is only limited by the software for designing custom genomic sequences.

Considerations for the use of genetically modified microalgae

We have proposed, above, that genetic modification of microalgae may be a significant tool in the development of new strains that will increase productivity and lower costs. We believe that this work needs to be carried out responsibly. The following four paragraphs summarize our approach to GM development of microalgal strains.

GM technology is part of the solution. GM technology is not the only solution. There are other activities that can help develop more robust and productive processes for the cultivation, harvest, and processing of microalgae. But, these may produce relatively small improvements; our goal is to apply GM technology wherever game-changing improvements are needed.

Safety. It is critically important to demonstrate the safety of the products and services provided by GM microalgae and these microorganisms must be deployed in an environmentally responsible manner. This might include physical control measures (such as deployment in enclosed PBRs) or genetic control measures that would render the microalga unable to compete with native species if accidentally released. Our goal is to prevent the spread of GM microalgae outside of the cultivation area.

Food and feed applications. We believe that consumer educations is critical. To help with this task, we support appropriate labeling of the products and services and that information is made widely available. Our goal is to offer the consumer choices, GM and non-GM produced.

Development of guidelines. To develop guidelines on the use of GM microalgae it is imperative to collaborate with stakeholders such as governments, universities, consumers, and others. Synthetic Genomics has proactively worked with policymakers to establish guidelines. One example is our participation in a 20 month study funded by the Afred P. Sloan Foundation on the safety and security concerns of this technology (Garfinkel et al. 2007). Another example is the work carried out by the 2010 Presidential Commission for the Study of Bioethical Issues (http://bioethics.gov/cms/synthetic-biology-report). Also, the US Environmental Protection Agency already has a framework for reviewing and approving geneticacally modified organisms for cultivation, which applies to microalgae (Environmental Release Application under the Toxic Substances Control Act, TERA). Our goal is to develop and establish guidelines for responsible development and deployment of this technology.

Predicting economic and environmental impacts: examples from Life Cycle Assessments

On several occasions throughout this chapter we have noted that that the economic and environmental impacts of microalgal technology are small (whether positive or negative) because of the small scale of microalgal production at present. However, we predict that the scale of production will increase as new strains (wild type and GMO) and new production technologies are developed: we see a future in which more nutritional, environmental, and energy products from microalgae will be economically attractive. In the meantime, and even at small scale, localized positive economic benefits of microalgal technology have been documented (see section on Economic impacts, above).

Technoeconomic analysis and Life Cycle Assessments (LCAs) are modeling tools that can help us understand what those impacts might be at larger-than-present scale. One can ask, for the different proposed products and processes, whether they will be economically feasible, whether the various balances of energy, water, CO_2, and nutrients are positive or negative, whether the environmental impacts will be acceptable and whether the processes are sustainable. Several such studies have been published over the last few years to study different aspects of microalgal production although mostly regarding microalgal biofuel production. This is probably the case because biofuel production would be expected to occur at a scale where environmental and economic impacts would be expectedly large. Still, some of their findings are useful and applicable to microalgal production in general.

It is generally agreed that large scale microalgal production needs large amounts of nutrients, CO_2, and water. It will also require large amounts of energy (from mixing of cultures to dewatering and downstream processing of the biomass). In reviewing this chapter, one can see that we have described processes where microalgae can replace other products at large scale: e.g., microalgal proteins, oils, and whole biomass to replace fishmeal and nutritional oils and oils and biomass to replace fossil fuels. So the questions are not simply how much CO_2, water, nutrients, and energy microalgal products require but how that compares to that needed by the products and processes they replace. The assumption is that consuming more CO_2, less water, less energy, and fewer nutrients are desirable characteristics.

Detailed LCAs comparing microalgal feed ingredients versus those of agricultural and animal origin are lacking while there are numerous ones comparing microalgal fuel precursors to fossil fuels, including a recent effort in which Synthetic Genomics was a participant (Vasudevan et al. 2012). In that study, special attention was paid to to GHG (greenhouse gas) emissions, freshwater consumption, and energy inputs and outputs. Depending on the technology set chosen, that study determined that microalgal biofuels could have a negative or positive energy balance, emit more or less GHG than conventional fuels and, when using brackish or salt water, consume about as much freshwater as fossil fuels. Replacing dry extraction with wet extraction technology was identified as a leading hurdle that, if overcome, would result in large (> 50%) reductions in GHG emissions and a favorable energy balance.

A similar conclusion was reached by Sander and Murthy (2010). They identified the amount of energy needed to process microalgal biomass into useable components as a major obstacle. They considered that if the energy needed for dewatering could be minimized by using wet-based processes based on enzyme-driven degradation of the microalgal biomass the energy balance for microalgae based products would be favorable. Another option is that suggested by Taylor et al. (2013). In that study (that is, Taylor et al. 2013), the authors propose to use non-fossil renewable energy, specifically concentrated solar power (CSP) and solar drying, for the energy-requiring processes such as providing mixing to the cultures, dewatering, etc. By doing so, the process can be CO_2 negative as well as energetically positive. Shirvani et al. (2011) also reached a similar conclusion; they suggest the need to decarbonize heat, electricity, and other energy requirements to produce, for example, fertilizers. Vasudevan et al. (2012) noted that fresh water consumption is expected to be similar for microalgal biofuels when compared to fossil fuels. Yang et al. (2011) considered the fresh water footprint when different types of water were used (fresh, salt, and waste) under different levels of recycling. Their results indicate that microalgae are very competitive with other types of agricultural feedstocks. Clarens et al. (2010) found that other biofuels feedstocks may actually have a lower environmental impact than microalgae but that the ability to utilize flue gasses and waste waters would improve their performance. They also found that microalgae are superior to other feedstocks when land area use and eutrophication potential are considered.

A common conclusion in microalgal biofuels LCAs is the need for full utilization of coproducts in the biorefinery concept. In many cases, these coproducts represent animal feed opportunities. If the environmental and economic burden of microalgal production is shared between biofuels and feed products, the economic and environmental balance would, of course, tend to be more positive. The biofuels industry based on land plants has already produced large quantities of feed ingredients in use throughout the world (Makkar 2012). Microalgal biofuels are also expected to generate large quantities of "spent" biomass rich in proteins and minerals and other bioactive compounds (Ravishankar et al. 2012). When considering the production of feeds independently, we have already made the arguments, above, that microalgae can reduce the water and nutrient footprint of feed production assuming water is recycled and nutrients are supplemented with waste water streams. It can also replace feed ingredients which are slowly being reduced (such as wild caught fish) with unpredictable ecological implications.

Essentially, microalgal technology can replace fuels and feeds with (when using the proper technology sets) lower economic and environmental impact. So, why is the scale of microalgal technology so small, as was pointed out earlier in the chapter? Several studies (Davis et al. 2011; Haruna et al. 2011; Amer et al. 2011, Chapter 14 in this volume are just examples) have pointed out that biofuels production form microalgae is not yet economically feasible and several technological breakthroughs will need to occur before cost parity with fossil fuels is achieved. A similar situation exists with feed replacements. Microalgal production is still too expensive (Table 1). However, we are optimistic that the cost differential between microalgal feedstocks and agricultural and fossil feedstocks will narrow as new microalgae strains are developed, new processes are established, larger scale microalgae culture is achieved, and the full cost (economic and environmental) of microalgae, agricultural, and fossil feedstocks is considered. As these developments occur, we expect to realize the promise of microalgal products with a smaller environmental footprint than those available today. Our review suggests that based on price and scale considerations (e.g., Fig. 8), microalgal-based feed replacements will likely find commercial viability sooner than microalga-based fuels.

Acknowledgements

Much of the work presented here has been possible thanks to the efforts of many different scientists at Synthetic Genomics. We are grateful to Dr. Gerardo Toledo for sharing unpublished data and Dr. James Flatt, Dr. Teresa Spehar, and Mr. Charlie Witherspoon for reviewing an earlier version of the manuscript. Special thanks to Dr. Claudia Grewe for reviewing the manuscript.

References

Abdulqader, G., L. Barsanti and M.R. Tredici. 2000. Harvest of Arthrospira platensis from Lake Kossorom (Chad) and its household usage among the Kanembu. J. Appl. Phycol. 12: 493–498.
Ahsan, M., B. Habib, M. Parvin, T.C. Huntington and M.R. Hasan. 2008. A review on culture, production and use of Spirulina as foods for humans and feeds for domestic animals and fish. FAO Fisheries and Aquaculture Circular No. 1034. FAO, Rome. 33 pp.
Ajjawi, I., J. Verruto, M. Aqui, L.B. Soriaga, J. Coppersmith, K. Kwok, L. Peach, E. Orchard, R. Kalb, W. Xu, T.J. Carlson, K. Francis, K. Konigsfeld, J. Bartalis, A. Schultz, W. Lambert, A.S. Schwartz, R. Brown and E.R. Moellering. 2017. Lipid production in *Nannochloropsis gaditana* is doubled by decreasing expression of a single transcriptional regulator. Nature Biotechnol. 35: 647–652.
Amer, L., B. Adhikari and J. Pellegrino. 2011. Technoeconomic analysis of five microalgae-to-biofuels processes of varying complexity. Bioresour. Technol. 102: 9350–9359.
Batan, L., J. Quinn, B. Wilson and T. Bradley. 2010. Net energy and greenhouse emission evaluation of biodiesel derived from microalgae. Environ. Sci. Technol. 44: 7975–7980.
Becker, E. 2007. Micro-algae as a source of protein. Biotechnol. Adv. 25: 207–210.
Beckmann, J., F. Lehr, G. Finazzi, B. Hankamer, C. Posten, L. Wobbe and O. Kruse. 2009. Improvement of light to biomass conversion by de-regulation of light-harvesting protein translation in *Chlamydomonas reinhardtii*. J. Biotechnol. 142: 70–77.
Benemann, J. 1989. The future of microalgal biotechnology. pp. 317–337. *In*: R.C. Creswell, T.A.V. Rees and N. Shah (eds.). Algal and Cyanobacterial Biotechnology. Longman, London.
Benemann, J. 2010. Microalgae Aquafeeds. 4th Algae Biomass Summit, Phoenix, AZ, USA.
Breithaupt, H. 2004. GM plants for your health. EMBO Reports 5: 1031–1034.

Carlsson, A.S., J.B van Beilen, R. Möller and D. Clayton. 2007. EPOBIO: Realising the Economic Potential of Sustainable Resources – Bioproducts from Non-food Crops. CNAP, University of York. CLP Press, Newbury. 82 pp.

Carlson, S.R., G.W. Rudgers, H. Zieler, J.M. Mach, S. Luo, E. Grunden, C. Krol, G.P. Copenhaver and D. Preuss. 2007. Meiotic transmission of an *in vitro*–assembled autonomous maize minichromosome. PLoS Genet. 3: e179.

Chacón-Lee, T.L. and G.E. González-Mariño. 2010. Microalgae for "healthy" foods—possibilities and challenges. Comprehensive Rev. Food Sci. Food Saf. 9: 655–675.

Cho, C.Y. and D.P. Bureau. 2001. A review of diet formulation strategies and feeding systems to reduce excretory and feed wastes in aquaculture. Aquacult. Res. 32: 349–360.

Ciferri, O. 1983. *Spirulina*, the edible microorganism. Microbiol. Rev. 47: 551–578.

Clarens, A., E. Resurreccion, M. White and L. Colosi. 2010. Environmental life cycle comparison of algae to other bioenergy feedstocks. Environ. Sci. Technol. 44: 1813–1819.

Coll, J.M. 2006. Review. Methodologies for transferring DNA into eukaryotic microalgae. Span. J. Agric. Res. 4: 316–330.

Cong, L., F.A. Ran, D. Cox, S. Lin, R. Barretto, N. Habib, P.D. Hsu, X. Wu, W. Jiang, L.A. Marraffini and F. Zhang. 2013. Multiplex Genome Engineering Using CRISPR/Cas Systems. Science 339: 819–823.

Cordell, D., J. Drangert and S. White. 2009. The story of phosphorus; global food security and food for thought. Global Environ. Chang. 19: 292–305.

Craggs, R., D. Sutherland and H. Campbell. 2012. Hectare-scale demonstration of high rate algae ponds for enhanced wastewater treatment for biofuel production. J. Appl. Phycol. 24: 329–337.

Davis, R, A. Aden and P.T. Pienkos. 2011. Techno-economic analysis of autotrophic microalgae for fuel production. Appl. Energy 88: 3524–3531.

Diaz, R.J. and R. Rosenberg. 2008. Spreading dead zones and the consequences for marine ecosystems. Science 321: 926–929.

Elser, J.J., W.F. Fagan, R.F. Denno, D.R. Dobberfuhl, A. Folarin, A. Huberty, S. Interland, S.S. Kilham, E. McCauley, K.L. Schulz, E.H. Siemann and R.W. Sterner. 2000. Nutritional constraints in terrestrial and freshwater food webs. Nature 408: 578–580.

Fabregas, J. and C. Herrero. 1990. Vitamin content of four marine microalgae. Potential use as source of vitamins in nutrition. J. Ind. Microbiol. 5: 259–263.

Falkowski, P.G. and Y.-B. Chen. 2003. Photoacclimation of light harvesting systems in eukaryotic algae. pp. 423–447. *In*: B.R. Green and W.W. Parsons (eds.). Light-Harvesting Antennas in Photosynthesis. Kluwer Academic Publishers, The Netherlands.

[FAO] Food and Agriculture Organization. 2016. The state of world fisheries and aquaculture. FAO Fisheries and Aquaculture Department. Food and Agriculture Organization of the United Nations, Rome. 200 pp.

Farr, C., J. Fantes, P. Goodfellow and H. Cooke. 1991. Functional reintroduction of human telomeres into mammalian cells. PNAS 88: 7006.

Fisher, A.W. and J.S. Burlew. 1953. Nutritional value of microscopic algae. pp. 303–310. *In*: J.W. Burlew (ed.). Algal Culture from Laboratory to Pilot Plant. Carnegie Institution of Washington, Washington.

Garfinkel, M.S., D. Endy, G.L. Epstein and R.M. Friedman. 2007. Synthetic genomics; Options for Governance. JVCI-MIT-CSIS Report. 57p. (http://www.jcvi.org/cms/fileadmin/site/research/projects/synthetic-genomics-report/synthetic-genomics-report.pdf).

García, J., R. Mujeriego and M. Hernández-Mariné. 2000. High rate algal pond operation strategies for urban wastewater nitrogen removal. J. Appl. Phycol. 12. 331–339.

Garofalo, R. 2010. AquaFUEL s Roundtable Proceedings, October 21–22, 2010, Brussels.

Gonzalez, L.E., R.O. Canizare and S. Baena. 1997. Efficiency of ammonia and phosphorus removal from a Colombian agroindustrial wastewater by microalgae *Chlorella vulgaris* and *Scenedesmus dimorphus*. Bioresour. Technol. 60: 259–262.

Grewe, C.B. and O. Pulz. 2012. The biotechnology of Cyanobacteria. pp. 707–739. *In*: B.A. Whitton (ed.). Ecology of Cyanobacteria II: Their Diversity in Space and Time. Springer, Dordrecht.

Gregory, J.A., F. Li, L.M. Tomosada, C.J. Cox, A.B. Topol, J.M. Vinetz and S. Mayfield. 2012. Algae-produced Pfs25 elicits antibodies that inhibit malaria transmission. PLoS ONE 7: e37179.

Habib, M.A.B., M. Parvin, T.C. Huntington and M.R. Hasan. 2008. A review on culture, production and use of Spirulina as food for humans and feed for domestic animals and fish. FAO Fisheries and Aquaculture Circular No. 1034. Rome. 33p.

Hallmann, A. 2007. Algal Transgenics and Biotechnology. Algae 1: 81–98.

Harris, W.S., S.L. Lemke, S.N. Hansen, D.A. Goldstein, M.A. DiRienzo, H. Su, M.A. Nemeth, M.L. Taylor, G. Ahmed and C. George. 2008. Stearidonic Acid-Enriched Soybean Oil Increased the Omega-3 Index, an Emerging Cardiovascular Risk Marker. Lipids 43: 805–811.

Haruna, R., M. Davidson, M. Doyle, R. Gopiraj, M. Danquah and G. Forde. 2011. Technoeconomic analysis of an integrated microalgae photobioreactor, biodiesel and biogas production facility. Biomass Bioenergy 35: 741–747.

Hintz, H.F, H. Heitman, W.C Weir, D.T. Torell and J.H. Meyer. 1966. Nutritive value of algae grown on sewage. J. Anim. Sci. 25: 675–681.

Hill, J., E. Nelson, D. Tilman, S. Polasky and D. Tiffany. 2006. Environmental, Economic and energetic costs and benefits of biodiesel and ethanol biofuels. PNAS. 103: 11206–11210.

Hoffman, J. 1998. Wastewater treatment with suspended and nonsuspended algae. J. Phycol. 34: 757–763.

Houben, A., R.K. Dawe, J. Jiang and I. Schubert. 2008. Engineered plant minichromosomes: A bottom-up success? The Plant Cell Online 20: 8–10.

Huesemann, M.H., T.S. Hausmann, R. Bartha, M. Aksoy, J.C. Weissman and J.R. Benemann. 2009. Biomass productivities in wild type and pigment mutant of *Cyclotella sp.* (Diatom). Appl. Biochem. Biotechnol. 157: 507–526.

[IAFMM] International Association of Fish Meal Manufacturers. 1970. Available amino acid content of fish meals. IAFMM Report 1: 11 pp.

[IEA] International Energy Agency. 2012. CO_2 emissions from fuel combustion 2012 edition. Paris 138 pp.

Iqbal, A., I.A. Khalil, N. Ateeq and M.S. Khan. 2006. Nutritional quality of important food legumes. Food Chem. 97: 331–335.

Kannan, N., S.S. Taylor, Y. Zhai, J.C. Venter and G. Manning. 2007. Structural and functional diversity of the microbial kinome. PLOS Biology S91–S102.

Khoshmanesh, A.F. Lawson and I.G. Prince. 1996. Cadmium uptake by unicellular green microalgae. Chem. Eng. J. Biochem. Eng. J. 62: 81–88.

Kiron, V., W. Phromkunthong, M. Huntley, I. Archibald and G. De Scheemaker. 2012. Marine microalgae from biorefinery as a potential feed protein source for Atlantic salmon, common carp and whiteleg shrimp. Aquac. Nutr. 18: 521–531.

Kleivdal, H. 2012. CO_2 sequestration and aquafeed development by industrial microalgae production. 6th International Algae Congress, Rotterdam, December 4–5, 2012.

Lipinsky, E.S., J.H. Litchfield and D.I.C. Wang. 1970. Algae, bacteria and yeasts as food or feed. CRC Crit. Rev. Food Technol. 1: 581–618.

Lopez, D., D. Casero, S.J. Cokus, S.S. Merchant and M. Pellegrini. 2011. Algal functional annotation tool: a web-based analysis suite to functionally interpret large gene lists using integrated annotation and expression data. BMC Bioinformatics 12: 282.

Makkar, H.P.S. 2012. Biofuel co-products as livestock feed: Opportunities and challenges. FAO, Rome, 533 p.

Mekonnen, M.M. and A.Y. Hoekstra 2012. A global assessment of the water footprint of farm animal products. Ecosystems 15: 401–415.

Melis, A. 2009. Solar energy conversion efficiencies in photosynthesis: Minimizing the chlorophyll antennae to maximize efficiency. Plant Sci. 177: 272–280.

Miller, E.L. 1970. Available amino acid content of fish meals. FAO Fish. Rep. 92: 66 p.

Mitra, M. and A. Melis. 2010. Genetic and biochemical analysis of the TLA1 gene in *Chlamydomonas reinhardtii*. Planta 231: 729–740.

Montague, M.G., C. Lartique and S. Vashee. 2012. Synthetic genomics: potential and limitations. Curr. Opin. Biotechnol. 5: 659–665.

Mussatto, S.I., G. Dragone, P.M.R. Guimarães, J.P.A. Silva, L.M. Carneiro, I.C. Roberto, A. Vicente, L. Domingues and J.A. Teixeira. 2010. Technological trends, global market, and challenges of bio-ethanol production. Biotechnol. Adv. 28: 817–830.

Murray, A.W. and J.W. Szostak. 1983. Construction of artificial chromosomes in yeast. Nature 305: 189–193.

Nakajima, Y., M. Tsuzuki and R. Ueda. 2001. Improved productivity by reduction of the content of light-harvesting pigment in *Chlamydomonas perigranulata*. J. Appl. Phycol. 13: 95–101.

Nakajima, Y. and R. Ueda. 1997. Improvement of photosynthesis in dense microalgal suspension by reduction of light harvesting pigments. J. Appl. Phycol. 9: 503–510.

Nakajima, Y. and R. Ueda. 1999. Improvement of microalgal photosynthetic productivity by reducing the content of light harvesting pigment. J. Appl. Phycol. 11: 195–201.

Nakajima, Y. and R. Ueda. 2000. The effect of reducing light-harvesting pigment on marine microalgal productivity. J. Appl. Phycol. 12: 285–290

Nakamura, C.E. and G.M. Whited. 2003. Metabolic engineering for the microbial production of 1, 3-propanediol. Curr. Opin. Biotechnol. 14: 454–459.

Nakamura, T., C.L. Senior, M. Olaizola, T. Bridges, S. Flores, L. Sombardier and S.M. Masutani. 2005. Recovery and sequestration of CO_2 from stationary combustion systems by photosynthesis of microalgae. US Department of Energy (Contract #DE-FC26-00NT40934), 220 pp.

Naylor, R.L., R.W. Hardy, D.P. Bureau, A. Chiu, M. Elliott, A.P. Farrell, I. Forster, D.M. Gatlin, R.J. Goldburg, K. Hua and P.D. Nichols. 2009. Feeding aquaculture in an era of finite resources. PNAS 106: 15103–15110.

Neori, A. 2011. "Green water" microalgae: the leading sector in world aquaculture. J. Appl. Phycol. 23: 143–149.

[NREL] National Renewable Energy laboratory. 2012. Irradiance data: NREL location USAF #747185 - IMPERIAL, CA (Class II), website: http://rredc.nrel.gov/solar/old_data/nsrdb/1991-2005/hourly/siteonthefly.cgi?id=747185.

O'Neill, B.M., K.L. Mikkelson, N.M. Gutierrez, J.L. Cunningham, K.L. Wolff, S.J. Szyjka, C.B. Yohn, K.E. Redding and M.J. Mendez. 2011. An exogenous chloroplast genome for complex sequence manipulation in algae. Nucleic Acids Res. 40: 2782–2792.

Olaizola, M. 2000. Commercial production of astaxanthin from *Haematococcus pluvialis* using 25,000 liter outdoor photobioreactors. J. Appl. Phycol. 12: 499–506.

Olaizola, M. 2003a. Commercial development of microalgal biotechnology: From the test tube to the marketplace. Biomol. Eng. 20: 459–466.

Olaizola, M. 2003b. Microalgal removal of CO_2 from flue gases: Changes in medium pH and flue gas composition do not appear to affect the photochemical yield of microalgal cultures. Biotechnol. Bioprocess Eng. 8: 360–367.

Ott, D.W. and C.K. Oldham-Ott. 2003. Eustigmatophyte, raphidophyte, and tribophyte algae. pp. 423–470. *In*: J.D. Wehr and R.G. Sheath (eds.). Freshwater Algae of North America. Elsevier Sciences, San Diego .

Pate R., G. Klise and B. Wu. 2011. Resource demand implications for US algae biofuels production scale-up. Appl. Energy 88: 3377–3388.

Pettersen, A.K., G.M. Turchini, S. Jahangard, B.A. Ingram and C.D.H. Sherman. 2010. Effects of different dietary microalgae on survival, growth, settlement and fatty acid composition of blue mussel (*Mytilus galloprovincialis*) larvae. Aquaculture 309: 115–124.

Picot B., H. El Halouani, C. Casellas, S. Moersidik and J. Bontoux. 1991. Nutrient removal by high rate pond system in a Mediterranean climate (France). Water Sci. Technol. 23: 1535–1541.

Pike, I., G. Andorsdóttir and H. Mundheim. 1990. The role of fish meal in diets for salmonids. International Association of Fish Meal Manufacturers.

Polle, J.E.W., S.D. Kanakagiri and A. Melis. 2003. Tla1, a DNA insertional transformant of the green alga *Chlamydomonas reinhardtii* with a truncated light-harvesting chlorophyll antenna size. Planta 217: 49–59.

Pyle, D.J, R.A. Garcia and Z. Wen. 2008. Producing docosahexaenoic acid (DHA)-rich algae from biodiesel-derived crude glycerol: Effects of impurities on DHA production and algal biomass composition. Journal of Agricultural and Food Chem. 56: 3933–3939.

Radakovits, R., R.E. Jinkerson, A. Darzins and M.C. Posewitz. 2010. Genetic engineering of algae for enhanced biofuel production. Eukaryotic Cell 9: 486–501.

Ravishankar, G.A., R. Sarada, S. Vidyashankar, K.S. VenuGopal and A. Kumudha. 2012. Cultivation of micro-algae for lipids and hydrocarbons, and utilization of spent biomass for livestock feed and for bio-active constituents. pp. 423–446. *In*: H.P.S. Makkar (ed.). Biofuel Co-Products as Livestock Feed: Opportunities and Challenges. FAO, Rome,

Rusch, D.B., A.L. Halpern, G. Sutton, K.B. Heidelberg, S. Williamson, S. Yooseph, D. Wu, J.A. Eisen, J.M. Hoffman, K. Remington, K. Beeson, B. Tran, H. Smith, H. Baden-Tillson, C. Stewart, J. Thorpe, J. Freeman, C. Andrews-Pfannkoch, J.E. Venter, K. Li, S. Kravitz, J.F. Heidelberg, T. Utterback, Y.-H. Rogers, L.I. Falcón, V. Souza, G. Bonilla-Rosso, L.E. Eguiarte, D.M. Karl, S. Sathyendranath, T. Platt, E. Bermingham, V. Gallardo, G. Tamayo-Castillo, M.R. Ferrari, R.L. Strausberg, K. Nealson, R. Friedman, M. Frazier and J.C. Venter. 2007. The Sorcerer II Global Ocean Sampling Expedition: Northwest Atlantic through Eastern tropical Pacific. PLOS Biology S22–S55.

Rust, M.B., F.T. Barrows, R.W. Hardy, A. Lazur, K. Naughten and J. Silverstein. 2011. The future of aquafeeds. NOAA Technical Memorandum NMFS F/SPO-124. 93 pp.

Sander, K and G.S. Murthy. 2010. Life cycle analysis of algae biodiesel. Int. J. Life Cycle Assess. 15: 704–714.

Samori, G., C. Samori, F. Guerrini and R. Pistocchi. 2013. Growth and nitrogen removal capacity of *Desmodesmus communis* and of a natural microalgal consortium in a batch culture system in view of urban wastewater treatment Part 1. Water Res. 47: 791–801.

Sargent, J., G. Bell, L. McEvoy, D. Tocher and A. Estevez. 1999. Recent developments in the essential fatty acid nutrition of fish. Aquaculture 177: 191–199.

Singh, A., P. Nigam and J. Murphy. 2011. Renewable fuels from algae: An answer to debatable land based fuels. Bioresour. Technol. 102: 10–16.

Sheehan, J., T. Dunahay, J. Benemann and P. Roessler. 1998. A look back at the US Department of Energy's Aquatic Species Program – Biodiesel from Algae. National Renewable Energy Laboratory, NREL/TP-580-24190.

Shirvani, T., X. Yan, O.R. Inderwildi, P.P. Edwards and D.A. King. 2011. Life cycle energy and greenhouse gas analysis for algae-derived biodiesel. Energ. Environ. Sci. 4: 3773–3778.

Slocombe, S.P. and J.R. Benemann. 2016. Microalgal Production for Biomass and High-Value Products. CRC Press, Boca Raton.

Spencer, K.G. 1989. Lipids and polyols from microalgae. pp. 237–254. *In*: C.A. Lembi and J.K. Waaland (eds.). Algae and Human Affairs. Cambridge University Press, Cambridge.

Tacon, A.G.J. and M. Metian. 2008. Global overview on the use of fish meal and fish oil in industrially compounded aquafeeds: trends and future prospects. Aquaculture 285: 146–158.

Tacon, A.G.J., M.R. Hasan and M. Metian. 2011. Demand and supply of feed ingredients for farmed fish and crustaceans: trends and prospects. FAO Fisheries and Aquaculture Technical Paper No. 564. Rome, FAO. 87 pp.

Tam, N.F.Y. and Y.S. Wong. 1989. Wastewater nutrient removal by *Chlorella pyrenoidosa* and *Scenedesmus* sp. Environ. Microbiol. 58: 19–34.

Taylor, B., N. Xiao, J. Sikorski, M. Yong, T. Harris, T. Helme, A. Smallbone, A. Bhave and M. Kraft. 2013. Tech no-economic assessment of carbon-negative algal biodiesel for transport solutions. Appl. Energy 106: 262–274.

Tran, M., C. Van, D.J. Barrera, P.L. Pettersson, C.D. Peinado, J. Bui and S.P. Mayfield. 2013. Production of unique immunotoxin cancer therapeutics in algal chloroplasts. PNAS 110: E15–E22.

Treangen, T.J. and S.L. Salzberg. 2012. Repetitive DNA and next-generation sequencing: computational challenges and solutions. Nat. Rev. Genet. 13: 36–46.

[UNESCO] United Nations Educational, Scientific and Cultural Organization. 2009. Water in a changing world. The United Nations World Water Development Report 3. Paris: UNESCO Publishing, Earthscan. 318 p.

[UNESCO] United Nations Educational, Scientific and Cultural Organization. 2017. Wastewater the untapped resource. The United Nations World Water Development Report 2017. Paris: Unesco. 198 p.

US Bureau of Reclamation, 2004. Data obtained from US Department of the Interior, Bureau of Reclamation, Lower Colorado Region, Boulder City, Nevada and the Salton Sea Authority La Quinta, California "Salton Sea Salinity Control Research Project" 2004 Bureau of Reclamation Technical Service Center, Denver, Colorado.

Valenzuela, C.L. 2011. Imperial County agricultural and livestock report 2011. Office of Agricultural Commissioner. Available at http://www.co.imperial.ca.us/ag/Crop_&_Livestock_Reports.

Vasudevan, V., R.W. Stratton, M.N. Pearlson, G.R. Jersey, A.G. Beyene, J.C. Weissman, M. Rubino and J.I. Hileman. 2012. Environmental performance of algal biofuel technology options. Environ. Sci. Technol. 46: 2451–2459.

Venter, J.C., K. Remington, J.F. Heidelberg, A.L. Halpern, D. Rusch, J.A. Eisen, D. Wu, I. Paulsen, K.E. Nelson, W. Nelson, D.E. Fouts, S. Levy, A.H. Knap, M.W. Lomas, K. Nealson, O. White, J. Peterson, J. Hoffman, R. Parsons, H. Baden-Tillson, C. Pfannkoch, Y.H. Rogers and H.O. Smith. 2004. Environmental Genome Shotgun Sequencing of the Sargasso Sea. Science 304: 66–74.

Verdelho, V. 2012. Scaling-up process from 0.001 L to 1100000 L culture volume in photobioreactors. 6th International Algae Congress, Rotterdam, December 4–5, 2012.

Walker A.B. and D.L. Berlinsky. 2011. Effects of partial replacement of fish meal protein by microalgae on growth, feed intake, and body composition of Atlantic cod. N. Am. J. Aquac. 73: 76–83.

Woertz, I., A. Feffer, T. Lundquist and Y. Nelson. 2009. Algae grown on dairy and municipal wastewater for simultaneous nutrient removal and lipid production for biofuel feedstock. J. Environ. Eng. 135: 1115–1122.

Yamada, T., Y. Fujimoto, Y. Yamamoto, K.I. Machida, M. Oda, M. Fujie, S. Usami and H. Nakayama. 2003. Minichromosome formation in *Chlorella* cells irradiated with electron beams. J. Biosci. Bioeng. 95: 601–607.

Yang J., M. Xu, X. Zhang, Q. Hu, M. Sommerfeld and Y. Chen. 2011. Life-cycle analysis on biodiesel production from microalgae: Water footprint and nutrients balance. Bioresour. Technol. 102: 159–165.

Yooseph, S., G. Sutton, D.B. Rusch, A.L. Halpern, S.J. Williamson, K. Remington, J.A. Eisen, K.B. Heidelberg, G. Manning, W. Li, L. Jaroszewski, P. Cieplak, C.S. Miller, H. Li, S.T. Mashiyama, M.P. Joachimiak, C. van Belle, J.M. Chandonia, D.A. Soergel, Y. Zhai, K. Natarajan, S. Lee, B.J. Raphael, V. Bafna, R. Friedman, S.E. Brenner, A. Godzik, D. Eisenberg, J.E. Dixon, S.S. Taylor, R.L. Strausberg, M. Frazier and J.C. Venter. 2007. The Sorcerer II Global ocean Sampling Expedition: expanding the universe of protein families. PLOS Biology S56–S90.

Yun-Ming, L., X. Wen-Zhou and W. Yong-Huang. 2011. *Spirulina* (*Arthrospira*) industry in Inner Mongolia of China: current status and prospects. J. Appl. Phycol. 23: 265–269.

Zieler, H., R.C. Brown, T. Richardson and D.G. Smith. 2009. Identification of centromere sequences and uses thereof. WO2009/134814 A2.

Index

A

A/R CDM 25, 27, 30, 31, 33–35, 38
Agar 146–150, 154
algal biofuel 302, 304–305, 309, 316, 324–325
algal form
    Size and shapes 2
algal physiology 53–55
Alginate 147–149, 152–154, 204, 206–208
Alveolata 217
animal feed 166, 168, 173, 174
Anti-cancer 129
Anti-inflammatory 117, 120, 121, 201, 203, 205, 206
Antileishmanial 132, 133, 136–138
Antimicrobial 16, 90, 178, 183
antimicrobial compounds 178
Antioxidant 116–120, 128, 129, 199–203, 205, 206,
    209–211
Antiplasmodial 137, 138
Anti-proliferative 120, 124, 126, 128
Anti-protozoal 131, 132
Antitrypanosomal 133, 137
aquaculture 162, 168–170, 172, 174
Asian Network of Algae as Mitigation and Adaptation
    Measures (ANAMAM) 21, 22
astaxanthin 43, 49–52

B

β-carotene 43, 49–51
Benthic algae 2, 12
bioactive 243, 244
bioactive compounds 216, 217
bioactive seaweeds 178
Bioactivity 117, 128, 132
Biodiesel 249–251, 257, 258, 260, 261, 263, 277, 282,
    283
Bioethanol 249–251, 257, 258, 262, 263, 284
biofilter 237
biofuels 52, 53, 248–252, 254, 257, 262, 263
Biogas 250, 258, 262, 263
biohydrogen 249, 258, 262
Biomass 63, 64, 66–70, 73, 74, 249–259, 262, 263
biostimulants 177, 178, 190
blue carbon 20, 21

C

carbon fixation 279, 281, 282, 284
Carbon partitioning 300, 320, 321

Carbon Zero Seaweed Town (CØST) 21, 22
Carrageenan 146–157
Cell disruption 100–103, 105
Centrifugation 90, 92, 100, 105
Cercozoa (Chlorarachniophyceae) 11
challenges 243, 245
Chemical defenses 15
chemical synthesis 216, 222, 228
Chlorella 165–171, 174
Chlorophyta 6–9, 12, 13
circadian cycle 223, 228
Clean Development Mechanism (CDM) 22
climate change 20–22, 26, 40
Coastal CO_2 Removal Belt (CCRB) 22
Commercial scale 302, 316
cosmeceutical 198, 199, 203, 204, 206, 209, 211
Cryptophyta 6, 7, 10
Cultivation 252–255, 258, 262, 263
Cyanobacteria 1–3, 7, 11–14

D

Dinoflagellates 216–228
dehydrogenase 108, 137, 282–283, 285–286, 290, 321
disease management 178, 184, 190, 191
Downstream processing 90–92, 96
Drying 90, 92, 100
Dunaliella salina 49, 51

E

Economic impact 302, 303, 305, 306, 308, 324
ecosystem services 244
Environmental impact 300, 302, 307, 309, 310, 324, 325
Euglenophyta 6–8, 11
eutrophication 236, 237
Extraction 90, 100–109

F

fed aquaculture 243
feed ingredients 161, 174
feed market 162, 169, 172
Filtration 90, 99, 100
Flocculation 90, 92–99
Flotation 90, 98
Foaming 98
food ingredients 243
FTIR-ATR 153–155
FT-Raman 153–153
Fucoidans 203–206, 211

G

Gelling 146, 147, 149
Genetic engineering 274–276, 283, 286, 287, 291
Genetic resources 315–317
Glaucophyta 6–8
Greenhouse gas (GHG) 22, 26
Green water 301–303

H

Haematococcus 47, 50, 51
Haptophyta 6, 7, 10, 11
harmful algal blooms (HABs) 216, 219
Harvesting 90–93, 96–99, 108, 109, 253, 255–257, 259
husbandry 169, 171
Hydrocolloids 147, 152
hydrogenase 284–290

I

induced plant defences 188
inflammation 202, 206
Inorganic carbon utilization 45
integrated multi-trophic aquaculture 79, 84, 86
Integrated Multi-Trophic Aquaculture (IMTA) 21, 235
irradiance 44–46, 49–54
light capture 276, 291

l

Lipids 50, 52, 53, 251–253, 257–259, 262

M

macroalgae 177, 178, 180, 183, 186, 188, 236
mariculture 236, 237
marine toxins 216
matrix metalloproteinases 200
mechanisms 180, 183, 187–189
melanogenesis 201, 202, 211
Metabolic engineering 217, 222, 228
metabolic engineering 217, 222, 228, 321, 322
Microalgae 63–74, 90–109, 161, 162, 165–175, 248, 249, 251–260, 262, 263
Microalgae value 308, 315
Microalgal biotechnology 315
Microalgal commercialization 302
mycosporine-like amino acids 209–211

N

Nationally Determined Contribution (NDC) 21
Neuroprotective 130, 131
Nitrogen 44, 45, 48, 49, 52
NMR 155–157
Nutrients 48, 50, 53, 63, 66–69, 71–73

O

Ochrophyta 6–10
oil 274, 275, 277, 282, 283
open raceway ponds 67, 68
Organic carbon utilisation 45, 47
outdoor cultures 46, 53, 54

P

pet feed 172
Phlorotannins 202
Phosphorous 49
photobioreactor 52–54, 65–68, 73, 74
Photo-oxidative stress 202, 210
Photosynthesis 43–48, 50, 53, 54
photosynthetic efficiency 274, 278, 281
Phytoplankton 2, 10, 11
plant pathogens 177–183, 185–186, 190–191
polyculture 236, 237, 242
Polyunsaturated fatty acids 208
productivity 45, 50, 52–54
project design document (PDD) 23
purification 90, 105, 108, 109

R

regulatory constraints 243
Renewable 248–250, 257, 258, 261, 263
Respiration 43, 45, 54
Rhodophyta 6–8, 10, 147
rubisco 279–281, 283, 286, 289

S

Seaweed Aquaculture Beds (SABs) 21
seaweed solution 20–22, 40
seaweeds 235–237, 241, 244
Sedimentation 91–93, 96, 98
separation 90, 92, 97–100, 105, 109
shear stress 113
Silicon 49
skin ageing 205, 211
Sonication 97, 101–103, 107
space availability 243
spatial planning 242–244
Spirulina 166, 168–171
Stabilizing 149, 153
Strain improvement 316
Sulfated galactans 147
sustainable aquaculture 79, 83

T

terpenoids 200
Theca 218, 224, 228
Thickening 149, 153
Toxin biosynthesis 220

About the Editors

F. Xavier Malcata, 2018

PhD in Chemical Engineering, with Minor in Food Science, Biochemistry and Statistics, at University of Wisconsin – Madison, USA. Full Professor at Department of Chemical Engineering, University of Porto, Portugal. Principal Investigator at LEPABE – Laboratory of Process Engineering, Biotechnology and Energy – responsible for research line *PHOTOBEAM: PHOTO-Bioprocess Engineering with Aquatic Microorganisms*. Main research interests focused on marine bioengineering – toward synthesis and extraction of added-value functional metabolites (with pharmaceutical, cosmetic and food applications), via novel designs of photobioreactors and sensors for microalga-mediated processes. Author of 400+ scientific papers in top-ranking scientific journals, with 12,000+ cites by the peers.

Isabel Sousa Pinto, 2018

PhD in Marine Biology (phycology) on ecophysiology and cultivation of seaweeds, at the UC Santa Barbara USA. Associate Professor, University of Porto and Director of Coastal Biodiversity Laboratory at the Interdisciplinary Centre for Marine and Environmental Research (CIIMAR-UP). Her main research has been on biodiversity and ecology of rocky shore and reefs and particularly on the seaweed flora as well as on algal ecophysiology, cultivation and use. She is also working on the science-policy- society interfaces and on promoting ocean literacy.

A. Catarina Guedes, 2018

PhD in Biotechnology – specialty Microbiology (2010) at the Portuguese Catholic University. Researcher in Algal Biotechnology at CIIMAR/UP - Interdisciplinary Centre of Marine and Environmental Research, since 2011. Main research interests focused on algal biotechnology – namely production, extraction and characterization of added-value functional metabolites from micro- and macroalgae, with pharmaceutical, cosmetic and food/feed applications. Author of more than 20 research papers in scientific journals, co-authored 14 book chapters in internationally published books and c. 30 communications in scientific meetings.